B. KEDEM

MATH. DEPT.

APRIL 1989

TROPICAL RAINFALL MEASUREMENTS

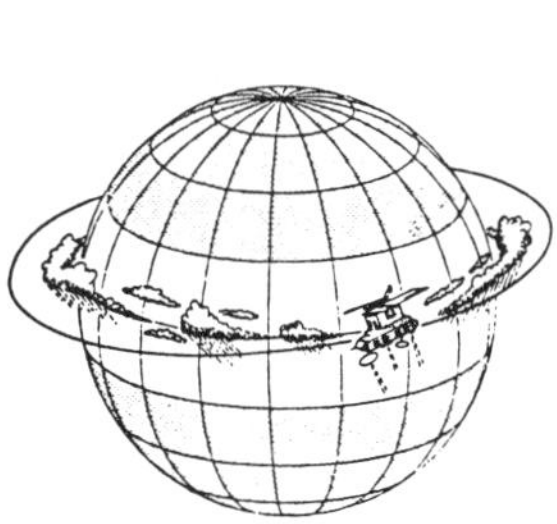

Studies in
Geophysical Optics and Remote Sensing
Series Editor: Adarsh Deepak
Published Volumes and Volumes in Preparation

Advances in Remote Sensing Retrieval Methods
Adarsh Deepak, Henry E. Fleming, and Moustafa T. Chahine (Eds.)

Aerosols and Climate
Peter V. Hobbs and M. Patrick McCormick (Eds.)

Aerosols and Their Climatic Effects
Hermann E. Gerber and Adarsh Deepak (Eds.)

Applications of Remote Sensing for Rice Production
Adarsh Deepak and K.R. Rao (Eds.)

Atmospheric Aerosols: Their Formation, Optical Properties, and Effects
Adarsh Deepak (Ed.)

Hygroscopic Aerosols
Lothar H. Ruhnke and Adarsh Deepak (Eds.)

IRS '84: Current Problems in Atmospheric Radiation
Giorgio Fiocco (Ed.)

IRS '88: Current Problems in Atmospheric Radiation
J. Lenoble and J.P. Geleyn (Eds.)

Light Absorption by Aerosol Particles
Hermann E. Gerber and Edward E. Hindman (Eds.)

Earth Microwave Remote Sensing
A. Chedin (Ed.)

Optical Properties of Clouds
Peter V. Hobbs and Adarsh Deepak (Eds.)

Polarization and Intensity of Light in the Atmosphere
Kinsell L. Coulson

Proceedings in Atmospheric Electricity
Lothar H. Ruhnke and John Latham (Eds.)

Proceedings of Quadrennial Ozone Symposium '88
R.D. Bojkov and P. Fabian (Eds.)

Radiative Transfer in Scattering and Absorbing Atmospheres: Standard Computational Procedures
Jacqueline Lenoble (Ed.)

RSRM '87: Advances in Remote Sensing Retrieval Methods
Adarsh Deepak, Henry E. Fleming, and John S. Theon (Eds.)

Spectral Line Shapes, Volume 4
R.J. Exton (Ed.)

Tropical Rainfall Measurements
John S. Theon and Nobuyoshi Fugono (Eds.)

Proceedings of the
International Symposium on Tropical Precipitation Measurements, held at Tokai University, Tokyo, 28–30 October 1987.

Sponsored by
National Aeronautics and Space Administration, USA; National Oceanic and Atmospheric Administration, USA; Science and Technology Agency, Japan; National Space Development Agency, Japan; Japan Meteorological Agency; Radio Research Laboratory, Japan; and Tokai University, Japan

Cohosted by
Horizon of Observation Platform for Earth and Space Group, Japan; and Tropical Rainfall Measuring Mission Science Steering Group, USA

Organized by
Science and Technology Corporation under contract NAS1-18397.

TROPICAL RAINFALL MEASUREMENTS

Edited by

John S. Theon
NASA Headquarters, Washington, DC

and

Nobuyoshi Fugono
Radio Research Laboratory, Tokyo

A. DEEPAK Publishing 1988
A Division of Science and Technology Corporation
Hampton, Virginia USA

A. DEEPAK Publishing
A Division of Science and Technology Corporation
101 Research Drive
Hampton, Virginia 23666–1340 USA

Library of Congress Cataloging-in-Publication Data

International Symposium on Tropical Precipitation Measurements (1987 : Tokai University)
Tropical rainfall measurements / edited by John S. Theon and Nobuyoshi Fugono.
p. cm.—(Studies in geophysical optics and remote sensing)
"Proceedings of the International Symposium on Tropical Precipitation Measurements, held at Tokai University, Tokyo, 28–30 October 1987. Sponsored by National Aeronautics and Space Administration, USA . . . [et al.]"—P.
Bibliography: p.
Includes index.
ISBN 0–937194–14–X
1. Rain and rainfall—Tropics—Measurement—Congresses. 2. Microwave detectors—Congresses. 3. Radar meteorology—Congresses. 4. Satellite meteorology—Congresses. I. Theon, J. S. II. Fugono, Nobuyoshi. III. United States. National Aeronautics and Space Administration. IV. Title. V. Series.
QC925.82.I58 1987 88–32080
551.57′81213—dc19 CIP

Printed in the United States of America

CONTENTS

SESSION A. REQUIREMENTS FOR MEASURING TROPICAL RAINFALL, ITS NATURE AND ITS ROLE IN THE GENERAL CIRCULATION

SESSION B. SPACEBORNE AND AIRBORNE TECHNIQUES FOR REMOTE SENSING OR RAINFALL

SESSION C. RETRIEVAL ALGORITHMS AND DATA PROCESSING AND ANALYSIS

SESSION D. SAMPLING STRATEGIES

SESSION E. MISSION REQUIREMENTS AND DATA MANAGEMENT

SESSION F. GROUND-BASED MEASUREMENTS

PREFACE

This volume records the papers presented at the International Symposium on Tropical Precipitation Measurements that was held in Tokyo, Japan, on October 28–30, 1987, at the facilities of Tokai University. It was the first International Symposium specifically focused on the problems of observing tropical rainfall from space and analyzing and interpreting the data. Although many aspects of tropical rainfall related to space observations were discussed, the symposium also provided a forum to discuss the present state of knowledge on available rainfall measurement techniques, the analysis and interpretation of available data, and proposed new observing techniques. More than 170 scientists and engineers from universities, government laboratories, and the private sector in Australia, Canada, France, Indonesia, Japan, The People's Republic of China, the United Kingdom, the United States, and Taiwan participated. Seventy papers were presented, and are contained herein.

Significant progress in the area of tropical rainfall measurements has been made during the past few years. This has been spurred by the development of new measuring techniques that make it feasible, for the first time, to observe precipitation from space. This, of course, presents the potential for measuring rainfall not only over well-instrumented continental areas, but over the vast oceanic regions as well.

It has been known for many years that tropical rain systems constitute a life-giving resource for billions of people, but their large variability represents a serious threat to those same people in the form of drought or flood extremes. The atmospheric circulation has been compared with a heat engine in which the process of low-level convergence, convection, and precipitation production in the equatorial regions releases latent heat of condensation in a "firebox" effect. The resulting vertical transports of heat, moisture, and momentum have a pronounced effect on the large-scale circulations of the atmosphere. The tropics produce occasional severe weather (typhoons, cyclones, and hurricanes), which has a very large impact on the areas directly affected. The tropical rainfall process provides an important coupling between the ocean and the atmosphere and a key link in the quasi-periodic phenomenon known as the El Niño/Southern Oscillation. Of course, convective tropical rain systems provide an important mechanism for fluxes from the troposphere to the stratosphere. Inherent in these precipitation-producing processes is a mechanism for charge separation between the earth and the ionosphere and, finally, precipitation provides an important scavenging mechanism for removing gaseous and aerosol chemicals from the lower atmosphere.

Certainly, observing precipitation distributions and their variations is a required step in attempting to quantify the hydrological cycle that is identified as a key process to be studied if we are to develop an understanding of global change.

The symposium was divided into six topical areas, namely: Requirements for Measuring Tropical Rainfall, Its Nature and Its Role in the General Circulation; Spaceborne and Airborne Techniques for Remote Sensing of Rainfall; Retrieval Algorithms and Data Processing and Analysis; Sampling Strategies; Mission Requirements and Data Management; and Ground-Based

Measurements. These areas were chaired by the following scientists, respectively: Dr. Taroh Matsuno, University of Tokyo; Dr. J. Michael Hall, National Oceanic and Atmospheric Administration; Dr. Marvin A. Geller, NASA Goddard Space Flight Center; Dr. John S. Theon, NASA Headquarters; Dr. Nobuyoshi Fugono, Radio Research Laboratory; Dr. Tasuku Tanaka, National Space Development Agency; and Dr. Tsuyoshi Nitta, Meteorological Research Institute.

To ensure proper representation of major disciplines, a workshop program committee composed of the following scientists was set up: Dr. John S. Theon, Dr. Nobuyoshi Fugono, and Dr. Ramesh Kakar.

The editors wish to acknowledge the enthusiastic support and cooperation of the members of the Technical Program Committee, Session Chairmen, speakers, and participants for making this a stimulating and valuable symposium. Special recognition must go to Professor T. Sakata, Director, Tokai University Research and Information Center, for his valuable support and hospitality before and during the Symposium and Dr. A. Deepak, Science and Technology Corporation (STC), and Ms. Carolyn Keen, STC Meetings Division, for their organization of the Symposium.

The symposium was cohosted by: Horizon of Observation Platform for Earth and Space Group (Japan); and Tropical Rainfall Measuring Mission Sciences Steering Group (USA). The cosponsors of the symposium were: National Aeronautics and Space Administration; National Oceanic and Atmospheric Administration; Science and Technology Agency; National Space Development Agency; Japan Meteorological Agency; Radio Research Laboratory; and Tokai University.

It is our hope that the research findings presented and compiled in this volume will provide an important source of information for many researchers in the field.

John S. Theon
Nobuyoshi Fugono
Editors

PARTICIPANTS LIST

Robert F. Adler, NASA/GSFC, Laboratory for Atmospheres, Code 612, Greenbelt, MD 20771, USA

K. Akaeda, Meteorological Research Institute, 1-1 Nagamine, Yatabe-machi, Tsukuba-gun, Ibaraki-ken 305, JAPAN

Jiro Aoyagi, Meteorological Research Institute, 1-1 Nagamine, Yatabe-machi, Tsukuba-gun, Ibaraki-ken 305, JAPAN

Kohei Arai, National Space Development Agency, Earth Observation Center, 1401 Ohashi, Hatoyama-Machi, Hikigun, Saitama 350-03, JAPAN

Phillip A. Arkin, NOAA/NWS/NMC, Climate Analysis Center-W/NMC52, 5200 Auth Road, Washington, DC 20233, USA

James Arnold, NASA/MSFC, Mail Code ED43, Huntsville, AL 35812, USA

Tadashi Aso, Meteorological Satellite Center, JMA, 3-235 Nakakiyoto, Kiyose-shi, Tokyo 204, JAPAN

David Atlas, 7420 Westlake Terrace, Apartment 810, Bethesda, MD 20817, USA

G.L. Austin, McGill Radar Weather Observatory, Box 241, Macdonald College, Ste. Anne de Bellevue, Quebec, H9X ICO, CANADA

Jun Awaka, Radio Research Laboratory, 4-2-1 Nukui-Kita-Machi, Koganei-shi, Tokyo 184, JAPAN

Koichi Ayabe, National Space Development Agency, 1401 Ohashi, Hatoyama-Machi, Saitama 350-03, JAPAN

Jakondar Bakara, Tokai University, 2-28-4 Tomigaya, Shibuya-ku, Tokyo 151, JAPAN

Eric C. Barrett, University of Bristol, Remote Sensing Unit, Dept. of Geography, University Road, Bristol, BS8 1SS, UNITED KINGDOM

James Bilbro, NASA/MSFC, Mail Code EB23, Huntsville, AL 35812, USA

Priscilla Cehelsky, Tokyo University, Geophysical Institute, Bunkyo-ku, Tokyo 113, JAPAN

C.Y. Chang, Oceanographical Inst. of Tokyo University, Minamidai 1-15-1, Nakanoku, Tokyo, JAPAN

Long-Nan Chang, National Central University, Department of Atmospheric Physics, Chung-Li, Taiwan 32054, REPUBLIC OF CHINA

Long S. Chiu, Applied Research Corp., Landover, Maryland 02139, USA

Kohei Cho, NASDA, Earth Observation Program Office, 2-4-1 Hamamatucho, Minatoku, Tokyo 105, JAPAN

Li Chong-yin, Institute of Atmospheric Physics, Academia Sinica, Beijing, PEOPLE'S REPUBLIC OF CHINA

Robert Curran, NASA Headquarters, Code EE, 600 Independence Avenue, SW, Washington, DC 20546, USA

Adarsh Deepak, Science and Technology Corporation, 101 Research Drive, Hampton, VA 23666, USA

James C. Dodge, NASA Headquarters, Mail Code EET, 600 Independence Avenue, SW, Washington, DC 20546, USA

Toshio Doura, National Space Development Agency of Japan, Tsukuba Space Center, 2-1-1, Sengen, Sakura-Mura, Niihari-Gun, Ibaraki-Ken 305, JAPAN
Takashi Eguchi, University of Tokyo, Department of Geography, 7-3-1, Hongo, Bunkyo-Ku, Tokyo 113, JAPAN
Charles Elachi, Jet Propulsion Laboratory, Mail Stop 183-335, 4800 Oak Grove Drive, Pasadena, CA 90119, USA
Paul C. Etter, ODSI Defense Systems, Inc., Rockville, Maryland 20852, USA
Daniel Fitzjarrald, NASA/MSFC, Mail Code ED43, Huntsville, AL 35812, USA
Nobuyoshi Fugono, Radio Research Laboratory, Planning Division, Ministry of Posts & Telecommunications, 4-2-1 Nukui-kita-machi, Koganei-shi, Tokyo 184, JAPAN
Masaharu Fujita, Radio Research Laboratory, Kashima Space Research Center, 893-1 Hirai, Kashima-Machi, Ibaraki-Ken 314, JAPAN
Yoji Furuhama, ATR Optical & Radio Communications Res. Lab, Twin 21 Bldg. MID Tower, 2-1-61 Shiromi Higashi-ku, Osaka 500, JAPAN
Oswaldo Garcia, NOAA/ERL/CRD, R/E/AR 6, 325 Broadway, Boulder, CO 80303, USA
Michael Garstang, U. of Virginia, Department of Environmental Sciences, Clark Hall, Charlottesville, VA 22903, USA
Marvin A. Geller, NASA/GSFC, Code 610, Greenbelt, MD 20771, USA
Ruslan Gintinl, Tokai University, Research & Information Center, 2-28-4 Tomigaya, Shibuya-ku, Tokyo 151, JAPAN
J. Michael Hall, NOAA, R/CAR, Room 825, Rockville, MD 20852, USA
Tadaaki Hamada, Meteorological Satellite Center, 3-235 Nakakiyoto, Kiyoseshi, Tokyo 204, JAPAN
Daesoo Han, NASA/GSFC, Code 636, Greenbelt, MD 20771, USA
Bidawi Hasyim, Tokai Regional Fisheries, Research Lab, Oceanographic Division, 5-5-1 Kachidoki, Chuo-Ku, Tokyo 104, JAPAN
S. Hayashida-Amano, National Institute for Environmental Studies, 16-2 Onogawa, Yatabe-machi, Tsukuba, Ibaraki 305, JAPAN
Kotaro Higuchi, Fujitsu Limited, Space Systems Department, 1-17-25, Shinkamata, Ota-ku, Tokyo 144, JAPAN
B.B. Hinton, Space Science and Engineering Center, University of Wisconsin-Madison, Madison, Wisconsin 53706, USA
Kinsaburo Hirano, Obunsho Publishing Co., 55 Yokodera-cho, Shinjuku-ku, Tokyo 162, JAPAN
Mitsuaki Horiguchi, Gokasho, Uji-shi, Kyoto-hu 611, JAPAN
Robert A. Houze, Jr., University of Washington, Department of Atmospheric Sciences, AK40, Seattle, WA 98195, USA
R.J. Hung, The University of Alabama in Huntsville, Huntsville, Alabama 35899, USA
Toshio Ihara, Radio Research Laboratory, Ministry of Posts and Telecommunications, 4-2-1 Nukui-kita-machi, 4-chome, Koganei-shi, Tokyo 184, JAPAN
K. Eastwood Im, Jet Propulsion Laboratory, 4800 Oak Grove Drive, MS T-1206D, Pasadena, CA 90119, USA
Toshiro Inoue, Meteorological Research Institute, 1-1 Nagamine, Yatabe-machi, Tsukuba-gun, Ibaraki-ken 305, JAPAN

Takashi Ishiyama, Chiba University, 1-33 Yayoicho, Chiba City 260, JAPAN
Yoshiako Jakewcki, Meteorological Satellite Center, 3-235 Nakakiyoto, Kiyoseshi, Tokyo 204, JAPAN
Arthur R. Jameson, Applied Research Corporation, 8201 Corporate Drive, Landover, MD 20785, USA
Ramesh Kakar, NASA Headquarters, Code EE, 600 Independence Avenue, SW, Washington, DC 20546, USA
Shogo Karitani, NASA/MSFC, Mail Code ED43, Huntsville, AL 35812, USA
Hiromu Kashihara, NEC Corporation, 10 Nisshincho 1-Chome, Fuchu City, Tokyo 183, JAPAN
Kuranoshin Kato, Nagoya University, Water Research Institute, Furoh-cho, Chikusa-ku, Nagoya 464, JAPAN
Susumu Kato, Radio Atmospheric Science Center, Kyoto University, UJI, Kyoto 611, JAPAN
Riona Kawashima, Toshiba Corporation, 1 Komukai, Toshiba-cho, Saiwai-ku, Kawasaki 210, JAPAN
Mitsu Kayama, Hiroshima University, 2-17 Midori-machi, Fukuyama-shi, Hiroshima 720, JAPAN
Thomas Keating, NASA/GSFC, Code 402, Greenbelt Road, Greenbelt, MD 20771, USA
Akio Kitoh, Meteorological Research Institute, 1-1 Nagamine, Yatabe-machi, Tsukuba-gun, Ibaraki-ken 305, JAPAN
Nobuhiko Kodaira, Remote Sensing Technology Center, 7-15-17, Roppongi, Minato-Ku, Tokyo 106, JAPAN
Y. Kodama, University of Tokyo, Ocean Research Institute, Nakanoku, Tokyo 154, JAPAN
Kunihiko Kodera, Meteorological Research Institute, 1-1 Nagamine, Yatabe-machi, Tsukuba-gun, Ibaraki-ken 305, JAPAN
Tadashi Kondo, Tokai University, 1117 Kitakaname, Hiratsuka, Tokyo 259-12, JAPAN
Hajime Koshiishi, National Aerospace Laboratory, 7-44-1 Higashi-machi, Jindaiji, Chofu-shi, Tokyo 182, JAPAN
Witold F. Krajewski, University of Iowa, Institute of Hydraulic Research, Iowa City, IA 52242, USA
Ken'ichi Kuma, Japan Meteorological Agency, Numerical Prediction Division, Ote-machi 1-3-4, Chiyoda-ku, Tokyo 100, JAPAN
Toshiyuki Kurino, Meteorological Satellite Center, 3-235 Nakakiyoto, Kiyoseshi, Tokyo 204, JAPAN
Dony Kushardono, Indonesian National Institute of Aeronautics and Space, Shonan Campus, Kaname, Hiratsuka City, INDONESIA
William K.-M. Lau, NASA/GSFC, Laboratory for Atmospheres, Code 613, Greenbelt, MD 20771, USA
Pay-Liam Lin, National Central University, Department of Atmospheric Physics, Chung-Li, Taiwan 32054, REPUBLIC OF CHINA
Roland List, University of Toronto, Department of Physics, Toronto, Ontario, M5S 1A7, CANADA

Takeshi Manabe, Radio Research Laboratory, 4-2-1 Nukui-Kita-Machi, Koganei-shi, Tokyo 184, JAPAN
M. J. Manton, BMRC, Bureau of Meteorology, P.O. Box 1289K, Melbourne, Victoria 3001, AUSTRALIA
David W. Martin, University of Wisconsin-Madison, Space Science and Engineering Center, 1225 West Dayton Street, Madison, WI 53706, USA
Keiji Maruo, Remote Sensing Technology Center, UNI-Roppongi Building 6F, 7-15-17 Roppongi, Minato-ku, Tokyo 106, JAPAN
Mongi Marzoug, CRPE-CNET, 38-40, Rue du General Leclerc, 92131 Issy-les-Moulineaux, FRANCE
Kooiti Masuda, University of Tokyo, Geophysical Insitute, 2-11-16 Yayoi, Bunkyo-ku, Tokyo 113, JAPAN
Harunobu Masuko, Radio Research Laboratory, 4-2-1 Nukui-Kita-Machi, 4-Chome, Koganei-shi, Tokyo 184, JAPAN
Tatsuro Matsumae, Tokai University, 2-28-4 Tomigaya, Shibuya-ku, Tokyo 151, JAPAN
Jun Matsumoto, University of Tokyo, Department of Geography, 7-3-1, Hongo, Bunkyo-Ku, Tokyo 113, JAPAN
Taroh Matsuno, University of Tokyo, 2-11-16 Yayoi, Bunkyo-ku, Tokyo 113, JAPAN
George McDonough, NASA/MSFC, Mail Code ED01, Huntsville, AL 35812, USA
Robert Meneghini, NASA/GSFC, Code 675, Greenbelt, MD 20771, USA
Sumio Miida, National Space Development Agency, 2-4-1, Hamamatsu-Cho, Minato-Ku, Tokyo 105, JAPAN
Takehiko Mikami, Tokyo Metropolitan University, 2-1-1 Fukazawa, Setagaya-ku, Tokyo 158, JAPAN
Richard K. Moore, University of Kansas, Radar Systems and Remote Sensing Lab, 2291 Irving Hill Road, Lawrence, KS 66045, USA
Masato Mukakami, Meteorological Research Institute, 1-1 Nagamine, Yatabe-machi, Tsukuba-gun, Ibaraki-ken 305, JAPAN
Nobuhiko Murayama, Meteorological Satellite Center, 3-235 Nakakiyoto, Kiyoseshi, Tokyo 204, JAPAN
Tsugunobu Nagai, Meteorological Research Institute, 1-1 Nagamine, Yatabe-machi, Tsukuba-gun, Ibaraki-ken 305, JAPAN
Y. Nakahara, 1 Komukai, Toshiba-cho, Saiwai-ku, Kawasaki 210, JAPAN
Eiichi Nakakita, Kyoto University, Disaster Prevention Res. Inst., Gokasho, Uji-shi, Kyoto 611, JAPAN
Kenji Nakamura, Radio Research Laboratory, 4-2-1 Nukui-Kita-Machi, Koganei-shi, Tokyo 184, JAPAN
Tetsuo Nakazawa, Meteorological Research Institute, 1-1 Nagamine, Yatabe-machi, Tsukuba-gun, Ibaraki-ken 305, JAPAN
Kazuro Nakome, National Institute of Resources S.T.A, 3-1 Chiyoda-ku, Tokyo, JAPAN
Katsuhiko Nishikawa, NEC Corporation, 4035 Ikebe-Cho, Midori-ku, Yokohama 226, JAPAN
Tsuyoshi Nitta, Meteorological Research Institute, 1-1 Nagamine, Yatabe-machi, Tsukuba-gun, Ibaraki-ken 305, JAPAN

Akira Noda, Meteorological Research Institute, 1-1 Nagamine, Yatabe-machi, Tsukuba-gun, Ibaraki-ken 305, JAPAN
Julia Nogues-Paegle, University of Utah, Department of Meteorology, 819 Wm. B. Browning Building, Salt Lake City, UT 84112, USA
Gerald R. North, Texas A and M University, College of Geosciences, Dept. of Meteorology, Climate Systems Research Program, College Station, TX 77843, USA
Atsusi Numaguchi, University of Tokyo, Geophysical Institute, 2-11-16 Yayoi-cho, Bunkyo-ku, Tokyo 113, JAPAN
Yoshio Ohgushi, NEC Corporation, 4035 Ikebe-Cho, Midori-ku, Yokohama 226, JAPAN
Takeyuki Ojima, Radio Research Laboratory, 4-2-1 Nukui-Kita-Machi, Koganei-shi, Tokyo 184, JAPAN
Ken'ichi Okamoto, Radio Research Laboratory, 4-2-1 Nukui-Kita-Machi, Koganei-shi, Tokyo 184, JAPAN
Shingo Osano, Meteorological Satellite Center, 3-235 Nakakiyoto, Kiyoseshi, Tokyo 204, JAPAN
Tomoaki Ose, Meteorological Research Institute, 1-1 Nagamine, Yatabe-machi, Tsukuba-gun, Ibaraki-ken 305, JAPAN
Ayako Ouchi, University of Tokyo, Geophysical Institute, 2-11-16 Yayoi-cho, Bunkyo-ku, Tokyo 113, JAPAN
Akira Ozeki, The Asahi Shimbun, Tsukiji, Tokyo, JAPAN
Garth Paltridge, CSIRO, Division of Atmospheric Research, Private Bag No. 1, Mordialloc, VIC. 3195, AUSTRALIA
Xie Ping-ping, Gokasho, Ugi-shi, Kyoto-hu 611, JAPAN
Eugene M. Rasmusson, University of Maryland, Department of Meteorology, College Park, MD 20742, USA
George T. Roach, NASA Goddard Space Flight Center, Greenbelt, MD 20771, USA
Huang Ronghui, Institute of Atmospheric Physics, Academia Sinica, Beijing, PEOPLE'S REPUBLIC OF CHINA
B.F. Ryan, CSIRO Division of Atmospheric Research, Aspendale, Victoria, AUSTRALIA
Toshibumi Sakata, Tokai University, Research & Information Center, 2-28-4 Tomigaya, Shibuya-ku, Tokyo 151, JAPAN
Hardjito Saroso, Tokai University, Research & Information Center, 2-28-4 Tomigaya, Shibuya-ku, Tokyo 151, JAPAN
Hideyuki Sasaki, Meteorological Satellite Center, 3-235 Nakakiyoto, Kiyoseshi, Tokyo 204, JAPAN
H. Sato, Tokyo Astronomical Observatory, Osawa, Mitaka 181, JAPAN
Hideo Sato, National Space Development Agency, Earth Observation Center, 1401 Ohashi, Hatoyama-Machi, Hikigun, Saitama 350-03, JAPAN
Robert A. Schiffer, NASA Headquarters, Code EE, 600 Independence Avenue, SW, Washington, DC 20546, USA
Shigeko Segawa, Newton Magazine, 1-20 Nishishinjuku, Horai Bldg. 6F, Tokyo 160, JAPAN
Akira Shibata, Meteorological Research Institute, 1-1 Nagamine, Yatabe-machi, Tsukuba-gun, Ibaraki-ken 305, JAPAN
Itsuro Shimizu, NASDA, 2-4-1 Hamamatucho, Minatoku, Tokyo 105, JAPAN

Masato Shinoda, University of Tokyo, Department of Geography, 7-3-1 Hongo, Bunkyo-ku, Tokyo 113, JAPAN

Jagadish Shukla, U. of Maryland - COLA, Department of Meteorology, College Park, MD 20742, USA

Joanne Simpson, NASA/GSFC, Head, Severe Storms Branch, Code 612, Greenbelt, MD 20771, USA

Suryadi Siregar, Institute Technology of Bandung, Department of Astronomy, Bandung, West Java 04132, INDONESIA

Roy W. Spencer, NASA/MSFC, ED43, Huntsville, AL 35812, USA

Akimasa Sumi, University of Tokyo, Geophysical Institute, Faculty of Science, Tokyo, JAPAN

Verner E. Suomi, University of Wisconsin-Madison, Space Science and Engineering Center, 1225 W. Dayton Street, Madison, WI 53706, USA

R. Suppiah, University of Tsukuba, Institute of Geoscience, Ibaraki 305, JAPAN

Totok Suprapto, LAPAN, Jakarta, INDONESIA

Kenji Takano, Courant Institute of Mathematical Sciences, New York University, 251 Mercer Street, New York, NY 10012, USA

Hitoshi Takatsuka, National Space Development Agency, 2-4-1, Hamamatsu-Cho, Minato-Ku, Tokyo 105, JAPAN

Yukari Takayabu, National Institute for Environmental Studies, 16-2 Onogawa, Yatabe-machi, Tsukuba, Ibaraki 305, JAPAN

Atsushi Takeda, National Research Center for Disaster Prevention, 9-2 Nijigahama, Hiratsuka, Kanagawa 254, JAPAN

Takao Takeda, Nagoya University, Water Research Laboratory, Furoh-cho, Chikusa-ku, Nagoya City 464, JAPAN

Yukio Takemura, Japan Meteorological Agency, Forecast Division, 1-3-4 Ohte-machi Chiyoda-ku, Tokyo 100, JAPAN

Nobuo Takeuchi, National Institute for Environmental Studies, 16-2 Onogawa, Yatabe-machi, Tsukuba, Ibaraki 305, JAPAN

Hiroshi Tanaka, Nagoya University, Water Research Institute, Furoh-cho, Chikusa-ku, Nagoya City 464, JAPAN

Tasuku Tanaka, National Space Development Agency, 2-4-1 Hamamatsu-cho, Minato-ku, Tokyo 105, JAPAN

Toshihiko Teramoto, 1-37-9 Wakaba-cho, Chofu-shi, Tokyo 182, JAPAN

John S. Theon, NASA Headquarters, Code EE, 600 Independence Avenue, S.W., Washington, DC 20546, USA

Otto W. Thiele, NASA/GSFC, Laboratory for Atmospheres, Code 613, Greenbelt, MD 20771, USA

S.F. Tian, University of Tsukuba, Institute of Geoscience, Sakurawura, Niiharigu, Ibaraki 305, JAPAN

Tatsushi Tokioka, Meteorological Research Institute, 1-1 Nagamine, Yatabe-machi, Tsukuba-gun, Ibaraki-ken 305, JAPAN

K. Tsuchiya, Chiba University, 1-33 Yayoicho, Chiba City 260, JAPAN

Ken-ichi Tsukamoto, Radio Research Laboratory, 4-2-1 Nukui-Kita-Machi, Koganei-shi, Tokyo 184, JAPAN

Yasufumi Tsukamura, Tokyo Metropolitan University, Department of Geography, 2-1-1 Fukazawa, Setagaya-ku, Tokyo 158, JAPAN
Tadashi Tsuyuki, Japan Meteorological Agency, Ote-machi 1-3-4, Chiyoda-ku, Tokyo 100, JAPAN
Robert Turner, 7501 Cadillac Drive, Huntsville, AL 35802, USA
William W. Turner, NASA Headquarters, Code XID, Washington, DC 20546, USA
Hiroyuki Uchida, Meteorological Satellite Center, 3-235 Nakakiyoto, Kiyoseshi, Tokyo 204, JAPAN
Shigetsugu Uehara, National Research Center for Disaster Prevention, S.T.A., 3-1 Tennodai, Sakura-mura, Ibaraki 305, JAPAN
Mitsuro Ueno, Japan Meteorological Agency, Ote-machi 1-3-4, Chiyoda-ku, Tokyo 100, JAPAN
Noboru Wakai, Tokai University, 2-28-4 Tomigaya, Shibuya-ku, Tokyo 151, JAPAN
Fumio Watanabe, Japan Meteorological Agency, Long Forecasting Division, 1-3-4 Ohtemachi Chiyoda-ku, Tokyo 100, JAPAN
Kazuo Watanabe, Remote Sensing Technology Center, 7-15-17, Roppongi, Minato-Ku, Tokyo 106, JAPAN
Peter J. Webster, Pennsylvania State University, Department of Meteorology, 518 Walker Building, University Park, PA 16802, USA
James Weinman, University of Wisconsin, Department of Meteorology, 1225 West Dayton Street, Madison, WI 53702, USA
Thomas T. Wilheit, NASA/GSFC, Laboratory for Oceans, Code 675, Greenbelt, MD 20771, USA
Gregory S. Wilson, NASA/MSFC, Mail Code ED41, Huntsville, AL 35812, USA
William J. Wilson, Jet Propulsion Laboratory, Building 168-327, 4800 Oak Grove Drive, Pasadena, CA 90119, USA
Isamu Yagai, Meteorological Research Institute, 1-1 Nagamine, Yatabe-machi, Tsukuba-gun, Ibaraki-ken 305, JAPAN
Shuji Yamakawa, University of Tsukuba, Institute of Geoscience, Sakura, Tsukuba, Ibaraki, JAPAN
Koji Yamazaki, Meteorological Research Institute, 1-1 Nagamine, Yatabe-machi, Tsukuba-gun, Ibaraki-ken 305, JAPAN
Peicai Yang, Institute of Atmospheric Physics, Academia Sinica, Beijing, PEOPLE'S REPUBLIC OF CHINA
Tetsuzo Yasunari, Tsukuba University, Sakura-mura, Niihari-gun, Ibaraki-ken 105, JAPAN
Tatsuo Yokoyama, Meteorological Research Institute, 1-1 Nagamine, Yatabe-machi, Tsukuba-gun, Ibaraki-ken 305, JAPAN
Fumio Yoshino, Public Works Research Institute, Asahi 1, Tsukuba, Ibaraki-ken, JAPAN
Wang Yunkuan, Institute of Atmospheric Physics, Academia Sinica, Beijing, PEOPLE'S REPUBLIC OF CHINA
Edward J. Zipser, NCAR, Box 3000, Boulder, CO 80307, USA

KEYNOTE ADDRESS

THE AGE OF EARTH OBSERVATION TECHNOLOGY

Dr. Tatsuro Matsumae
President, Tokai University
2–28–4 Tomigaya, Shibuya-ku
Tokyo 151, Japan

The starting point of the development of the present day science, I believe, was the conversion of the concept men held with regard to the cosmos. It might be said that when Nicolaus Copernicus published *On the Revolutions of the Heavenly Spheres* in 1543, the conflict began between his new concept of the universe and the conventional Ptolemaic geocentric cosmology, which was deeply believed in by contemporary, devout Roman Catholic church scholars.

When his publication of *On the Revolutions of the Heavenly Spheres* was completed, Copernicus was on his deathbed. The Roman Church put on his tombstone the following inscription, "I shall not beseech either the grace given to Paul, or that bestowed on Peter. I only implore the same blessing Thou gave to the thief on the cross."

Thereafter, Johannes Kepler, who studied under Tycho Brahe, also supported the Copernican theory, and established the law of the motion of the planets in the solar system in his works *New Astronomy* (1609), and *Harmonice Mundi* (1619). He first studied theology as a Protestant, and then put most of his efforts into studying astronomy. Being a Protestant, he moved in the wake of religious persecutions from the University of Graz in Austria to Prague to continue his work on those volumes, and made them public there.

Galileo Galilei made astronomical observations, by means of the telescope he invented, of Jupiter, Saturn, the Moon, and the Sun, and discovered Jupiter's moons, Saturn's rings, and the Sun's black spots. He also supported the moving-Earth theory, and because of that, was summoned before the Inquisition in Rome, and sentenced to prison for his publication, *Dialogue on the Two Chief World Systems*.

Nobody would have believed at that time that the Earth on which one stood was moving, and in 1616, official theologians and the Inquisition in Rome declared the doctrine that the Earth moves around a stationary Sun was contrary to the Holy Scriptures, and any books related to the moving-Earth theory were prohibited by the Pope.

The prohibition of the books published by Copernicus was lifted, finally, as late as 1835. In the case of Galileo Galilei, more than 120 years had to pass before a trial for his reinstatement was held, and he was exonerated in 1757.

Isaac Newton, following in their footsteps, succeeded in the calculation of the orbital movement of heavenly bodies and published it in his work, *Philosophiae naturalis principia mathematica*, or the *Principia*. He thus gave the theoretical foundation of dynamics to the moving-Earth theory contended by Copernicus, Kepler, and Galileo, and made the world at large recognize the fact that the Earth revolves around the Sun.

Why did such harsh treatment of the moving-Earth theory not attract society's attention before? It is now assumed that for human beings who lived on the Earth at that time, such discussion had no relevance to their daily lives and was also too vague a concept for them to grasp or visualize.

And now, after almost 400 years since Copernicus first contended his new theory, mankind can see with their own eyes that the Earth is spherical in shape. The first artificial satellite, *Sputnik 1,* was launched in 1957, and the first astronaut, Gagarin, actually saw the round and blue Earth.

The dream world depicted in the modern science fiction of Jules Verne has become a reality. The Earth observed from the Moon is as beautiful a sphere as Verne imagined and described to his readers.

The desire of man, who has miniscule mass, to liberate himself from the surface of the Earth, which has huge mass and is in continual revolution, and to overcome Earth's gravity instigated the development of airplanes and rockets. In 1902, a Russian school teacher, Tiolkovsky, wrote a book entitled *Exploration of Space by Means of Reaction Apparatus.* In the book he depicted the means of travel into the space of the universe, and contended that navigation in contravention to gravity could be made possible if one could make a rocket vehicle, generating sufficient power to overcome the gravity of the Earth. With such a carrier, called a rocket, an object riding on it could become an artificial star.

If an object is pushed by a rocket to the veolocity of 7.91 km per second, and hauled out of the Earth's atmosphere, such an object will rotate around the Earth. With further acceleration, the object can move out of the Earth's gravitational sphere in a parabolic trajectory and begin circling around the Sun. If the object is accelerated still further, it will move in a hyperbolic curve out of the solar system to the galactic universe. The velocity required to move out of the gravitational zone of the Earth is 11.8 km per second (the second astronautical velocity), and that needed to move out of the solar system is 16.5 km per second (the third astronautical velocity).

Now let's consider a flying object staying within the gravitational sphere of the Earth. The speed of revolution of the Earth is 460 meters per second at the equator, 380 meters at the central part of Japan, which is at 35 degrees north latitude, and 220 meters per second at 60 degrees north latitude (around Stockholm or Anchorage), equivalent to the speed of a jet-engine aircraft (about 800 km/h). If a flying object is launched above the equator to an altitude of 36,000 km and set to circle around the Earth on a fixed position on the longitudinal axis, it will be seen by people looking up from the surface of the Earth as if it remains stationary.

A flying object set at such conditions, in the case of an artificial satellite, is called a geostationary satellite. If a similar object is circled on the polar axis, it is called a polar-orbiting satellite, and it will be floating at an altitude between 700 and 1,000 km.

The photo of the Earth taken and transmitted from the Apollo spaceship heading toward the Moon was impressive. The outer solar system probe, Voyager, transmitted a photo image of the Earth and the Moon together. While Gagarin conveyed his impressions of the "blue Earth," we could identify in the picture of the Earth transmitted by the satellites, the clouds, and oceans of the Earth with surprising clarity. The Earth's image the Voyager transmitted was, on the other hand, whitish and blurred.

The age has become one of observing the Earth from outer space. In 1972 the U.S. National Aeronautics and Space Administration (NASA) launched a satellite exclusively for the purpose of observing the Earth. This project of the "LANDSAT" series has put five satellites into orbit now. Various types of satellites launched for observation of the Earth, having such acronyms as ESSA, NIMBUS, TIROS, NOAA, and so forth, have contributed new and highly valuable information to mankind on Earth.

Rapid advancement has been made in the sensor equipment placed aboard satellites. The infrared sensors carried by the LANDSATs can cover the near and far ends of the infrared range as well as infrared thermal waves, and thus can make surveys encompassing the degree of activity in forests and the humidity of ocean surfaces. The NOAA series of meteorological observation satellites has facilities to cover more than the graphic images of the surface of the Earth. France launched its Earth observation satellite "SPOT" recently, and Japan followed it with a marine observation satellite "MOS-1," nicknamed "Momo." The information sent from these satellites enables us to analyze the conditions of the forests and agricultural products, to detect abnormalities in the temperature distribution of ocean water mass, such as during an El Niño event, coldwater masses, and other abnormal phenomena in oceanography. NOAA and NIMBUS keep us informed of the reduction of ozone in the antarctic atmosphere, the diminution of forests, and expansion of deserts, warning men of their precarious future.

Many countries have put forth a large amount of funds for their space development activities. Yet in view of the fact that space itself is to be commonly shared by mankind, and that this new territory should not belong to any particular men or nation, human activities in space must reflect the common themes or subjects mankind must resolve. The activities in space, for that reason, must be participated in by all mankind. The natural environment, which has kept its trim through natural action and reaction, must be maintained in stable condition, for the Earth is irreplaceably precious for humans as well as other fauna and flora. Therefore, this environment should not be despoiled either by the excessive activities of mankind or the national egotism of specific countries.

The satellites' mission of observing the Earth is, therefore, extremely important, and I believe the cooperation and interchange among the personnel engaged in related activities, irrespective of their field of study or nationality, will play a very significant role in the future of man on Earth.

TROPICAL RAINFALL FROM SPACE

Joanne Simpson
NASA Goddard Space Flight Center
Greenbelt, Maryland 20771, USA

ABSTRACT

Tropical precipitation is important in maintaining atmospheric circulations on many scales, from the mesoscale up to planetary. There is evidence that tropical rainfall variability, apparently coupled with changes in the underlying surface (particularly sea surface temperatures), is associated with significant alterations in short-term climate. This variability may lead to hemisphere-wide alterations in wind patterns and rainfall. Unfortunately, the difficulty of measuring tropical precipitation is an obstacle in advancing understanding and prediction in this crucial portion of the earth system. Because of inaccessibility to ground-based sensors and the high costs of surface radar networks, space measurements offer the only means to obtain rainfall and its variability over the global tropics. While useful qualitative descriptions of the linkages have been derived using proxy variables for rain from cold IR signals (related to high cold cloud tops) on several satellites, more quantitative measurements are needed in the nearest possible future, confirmed by a substantial ground truth program. A relatively inexpensive satellite tropical rainfall measuring mission (TRMM) is mentioned, which, by means of international cooperation, can be launched by the mid-1990's. A low-altitude, low-inclination space vehicle is intended to carry the first spaceborne precipitation radar, a multi-channel passive microwave radiometer, and high-resolution visible and infrared sensors. The last part of the report presents the goals and priority science questions for TRMM.

1. PRECIPITATION: WHAT WE KNOW AND WHAT WE NEED TO KNOW

1.1 THE IMPORTANCE OF PRECIPITATION IN EARTH SYSTEM SCIENCE

The sciences involving the individual parts of the earth, such as the atmosphere, the oceans and the biosphere, have advanced to the point where it is both possible and necessary to treat the whole earth as a system with mutually interacting components. In particular, to better understand and predict the large variations in weather and climate, we must explicitly incorporate into our models the coupling between atmosphere and the surface below, which is ocean over 75% of the globe. The hydrological cycle is the primary coupling between land, sea and air. The phase changes of water substance are unique to this planet and are necessary for life as we know it. Precipitation is both one of the key products of the hydrological cycle and, at the same time, its latent heat release is a main driver of circulations in both the atmosphere and the oceans.

A. DEEPAK Publishing
ISBN 0-937194-14-4

1.2 TROPICAL PRECIPITATION: ITS ROLE IN ATMOSPHERIC AND OCEANIC CIRCULATION

The majority of global precipitation occurs in the tropics, between latitudes 30° N and 30° S. It has been long known that the latent heat released by tropical precipitation plays a vital role in driving low-latitude circulations and in supplying energy to balance the radiational heat losses and fuel the wind systems of middle and high latitudes. In the context of the meridional circulation of the Hadley cells on either side of the equator, the important role of the "hot towers" in their equatorial ascending portions was postulated and confirmed (Riehl and Simpson, 1979). In fact, the conversion of energy by precipitating cloud systems in this narrow zone is a major term in the global heat energy budget (Malkus, 1962). More recently, attention has been focused on zonal circulations, which have more dramatic variability, particularly the Walker cell in the tropical Pacific and its portentous break down every two to seven years in an El Niño/Southern Oscillation (ENSO). The Walker cell (W) and its relationship to the meridional Hadley cell (H) under normal conditions are illustrated schematically in Fig. 1. The larger-scale impacts of its anomalies in an ENSO condition are sketched in Fig. 2.

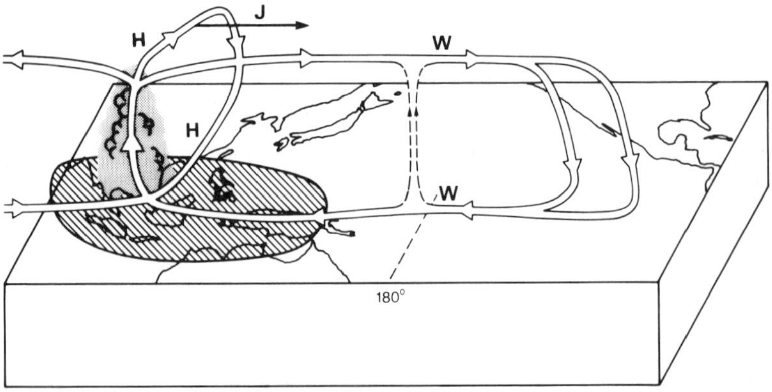

Figure 1. Schematic illustration of normal interaction between the Hadley (H) and Walker(W) circulations in the Pacific basin in the northern hemisphere winter. The descending branch of the Walker cell is in the eastern Pacific, offshore from California, Ecuador and Chile, where cold ocean surface temperatures prevail due to upwelling. The thin straight arrows are upper-tropospheric jets, the easternmost being much stronger. Note the mutual reinforcement between the ascending branches of both cells in the region of the "maritime continent". In the southern (summer) hemisphere, heavy monsoon rainfall extends as far south as northern Australia. The dashed lines just east of the dateline show where the Walker cell breaks down, and where ascent occurs during an ENSO event (Modified from Lau and Li, 1984).

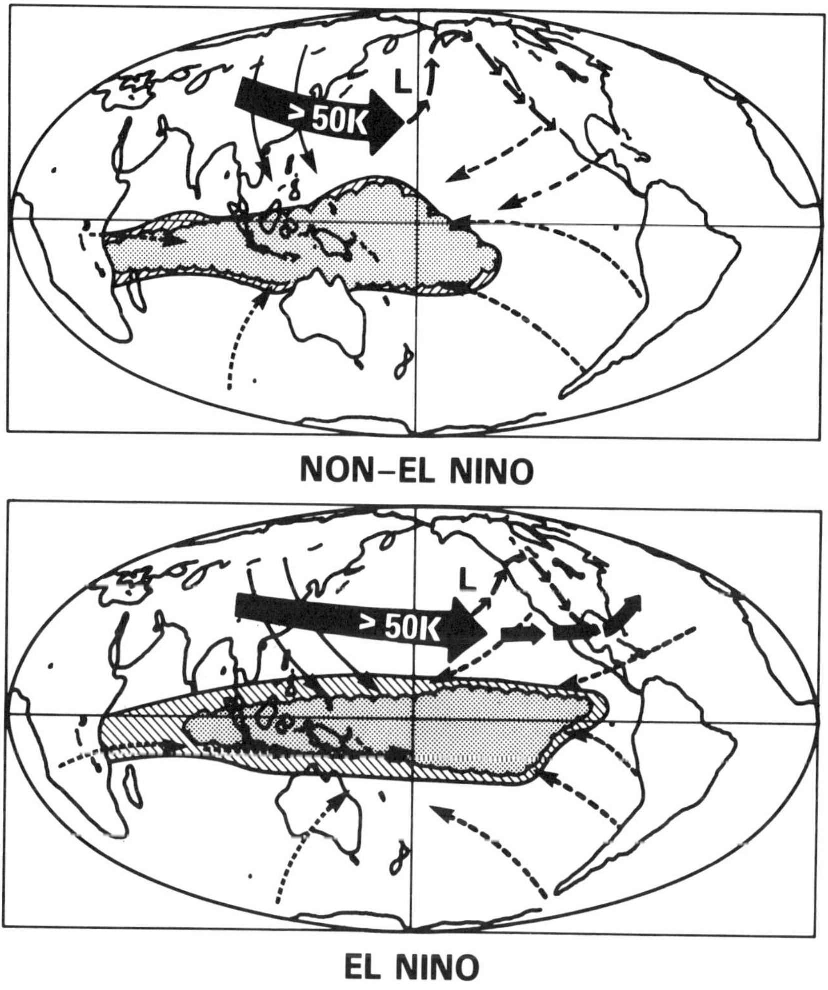

Figure 2. Comparison between normal, non-El Niño, conditions with El Niño conditions. The areas with scalloped outlines are those of high cold clouds, assumed to be the locations of precipitation heat release. The underlying smooth outlined areas are warm sea surface temperature anomalies. Note how the warm, convective "boiler box" moves eastward when the ENSO occurs, indicating that the Walker circulation has broken into two parts with ascent in the central Pacific. The thin arrows denote low-level windflow; the northeast and southeast trades are dashed, while the cold surges in the winter hemisphere are the slightly heavier solid arrows. The upper level outflow from the boiler box (not shown) has pronounced impacts on the high-tropospheric jet streams (heavy solid arrows). In an ENSO situation, the altered upper-level outflow from the boiler box changes the jet configuration. Thereby the extra-tropical weather is altered significantly, frequently with disastrous consequences. (After Webster, 1987)

1.3 A NEED FOR MORE ACCURATE DETERMINATION OF TROPICAL RAIN

Many models and theories have been advanced to explain the ENSO and its widespread teleconnections (Rasmusson, 1987). Verification and improvement of these theories suffer from lack of quantitative knowledge of the latent heat release and how it varies. Lau and colleagues (Lau, 1987) show evidence of a relationship with the 30-60 day tropical oscillation and the propagation of Kelvin-type waves which grow by a positive feedback with precipitation (the so-called wave CISK). To test these concepts, improved knowledge of the vertical profile of latent heat is required. Houze (1987) and Zipser (1987) infer from intensive (but time-and space-limited) field measurements that the heating profile will depend upon the ratio of convective core and stratiform rain within the tropical cloud clusters. However, the latent heat profile over a substantial area has not yet been quantitatively related to the ratio of convective to anvil rain. Therefore, reliable deductions of latent heat profiles from routinely made observations await 1) improved measurements of tropical rain, including its vertical distribution, combined with 2) cloud population models to relate the rain amount and vertical distribution to the actual condensation heating and evaporational cooling by cloud processes, averaged over the desired area.

Models of large-scale tropical circulations (Shukla, 1987) have progressed to the point where they can predict precipitation amounts and locations, but measurements are lacking for initializing the models and testing how they parameterize rain-producing processes. Many of the predictive problems of GCM's in midlatitudes have been attributed to tropical forecasts which are faulty owing to lack of rain and wind data (Paegle and Baker, 1983). Moreover, near the equator, the atmosphere and oceans are closely coupled systems (Philander, 1979). The fresh water supplied to the upper ocean by rain has been estimated to create thermohaline driving forces as strong as those contributed by wind stress (Webster, 1987). Monthly rainfalls over the global tropics, with spatial resolution of about 5°, are greatly needed to address these problems.

2.0. APPROACH TO A TROPICAL RAINFALL MEASURING MISSION (TRMM) FROM SPACE

2.1. COMBINATION PASSIVE AND ACTIVE MICROWAVE (RADAR), WITH VISIBLE AND INFRARED SENSORS IN A LOW-ELEVATION, LOW-INCLINATION ORBIT

A relatively inexpensive satellite measuring mission for tropical rainfall is planned for launch in 1993 or 1994 as a cooperative effort between the United States and Japan, with other nations involved in the surface validation sites. It is now in the design stage. The sensors, instruments, and sampling accuracy, and rainfall retrieval methods will be described later in this volume. The goals and priority science questions are outlined in Tables 1 and 2.

2.2. TRMM GOALS AND EXPECTED ADVANCES

TABLE 1 - TRMM GOALS

I. TO ADVANCE THE EARTH SYSTEM SCIENCE OBJECTIVE OF UNDERSTANDING THE GLOBAL ENERGY AND WATER CYCLES BY MEANS OF PROVIDING DISTRIBUTIONS OF RAINFALL AND LATENT HEATING OVER THE GLOBAL TROPICS

II. TO UNDERSTAND THE MECHANISMS THROUGH WHICH CHANGES IN TROPICAL RAINFALL INFLUENCE GLOBAL CIRCULATION, AND TO IMPROVE OUR ABILITY TO MODEL THESE PROCESSES IN ORDER TO PREDICT GLOBAL CIRCULATION AND RAINFALL VARIABILITY AT MONTHLY AND LONGER TIME SCALES

III. TO EVALUATE A SPACE-BASED SYSTEM FOR RAINFALL MEASUREMENT

TABLE 2 - EXPECTED SCIENTIFIC ADVANCES

- VALIDATION OF GLOBAL CIRCULATION MODELS
- VERTICALLY INTEGRATED LATENT HEAT RELEASE AND ITS VERTICAL DISTRIBUTION FOR MODEL INITIALIZATION
- CALIBRATION OF IR RAINFALL ESTIMATION TECHNIQUES
- DIRECT OCEANIC PRECIPITATION MEASUREMENTS TO VALIDATE MODELS OF THERMOHALINE CIRCULATIONS
- BIOMASS RELATIONSHIPS - DIAGNOSTIC STUDIES - DROUGHT AND/OR ANOMALOUS RAINFALL
- CLIMATE VARIATIONS: PREDICTABILITY STUDIES AND IMPROVED MODELS

2.3. TRMM PRIORITY SCIENCE QUESTIONS

The TRMM Science Steering Group (Rasmusson, 1988) has agreed on the priority science questions in Table 3.

TABLE 3 -TRMM PRIORITY SCIENCE QUESTIONS

1) WHAT IS THE 4-DIMENSIONAL STRUCTURE OF LATENT HEATING IN THE TROPICAL ATMOSPHERE? HOW DOES IT VARY DIURNALLY, INTRASEASONALLY, SEASONALLY AND ANNUALLY?

2) WHAT IS THE ROLE OF LATENT HEAT RELEASED IN THE TROPICS IN BOTH TROPICAL AND EXTRA-TROPICAL CIRCULATIONS?

3) WHAT IS THE MONTHLY AVERAGE RAINFALL OVER TROPICAL OCEAN AREAS OF ABOUT 10^5 KM^2 AND HOW DOES THIS RAIN AND ITS VARIABILITY AFFECT THE STRUCTURE AND CIRCULATION OF THE TROPICAL OCEANS?

4) WHAT IS THE RELATIONSHIP BETWEEN PRECIPITATION AND CHANGES IN THE BOUNDARY CONDITIONS AT THE EARTH'S SURFACE (E.G. SEA SURFACE TEMPERATURES, SOIL PROPERTIES, VEGETATION)?

5) WHAT IS THE DIURNAL CYCLE OF TROPICAL RAINFALL AND HOW DOES IT VARY IN SPACE?

6) WHAT IS THE RELATIVE CONTRIBUTION OF CONVECTIVE AND STRATIFORM PRECIPITATION AND HOW DOES THE RATIO VARY IN DIFFERENT PARTS OF THE TROPICS AND IN DIFFERENT SEASONS?

7) HOW CAN IMPROVED DOCUMENTATION OF RAINFALL IMPROVE UNDERSTANDING OF THE HYDROLOGICAL CYCLE IN THE TROPICS?

3. CONCLUDING REMARKS

This paper has emphasized the TRMM science needs and key science questions. In the remainder of this volume, it will be shown that the TRMM participants (NASA, with university colleagues and with the cooperation of other nations, particularly Japan) are ready to conduct this mission and that they, together with the global change and earth system science communities, are eager to analyze and utilize the measurements.

ACKNOWLEDGMENT

The TRMM Science Steering Group has jointly worked out much of the material, particularly the goals and science questions presented.

REFERENCES

Houze, R. A., Jr., 1987: Convective and stratiform precipitation in the tropics. This vol.

Lau, K.-M., 1987: Quasi-stationary and transient fluctuations in latent heat release from tropical precipitation: Their role in the atmospheric circulation. This vol.

Lau, K.-M. and M.-T. Li, 1984: The monsoon of East Asia and its global associations--A survey. Bull. Amer. Meteor. Soc.,65, 114-125.

Malkus, J. S., 1962: Large-Scale Interactions. Chap. 4 in The Sea: Ideas and Observations on Progress in the Study of the Seas. John Wiley and Sons, New York, 88-294.

Paegle, J. and W. E. Baker, 1983: The influence of the tropics on the prediction of ultralong waves. Part II: Latent heating. Mon. Wea. Rev., 111, 1356-1371.

Philander, S, G. F., 1979: Variability of the tropical oceans. Dyn. Oceans Atmos., 3, 191-208.

Rasmusson, E. M., 1987: Time-space scales of low frequency tropical rainfall. This vol.

Rasmusson, E. M., Ed.,1988: Report of the TRMM Science Steering Group. In progress.

Riehl, H. and J. Simpson, 1979: The heat balance of the equatorial trough zone, revisited. Contr. to Atmos. Phys., 52, 287-305.

Shukla, J. , 1987: Simulation of rainfall variability over the tropical oceans. This Vol.

Webster, P.J., 1987: The role of hydrological processes in ocean/atmosphere interaction. This vol.

Zipser, E. J., 1987: The evolution of mesoscale convective systems: Evidence from radar and satellite observations. This vol.

OVERVIEW OF JAPAN TRMM PROGRAM

A. Sumi
Geophysical Institute, University of Tokyo
Bunkyo-ku, Tokyo 113, Japan

ABSTRACT

Present status of TRMM activities by the Japanese meteorological community is reviewed and the requirement for TRMM is proposed.

1. PRESENT STATUS OF ACTIVITIES FOR TRMM IN JAPAN

At first, the TRMM project was proposed personally to Professor T. Matsuno (Univ. of Tokyo) by Dr. Geller (NASA). After that, through frequent meetings between US and Japanese specialists, an atmosphere promoting the TRMM project has been gradually generated in the meteorological community. However, the attitude of Japanese meteorologists was not yet definite.

The decision was made when Professor Takeda returned from the Global Energy and Water Cycle Experiment (GEWEX) Scientific Steering Group Meeting. Japan was expected to promote the TRMM project, because the measurement of precipitation is crucial for the success of the GEWEX project and because satellite observation is the only possible way to observe the precipitation over the ocean. Then, a working group for the TRMM project, which is chaired by Professor T. Matsuno, was established under the committee for the World Climate Research Program (WCRP) of the Japanese Scientific Council. People in both universities and government agencies are participating in this group. At the same time, the working group was established in the JMA to discuss the TRMM project.

To push the TRMM project a booklet was published and distributed to the people of the various fields in Japan. We are expecting that the booklet and this symposium will accelerate the TRMM project.

2. SCIENTIFIC OBJECTIVES FOR TRMM

The scientific objectives for TRMM that were discussed in Japan are summarized as follows:

(A) Quantitative measurement of rainfall in the tropics and GCM validation.

The importance of measuring the rainfall in the tropics is obvious, and it is clear that GCM validation by using rainfall data will enhance our understanding of the climate system. However, it should be noted that there is a difference between the heating and the precipitation, although they are closely related. Heating profiles in the tropics are essential to validate our GCM results. It should be investigated how to retrieve the heating profile from the precipitation data.

TROPICAL RAINFALL MEASUREMENTS
John S. Theon and Nobuyoshi Fugono (Eds.)

ISBN 0-937194-14-4

(B) Documentation of intraseasonal variability (30-60 day oscillation).

The discussion of (A) also holds for this item. Recently, the importance of the heating profile has been proposed for the 30-60 day oscillation (M. Takahashi, 1987; C.P. Chang and H. Lim, 1986). Direct estimation of the heating profile of 30-60 day oscillation is crucial to understanding the dynamic aspect of this oscillation.

(C) Monsoon - Baiu and Meiu front

Monsoonal rainfall is a typical rainfall in the tropics. However, monsoon activity in the northern hemisphere summer lies in the subtropics. Rainfall in the Baiu - and Meiu - front is also a typical example for this subtropical rainfall system. Furthermore, rainfall is realized in the mesoscale clusters in the Baiu front. In order to realize the behavior of the mesoscale systems, finer space and time resolution is required. To achieve this requirement, TRMM alone is not sufficient but the integration of existing satellite systems (GMS, polar-orbiting satellite, and TRMM satellite) will be necessary.

(D) Improvement of initial fields for numerical prediction.

In the tropics, the balance between flow field and forcing is important, and this imbalance rapidly deteriorates the forecasted fields. If we can observe forcing fields with a reasonable accuracy, we can improve the initial fields by using the diabatic nonlinear normal mode initialization. This improvement will possibly extend the predictability.

(E) Cloud clusters, tropical disturbances and the typhoons in the tropics.

Recently, much attention has been paid to the detailed structure of cloud clusters and tropical disturbances in Japan. The integrated rainfall evaluation system composed of GMS, polar-orbiting satellite and TRMM satellite will contribute to the understanding of the dynamics of cloud clusters, and tropical disturbances (typhoons) in the tropics and the subtropics.

3. REQUIREMENTS AND RECOMMENDATIONS

In order to achieve the scientific objectives described in section 2, the following requirements are necessary in addition to the original requirements proposed by NASA.

1) As to TRMM orbits, they should cover at least from 37°S to 37°N in order to achieve the scientific objective (C).
2) Special efforts should be made to establish the "integrated rainfall measurement system based on the space platform (GMS, polar-orbiting satellite and TRMM satellite)."

In order to achieve the success in the TRMM project, the ground truth experiments are indispensable. With respect to this, Japan will contribute the following:

1) Over the area of Japan, the composite rainfall measurements using the surface rain-gauge network (AMEDAS) and the digitized radar system are routinely taken. In the future, GMS radiance will be included. These data are useful for calibration of TRMM observations.

2) A special observational project is expected for the mesoscale phenomena in the Baiu-front over the Ryukyu islands. This project will be useful as a ground truth experiment for TRMM.

3) Japan is now considering construction of the International Equatorial Radar Observatory at Pontianak (West Borneo), which is composed of the huge MST radar, two microwave doppler radars and two wind profilers. This facility will be useful for a calibration for TRMM.

4. SUMMARY

The TRMM project proposed by NASA has thus far been welcomed by the Japanese scientists, and they are trying to realize its objectives in Japan. The basic idea of launching the active sensor on the space platform is exciting and its possibilities for atmospheric sciences are immeasurable. We are very pleased to participate in this project. Although this project has not yet been officially sanctioned in Japan, we believe it will be. We believe that close cooperation between Japanese and U.S. scientists will help this project achieve success.

REFERENCES

Chang, C.P., and H. Lim, 1986: Kelvin wave-CISK: A possible mechanism for the 30-50 day oscillations, International Conference on Monsoon and Mesoscale Meteorology with MSROC, Taipei, Taiwan.

Takahashi, M., 1987: A theory of the slow phase speed of the intraseasonal oscillation using wave-CISK, J. Meteo. Soc. Japan, 65, 43-49.

TIME-SPACE SCALES OF LOW FREQUENCY TROPICAL RAINFALL VARIABILITY

E.M. Rasmusson
University of Maryland
College Park, Maryland 20742, USA
and
C.F. Ropelewski and M.S. Halpert
Climate Analysis Center, NMC, NWS, NOAA
Washington, DC 20233, USA

ABSTRACT

Unlike the small-scale synoptic features which appear on daily weather maps, the long-term, time-averaged circulation and precipitation fields exhibit large-scale features associated with the annual evolution of the monsoons, tradewind systems and oceanic convergence zones. Monthly-seasonal anomalies also exhibit large-scale features. The dominant intraseasonal pattern is associated with the global-scale Madden-Julian (30-60 day) oscillation. Interannual variability is dominated by tropical Pacific ocean-atmosphere interactions. These interactions are associated with the El Nino/Southern Oscillation(ENSO) cycle, whose major swings, which occur at irregular intervals of 2-7 years are associated with pronounced year-to-year precipitation variations over large areas of the tropics. The large-scale spatial coherence and systematic evolution of both intraseasonal and interannual tropical precipitation anomalies is of great significance to climate prediction. They are also important to the development of optimum sampling strategies for estimating accumulated precipitation over periods of a few weeks to a few months.

1. INTRODUCTION

The atmospheric forcing that results from the spatial variations in latent heating is a fundamental factor in the maintenance of the large-scale circulation systems of the tropics. It is also a key element in the dynamics of coupled ocean-atmosphere-land surface interactions. A quantitative knowledge of the time-averaged rainfall distribution is, therefore, of crucial importance for understanding and possibly predicting tropical climate variability. Satellite "snapshots" reveal the basic mesoscale character of convective precipitation associated with tropical synoptic disturbances. When these "high frequency" variations are averaged over several weeks or longer, the large-scale, coherent rainfall patterns associated with the slowly varying planetary scale features of the tropics emerge. Intraseasonal and interannual variations of these features, i.e. in the seasonal monsoons, tradewind systems and oceanic convergence zones, are also associated with large-scale circulation and rainfall anomaly patterns.

ISBN 0-937194-14-4

2. ANNUAL CYCLE

Figure 1 shows average outgoing longwave radiation (OLR) values derived from NOAA operational polar orbiter satellite measurements for the two solstice seasons. Experience suggests that areas in the tropics where time-averaged OLR values are less than 240 W/m^{-2} , i.e. areas where the largest amounts of OLR originate from the cold cloud tops of convective systems, bear a general qualitative correspondence with the regions of heavy convective rainfall and mean upward motion. One large-scale OLR minimum (implied precipitation maximum) is located over Africa, another over South America/Central America; both migrate north-south with the high sun season. The third and most extensive area is centered over the land areas and adjacent warm waters of the Indonesian-Australian-East Asian monsoon region, with eastward extensions along the the ITCZ north of the equator, and the South Pacific Convergence Zone (SPCZ) in the southwest Pacific. The implied regions of heavy precipitation mark the upward branches of the time-averaged direct thermal circulations (warm, moist air rising; cool, dry air sinking). These provide the potential to kinetic energy conversion that serves as a major source of energy maintaining the circulation of the tropics (Rasmusson and Arkin, 1985).

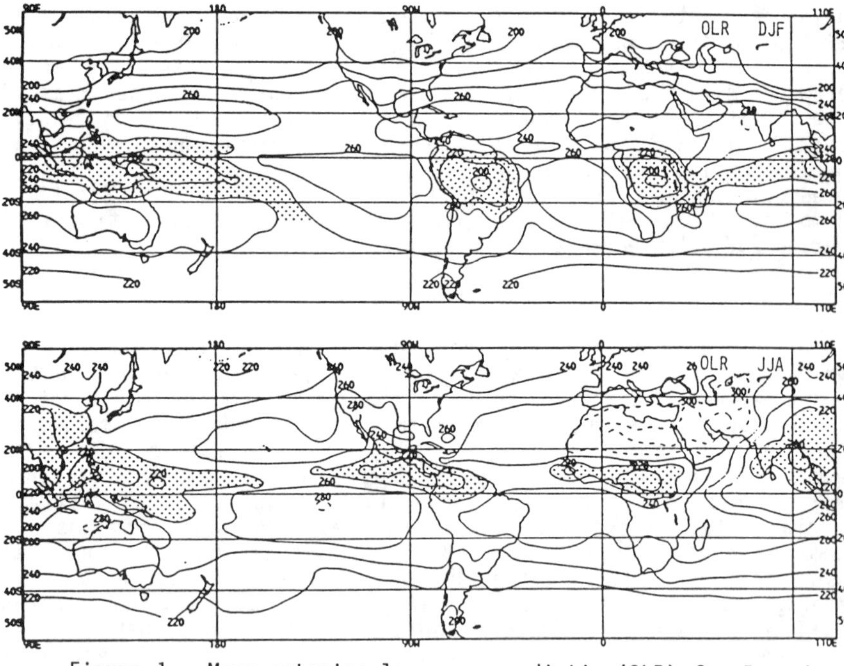

Figure 1. Mean outgoing long-wave radiation(OLR) for December-February (upper) and June-August (lower). Areas where OLR is less than 240 W/m^{-2} are stippled. Contour interval is 20 W/m^{-2}.

The annual cycle in tropical circulation and rainfall is closely related to the annual cycle of surface temperature, which, for most of the tropics, is sea surface temperature(SST). Even though the amplitude of the annual cycle of SST is relatively small, the large-scale pattern of tropical convection is quite sensitive to these changes. Specifically, the tropical oceanic regions of OLR less than 240 W/m^{-2} lie mostly within the 28°C SST isotherm. The east-west SST difference between the upwelling region of the eastern equatorial Pacific, and the huge "warm pool" that overlays the west Pacific-eastern Indian Ocean monsoon region is particularly important. Comparison of Figure 1 with mean SST charts (not shown) reveals a good correspondence between the location and seasonal migration of the west Pacific-Indonesian warm pool and the monsoon precipitation regime. Thus, the time-averaged planetary-scale thermal circulations not only reflect the classical land-sea monsoon forcing, but also reflect a pronounced forcing due to zonal SST gradients. This gives rise to the idealized east-west "Walker Circulation" that spans the equatorial Pacific basin (Bjerknes, 1969).

For a full understanding of both interannual and intraseasonal variability, it seems essential to view the annual cycle as a low-frequency mode of global climate variability. Although fundamentally driven by the annual variation in incoming radiation, the direct solar forcing is significantly modified by the inertia of the coupled ocean-atmosphere-land system. Year-to-year variations in the annual cycle are due in part to low frequency modes of ocean-atmosphere interaction.

The circulation systems of the atmosphere and the ocean are coupled through radiation and heat exchange at the air-sea interface. However, in the deep tropics, where the time scale of the oceans is relatively short, and that of the slowly evolving planetary-scale atmospheric circulations is relatively long, there is, in addition, a close dynamical coupling on time scales as short as a few months (Philander,1979).

The three key parameters in the coupling of the tropical ocean and atmosphere are SST, surface wind stress, and rainfall. What would otherwise be a weak thermal coupling is strongly amplified by the atmospheric latent heating associated changes in the large-scale rainfall regime. They produce significant changes in the low-level atmospheric circulation, that in turn leads to changes in the ocean upwelling and current systems. These, in turn, produce further changes in the SST, thus closing an interactive feedback loop. Large-scale convection is therefore a key element in this interactive process.

We now have reasonably good estimates of mean monthly SST. There is optimism that scatterometer data will lead to substantial improvement in the estimation of surface stress. Present estimates of precipitation over the tropical oceans are inadequate for a quantitative description of the coupled interactions.

3. INTERANNUAL VARIABILITY

The amplitude of the annual cycle of SST is quite small over the western and central equatorial Pacific. The interannual variance is of the same order of magnitude as the variance of the mean annual cycle itself. Furthermore, the zonal SST gradient is so weak that an anomaly of 1°C-2°C is sufficient to produce a shift of several tens of degrees longitude in the 28°C SST isotherm and related region of heavy rainfall. The associated change in atmospheric heating excites teleconnections both within the tropics and between the tropics and higher latitudes resulting in a global pattern of climate anomalies. (Ropelewski and Halpert, 1987). This systematic pattern of climate fluctuations is associated with the El Niño/Southern Oscillation (ENSO) phenomenon (Rasmusson, 1985).

ENSO is the most notable and pronounced example of global climate variability on the interannual time scale. The nearly simultaneous appearance of pronounced climate anomalies around the world, together with the disruption of the marine ecosystems along the west coast of the Americas, can have serious weather and climate consequences and major adverse effects on regional and global food production.

The "warm" phase of the ENSO Cycle has received the most attention (Rasmusson and Carpenter, 1982). These episodes often begin early in the year, reach their mature phase around the end of that year, and decay during the following six months (Figure 2).

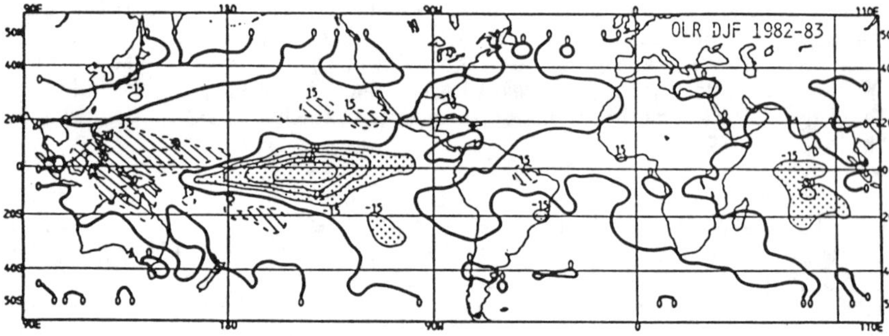

Figure 2. Average OLR anomaly for December-February, 1982/83. Contour interval 15 W/m^{-2}. Areas less than -15 W/m^{-2} stippled; areas greater than + 15 W/m^{-2} hatched. This diagram illustrates the scale and distribution of precipitation anomalies during the mature phase of the 1982/83 ENSO warm episode. The Pacific anomaly pattern developed during the middle of 1982, migrated eastward across much of the Pacific basin and weakened during the middle of 1983.

While the tropical circulation response to heating is quite well understood in terms of simple equatorial wave dynamics, e.g., Gill(1980), the way in which tropical heating influences extratropical climate is rather complex, being strongly dependent on the climatological mean circulation and the location and distribution of the heating itself. The teleconnection process, through which an atmospheric response is generated outside the region of direct forcing, has been the subject of intense study and lively debate during the past few years, but the issue is yet to be resolved (Horel and Wallace, 1981; Simmons et al., 1983; Palmer and Owen, 1986).

Many aspects of tropical rainfall variability are related to factors other than ENSO. This includes the multi-year droughts of the Sahel, and the recurrent droughts of northeast Brazil. For example, although the typical ENSO-related drought pattern which occurred over southeast Africa during the rainy season of 1982-83 was among the worst of the century, it was only one aspect of a more pervasive and longer lasting period of drought that afflicted much of sub-Saharan Africa between 1982-84 (Rasmusson,1987). For these and other features of tropical variability one must adopt an even broader view, which involves the entire tropics, both land and ocean, as well as a possible links with the extratropics.

4. INTRASEASONAL VARIABILITY

There is also a high degree of coherence in tropical variations on time-scales centered around 40-50 days. The first evidence of organized large amplitude intraseasonal oscillations was reported by Madden and Julian (1971; 1972). The oscillations are particularly evident in rainfall and wind fluctuations over the Indian Ocean and the Indonesian-North Australian sector of the low-latitude tropics (Figures 3 and 4).

The period of the oscillation generally ranges between 30-60 days. Studies with OLR data indicate that the oscillation typically first appears as a convective outburst in the western equatorial Indian Ocean, then migrates eastward, reaching maximum amplitude in the Indonesian-north Australian region after which it weakens or disappears over the central and eastern Pacific. However, the oscillation can often be traced around the entire equatorial belt in the upper tropospheric wind field. The oscillation may play an important role in intraseasonal monsoon variability. There is evidence that it also affects the extratropical winter hemisphere (see, for example, Knutson and Weickmann, 1987). Thus, this mode of variability is of vital importance to medium-range weather prediction and short-term climate variability.

There has been significant progress in our theoretical understanding of the 30-60-day oscillation during the past two years. GCM experiments (Lau and Lau, 1986; Hayashi and Sumi, 1986) have pointed to internal atmospheric dynamics associated with tropical convection as the basic cause of these oscillations. Recent studies suggest that the details of the vertical heating profile from convection and low-level heating from boundary layer processes and surface evaporation may be important in modifying the time scale of the oscillation (Lau and Chan, 1987).

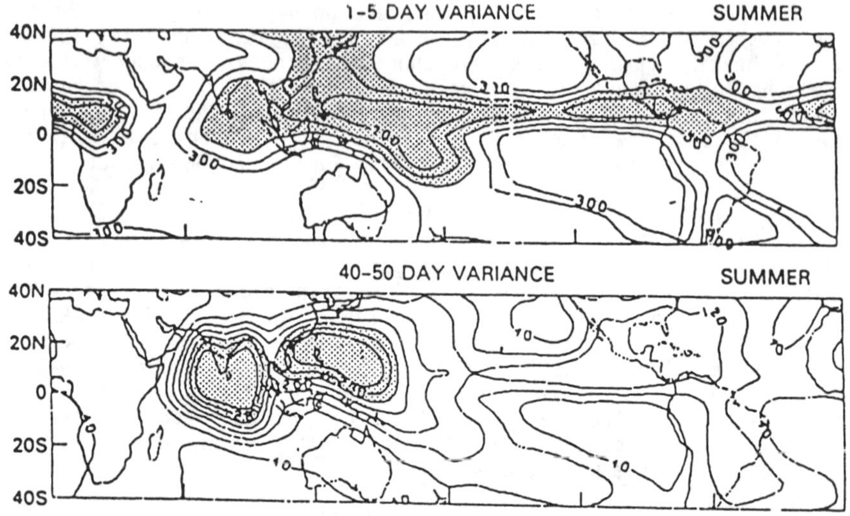

Figure 3. 1 to 5 day (upper) and 40-50 day (lower) components of OLR variance (summer). Contour interval: 200 $(W\ m^{-2})^2$ for 1-5 day; 40 $(W\ m^{-2})^2$ for 40-50 day. Note in this and Figure 4 the different patterns of synoptic and intraseasonal variability.

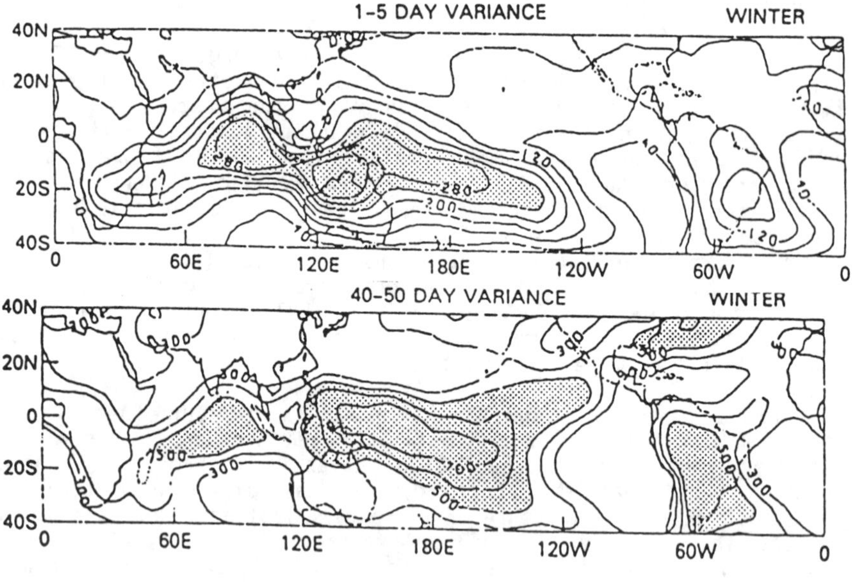

Figure 4. Same as Figure 3 but for winter.

One of the most intriguing observations of the 30-60- day oscillation is the remarkable similarity between the spatial structure and evolution of the atmospheric anomalies with those observed during ENSO (Lau, this meeting). While the 30-60- day oscillations show little dependence on coupled ocean-atmosphere dynamics, they are nevertheless related to ENSO variability. The regions of intense convection are largely positioned by the SST field: thus, the amplitude of the 30-60- day mode in the convection and low-level circulation is zonally asymmetric. The maximum response typically occurs in the Indian Ocean-western Pacific monsoon sector, but during major ENSO episodes, the area of maximum amplitude shifts eastward with the eastward migration of the 28° C SST isotherm.

5. CONCLUDING REMARKS

For an understanding of the dynamics of low frequency variability and coupled ocean-atmosphere land surface interactions in the tropics, it is crucial to have a quantitative knowledge of the time averaged distribution of precipitation. While adequate resolution of the synoptic variations requires almost continuous observations in time and space, the large-scale coherence of the low-frequency anomaly pattern eases the sampling problem (North, 1987).

REFERENCES

Bjerknes, J., 1969: Atmospheric teleconnections from the equatorial Pacific. Mon. Wea. Rev., 97, 820-829.

Gill, A. E., 1980: Some simple solutions for heat induced tropical circulation. Quart. J. Roy. Meteor. Soc., 106, 447-462.

Hayashi, Y. Y., and A. Sumi, 1986: The 30-40 day oscillations simulated in an "aqua planed" model. J. Meteor. Soc. Japan, 61, 812-828.

Horel, J. D., and J. M. Wallace, 1981: Planetary scale atmospheric phenomena associated with the Southern Oscillation. Mon. Wea. Rev., 109, 813-829.

Knutson, T. R., and K. M. Weickmann, 1987: 30-60 day atmospheric oscillations: composite life cycles of convection and circulation anomalies. Mon. Wea. Rev., 115, 1407-1436.

Lau, N-C, and K.-M. Lau, 1986: The structure and propagation of intraseasonal oscillations appearing in a GFDL general circulation model. J. Atmos. Sci., 43, 2023-2047.

Lau, K.-M., and P. H. Chan, 1987: Intraseasonal and interannual variations of tropical convection: A possible link between the 40- day mode and ENSO?. Mon. Wea. Rev, 115, (in press).

Madden, R. A., and P. R. Julian, 1972: Description of global scale circulation cells in the tropics with a 40-50 day period. J. Atmos. Sci., 29, 1109-1123.

Madden, R. A., and P. R. Julian, 1971: Detection of a 40-50 day oscillation in the zonal wind in the tropical Pacific. J. Atmos. Sci., 28, 702-708.

North, G. R., 1987: Survey of sampling problems for TRMM. Proceedings, International Symposium on Tropical Precipitation Measurements. Tokai University, Tokyo, Japan, 28-30 Oct. 1987.

Palmer, T. N., and J. S. Owen, 1986: A possible relationship between some "severe" winters in North America and enhanced convective activity over the tropical west Pacific. Mon. Wea. Rev., 114, 648-651.

Philander, S.G.F., 1979: Variability of the tropical oceans. Dyn. Oceans Atmos., 3, 191-208.

Rasmusson, E. M., 1987: Global climate change and variability: effects on drought and desertification in Africa. Drought and Hunger in Africa: Denying Famine a Future. M. H. Glantz, Editor. Cambridge University Press, Cambridge, U.K. pp. 3-22.

Rasmusson, E. M., 1985: El Nino and variations in climate. American Scientist, 73, 168-178.

Rasmusson, E. M. and P. A. Arkin, 1985: Interannual climate variability associated with the El Nino/Southern Oscillation. Coupled Ocean-Atmosphere Models., Elsevier Science Publishers B.V., Amsterdam, 697-725.

Rasmusson, E. M. and T. H. Carpenter, 1982: Variations in tropical sea surface temperature and surface wind fields associated with the Southern Oscillation/El Nino. Mon. Wea. Rev., 110, 354-384.

Ropelewski, C. F., and M. S. Halpert, 1987: Global and regional scale precipitation patterns associated with the El Nino/Southern Oscillation. Mon. Wea. Rev., 115, 1606-1626.

Simmons, A. J., Wallace, J. M. and G. W. Branstator, 1983: Barotropic wave propagation and instability and atmospheric teleconnection patterns. J. Atmos. Sci., 40, 1363-1392.

GLOBAL ENERGY AND WATER CYCLE EXPERIMENT (GEWEX)

Robert A. Schiffer
Office of Space Science & Applications
National Aeronautics and Space Administration
Washington, DC 20546, USA
and
Joint Planning Staff for the WCRP
World Meteorological Organization
Geneva, Switzerland

1. INTRODUCTION

The atmospheric circulation transports both energy and water. Heat from the tropics is transferred to polar regions. Water, evaporated from the oceans and the land surface, falls as rain or snow often in places far removed from the major evaporation areas. The changing distribution of rain and snowfall dominates the impact of weather and climate on man and his activities.

The Global Energy and Water Cycle Experiment (GEWEX), a proposed new initiative of the World Climate Research Program (WCRP), will concentrate on improving our knowledge of the transport of water and energy within the atmospheric system. Its scientific objectives are:

- To describe and understand the transport of water (vapor, liquid and solid) and energy in the global atmosphere and at the underlying surface.

- To develop methods of predicting changes in the distribution of water (vapor, liquid and solid) within the global atmosphere and at the underlying surface, which may occur naturally or through the influence of man's activities.

A further objective of GEWEX is to promote scientific advances and technical developments, such as the deployment of new satellite systems and orbital testing of new remote–sensing instruments, which would find operational applications for improved weather forecasting and climate prediction. In addition, GEWEX will also promote the development of time-dependent representations and predictive models of the global transport and ground storage of water, and validation of these models against the observed annual and multi-annual evolution of available water resources and river discharge in several selected continent-scale domains.

The overall research strategy of the WCRP is formulated into three specific objectives, or "streams".

- Stream 1: Establishing the physical basis of long-range weather prediction (Atmospheric Prediction Research)

- Stream 2: Understanding the predictable aspects of global climate variations over periods of several months to several years (TOGA)

- Stream 3: Assessing the response of climate to natural or man-made influences over periods of several decades (WOCE)

A. DEEPAK Publishing
ISBN 0–937194–14–4

The TOGA (Tropical Oceans and Global Atmosphere) and WOCE (World Oceans and Global Atmosphere) programs are already underway and we can expect, by the mid-1990's, significant progress in our knowledge of the oceans and their interaction with the atmosphere. One of the largest remaining problems will then be the transports of water and energy within the atmosphere.

The present operational World Weather Watch (WWW) observing system does not meet the observational requirements of GEWEX with respect to describing the global distribution of precipitation, evaporation and water transport. A number of special WCRP data projects, based on existing observations, have been implemented in an attempt to improve the data coverage but the total volume of available information remains inadequate. The situation is further aggravated by the fact that the global hydrological processes are not adequately described by present general circulation models. Thus, GEWEX has been proposed to be the central project in WCRP Stream 1 to address these problems.

2. OBSERVATIONAL SYSTEMS

Most studies of global environmental problems rely, at present, on information provided by the worldwide network of surface and upper-air sounding stations organized by the WWW, and the system of 4 - 5 geostationary and 2 polar-orbiting meteorological satellites, which has been in operation since the GARP Global Weather Experiment in 1979. This near-real-time operational observing and data management system is complemented by other sources of mainly delayed-time oceanic data collected by other programs. The observing system suffers from several deficiencies which include:

• Serious gaps in the three-dimensional distribution of global wind observation, especially over the Indian Ocean

• Large uncertainties, mainly due to sampling errors, in the estimation of precipitation over land areas and total lack of information over the oceans

• Slow and inefficient transmission of essential marine meteorological and oceanographic observations

• Significant uncertainties in the three-dimensional global distribution of water vapor and clouds, and almost total lack of knowledge of the surface radiation fluxes and optical properties

• Widespread lack of detailed and quantitative geographical information on land use, vegetation types and properties

Major national and international efforts are underway in planning the next generation global operational meteorological satellite system. It is timely for the earth science community to introduce its own input, ideas and requirements into the debate on future space platform plans. It is envisioned that the observational requirements for a global atmospheric energy and water research program could be met with the following (ideal) observing system:

• A constellation of five (essentially) identical geostationary satellites, each equipped with:

(i) A multi-spectral (visible and infrared), high-resolution imaging radiometer to observe cloud developments and cloud drifts within several atmospheric layers;

(ii) A medium resolution, infrared/ microwave atmospheric sounder to observe the time-dependent three-dimensional distribution of temperature and moisture.

The geostationary component is considered essential for determining the basic atmospheric variables (wind, temperature and moisture) as well as providing reasonably frequent observations of rapidly varying atmospheric processes (in particular, precipitation and various boundary layer phenomena with strong diurnal variations).

• An array of observing components, or packages, on a number of satellites in low-altitude earth orbit (LEO). These include:

(i) A basic operational meteorological package on one or two polar LEO spacecraft. This package includes an advanced visible and infrared radiometer, e.g., a later version of the AVHRR instrument with improved calibration of the radiometric channels, a temperature and moisture sounder combining the infrared and microwave spectrometric capabilities of AMSU and HIRS, and a data collection and location facility such as the ARGOS system.

(ii) A cloud structure and wind package to complement the basic geostationary observing system by providing quantitative information on the three-dimensional distribution of clouds, additional cloud-drift wind data and, if feasible, Doppler lidar wind data. The package would ideally be flown in both low-inclination and polar orbits, and include:

- a ranging altimetric lidar,

- a multi-spectral stereographic cloud imaging radiometer, providing time-lapse pictures of the same cloud scene to estimate cloud drift vectors,

- a microwave scatterometer to measure the surface wind or wind stress over the ocean,

- within the constraints of available technology, a suitable Doppler lidar which could constitute a first step in the development of the Lidar Atmospheric Wind Sounder (LAWS) concept, e.g., measuring one component of the wind only.

(iii) A precipitation package, comprised of two main instruments, to be flown on both low-inclination and polar platforms.

- multichannel passive microwave imaging radiometer including, to the extent possible, a low-frequency (C-band) radiometric channel for observation of ground moisture, and featuring a relatively wide imaging swath for global coverage;

- a two- frequency (10 - 15 and 35 GHz) high-resolution rain-measuring radar, possibly a full-size instrument with a relatively large antenna on the Space Station to provide a 200 km wide swath, without compromising the 0.5 km vertical resolution of the system, and more compact instruments on the polar platforms, e.g., using a single frequency only for the detection of the rain cloud brightband.

(iv) A radiation and surface flux package on two or more spacecraft in polar or high inclination orbits, and consisting of:

- an upgraded Earth Radiation Budget Experiment (ERBE) type instrument to measure the short- and long-wave components of the radiation budget at the top of the atmosphere, with improved absolute calibration, narrow field-of-view and adequate spectral discrimination for separating the effects of clouds, water vapor, carbon dioxide and other greenhouse gases;

- an aerosol ranging lidar capable of detecting the top of the boundary layer as well as measuring cloud altitudes and providing backscatter measurement of surface characteristics;

- a moderate resolution, multi-spectral (visible and infrared) imaging radiometer;

- a multi-spectral, multiple field-of-view thermographic radiometer to measure the surface temperature, e.g., the Along-Track Scanning Radiometer (ATSR).

A number of technical studies are currently underway on potential space-borne instruments which could contribute measurements needed for GEWEX. These include a Lidar Atmospheric Sounder and Altimeter (LASA), a precipitation radar for the Tropical Rainfall Measurement Mission (TRMM), a Laser Atmospheric Wind Sounder (LAWS), an Along-Track Scanning Radiometer (ATSR), and a MODerate-resolution Imaging Spectrometer (MODIS).

The success of GEWEX will be determined by the degree of enthusiasm and participation of a wide and heterogeneous scientific community, as much as by the availability of new observing techniques in space. The overall planning for GEWEX envisions a build-up phase for technological development (instruments, satellites, data handling, model improvements), an operational phase targeted for the period 1995 - 2000 when the new space missions supporting GEWEX are scheduled to be launched, and a scientific exploitation phase which would begin with exploratory research even before the operational phase and continue for many years thereafter.

Bibliography

National Aeronautics and Space Administration (NASA). 1986. Moderate-Resolution Imaging Spectrometer (MODIS) - Instrument Panel Report. Earth Observing System Report Vol. IIb. Washington, DC.

National Aeronautics and Space Administration (NASA). 1987. Lidar Atmospheric Sounder and Altimiter (LASA) - Instrument Panel Report. Earth Observing System Report Vol. IId. Washington, DC.

National Aeronautics and Space Administration (NASA). 1987. Laser Atmospheric Wind Sounder (LAWS) - Instrument Panel Report. Earth Observing System Report Vol. IIg. Washington, DC.

National Aeronautics and Space Administration (NASA). 1987. On Requirements for a Satellite Mission to Measure Tropical Rainfall. Reference Publication 1183. Washington, DC.

World Climate Research Program (WCRP). 1987. Space System Possibilities for a Global Energy and Water Cycle Experiment. (WCP-XX). Columbia, Maryland, January 19 - 23 1987.

CONVECTIVE AND STRATIFORM PRECIPITATION IN THE TROPICS[1]

Robert A. Houze, Jr.
University of Washington
Seattle, Washington 98195, USA

ABSTRACT

Tropical precipitation systems are characterized by extensive stratiform clouds directly associated with deep convection. Ice particles grown by riming in the mid-to-upper regions of the convective cells are detrained into the stratiform cloud where they travel great horizontal distances while growing slowly by vapor deposition. These particles undergo aggregation before reaching the melting layer. The horizontal patterns of convective and stratiform precipitation are complex and vary from case to case in tropical cloud systems. However, the distinct vertical and horizontal structures of the convective and stratiform regions make them amenable to identification from space, especially if the observations can distinguish both horizontal and vertical structure.

1. INTRODUCTION

Atmospheric precipitation patterns typically exhibit complicated patterns in time and space. Consequently, sophisticated sampling and measuring techniques, such as those proposed for the Tropical Rainfall Measuring Mission (TRMM), are required to describe precipitation accurately. The purpose of this paper is to show that despite its complexity, tropical precipitation structure remains understandable and explainable in terms of rather basic precipitation mechanisms. Awareness of these physical mechanisms is important, if not essential, to the proper design of technologies for determining amounts and patterns of precipitation on a global scale. In the following sections, a conceptual model of the basic mechanisms of tropical precipitation based on recent research results is presented. This model implies that observational technologies should take careful account of the vertical as well as the horizontal structure of tropical precipitation systems.

2. BASIC PRECIPITATION MECHANISMS IN THE TROPICS

Most elementary textbooks on meteorology, as well as advanced texts in cloud physics, recognize two types of mechanisms for the growth and fallout precipitation — stratiform and convective. It has long been recognized by tropical meteorologists, such as Malkus and Riehl (1964) and Ramage (1971), that both stratiform and convective precipitation mechanisms are important in tropical precipitation. The primary precipitation producers in the tropics are large mesoscale convective systems in equatorial regions ("cloud clusters"), tropical cyclones, which produce some 25% of the rain in certain regions of the tropics (Simpson and Simpson, 1966), and orographic cloud systems. Houze and Hobbs (1982) and Leary (1984) estimate that 25-50% of cloud-cluster rainfall is stratiform. Marks (1985) and Marks and Houze (1987) find that about 60% of hurricane inner-core region rain is stratiform. Estimates do not exist for the orographic rains, but similar proportions of stratiform rain would not be surprising since tropical

[1]This research was supported by NASA Grant No. NAG5-784

ISBN 0-937194-14-4

mountains probably trigger mesoscale systems rather similar to cloud clusters, but anchored to the topography.

3. A CONCEPTUAL MODEL OF PRECIPITATION MECHANISMS IN TROPICAL CLOUD CLUSTERS AND HURRICANES

In tropical cloud systems, stratiform precipitation typically occurs in direct association with deep intense convection—not as a separate phenomenon. The structure of the cloud systems is illustrated in vertical cross section in Fig. 1. The stratiform region is connected to a group of deep convective cells, which in the horizontal plane are often arranged in a line extending out of the plane of Fig. 1. Precipitation ice is generated as growing drops are carried up past the 0°C level by the convective updrafts. Thereafter the ice particles grow by riming as they accrete supercooled cloud drops forming in the updraft at mid-to-upper levels. Some of the particles growing by riming become dense

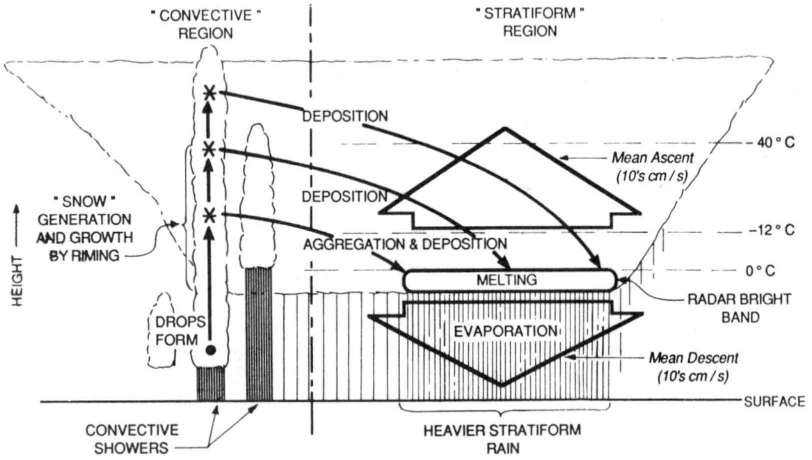

FIGURE 1. Schematic diagram of the precipitation mechanisms in a tropical cloud system. Solid arrows indicate particle trajectories.

graupel particles (terminal fall-speeds 2-6 m/s), while others, referred to as *snow* , become only moderately dense and fall more slowly (≈ 0.5-1 m/s). Substantial amounts of both graupel and snow are detrained to the environment during the lifetime of one of these cells (Fig. 2; from Ferrier and Houze — manuscript in preparation). Because of their large fall speed, the graupel particles fall quickly to the surface, within the convective region. The more slowly falling snow particles, however, are spread laterally through the stratiform cloud region by the horizontal wind as they slowly drift downward. The detrainment of the snow from the convective cells is thus the mechanism by which precipitating ice particles are introduced into the stratiform portion of the cloud system.

The most intense radar bright band and the heaviest stratiform rain at the surface occur where the convectively generated snow particles reach the 0°C level, after their passage through the stratiform cloud (Smull and Houze, 1985; Rutledge and Houze, 1987). The stratiform cloud through which the snow particles pass is characterized by mean upward air motion of 10s of cm/s (Leary and Houze, 1980; Gamache and Houze, 1982, 1983,

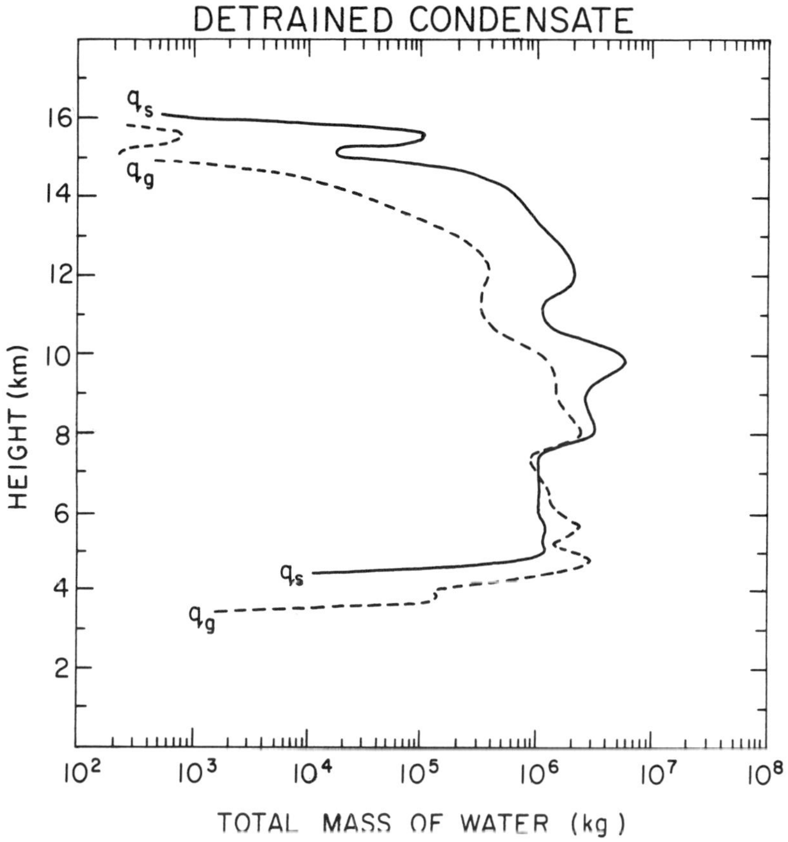

FIGURE 2. Mass of condensate detrained from a one-dimensional time-dependent model of the convective cells in the tropical oceanic squall line studied by Houze (1977). Solid line (q_s) indicates condensate in the form of snow. Dashed (q_g) indicates condensate in the form of graupel. Model calculations are from current work of Ferrier and Houze (manuscript in preparation).

1985; Johnson, 1982; Houze and Rappaport, 1984; Churchill and Houze, 1984b; Chong et al., 1987). In this environment of widespread but moderate vertical motion, the snow particles grow by vapor deposition but not by riming. These snow particles are typically of very irregular shape; however, sometimes their crystalline habits can be discerned in particle images sampled by aircraft. Figure 3, from Houze and Churchill (1987), shows ice particle types observed by aircraft in tropical stratiform regions at flight-level temperatures of + 5 to -25°C. Pristine crystal shapes, including needles, columns, plates and dendrites, were observed in maximum concentration about a kilometer below the altitude where their growth habit was determined. These shapes are a clear indication of growth by vapor deposition. The larger dendritic crystals apparently aggregated to form large snowflakes that were in maximum concentration about 1 km above the 0°C level. Such aggregation is a further characteristic of the stratiform precipitation process.

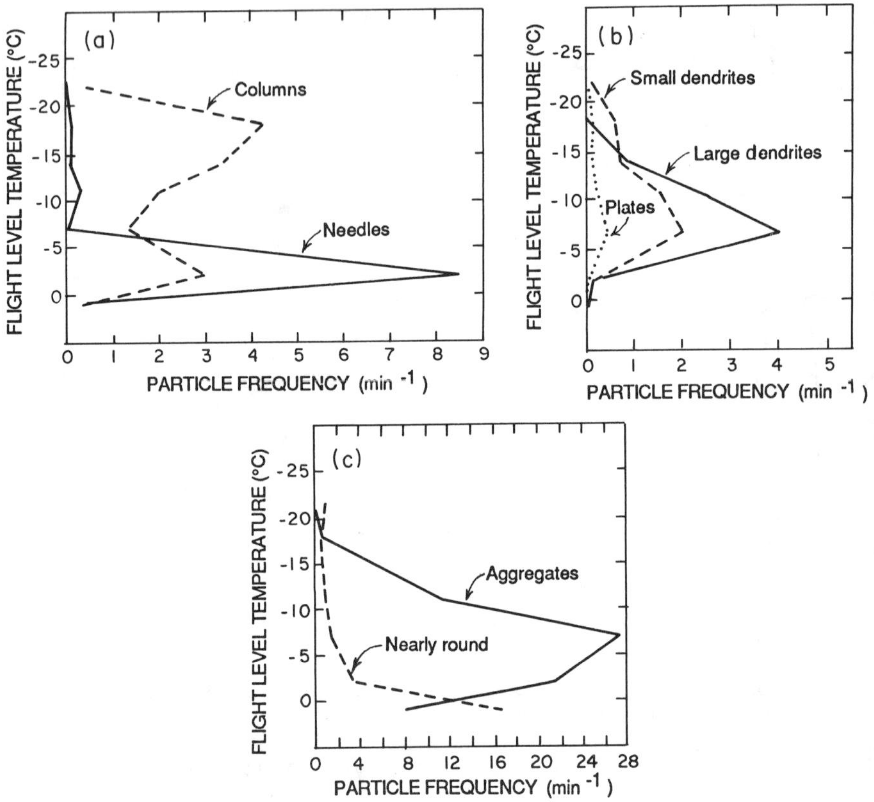

FIGURE 3. Vertical distribution of observed frequencies of ice-phase hydrometeors in the rain areas of the Bay of Bengal depression of 3-8 July 1979. For each particle type the number of hydrometeors observed for each minute of in-cloud flight time are plotted against the flight-level temperatures at which they were observed. (a) Columns and needles. (b) Small dendrites, large dendrites and plates. (c) Aggregates of ice particles and nearly round particles. From Houze and Churchill (1987).

From diagnostic modelling of the water budget and microphysical processes in a stratiform region similar to those in tropical rain systems, Rutledge and Houze (1987) found that both the influx of snow into the stratiform region from the convective cells and the growth of the snow as it passed through the region of the mesoscale updraft contributed strongly to the stratiform precipitation process. Without the lateral influx of snow, their model stratiform cloud was incapable of producing significant rain. On the other hand without the mesoscale ascent within the stratiform cloud, only about one-fourth as much stratiform rain reached the surface. Thus a kind of feeder-seeder process is at work; the deep convective cells feed snow particles into the stratiform cloud, where mean ascent provides moisture for vigorous depositional growth of the snow particles.

4. HORIZONTAL DISTRIBUTIONS OF CONVECTIVE AND STRATIFORM PRECIPITATION IN CLOUD CLUSTERS AND HURRICANES

The conceptual model in Fig. 1 envisages the vertical structure of a tropical cloud system. The convective cells and attendant stratiform structure depicted there occur in various horizontal patterns. The horizontal arrangement of stratiform and convective rain has proven to be a very useful indicator of the mesocale structure and organization of both cloud clusters (Houze and Betts, 1981; Johnson and Houze, 1987) and hurricanes (Jorgensen, 1984; Marks and Houze, 1987).

In cloud clusters, the stratiform and convective precipitation occur in a wide variety of patterns. Sometimes, the convection is arranged horizontally in a propagating arc-shaped line, with the stratiform cloud and precipitation trailing behind (Zipser, 1969, 1977; Houze, 1977; and others). In other clusters, the convection may be in less rapidly moving lines (Leary and Houze, 1979; Barnes and Sieckman, 1985) or groups of cells tied to some topographic feature such as a coastline (Churchill and Houze, 1984a).

In hurricanes, convective-stratiform structure is seen in the outer extremities of the storm, where convective cells can be located on the upwind end of a spiral band, while ice particles detrained from the cells are carried downwind into a stratiform region of the band (Atlas et al., 1963). In the inner-core region of the hurricane, which is dominated by the circular eyewall, radial cross sections typically show stratiform precipitation structure in an annular region just outside the eyewall convection (Marks and Houze, 1987).

In all these cases, the stratiform region occurs adjacent to the convection with the heaviest stratiform rain in a location consistent with the trajectories of the snow particles relative to the cells, as shown in Fig. 1. Sometimes the ice particles from the convective cells are carried great distances by the horizontal wind. For example, in the inner-core region of the hurricane, the trajectories of ice particles ejected from a spot within the eyewall swirl around the storm up to 1 1/2 times before they reach the melting layer (see Fig. 4; from Marks and Houze, 1987).

Several additional factors can contribute to making the horizontal pattern of stratiform and convective precipitation in a cloud cluster or hurricane complex: (1) Since the ice particles generated in convective regions take a long time to fall out, the convection may actually disappear before the detrained ice particles reach the melting level and fall out as stratiform rain. Thus, an instantaneous pattern of precipitation in a cloud cluster or hurricane may contain either stratiform precipitation associated with earlier convection, which is no longer present, or active convection may be present whose detrained snow has not yet had time to influence a stratiform precipitation region. (2) A cloud cluster may contain several groups or lines of convection each affecting different stratiform regions (e.g., Leary and Houze, 1979). (3) Several precipitation features, each consisting of some convection and some stratiform precipitation, may overlap, intersect or merge.

5. DETERMINATION OF CONVECTIVE AND STRATIFORM COMPONENTS OF TROPICAL RAINFALL FROM SPACE

The complexities of the horizontal precipitation patterns in cloud clusters and hurricanes lead to the necessity of careful sampling of rainfall from space in a program such as TRMM. However, the discussion in Sec. 5 indicates that these complexities appear primarily to be superpositions and arrangements of the basic convective and stratiform precipitation mechanisms depicted in Fig. 1. This result implies that, despite the

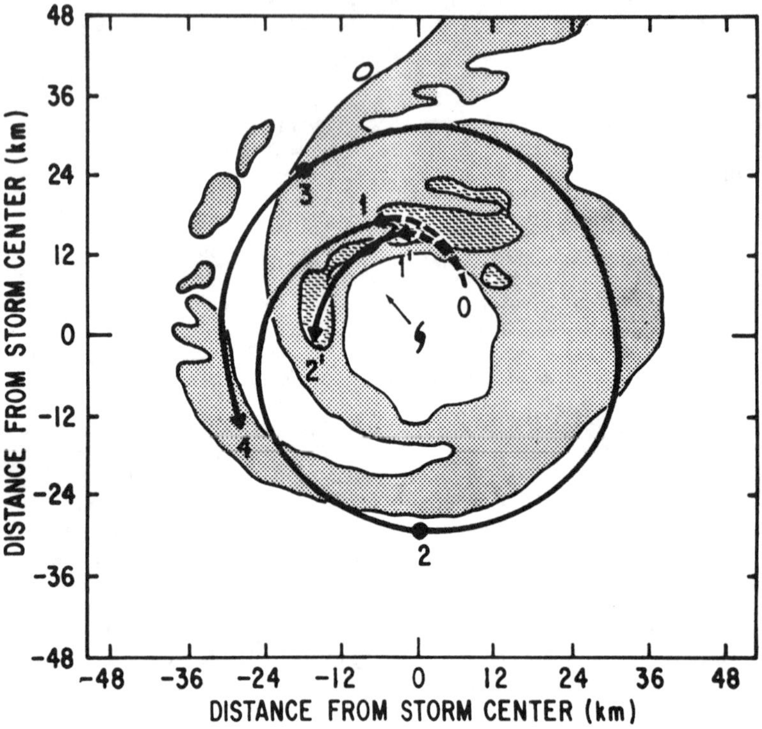

FIGURE 4. Plan view of the low-level reflectivity field in the inner core of Hurricane Alicia. Superimposed are horizontal projections of calculated trajectories of precipitation particles falling from 12 to 1 km altitude. Reflectivity contours are for 20 and 35 dBZ. Storm center and direction of motion are also shown. From Marks and Houze (1987).

complexity of the horizontal structure of tropical rain patterns, the building blocks of the patterns remain relatively simple. A primary objective of TRMM should therefore be to distinguish the basic convective and stratiform components of tropical precipitation.

Attempts are presently being made to resolve stratiform and convective components of tropical cloud clusters from conventional satellite measurements. Adler and Negri (1987) have developed an algorithm for distinguishing convective and stratiform regions by using infrared images from geosynchronous satellites. Their technique initially identifies all relative minima in the cloud-top temperature field as potential convective centers. These points are then tested against empirically derived criteria designed to screen out minima associated with minor irregularities in the background cirrus shield. A temperature threshold for the stratiform areas is determined from the statistics of the cloud-top temperatures in the region surrounding the convective centers. Application of a slightly modified version of the Adler-Negri algorithm to data from Winter MONEX has been used to estimate the convective and stratiform areas of cloud clusters over the South China Sea (Goldenberg and Houze — work in progress). The results (see Fig. 5) are encouraging in that they agree well with estimates of the stratiform areas obtained from radar data by Churchill and Houze (1984a).

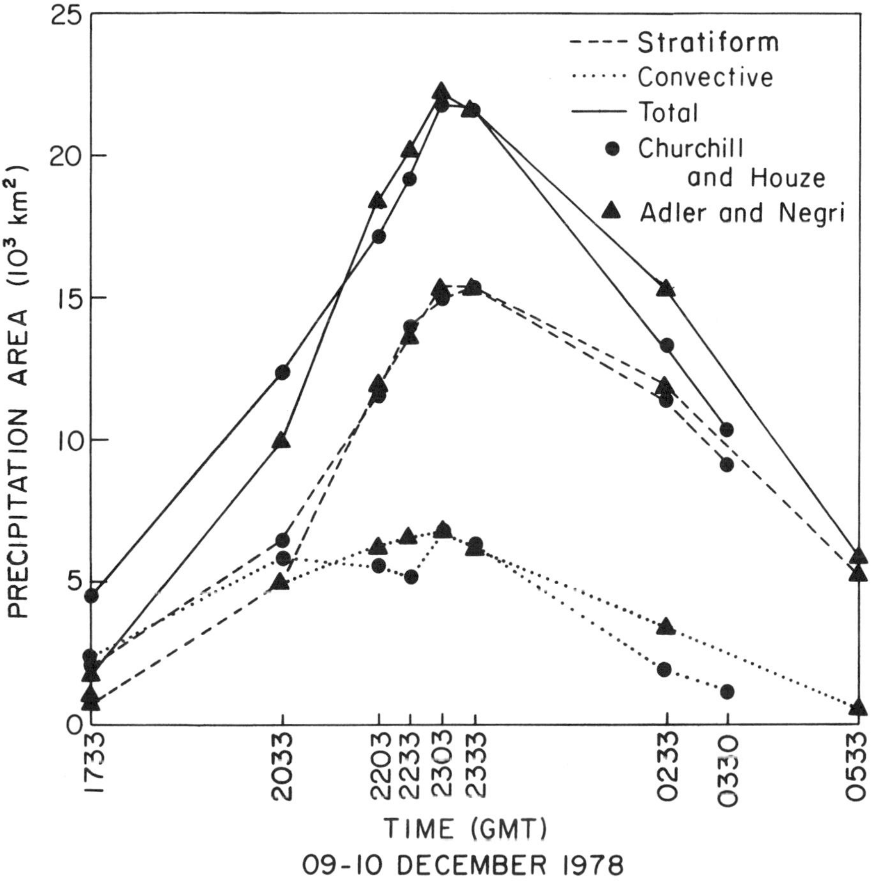

FIGURE 5. Estimates of the total, stratiform and convective precipitation areas for Churchill and Houze's (1984a) Winter MONEX cloud cluster "B". One set of curves was determined from radar data by Churchill and Houze. The other set was determined from infrared satellite data by Goldenberg and Houze (work in progress) using a modified version of the technique of Adler and Negri (1987).

While useful, this type of technique is limited in that it is based entirely on horizontal pattern recognition. A particular difficulty is that, while the method is able to recognize the existence of convective regions, a somewhat arbitrary assumption must be made regarding the horizontal size of the convective region surrounding a center. Consequently, satellite-derived results such as those in Fig. 5 have considerable uncertainty as a result of this assumption. From Fig. 1, it is evident that the stratiform and convective processes are most clearly distinguished by their differing vertical structure. It is therefore important for future measurements, such as those in TRMM, to have good observational capability in the vertical as well as the horizontal. In this case, it will be possible to use both vertical structure and horizontal variability to distinguish convective and stratiform regions in tropical precipitation, thereby removing much of the present uncertainty.

6. CONCLUSIONS

Although the horizontal patterns of precipitation in tropical cyclones and cloud clusters can be quite complex, they are made up of basic stratiform and convective components characterized by quite different and distinct precipitation mechanisms, which have been summarized in Fig. 1. Because of these different physical mechanisms, it is important that in an effort, such as TRMM, in which tropical precipitation is to be mapped globally, that the stratiform and convective components as well as the total amounts of rain be determined. If these components are clearly identified over the whole tropics, several important scientific results will accrue:

- The microphysical mechanisms of precipitation throughout the tropics will be known in addition to total rain amounts.

- The mesoscale internal organization of tropical cloud systems, which is often well indicated by the relative juxtaposition of convective and stratiform areas, will be sampled throughout the tropics, thus allowing a thorough census of squall-line and other types of cloud cluster types.

- The vertical distribution of diabatic heating associated with tropical cloud systems can be more accurately calculated. In the stratiform regions, the cooling that occurs in connection with melting and evaporation at low-to-mid levels strongly affect the vertical heating profile (Houze, 1982). Accurate determination of the relative amounts of stratiform and convective precipitation will allow these important stratiform effects to be included properly in the determination of the vertical distribution of heating in the tropics.

Current methods used to identify convective and stratiform precipitation regions in the tropics from detailed horizontal patterns are useful, but are limited by various arbitrary assumptions. Because stratiform and convective regions in tropical cloud systems can be distinguished physically by their vertical structure (as idealized in Fig. 1), identification of them will be improved by measurements that have the capability to distinguish the characteristic vertical structures of the convective and stratiform regions. This conclusion emphasizes the importance of measuring the vertical as well as the horizontal structure of tropical precipitation in TRMM.

ACKNOWLEDGMENTS

The author gratefully acknowledges information provided by J. Simpson, S. Rutledge, R. Adler, A. Negri, B. Ferrier and S. Goldenberg.

REFERENCES

Adler, R.F. and A.J. Negri, 1987: A satellite infrared technique to estimate tropical convective and stratiform rainfall. J. Clim. Appl. Meteor., In press.

Atlas, D., K.R. Hardy, R. Wexler, and R.J. Boucher, 1963: On the origin of hurrricane spiral bands, Geofisica Internacional, 3, 123-132.

Barnes, G.M., and K. Sieckman, 1985: The environment of fast-and slow-moving tropical mesoscale convective cloud lines. Mon. Wea. Rev., 112, 1782-1794.

Chong, M., P. Amayenc, G. Scialom and J. Testud, 1987: A tropical squall line observed during the COPT 81 Experiment in West Africa. Part I: Kinematic structure inferred from dual-doppler radar data. Mon.Wea. Rev., 115, 670-694.

Churchill, D.D., and R.A. Houze, Jr., 1984a: Development and structure of winter monsoon cloud clusters on 10 December 1978. J. Atmos. Sci., 41, 933 - 960.

Churchill, D.D., and R.A. Houze, Jr., 1984b: Mesoscale updraft magnitude and cloud-ice content deduced from the ice budget of a tropical cloud cluster. J. Atmos. Sci., 41, 1717-1725.

Gamache, J.F., and R.A. Houze, Jr., 1982: Mesoscale air motions associated with a tropical squall line. Mon. Wea. Rev., 110, 118-135.

Gamache, J.F., and R.A. Houze, Jr., 1983: Water budget of a mesoscale convective system in the tropics. J. Atmos. Sci., 40, 1835-1850.

Gamache, J.F., and R.A. Houze, Jr., 1985: Further analysis of the composite wind and thermodynamic structrure of the 12 September GATE squall line. Mon. Wea. Rev., 113, 1241-1259.

Houze, R.A., Jr., 1977: Structure and dynamics of a tropical squall-line system. Mon. Wea. Rev., 105, 1540-1567.

Houze, R.A., Jr., 1982: Cloud clusters and large-scale vertical motions in the tropics. J. Meteor. Soc. Japan, 60, 396-410.

Houze, R.A., Jr., and A.K. Betts, 1981: Convection in GATE. Reviews of Geophysics and Space Physics, 19, 541-576.

Houze, R.A., Jr., and D. D. Churchill, 1987: Mesoscale organization and cloud microphysics in a Bay of Bengal depression. J. Atmos. Sci., 44, 1845-1867.

Houze, R.A., Jr., and E.N. Rappaport, 1984: Air motions and precipitation structure of an early summer squall line over the eastern tropical Atlantic. J. Atmos. Sci., 41, 553-574.

Houze, R.A., Jr., and P.V. Hobbs, 1982: Organization and structure of precipitating cloud systems. Advances in Geophysics., 41, 3405-3411.

Johnson, R.H., 1982: Vertical motion of near-equatorial winter monsoon convection. J. Meteor. Soc. Japan, 60, 682-690.

Johnson, R.H., and R.A. Houze, Jr., 1987: Precipitating Cloud Systems of the Asian Monsoon. Reviews in Monsoon Meteorology, C.-P. Chang and T.N. Krishnamurti, Eds., In press.

Jorgensen, D.P, 1984: Mesoscale and convective-scale characteristics of mature hurricanes. Part I: General observations by research aircraft. J. Atmos. Sci., 41, 1268-1285.

Leary, C.A., 1984: Precipitation structure of the cloud clusters in a tropical easterly wave. Mon. Wea. Rev., 112, 313-325.

Leary, C.A, and R.A. Houze, Jr., 1979: The structure and evolution of convection in a tropical cloud cluster. J. Atmos. Sci., 36, 437-457.

Leary, C.A, and R.A. Houze, Jr., 1980: The contribution of mesoscale motions to the mass and heat fluxes of an intense tropical convective system. J. Atmos. Sci., 37, 784-796.

Malkus, J.S., and H. Riehl, 1964: Cloud Structure and Distributions over the Tropical Pacific Ocean. Univ. Calif. Press, Berkeley, 229 pp.

Marks, F.D., Jr., 1985: Evolution and structure of precipitation in Hurricane Allen (1980). Mon. Wea. Rev., 113, 909-930.

Marks, F.D., Jr., and R.A. Houze, Jr., 1987: Inner core structure of hurricane Alicia from airborne Doppler radar observations. J. Atmos. Sci., 44, 1296-1317.

Ramage, C.S., 1971: Monsoon Meteorology. Academic Press, New York, 271 pp.

Rutledge, S.A., and R.A. Houze, Jr., 1987: A diagnostic modeling study of the trailing stratiform region of a midlatitude squall line. J. Atmos. Sci., 44, in press.

Simpson, R.H., and J. Simpson, 1966: Why experiment on tropical hurricanes? Transactions of the New York Academy of Sciences, Series II, 28, 1045-1062.

Smull, B.F., and R.A. Houze, Jr., 1985: A midlatitude squall line with a trailing region of stratiform rain: radar and satellite observations. Mon. Wea. Rev., 113, 117-133.

Zipser, E.J., 1969: The role of organized unsaturated convective downdrafts in the structure and rapid decay of an equatorial disturbance. J. Appl. Meteor., 8, 799-814.

Zipser, E.J., 1977: Mesoscale and convective-scale downdrafts as distinct components of squall-line circulation. Mon. Wea. Rev., 105, 1568-1589.

ANALYSIS OF THE SUPER CLOUD CLUSTER BY GEOSTATIONARY METEOROLOGICAL SATELLITE (GMS) IR 3-HOURLY DATA

Tetsuo Nakazawa
Meteorological Research Institute
Tsukuba, Ibaraki 305, Japan

ABSTRACT

Using 3-hourly Japanese GMS IR data, the super cloud cluster was analyzed. The super cloud cluster is defined as an ensemble of cloud clusters with different life stages. It is observed in the tropics near the equator within the active phase of the intraseasonal variations (ISV). The hierarchy of ISV in the real atmosphere is shown; ISV, super cloud cluster, and cloud cluster. The ISV and the super cluster move eastward, however, each cloud cluster under the super cluster moves westward.

1. INTRODUCTION

Both numerical modeling (Y.Y. Hayashi and Sumi,1986) and analysis (Nakazawa, 1986b) suggest the hierarchic structure in the tropical convection. It is well known that the cloud cluster with its scale of about 100-500km consists of many deep cumulus clouds. In the 1960's and 70's vigorous studies about tropical disturbances were performed which showed that the lower-tropospheric disturbances frequently follow cloud clusters. On the other hand, Madden and Julian (1971,1972) found a 40-50 day oscillation in the tropics and speculated on the importance of large-scale convections for the maintenance of the oscillation. There is little doubt from the recent analysis that their speculation is correct (Murakami and Nakazawa,1985). However, the temporal and spatial scale on both ISV and cloud cluster is so different that much attention was not paid to the intermediate-scale phenomena. Most works about the ISV were done using the filtered data in time and/or space unable to see these fine features. By analyzing the non-filtered outgoing longwave radiation (OLR) data, Nakazawa (1986b) demonstrated that there are eastward moving active convective regions with a horizontal scale of several thousand kilometers under the planetary-scale ISV. Y.Y.Hayashi and Sumi (1986) simulated both phenomena; the super cluster and the planetary-scale 30-day oscillation.

The purpose of the present study is to analyze the super cluster using 3-hourly Japanese GMS IR data in order to see the fine structure in more detail.

2. DATA

The data used in this study is the following three types:
1) GMS IR fundamental histogram data, which was edited by Japan Meteorological Satellite Center (JMSC).

Spacial resolution:	0.25° longitude, latitude
Spacial coverage:	60°N-60°S, 80°E-160°W
Time resolution:	every 3 hours except for 16Z
Temperature resolution:	256 levels (~0.5°K)

2) OLR data obtained from NOAA polar-orbiting satellites.

ISBN 0-937194-14-4

Spatial resolution:	2.5° longitude, latitude
Spatial coverage:	global
Time resolution:	originally twice daily, but once a day after interpolation for missing data

3) global objective analysis datasets prepared by ECMWF/ WMO.

Spatial resolution:	2.5° longitude, latitude
Time resolution:	00Z and 12Z
Vertical resolution:	7 levels from 1000mb to 100mb

3. SUPER CLOUD CLUSTER UNDER INTRASEASONAL VARIATIONS

The term "super cluster" first appeared in the paper of Y.Y.Hayashi and Sumi (1986). They defined it as a group of cloud clusters. Using an aqua planet model, they found the hierarchy of ISV; under the planetary-scale eastward-moving wavenumber 1 mode with a period of 30 days, there are lots of active convective regions with a horizontal scale of several thousand kilometers. The active convective region, which they call super cluster, also moves eastward at the same speed as the 30-day mode.

From the analysis of the real atmosphere, Nakazawa (1986b) detected the hierarchic structure of the ISV using 8-year OLR data. Fig. 1 shows the time-longitude section of the transient (seasonal trend was subtracted) OLR data averaged between the equator and 5°N during May and July 1980. As contours were drawn where the value was less than -15Wm^{-2} with intervals of 30 Wm^{-2}, the closed contour region indicates above normal convective activity. The figure shows the clustering of the active convective regions over the eastern equatorial area from 90°E to 150°W during late May and early July. The convective region resembles the super cluster Y.Y. Hayashi and Sumi (1986) found in their model. We also call the active convective region a super cluster. This period coincides with the active phase of the ISV over the area. During this period we clearly define four super clusters, which move eastward with a phase speed of about 8-12°day^{-1} or 10-15ms^{-1} (It takes 45-30 days for moving around a globe). In late May a super cluster around 90°E moved eastward and reached the dateline in early June. We refer to the super cluster as SC1.

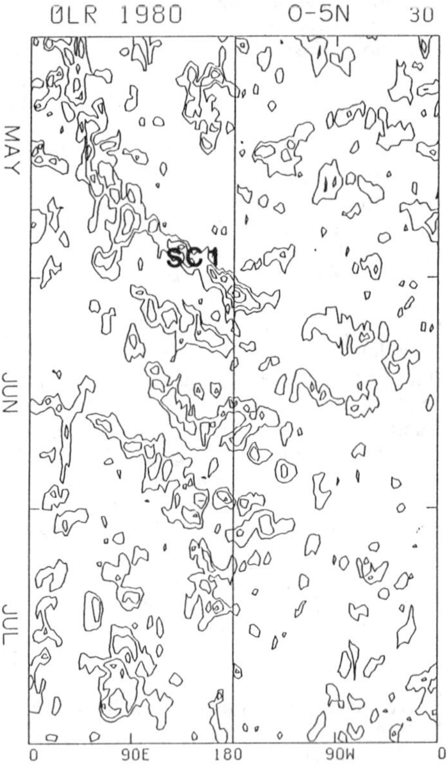

FIGURE 1. Time-longitude section of transient OLR averaged between the equator and 5°N during May and July 1980. Contours are drawn for values of -15,-45,-75 Wm^{-2}.

The spatial coverage of Japanese GMS mostly agrees with regions where the 30-60-day fluctuations are dominant; the western and central Pacific. The analysis using GMS data may give us more information about the ISV. Fig. 2

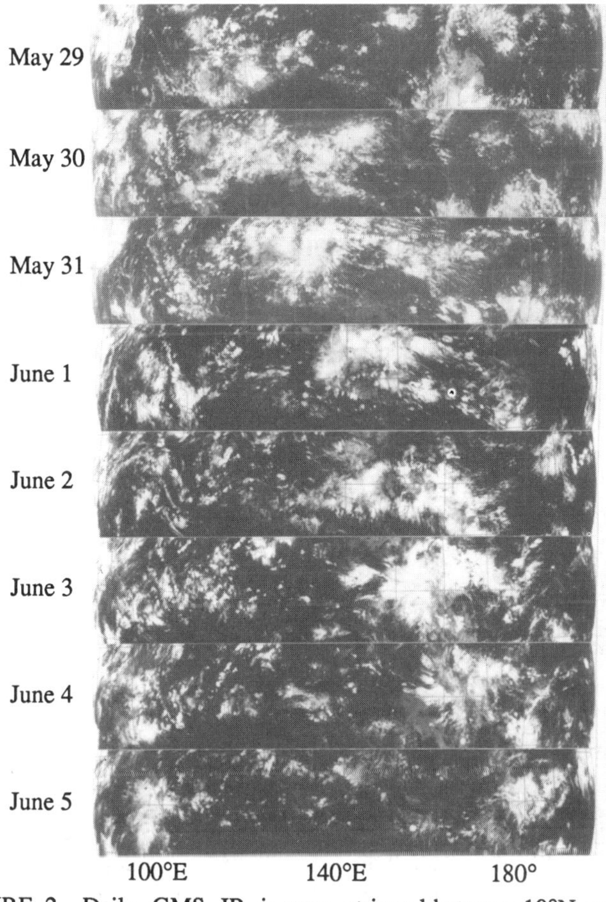

FIGURE 2. Daily GMS IR images stripped between 10°N and 10°S at 09Z from 29 May to 5 June in 1980. The center longigude and latitude is 140°E and the equator.

represents a sequence of daily GMS IR images stripped between 10°N and 10°S at 09Z from 29 May (top) to 5 June (bottom). On 29 May the center of the system was located around 110°E and moved eastward crossing 140°E, sub-satellite longitude of GMS,on 31 May or 1 June, then reached the dateline on 6 June. From Figs. 1 and 2, do we say that a group of whole cloud ensembles consisting of SC1 moved eastward directly ? In order to answer the question, the GMS IR 3-hourly data were used. Fig. 3 shows the equivalent blackbody temperature (TBB) distribution every 3 hours between the equator and 10°N from 00Z in 31 May to 21Z in 1 June. As is shown in Fig. 2, SC1 was located at 130° to 140°E at 00Z in 31 May, and at 135° to 155°E at 00Z in 1 June. The cloud cluster C1 located at 135°E at 00Z in 31 May of its mature stage at this time, moved westward and weakened until 00Z in 1 June. A different cloud cluster C2 in a mature stage took over at 00Z in 1 June around 140°E, which first evolved at 00Z in 31 May near 153°E about 2000km east of the mature-stage cloud cluster C1. C2 then developed with westward phase propagation and decayed around 135°E at 18Z in 1 June. At 21Z in 1 June, C3, the main portion of SC1, was located at 144°E. This cloud cluster was originally observed at 16Z in 31 May at around 155°E about 1000km east of the mature-stage cloud cluster C2. We find that within SC1 there are 2 or 3 cloud clusters with different life stages; a new cloud cluster is produced east of 1000-2000km away from a mature-stage cloud cluster, is located further east of a decay-stage cloud cluster.

The T_{BB} index I_{TBB} was introduced in order to pick up only very deep convective clouds and not to reflect the sea surface temperature and the lower clouds.

$$I_{TBB} = (225-[T_{BB}])/5 \quad , \text{if } [T_{BB}] \leq 225°K$$
$$= 0 \quad , \text{if } [T_{BB}] \geq 225°K$$

where $[T_{BB}]$ is the area mean T_{BB}.

To clearly see the phase direction of the two different time and space scale phenomena, Fig. 4 is shown. This figure indicates the time-longitude section for I_{TBB} summed up between the equator and 10°N from 00Z at 29 May to 21Z at 5 June between 90°E and the dateline. Although each cloud cluster with a horizontal scale of several hundred kilometers and with a time-scale of 1-2 days moves westward, a super cluster with a scale of several thousand kilometers moves eastward by the successive formation of a new cloud cluster east of a mature-stage cloud cluster. A schematic figure illustrating the hierarchic structure under ISV is shown in Fig. 5.

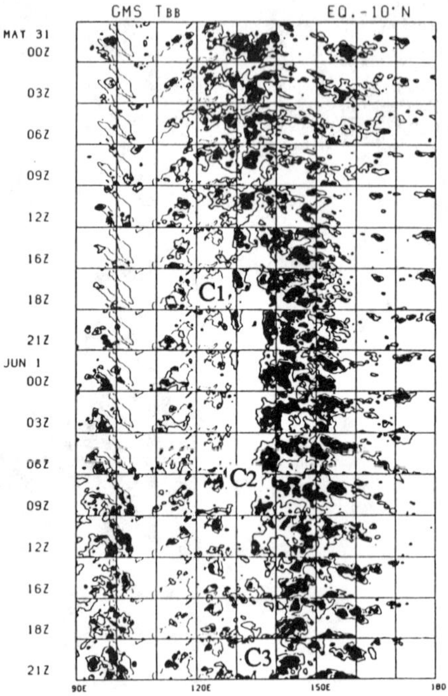

FIGURE 3. T_{BB} by GMS IR every 3 hours between the equator and 10°N from 00Z in 31 May to 21Z in 1 June. Contour interval is 25°K and shading denotes regions less than 225°K. The symbols, C1, C2, and C3 are the westward moving cloud clusters within SC1.

4. LOWER-TROPOSPHERIC DISTURBANCES WITHIN SUPER CLOUD CLUSTER

The successive formation of a new cloud cluster east of a mature-stage cloud cluster may correspond with the lower-tropospheric disturbances near the equator. The non-filtered wind field at 850 mb obtained by ECMWF/WMO global analysis dataset is shown in Fig. 6. The daily mean OLR data is also added in the figure by shading. Notice that the symbols A and C are counter-clockwise and clockwise, respectively. The daily OLR shows the eastward moving convective system SC1. Under the system SC1, it is seen that there exist several lower-tropospheric disturbances staggered on both sides of the equator. The monthly mean wind field at 850 mb in June 1980 (not shown) indicates that the generation area of the 4 super clusters was the region where the monsoon westerlies came across the easterly trade winds. Over the region low-level convergence along the equator enhances the convective activity and the formation or development of the disturbance. On 2 June (fourth from the top), the convection center of SC1 at the equator was located around 160°E where the westerlies prevailed. The westerlies came from the counterclockwise (clockwise) disturbance in the northern (southern) hemisphere. On 3 June, the location of westerly maximum shifted eastward at around 165°E in the vicinity of the counterclockwise disturbance located at 2°N,167°E where the

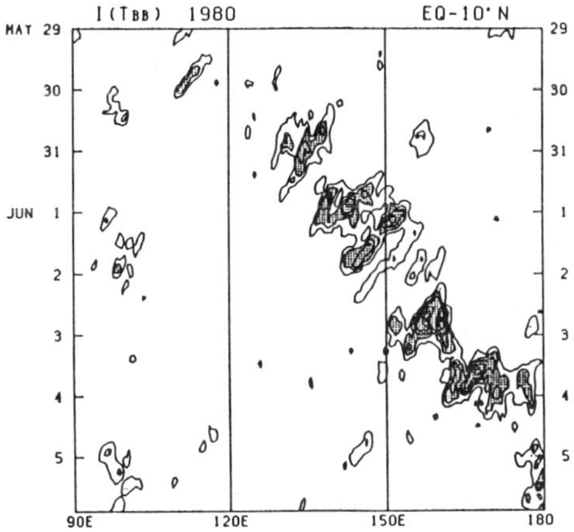

FIGURE 4. TBB index (ITBB) of the super cloud cluster SC1 summed up between the equator and 10°N. Contour interval is 10. Shading denotes value of greater than 20.

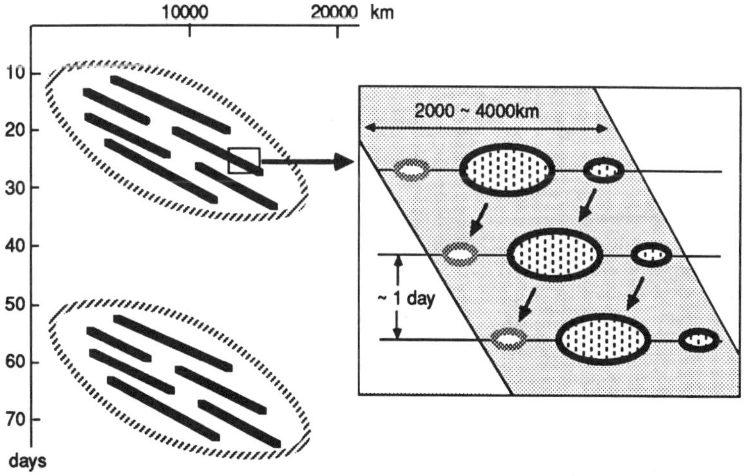

FIGURE 5. Schematic figure of 30-60 day oscillation (left; dotted), super cluster (left; heavy line), and cloud clusters (right).

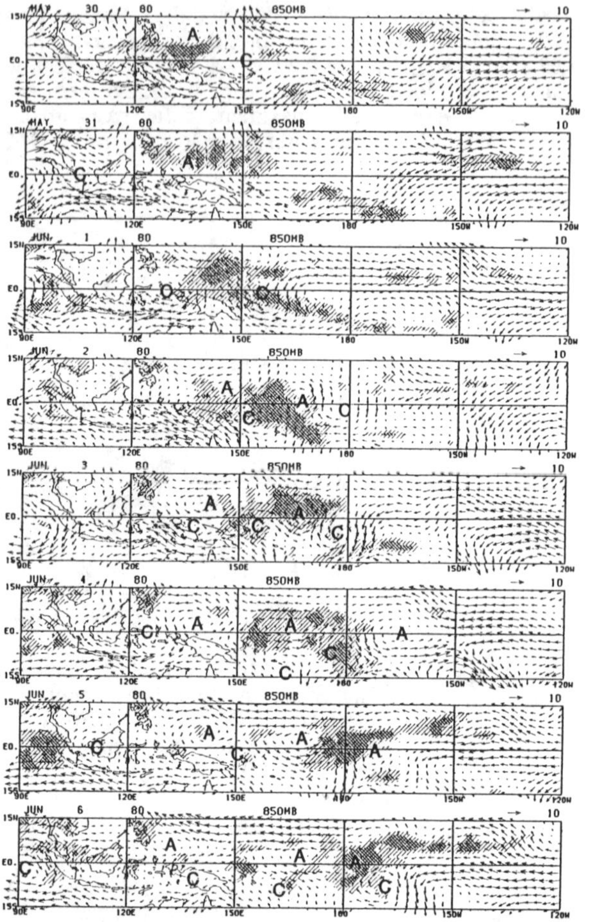

FIGURE 6. 850mb wind and OLR field from 30 May to 6 June in 1980. The symbols A and C denote counter-clockwise and clockwise, respectively. Light (dark) shading denotes less than 200 (160) Wm^{-2}.

convection center also shifted eastward and was located. In this study only the data at 12Z was available, so we could not examine the phase direction of the disturbances. However, it is speculated from Fig. 4 that the disturbances move westward following the cloud cluster. It is announced that 00Z dataset is available now. The phase direction must be checked using higher time resolution data.

Once the disturbance forms, equatorial westerlies tend to strengthen as the disturbance begins to develop. The westerlies shift the center of the low-level convergence eastward. How far the convergence center is shifted affects the east-west phase direction of the super cluster. That is, for example, super cloud cluster does not move when the new convergence occurs the same position as before. So far we do not know what determines the distance between the new and old convergence center. As is shown in Fig. 4, the super cloud cluster moves eastward about $8°\ day^{-1}$ and each cloud cluster moves westward about 8°/day^{-1}. As the average generation frequency of a cloud cluster is once a day, the new convergence occurs about 16° east of the previous convergence center. One possibility is that new convection can be organized by the downdraft of the old convection. A similar mechanism may work even for different time and space scale phenomena.

Another possibility of eastward movement of a super cloud cluster is the existence of some kind of eastward moving tropical disturbance. When baroclinic disturbance moves eastward, a new cloud cluster is generated east of the old cloud cluster.

5. DISCUSSION

After the paper by Y.Y. Hayashi and Sumi (1986) was published, several papers on the modeling of ISV followed. Lau and Peng (1987) proposed a theory of the origin of ISV. They explained the eastward propagation of ISV by eastward movement of the internal heat source in terms of the selective amplification of Kelvin waves. Using a simple shallow water equation model including the moisture equation, Yamagata (1987) showed the eastward movement of the heat source. Miyahara (1987) stressed that the model disturbance moved eastward and grew in the high SST region where the heating function is large, and suppressed in other regions. These experiments except for those by Y.Y. Hayashi and Sumi (1986) did not succeed in presenting the hierarchic structure of ISV. Only one super cluster prevailed in these models. Fine structure under the super cluster shown in this paper has not been simulated in any models. If the interaction between convection and dynamics through some kind of positive feed back mechanism is intrinsic for ISV, the analysis of the super cluster in the real atmosphere has to be done in the following direction; a composite structure of the super cluster, a triggering mechanism of the super cluster, the role of lower-tropospheric disturbances, the relationship between the super cluster and cloud clusters, seasonal dependency of the occurrence frequency of the super cluster.

Some analysis showed evidence of the super cluster. Zangvil (1975) analyzed the satellite brightness data and by performing the space-time spectral analysis developed by Hayashi (1973), he showed in his Figs. 7 and 8 the predominance of the power for eastward components along the equator with a phase speed of about 13ms^{-1} without regard to the wavenumbers. The power for ISV with a period of 40 days was also large in the figure. The space-time spectral analysis is very useful to separate the power into westward-moving and eastward moving components in each zonal wavenumber. However, if there is a correlation between the different period modes, such as our case when there is a hierarchic structure, we should return to examine the original data.

It is reported that a monsoon depression or tropical cyclone forms and develops during the active phase of ISV (Yasunari,1980; Murakami et al.,1985; Nakazawa,1986a). As Nitta and Motoki (1987) showed, in November 1986 an active convective region moved eastward from the Indian Ocean to the dateline. By using GMS IR 3-hourly data we are now analyzing the active convective region over the western Pacific during the onset phase of 1986/87 ENSO. From the analysis the active convective region is recognized as a super cluster and successive formation of tropical cyclones from the super cluster are shown. A detailed result will be reported in another paper.

6. CONCLUDING REMARKS

Using the GMS IR 3-hourly data, a super cloud cluster in the tropics was examined. The results show the existence of the hierarchic structure of ISV summarized in Table 1. Under ISV with zonal wavenumber 1 structure and a period from 30 to 60 days, several super clusters are observed with a horizontal scale of several thousand kilometers. A super cluster consists of 2 or 3 cloud clusters of several hundred kilometer width and 1-2 day life scale. A super cluster moves eastward, while each cloud cluster moves westward. New cloud cluster is produced at 1000-2000km east of the mature-stage cloud cluster which enhances the low-level westerlies and shifts the low-level convergence center eastward.

TABLE 1. Hierarchic structure of ISV shown in this study.

Hierarchy	Horizontal Scale	Time Scale	Phase Direction
ISV	zonal wavenumber 1	30-60 days	Eastward
Super Cluster	several 1000 km	< 10 days	Eastward
Cloud Cluster	several 100 km	1-2 days	Westward

REFERENCES

Hayashi, Y., 1973: A method of analyzing transient waves by space-time cross spectra. J. Appl. Meteor., 12, 404-408.

Hayashi,Y.Y. and A. Sumi, 1986: The 30-40 day oscillations simulated in an "aqua planet" model. J. Mereor. Soc. Japan, 64, 451-467.

Lau, K.M. and L. Peng, 1987: Origin of low-frequency (intraseasonal) oscillations in the tropical atmosphere. Part I: Basic theory. J. Atmos. Sci., 44, 950-972.

Madden, R.A. and P.R. Julian, 1971: Detection of a 40-50 day oscillation in the zonal wind in the tropical Pacific. J. Atmos. Sci., 28, 702-708.

Madden, R.A. and P.R. Julian, 1972: Description of global scale circulation cells in the tropics with a 40-50 day period. J. Atmos. Sci., 29, 1109-1123.

Miyahara, S., 1987: A simple model of the tropical intraseasonal oscillation. J. Mereor. Soc. Japan, 65, 341-351.

Murakami, T., T. Nakazawa and J. He, 1985: On the 40-50 day oscillations during the 1979 Northern Hemisphere summer. Part I: Phase propagation. J. Mereor. Soc. Japan, 62, 440-468.

Murakami, T. and T. Nakazawa, 1985: Tropical 40-50 day oscillations during the 1979 Northern Hemisphere summer. J. Atmos. Sci., 42, 1107-1122.

Nakazawa, T., 1986a: Intraseasonal variations of OLR in the tropics during the FGGE year. J. Meteor. Soc. Japan, 64, 17-34.

Nakazawa, T., 1986b: Mean features of 30-60 day variations as inferred from 8-year OLR data. J. Meteor. Soc. Japan, 64, 777-786.

Nitta, T. and T. Motoki, 1987: Abrupt enhancement of convective activity and low-level westerly burst during the onset phase of the 1986-87 El Niño. J. Mereor. Soc. Japan, 65, 497-506.

Yamagata, T., 1987: A simple moist model relevant to the origin of intraseasonal disturbances in the tropics. J. Mereor. Soc. Japan, 65, 153-165.

Yasunari, T., 1981: Structure of an Indian summer monsoon system with around 40-day period. J. Mereor. Soc. Japan, 59, 336-354.

Zangvil, A.,1975: Temporal and spatial behavior of large-scale disturbances in tropical cloudiness deduced from satellite brightness data. Mon Wea. Rev., 103, 904-920.

THE ROLE OF THE EQUATORIAL HEAT SOURCES ON THE ONSETS OF THE ASIAN SUMMER MONSOON OF 1986

Ken-ichi Kuma
Numerical Prediction Division
Japan Meteorological Agency
Tokyo, Japan

ABSTRACT

The main purpose of our study is to investigate the effect of the equatorial heat sources on the behavior of the upper anticyclone over South Asia during the early summer monsoon. We show that intensification of the equatorial heating generates the Rossby mode and intensifies the anticyclone over South Asia. We have performed prediction experiments, in which we can control the convective heating over the equatorial region, for two transition periods in 1986 using our nonlinear forecast model with full physics. The heat source, which is given over the domain from 30°S to 30°N and 90°E to 170°W, is estimated from the cloud top temperature measured by the GMS (Japanese Geostationary Meteorological Satellite). In these experiments, we found that the equatorial heat source, away from the monsoon region, plays an important role in generating the anticyclone at 200mb.

1. INTRODUCTION

Yasunari (1980) first pointed out that the Indian monsoon activity is well correlated with the 40-day oscillation detected by Madden and Julian (1971). Hayashi and Sumi (1986) showed that the equatorial convection plays an essential role in the eastward-moving equatorial mode with 30-50-day period. These results seem to suggest that the monsoon activity might be controlled by the equatorial heat sources.

The onset of the monsoon is one of the best indices of monsoon activity. He et al. (1987) show that there were two onsets during the early summer of 1979. The first one occurred over South East Asia during the middle of May. About 30 days later, the second one occurred over India. The movement and the development of the upper anticyclone over South Asia corresponds well to these onsets.

2.ANALYSIS FOR THE EARLY SUMMER OF 1986

The rainfall rate can be roughly estimated by the cloud top temperature measured by Japanese GMS. Fig.1 shows the time change of the area average of the rainfall rate from April 1 to July 31. The region between 10°N and 10°S, 90°E and 170°W is

TROPICAL RAINFALL MEASUREMENTS
John S. Theon and Nobuyoshi Fugono (Eds.)

ISBN 0-937194-14-

selected for the averaging to represent the equatorial rainfall. We can clearly see the intra-seasonal changes in the 30-40-day period.

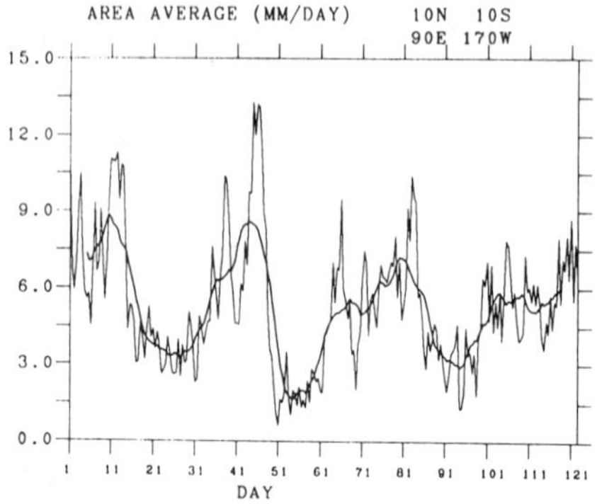

FIGURE 1. Time sequence from April 1 to July 31 in 1986 of the rainfall rate (mm/day) estimated from the GMS data averaged over the region between 10°N and 10°S, and 90°E and 170°W. (Thin line) original data with the 12-hour interval, (Heavy line) running mean by 8 days of the original data.

These changes are accompanied by the equatorial global-scale mode moving eastward with 30-40-day period. Fig.2 shows time sequence of the amplitude and the phase of the zonal wave number one of u (zonal wind averaged between 15°N and 15°S) from April 1 to July 31. When the convection shown in Fig.1 becomes intensive, the amplitude also becomes large. Fig.3 shows the same field as Fig.2 but the time mean for the whole analysed period subtracted to examine the transient component. The eastward propagation is dominant, though we can find some sudden transitions of the phase. 200mb features are shown in Fig.2 and Fig.3, and the phase of the wave number one at 850mb is just reverse of that at 200mb (not shown).

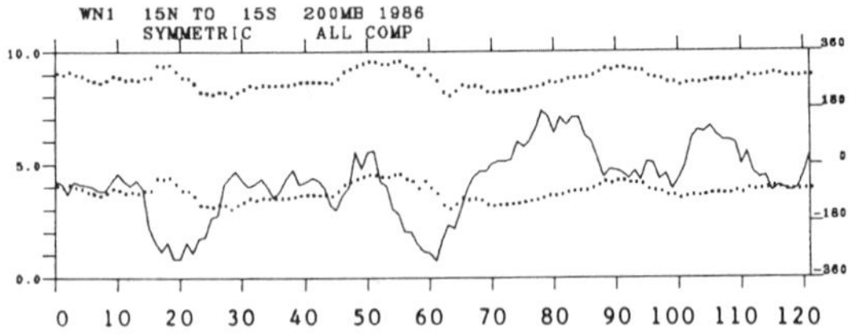

FIGURE 2. Time sequence from April 1 to July 31 in 1986 of the amplitude (m/s, solid) and the phase (degree, dashed) of the symmetric mode of the zonal flow for the zonal wave number one. The value of the phase corresponds to the longitude (+90 means 90°E and -90 mean 90°W) of the maximum westerly.

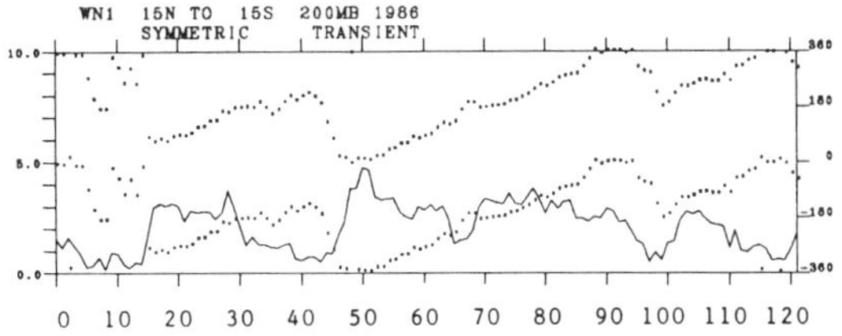

FIGURE 3. Same as Fig.2, but the deviation from the four months mean is shown.

When the convection in Fig.1 becomes active, the upper anticyclone over South Asia, which is closely related with the summer monsoon, develops with another anticyclone symmetrically appearing in the southern hemisphere. Ten-day mean stream function fields at 200mb for May and June are subsequently shown in Fig.4. The behavior of the anticyclone is similar to the one analysed by He et al. (1987). The development in May may correspond to the onset of the East Asia monsoon, while that in June corresponds to the onset of the Indian monsoon.

3. NUMERICAL EXPERIMENTS

The JMA specral global model currently used (see Kanamitu et al; 1983) tends to produce weaker convection over tropics than the observation. We used this model as the control run. In order to see the effect of the equatorial heating on the upper anticyclones, we performed the experiment run, in which we can fix the convective heating over the equatorial region. Again, we used the rainfall estimated from the satellite data. The rainfall is fixed over the region between 30°S and 30°N and 90°E and 170°W. Thus we can get much more rainfall in the experimental run over the region than in the control run. From the difference between the control and experimental run, we calculate the stream function shown in Fig.5. The 3- day average between day 6 and day 8 from the initial is shown. We can regard this difference as the impact of the rainfall over the equatorial region. The initial day is 1st of June and during the forecast period, the upper anticyclone over South Asia developed in the analysed fields. The impact at 850mb is the pair of cyclones over 90°E and that at 200mb is the pair of anticyclones at 90°E. These features seem to be similar to the stationary response to the symmetric heating (Gill,1980). We have performed similar experiments as to the onset in May. The results almost resemble Fig.5, though the anticyclone appears at 105°E.

4. DISCUSSION

This study suggests the importance of observing the convective heating over the tropics in predicting the onset of the monsoon. There may be many future works to confirm our hypothesis. Our estimate of the rainfall amount is quite rough.

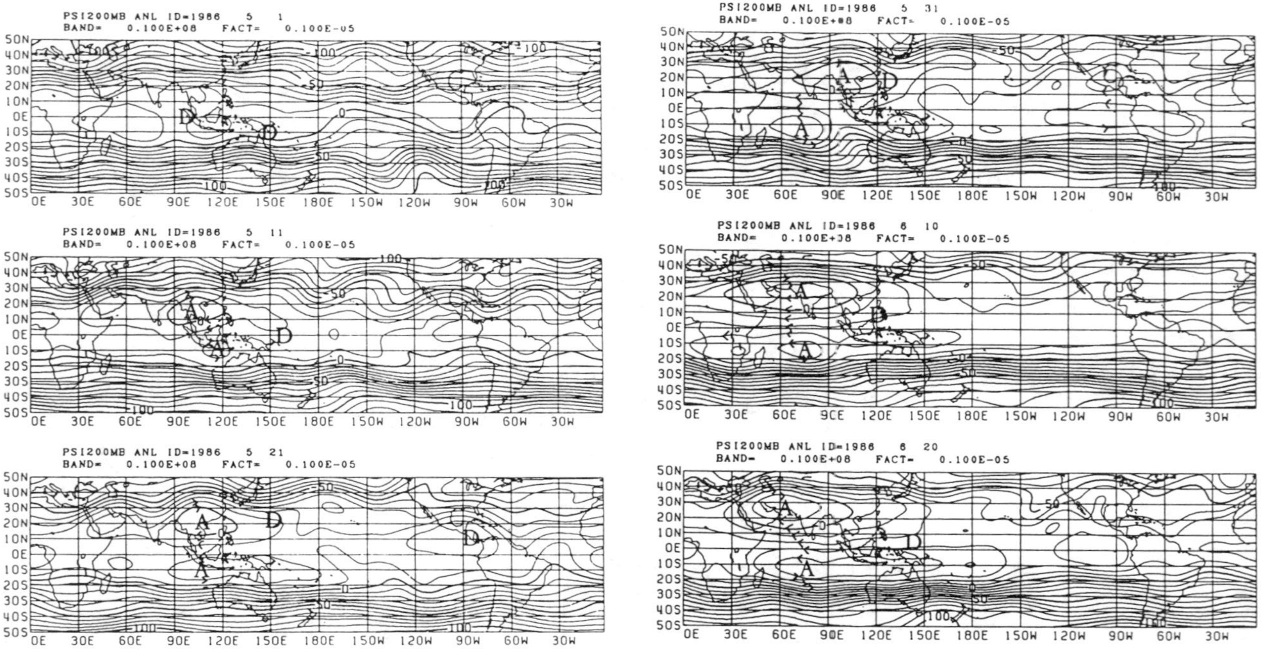

FIGURE 4. The analysed stream function at 200mb averaged for ten days. Contour interval is $10^7 m^2/s$. 'A' and 'D' indicate the center of the anticyclone and the center of the minimum velocity potential, respectively. Left hand side (Top) from 1 to 10, (middle) from 11 to 20, and (bottom) from 21 to 30, May 1986. Right hand side (Top) from 31 May to 9 Jun, (middle) from 10 to 19, and (bottom) from 20 to 29, June 1986.

We also assume the vertical profile of the convective heating in the experimental run (proportional to the difference between the moist adiabatic temperature and the grid temperature). The vertical profile of the heating has a significant effect on the circulation. There are many gaps in the available information which TRMM is expected to fill and thus help in increasing our understanding of nature.

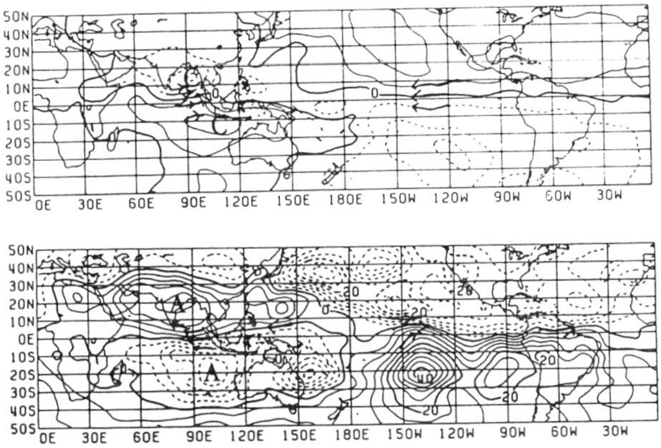

FIGURE 5. The stream function for the difference between the control run and experimental run at 850mb (top) and at 200mb (bottom). 3 days mean between day 6 and day 8 from the initial (1 Jun 1986). Contour interval is 4×10^6 m /s.

REFERENCES

Gill, A.E., 1980: Some simple solutions for heat induced tropical circulation. Quart. J. Roy. Met. Soc., 106 447-462

Hayashi, Yoshi., and A. Sumi, 1986: The 30-40 day oscillations simulated in an "Aqua Planet" model. J. Met. Soc. Japan, 64, 451-467.

He, H. W. McGinnis, Z. Song and M.Yanai, 1987: Onset of the Asian summer monsoon in 1979 and the effects of the Tibetan Plateau. Submitted to Mon. Wet. Rev.

Kanamitu, M., K.Tada, T. Kudo, N. Sato and S. Isa, 1983: Description of the JMA operational spectral model. J. Met. Soc. Japan, 61, 812-828.

Madden, R., and P. Julian, 1971: Detection of 40-50 day oscillation in the zonal wind. J. Atmos. Sci., 28, 702-708.

Yasunari, T., 1980: A quasi-stationary appearance of 30 to 40 day period in the cloudiness fluctuations during the summer monsoon over India. J. Met. Soc. Japan, 58, 225-229.

WORLD WATER BALANCE: ANALYSIS OF FGGE IIIb DATA

Kooiti Masuda
Geophysical Institute,
University of Tokyo
Bunkyo-Ku, Tokyo 113, Japan

ABSTRACT

The annual total precipitation minus evaporation over continents and ocean basins is estimated by the budget of water vapor in the atmosphere. The one-year data of the First GARP Global Experiment (FGGE) are used. The results are qualitatively realistic in the respect that the value is small and positive over dry land. Quantitative disagreements remain between different FGGE data sets.

1. INTRODUCTION

The precipitation process is so efficient that the average residence time of water vapor in the atmosphere is only on the order of ten days. Despite the short residence time, water vapor is carried over long distances, and evaporation and precipitation are not locally balanced. The hydrological cycle consists of (1) precipitation, (2) runoff (including river and underground runoff as well as net freshwater transport in the ocean), (3) evaporation and (4) advection in the atmosphere. The four components should have the same magnitude when globally averaged. Here we neglect the slow circulation of ice masses and deep groundwater. In addition, we will neglect the contribution of water in the liquid or solid phases in the atmosphere to the term of advection.

Let us consider the availability of data for the quantitative evaluation of the global hydrological cycle. Every term is difficult to obtain.

(1) Precipitation. Data coverage is good over large parts of the continents, though there are problems of representativeness and bias of rain gages (Korzun et al., 1974). Over the ocean, however, there are very few stations which observe precipitation continuously. Until some reliable methods of remote sensing are established, we have to rely on indirect estimation from reports of weather.

(2) Evaporation. Direct observations are rare over the land as well as over the ocean. Most of them are those of pan evaporation, and the relation between it and evaporation from natural surfaces is not simple over the land. It is because usually the humidity of the air contacting the ground cannot be approximated by the saturation humidity, and because the ground is often covered by vegetation. Usually bulk formulae are used for the estimation of areal evaporation. They include empirical constants (drag coefficients). The calibration of constants is effective over the ocean where the surface is relatively uniform, but difficult to apply over the land with varying topography and inhomogeneity of groundcover.

(3) Runoff. Runoff by many major rivers was gauged. Lvovitch (1973) used daily discharge data of rivers whose basins cover 35 % of

ISBN 0-937194-14-4

the total continental area. If incomplete data are added, the coverage extends to 60 %. For the freshwater transport in the ocean, only indirect estimates are possible, from the distribution of salinity or minor constituents, or from other terms of surface water balance.

(4) Advection of water vapor. The atmosphere is operationally observed every day because of the need for weather forecasts and aviation. Though it can be considered that the spatial representativeness of upper air variables is larger than surface meteorological variables, the network of radiosonde observations is coarser than that of rain gages.

The intention of this study is investigation of the feasibility of the approach from the term (4). Neglecting the small term of local time change, the annual mean convergence of vertically integrated water vapor transport in the atmosphere is equated to precipitation minus evaporation (P-E) at the same place (Peixóto, 1970). With the comprehensive statistics of 10-year radiosonde station data by Oort (1983), Bryan and Oort (1984) evaluated annual and seasonal mean P-E over ocean basins and continents. Their result, however, seems to have suffered from the coarseness of the observation network.

The recent advent of "four-dimensional assimilation" (4DA) of atmospheric data has changed the situation. Supported by numerical weather forecast models, values of atmospheric variables are nominally available at every gridpoint over the globe. Of course, the values at places where observations are absent are essentially model forecasts, which depend on the model parameterization of physical processes such as evaporation and precipitation. Despite this weakness, it is challenging to see what can be said about the global water balance with the product of 4DA.

2. DATA AND PROCEDURE

The data taken during the one-year period from December 1978 to November 1979 of the Global Weather Experiment or the First GARP Global Experiment (FGGE) are used. Both European Centre for Medium-Range Forecasts (ECMWF) (Bengtsson et al., 1982) and Geophysical Fluid Dynamics Laboratory (GFDL) (Stern et al., 1985) made gridpoint data sets for the one-year period, called "FGGE main IIIb data sets". In this study, two data sets are used: twice-daily data of ECMWF main IIIb, and monthly statistics compiled by N.-C. Lau (Lau, 1985) from twice-daily data of GFDL main IIIb.

Zonal and meridional moisture transports are evaluated at every gridpoint in every layer surrounded by two consecutive pressure levels where data are available, and summed up vertically. With ECMWF data, the transports are calculated twice-daily and time averages are made afterwards. With GFDL data, total transports are evaluated by adding transports by monthly mean fields and those by covariances of wind velocity and humidity. Horizontal convergence is calculated from the annual mean transport values. The procedure is described in the paper on the meridional energy transport (Masuda, 1988b). The lower boundary is set at the ground surface of the forecast model of ECMWF and GFDL, respectively. The wind field is adjusted in order to satisfy the relation of mass conservation. The moisture field in the 850 -- 1000 hPa layer of ECMWF data is adjusted to compensate for the bias due to interpolation from middle of the layer to the standard pressure levels (Masuda, 1988a).

3. ANNUAL WATER BALANCE OVER CONTINENTS AND OCEAN BASINS

Table 1 shows the estimates of annual total volume of runoff, or precipitation minus evaporation. This study (two columns labeled "FGGE") as well as Bryan and Oort (1984) evaluated convergence of water vapor transport in the atmosphere. Baumgartner and Reichel (1975), Korzun et al. (1974) and Lvovitch (1973) made surface-based hydrological estimates. In this study the boundaries of continents and ocean basins are defined as closely as possible to the verbal description by Baumgartner and Reichel. The Pacific and the Atlantic

TABLE 1. ANNUAL VOLUME OF PRECIPITATION MINUS EVAPORATION BY CONTINENTS AND OCEAN BASINS

(Total volume of liquid water, in the unit of 10^{12} m^3/year^{-1}.)

	Area	Aerological methods			Hydrological methods		
	$10^{12} m^2$	FGGE ECMWF	FGGE GFDL	BO84[1]	BR75[2]	K74[3]	L73[4]
Asia	47	4.4	4.7	1.5*	12.2	14.1	13.2
Europe	11	1.8	1.0	-2.0*	2.8	3.0	3.1
Africa	30	1.9	10.0	0.2	3.4	4.6	4.2
N.America	26	5.9	5.8	4.2	5.8	8.2	6.7
S.America	18	7.6	15.3	6.0	11.0	12.2	10.4
Australia	9	0.5	1.9	-3.6	2.4	2.5	2.0
Antarctica	14	1.4	1.5	0.6	2.0	2.3	2.2
Continents	155	23.6	40.3	6.5	39.7	47.0	41.7
N.Pacific	80	3.4	11.8	3.5	16.7	-9.7+	
S.Pacific	94	-1.3	-9.3	6.9	-0.9		
N.Atlantic	50	-9.9	-17.4	-2.4**	-17.8	-31.7+	
S.Atlantic	45	8.2	-15.4	-10.5	-18.7		
Indian Ocean	78	-8.8	-11.5	-4.1	-19.5	-7.6	
Arctic Ocean	8	1.3	1.4	---**	0.4	2.1	
Oceans	355	-23.6	-40.3	-6.5	-39.7	-47.0	-41.7

(1) Bryan and Oort (1984), Table 3.
(2) Baumgartner and Reichel (1975), Table VI (runoff).
(3) Korzun et al. (1974), Table 157 (land, runoff) and Table 185 (ocean, precip. minus evap.).
(4) Lvovitch (1973), Table 3 (runoff).
(*) In BO84, "Middle East" is included in Europe, not in Asia.
(**) In BO84, the Arctic Ocean is included in the North Atlantic.
(+) These values of K74 are for the whole Pacific and the whole Atlantic.

is further divided by the equator. The boundaries are not always the same in other studies.

A strange feature of the result of Bryan and Oort (1984) is the negative value, i.e. excess of evaporation, over Australia and Europe. Their definition of Europe includes most of so-called "Middle East", which is usually considered as the south-western part of Asia. The horizontal distribution of their estimation of P-E (their Fig. 2a) reveals that the excess of evaporation over Europe mainly come from that over the Middle East. Both Australia and the Middle East are mostly arid areas. It is true that in some arid areas negative runoff is possible by rivers flowing from more humid areas, or supply from groundwater through fountains and wells. It is unlikely, however, that a whole continent has negative runoff.

The two FGGE analyses result in excess of precipitation, or positive runoff, over those areas, and it is in qualitative agreement with hydrological estimates. The magnitude of excess is small, which is also plausible.

A possible cause of negative values in Bryan and Oort's estimate can be explained as follows. Around the arid continents, most of the observing stations are located in the coastal area. The observations there influence gridpoint values over the ocean as well as over the desert, so that the gradient of humidity across the coast is analyzed to be small. Suppose that the horizontal divergence in the lower troposphere is slightly positive. If there is no contrast, the divergence of moisture flux should be positive. With 4DA, on the other hand, the atmosphere is analyzed to be humid over the ocean and dry over the deserts. With this contrast, the divergence of moisture can be negative even if the divergence of mass in the lower troposphere is positive. In this respect, the 4DA seems to yield more realistic estimates of water balance than analyses of station observations. We may fall into a circular argument, however, since the contrast of moisture essentially comes from the model physics of evaporation and precipitation.

In the horizontal distribution of P-E (not shown), another remarkable qualitative difference between Oort's and FGGE analyses is found over western India. The annual balance shows a small excess of precipitation according to either of two FGGE analyses, while it is an excess of evaporation according to Bryan and Oort. Comparison of meridional moisture transport between the FGGE analyses and Fig. 8 of Peixóto and Oort (1983) reveals that the transport of moisture by the Indian summer monsoon circulation over the Arabian Sea is very different. Since observations were few before FGGE, it is suspected that Oort's analysis under-estimates moisture transport by the monsoon circulation, accordingly also the monsoon rainfall.

The quantitative agreement between the two FGGE analyses is not very good. For example, the runoff from all the land to all the ocean differs by a factor of two. From different approaches to 4DA of ECMWF and GFDL, the two data sets have different characters (see Masuda, 1988a and references therein). In essence, ECMWF tends to under-estimate the divergent component of wind, and the opposite is the case of GFDL. It is likely that the actual values of divergence lie between the two analyses. Improvements made after the assimilation of "main IIIb" made the discrepancy between 4DA products smaller.

Unfortunately, the production of the revised, or "final" FGGE IIIb data does not cover the whole year, so they cannot be directly used for the study of the annual mean water balance.

ACKNOWLEDGMENTS

I thank M. Kanamitsu, I. Kubota, M. Chiba, K. Yamazaki, T. Iwashima, N.-C. Lau, K. Miyakoda and J. Ploshay for helping me gain access to the data necessary for the study.

REFERENCES

Baumgartner, A. and E. Reichel, E., 1975: The World Water Balance: Mean Annual Global, Continental and Maritime Precipitation, Evaporation and Runoff. Ordenbourg, München (or Elsevier, Amsterdam).

Bengtsson, L., M. Kanamitsu, P. Kållberg and S. Uppala, 1982: FGGE 4-dimensional data assimilation at ECMWF. Bull. Amer. Meteor. Soc., 63, 29 - 43.

Bryan, F. and A. Oort, 1984: Seasonal variation of the global water balance based on aerological data. J. Geophys. Res., 89, 11717 - 11730.

Korzun, V.I. et al.(Eds.), 1974: World Water Balance and Water Resources of the Earth (in Russian), Gidrometeoizdat, Leningrad. English edition, 1978, Unesco Press, Paris, 663 pp.

Lau, N.-C., 1985: Publication of circulation statistics based on FGGE level III-b analyses produced by GFDL and ECMWF. Bull. Amer. Meteor. Soc., 66, 1293 - 1301.

Lvovitch, M.I., 1973: The global water balance. EOS, Transactions of Amer. Geophys. Union, 54, 28 - 42.

Masuda, K., 1988a: Problem notes about FGGE "main" IIIb datasets of ECMWF and GFDL. Tenki (Bulletin of the Meteorological Society of Japan), 35, 5 - 19 (in Japanese). English version is available from the author as Meteorological Research Report 87-01, Division of Meteorology, Geophysical Institute, University of Tokyo.

Masuda, K., 1988b: Meridional heat transport by the atmosphere and the ocean: Analysis of FGGE data. Tellus, 40A (in press).

Oort, A.H., 1983: Global Atmospheric Circulation Statistics, 1958 - 1973. NOAA Professional Paper, No. 14, 180 pp.+microfiches.

Peixóto, J.P., 1970: Pole to pole divergence of water vapor. Tellus, 22, 17 - 25.

Peixóto, J.P. and A.H. Oort, 1983: The atmospheric branch of the hydrological cycle and climate. In Street-Perrot, A., M. Beran and R. Ratcliffe (Eds.), Variations in the Global Water Budget, Reidel, Dordrecht, 5 - 65.

Stern, W.F., J.J. Ploshay and K. Miyakoda, 1985: Continuous data assimilation at GFDL during FGGE. In Seminar/Workshop 1984, Data Assimilation Systems and Observing System Experiments with Particular Emphasis on FGGE, European Centre for Medium-Range Weather Forecasts, Reading, U.K., Vol.2, 125 - 156.

SENSITIVITY OF CLIMATIC RESPONSE IN A GCM TO THE SELECTION OF A CUMULUS PARAMETERIZATION SCHEME

M.A. Geller, Y.C. Sud, and H.M. Helfand
Laboratory for Atmospheres
NASA Goddard Space Flight Center
Greenbelt, Maryland 20771, USA
and
K. Takano
Courant Institute of Mathematical Science
New York University
New York, New York 10012, USA

ABSTRACT

Results are shown for sea surface temperature anomaly atmospheric general circulation experiments in which the Kuo and Arakawa-Schubert cumulus parameterization schemes are compared within a single model. It is found that qualitatively similar results in the tropical precipitation anomaly fields are obtained with both parameterizations, but that there are important differences seen between them. With present observational capabilities, it would be difficult to use differences in observed tropical precipitation and nonradiative diabatic heating rates to distinguish between different cumulus parameterizations for use in a given model. TRMM would be a major step forward in this regard.

1. INTRODUCTION

The pioneering observational work of Bjerknes (1966, 1969) indicated that very strong changes in the global atmospheric circulation accompanied the occurrence of the warm El Nino sea surface temperature anomalies that are found to exist in some years in the equatorial eastern Pacific Ocean. In recent years, this has motivated a large number of general circulation modeling (GCM) groups to conduct simulation experiments with the objective of understanding the local and global implications of such SST anomalies. Some of the well known modeling studies include an early work by Rowntree (1972), and more recent works by Shukla and Wallace (1983), Cubasch (1985), Geisler et al. (1985), and Tokioka et al. (1986).

The physical chain of events by which these elevated sea surface temperatures affect the global circulation has been found to be the following. The region of deep cumulus convection with heavy rainfall, that is normally confined to the extreme western Pacific where the warmest sea surface temperatures are found in most years, shifts eastward following the shift in the region of warmest sea surface temperatures. This is accompanied by a corresponding shift in the pattern of the upper level (~200 mbar) divergence with maximum divergences overlying the region of heavy precipitation. Regions of low

 A. DEEPAK Publishing
ISBN 0-937194-14-4

level convergence lie below these regions of high level divergences. Accompanying these changes in the precipitation and divergence/ convergence patterns, a stationary wave pattern is set up which has global extent. This chain of events is extremely important to understand and predict since it gives rise to very persistent anomalous weather patterns globally.

Obviously, the energetics of this atmospheric response is far more complex than what would simply result from the enhanced vertical sensible and latent heat transports resulting directly from the warmer ocean temperatures. In fact, a large-scale convergence of water vapor takes place to produce the observed and/or modeled regions of heavy precipitation (see Mechoso et al., 1986, for example). These large changes in the moisture convergence pattern are a result of the warm tropical sea surface temperatures driving intense cumulus convective activity which in a GCM depend, to a great extent, on the chosen subgrid-scale parameterization processes. Miyakoda and Sirutis (1977) and Tiedtke (1984) have investigated the dependence of the climatologies of the Geophysical Fluid Dynamics Laboratory (GFDL) and the European Center for Medium-Range Weather Forecasting (ECMWF) GCM's to the choice of cumulus parameterization methods. The present investigation differs from these earlier works in that our goal is to look at the sensitivity of the Goddard Laboratory for Atmospheres' (GLA) Fourth-Order GCM to different choices of cumulus parameterization methods in modeling the atmospheric response to El Nino sea surface temperature anomalies.

2. DESCRIPTION OF MODEL EXPERIMENTS

The present model experiments were carried out using the GLA GCM, which has been described in detail by Kalnay et al. (1983), and has been used extensively for a number of weather prediction studies and data impact studies. Briefly, the version of the GLA GCM being used in the present study has a horizontal resolution of 4° latitude by 5° longitude and has nine sigma levels of equal thickness between the top of the model atmosphere at 10 mbar and the earth's surface. Prognostic equations are used for the horizontal wind components, temperature and specific humidity in the free atmosphere as well as for surface pressure and surface temperature over land. Sea surface temperature, ice conditions, and soil moisture were prescribed in these present experiments. The flux form equations of motion were solved using fourth-order finite difference equations in space and Matsuno (Euler backward) time differencing. A sixteenth-order Shapiro (1970) filter was used every 2 hours on the sea level pressure and potential temperature.

The physical parameterizations were essentially taken from the three layer UCLA model. They include bulk aerodynamic formulations for the PBL fluxes as given in Arakawa (1972) and further modified by Sud and Abeles (1980), dry and moist convection, vertical diffusion of heat and moisture, the solar radiation parameterization of Lacis and Hansen (1974) and the longwave parameterization of Wu (1980). The only new additions are the inclusion of a fractional cloudiness parameterization for cloud radiation interaction following Slingo and Ritter (1984). This is used for all simulations.

Two different cumulus cloud parameterizations were used. One is the scheme of Kuo (1974) in which both deep cumulus convection and shallow convection are included. This scheme was taken from NMC (personal communication, Dr. P. Caplan) and was implemented in the GLA GCM after making adjustments for the different vertical resolutions of the NMC and GLA GCM's. The Kuo scheme is very widely used, and its variations are being used in several operational and research models such as at ECMWF, NCAR, NMC, and FSU. Briefly speaking, the Kuo scheme uses the large-scale low level convergence of moisture and low level lifting to determine the onset of cumulus convection. The condensation rate is assumed to be proportional to both the large-scale moisture convergence and the temperature difference between the cloud air and the environmental air. One-dimensional cloud models are used to determine the in-cloud vertical motion and the in-cloud temperature.

The other cumulus parameterization scheme used is that of Arakawa and Schubert (1974), as modified by Sud and Molod (1987), which is referred to in the remainder of this paper as AS. The AS scheme allows for the coexistence of a spectrum of model clouds, such as a mixture of shallow and deep clouds, that interact with each other as well as with the large-scale environment. Each member of the spectrum has its own mass, momentum, and moisture budget and is characterized by its own fractional entrainment rate. The cloud bases are assumed to be located at the top of the boundary layer. Therefore, the AS scheme relies heavily on the relative humidity and moist static energy of the planetary boundary layer to infer the onset of convection. An integral component of the AS method is the "cloud work function" which is largely determined by the thermodynamic characteristics of the cloud ensemble with respect to the large-scale environment. By assuming approximate equilibrium of the cloud work function with respect to time, the AS scheme strives to maintain a balance between the generation of moist instability by the large-scale flow with the cumulus processes that destroy the large-scale instability. Sud and Molod (1987) modified the AS parameterization to include re-evaporation of raindrops. They also altered the GLA GCM to include a more realistic calculation of dry convective mixing as well as cloud radiation interaction.

Four simulations, two with the Kuo and two with the AS cumulus parameterizations, were carried out. The control simulation for each parameterization used the February climatological SST, whereas the corresponding anomaly simulations used the actual February 1983 SST that was produced by superposing the observed SST anomaly of Reynolds and Gimmel (1984) upon the climatological SST distribution. Both the climatological and anomaly sea surface temperatures are shown in Figure 1. For all four cases, the intitial conditions were those determined by the GLA analysis for January 15, 1983. A period of fifteen days was allowed for the initial adjustment, and the model-simulated Februaries were analyzed.

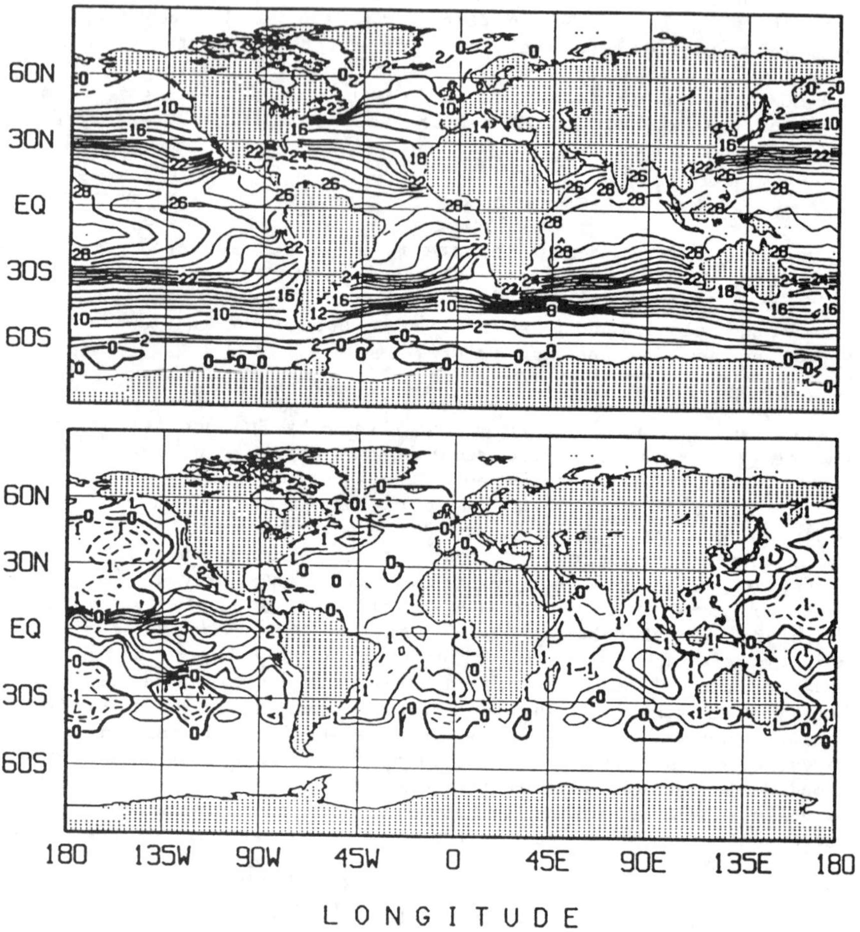

Figure 1. Top: Climatological sea surface temperature for the month of February in units of °C. Isotherms are drawn at 2 °C interval with the thick solid line indicating 0 °C and dashed lines indicating temperatures less than 0 °C. For temperatures 21, 23, . . . up to 29 °C, thin lines are added. Bottom: SST anomaly map for February 1983 taken from Reynolds and Gimmel (1984) at 0.5 °C intervals.

3. RESULTS

The primary variable that is examined in this paper for these various GCM runs is, in line with the topic of this Symposium, the tropical precipitation. Figure 2 shows the February averaged total precipitation in units of mm/day for the climatological runs with the AS and Kuo parameterizations. Both the AS and Kuo climatological precipitation fields show maximum rainfall in the Australasia equatorial area, over Equatorial Africa, and in a band extending from Equatorial South America to the northwest. This is consistent with the observational picture as given by Jaeger (1976) for climatological February. The level of agreement between the climatological results with the Kuo and AS parameterizations that we see is similar to what was found by Tiedtke (1984). The Kuo and AS precipitation anomalies in Figure 2 are qualitatively similar with both showing a large precipitation enhancement over the eastern equatorial Pacific and a general decrease in precipitation over the western Pacific that is in line with the eastward movement of the region of maximum precipitation that is expected to accompany the eastward motion of the warmest sea surface temperatures. There are differences though.

Perhaps the most striking difference is that in the Kuo case, the positive precipitation anomaly in the eastern Pacific is centered on the Equator whereas in the AS case, this anomaly is centered at a latitude that lies to the south of the Equator. Although there are other regions of sizeable precipitation anomalies, the Pacific region is the only one in which the precipitation responds directly to the altered sea surface temperatures. In other equatorial regions, large anomalies are seen, but they are produced by a rather more complicated combination of physical effects than just responding to the underlying sea surface temperatures. Since most of the precipitation anomalies occur over the tropical oceans and the pattern of the precipitation anomalies with the Kuo and AS schemes resemble each other quite closely, it would be difficult to use the differences in these modeled anomaly patterns to verify which of these parameterizations gave preferable results using just our present observing capability for tropical precipitation which is quite sparse over land and practically nonexistent over the oceans. It is interesting to note that Figure 3 (Prabhakara, unpublished figure derived by the methods of Prabhakara et al., 1986) shows a rainfall anomaly map over the oceans that was derived for February 1983 from Nimbus 7 SMMR data. It shows a region of positive rainfall anomaly that is centered at a latitude south of the Equator in better agreement with the AS results than with the Kuo results.

Another atmospheric parameter which is known to be very important in the physics of El Nino produced circulation anomalies is the pattern of the resulting diabatic heating anomalies. Figure 4 shows both the zonally averaged climatological nonradiative heating rates as well as the zonally averaged nonradiative heating rates for both the Kuo and AS El Nino simulations. One sees that the AS scheme gives larger climatological heating rates in the tropical middle troposphere than does the Kuo scheme. It is also seen that the AS

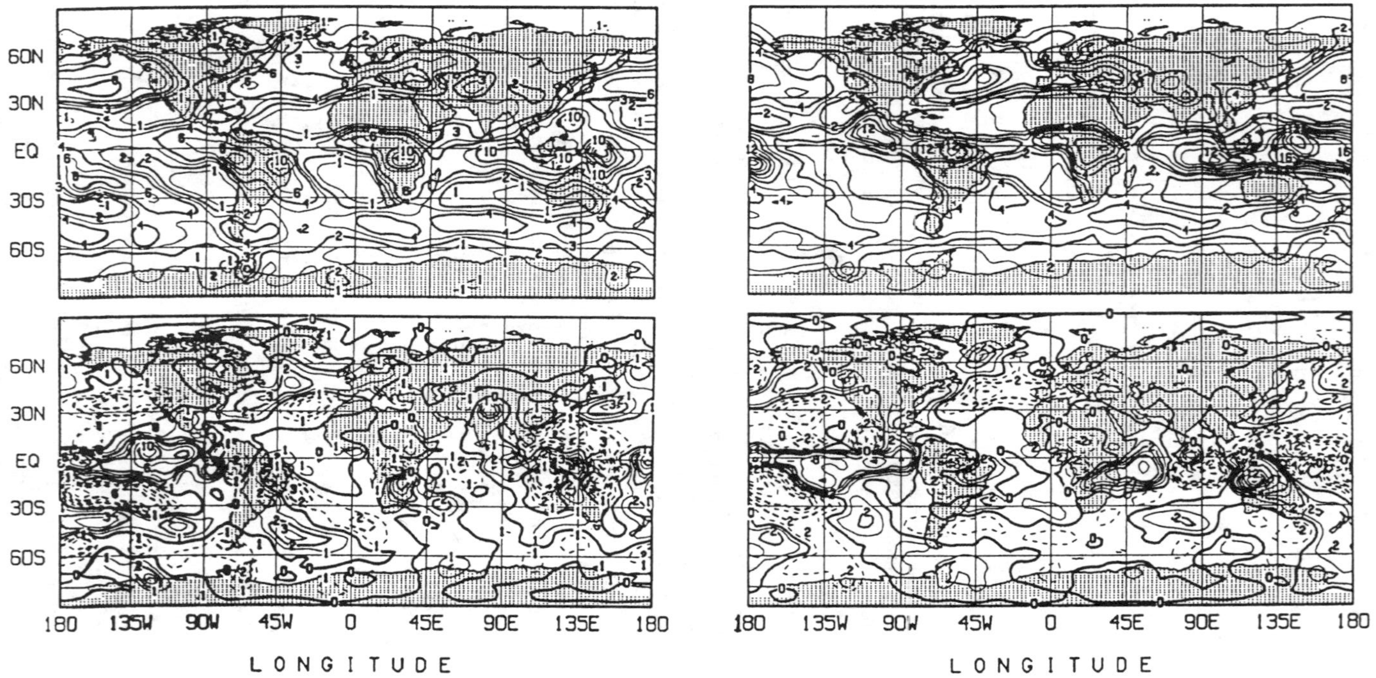

Figure 2. Top Left: Control simulation of precipitation rate with Kuo parameterization. Top Right: Control simulation of precipitation rate with AS parameterization. Bottom Left: Precipitation rate anomaly for SST anomaly simulation with Kuo parameterization. Bottom Right: Precipitation rate anomaly for SST anomaly simulation with AS parameterization. Contour intervals are 0, 1, 2, 3, 4, 6, 8, 10, 12 . . . mm/day.

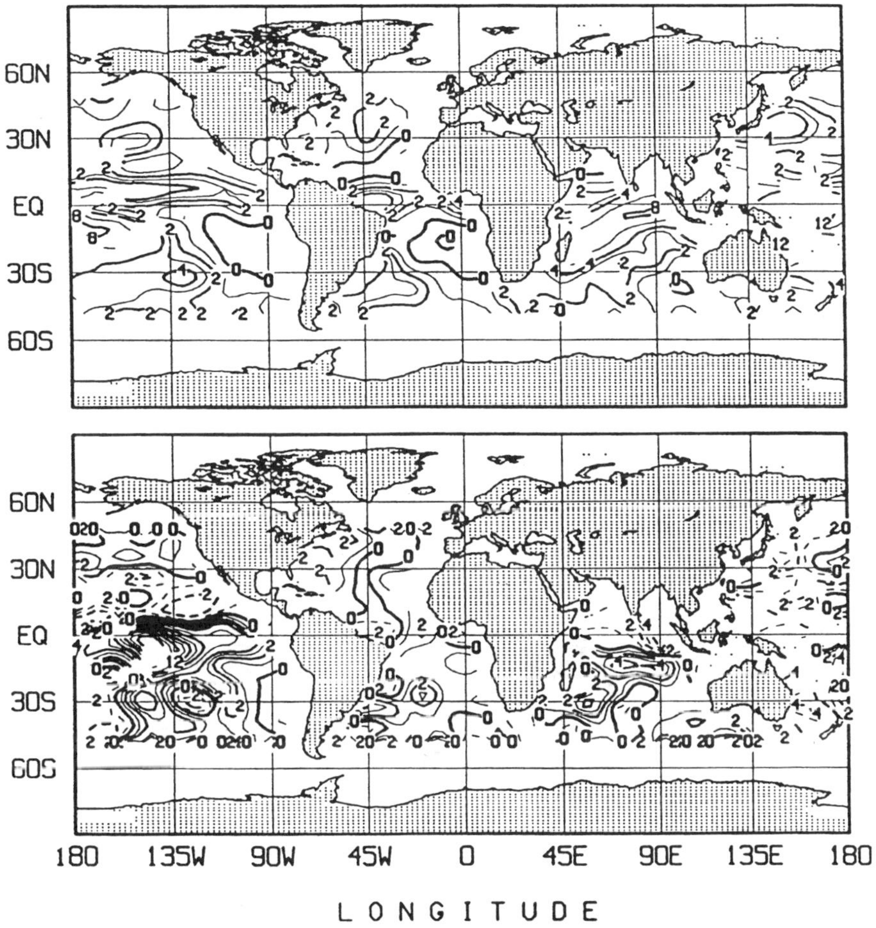

Figure 3. Top: Three-year mean February (1979, 1980, and 1981) precipitation. Bottom: Precipitation anomaly for February 1983 as a difference from the above three-year mean. Derived from Nimbus 7 SMMR data using the methods of Prabhakara et al. (1986). Contour intervals are same as in Figure 2.

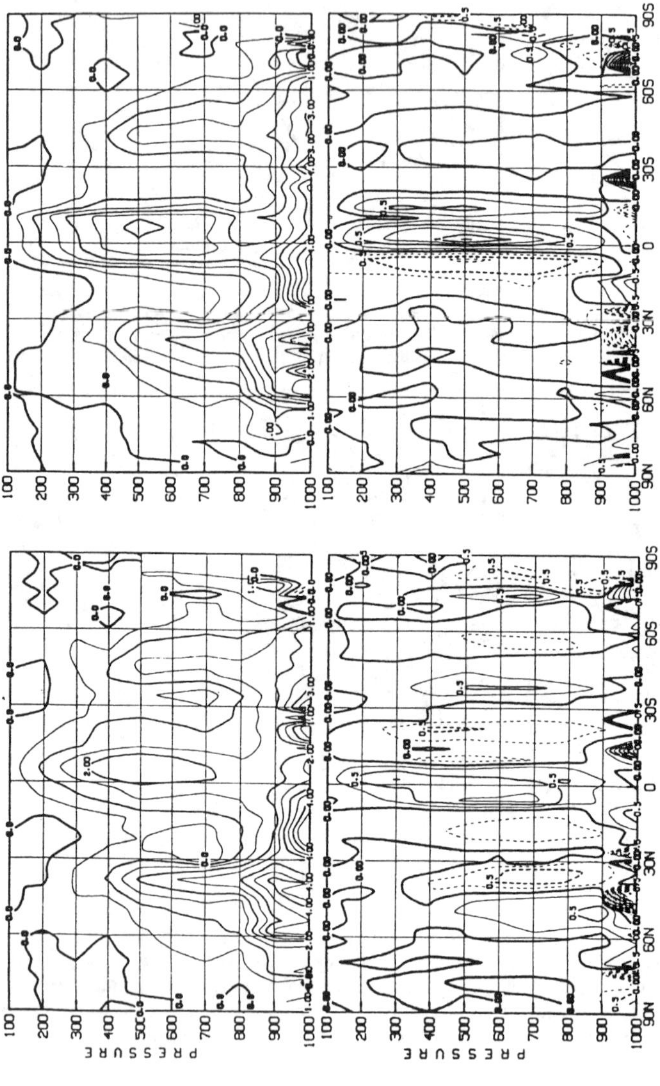

Figure 4: Top Left: Control zonally-averaged nonradiative heating rate for Kuo parameterization. Top Right: Control zonally-averaged nonradiative heating rate for AS parameterization. Bottom Left: Differences (SST anomaly minus control) in zonally-averaged nonradiative heating rate for Kuo parameterization. Bottom Right: Differences (SST anomaly minus control) in zonally-averaged nonradiative heating rate for AS parameterization. All have units of °C/day.

results give larger heating rates at very high levels. The AS El Nino results show a much greater enhancement in the nonradiative heating rates (almost all of this is by latent heat release) throughout the tropical troposphere. At the present time, we have no observations of the nonradiative heating profiles in the tropical atmosphere. This is one of the principal goals of the TRMM mission.

4. SUMMARY AND CONCLUSIONS

The understanding of the relationship between tropical sea surface temperature anomalies and changes in atmospheric circulation is generally recognized as being one of the most important topics in climate variability. In fact, it provides a good deal of the scientific justification for the TOGA program. GCM's are presently, and will continue to be used to model these effects toward the goal of having the ability to predict the atmosphere and ocean effects with coupled models. A crucial physical link in ENSO occurrences is the manner in which cumulus convective activity responds to anomalies in the sea surface temperatures since it is the resulting precipitation and accompanying condensational heating that drives the resulting global circulation anomalies.

We have used two of the most commonly used cumulus parameterizations within a single global atmospheric model to study how the resulting precipitation and circulation anomalies may depend on the choice of these parameterizations. We have seen that the most straightforwardly coupled atmospheric parameter to sea surface temperature anomalies is the change in the tropical precipitation distribution along with the resulting change in diabatic heating.

Given the present situation in which the distribution of precipitation is derived primarily from in situ land observations, it is not possible to compare the model-simulated rainfall fields with observations, particularly over the oceans. In other words, we cannot tell how accurate is the rainfall prediction of a GCM in response to different cumulus parameterizations but with observed SST anomalies. **It is, therefore, evident that TRMM, which would provide the tropical rainfall and diabatic heating rate data, will be of significant value for quantitative verifications of models as well as for future development of cumulus parameterization schemes.**

ACKNOWLEDGMENTS

We would like to thank Mr. G. K. Walker and Mr. W. E. Smith of Sigma Data for their help in setting up the model experiments and computer graphics.

REFERENCES

Arakawa, A., 1972: Design of the UCLA General Circulation Model . numerical simulation of weather and climate. Tech. Rep. No. 7, Dept. Meteorol., UCLA.

Arakawa, A. and W. H. Schubert, 1974: Interaction of cumulus cloud ensemble with the large-scale environment, Part I. J. Atmos. Sci., 31, 674-701.
Bjerknes, J., 1966: A possible response of the atmospheric Hadley circulation to equatorial anomalies of ocean temperature. Tellus, 18, 820-829.
Bjerknes, J., 1969: Atmospheric teleconnections from the equatorial Pacific. Mon. Wea. Rev., 97, 163-172.
Cubasch, U., 1985: The mean response of the ECMWF global model to the El-Nino anomaly in extended range prediction. Atmosphere-Ocean, 23, 43-66.
Geisler, J. E., M. L. Blackmon, G. T. Bates and S. Munoz, 1985: Sensitivity of January climate response to the magnitude and position of equatorial Pacific sea surface temperature anomalies. J. Atmos. Sci., 42, 1037-1049.
Jaeger, L., 1976: Monatskarten des Niederschlages fur die ganze Erde., Ber. Deutsch Wetterd., Ofenbach/Main, Nr. 139, Vol. 18, 38 pp.
Kalnay, E., et al., 1983: Documentation of the GLAS Fourth-Order General Circulation Model. NASA Tech. Memo. 86064, Vol. 1, NTIS Washington, DC.
Kuo., H. L., 1974: Further studies of the parameterization of the influence of cumulus convection on large-scale flow. J. Atmos. Sci., 31, 1232-1240.
Lacis, A. and J. E. Hansen, 1974: A parameterization for the absorption of solar radiation in the earth's atmosphere. J. Atmos. Sci., 31, 118-133.
Mechoso, C. R., A. Kitoh, S. Moorthi and A. Arakawa, 1986: Numerical experiments on the atmospheric response to a sea surface temperature anomaly over the equatorial eastern Pacific Ocean. In Workshop on Comparison of Simulations by Numerical Models of the Sensitivity of the Atmospheric Circulation to Sea Surface Temperature Anomalies, WMO/TD - No. 138, WCP - 121, World Meteorological Organization, Geneva, Switzerland.
Miyakoda, K. and J. Sirutis, 1977: Comparative integrations of global models with various parameterized processes of subgrid-scale vertical transports: description of the parameterizations. Beitr. Phys. Atmosph., 50, 445-487.
Prabhakara, C., D. A. Short, W. Wiscombe, R. S. Fraser and B. E. Vollmer, 1986: Rainfall over oceans inferred from Nimbus 7 SMMR: application to 1982-83 El Nino. J. Climate Appl. Meteor., 25, 1464-1474.
Reynolds, R. W. and W. H. Gimmel, 1984: An objective global monthly mean sea surface temperature analysis. Tropical Ocean-Atmosphere Newsletter, 18, 14-15.
Rowntree, P. R., 1972: The influence of tropical east Pacific Ocean temperatures on the atmosphere. Quart. J. Roy. Meteor. Soc., 98, 290-321.
Shapiro, R., 1970: Smoothing, filtering and boundary effects. Rev. Geophys. and Space Phys., 8, 359-387.
Shukla, J. and J. M. Wallace, 1983: Numerical simulation of the atmospheric response to equatorial Pacific sea surface temperature anomalies. J. Atmos. Sci., 40, 1613-1630.

Slingo, J. and B. Ritter, 1985: Cloud prediction in the ECMWF model. Tech. Rep. No. 46, ECMWF, England.

Sud Y. and J. Abeles, 1980: A non-iterative calculation for the surface temperature and surface fluxes in the GLAS GCM. NASA Tech. Memo 80650.

Sud, Y. and A. Molod, 1987: Influence of recent improvements in the parameterization of convection in the GLA GCM on circulation and rainfall, and the roles of dry convection, and cloud radiation feedback processes. Mon. Wea. Rev. (to appear).

Tiedtke, M., 1984: The effect of penetrative cumulus convection on the large-scale flow in a general circulation model. Beitrag. Phys. Atmosph., 57, 216-239.

Tokioka, T., A. Kitoh and A. Katayama, 1986: Atmospheric response to the sea surface temperature anomalies in the mature phase of El Nino: numerical experiment under the perpetual January condition. J. Meteor. Soc. Japan, 64, 347-362.

Wu., M. L., 1980: The exchange of infrared radiative energy in the troposphere. J. Geophys. Res., 85, 4084-4090.

LOW LATITUDE PRECIPITATION CHARACTERISTICS OF THE MRI GCM AND CUMULUS PARAMETERIZATION

Tatsushi Tokioka, Koji Yamazaki, Akio Kitoh, and Tomoaki Ose
Meteorological Research Institute
Tsukuba, Ibaraki 305, Japan

ABSTRACT

Numerical experiments with the use of the Meteorological Research Institute atmospheric general circulation model (MRI·GCM) demonstrate high sensitivity of both the time mean fields and low latitude intraseasonal variation fields to penetrative cumulus parameterizations, i.e. to the precipitation characteristics of the model. If we restrict occurrence of deep penetrative cumuli to the area with thick planetary boundary layer depth under the framework of the Arakawa-Schubert parameterization (1974), following changes occur in the model atmosphere. Temperature in the upper troposphere decreases in low latitudes. Precipitation increases there, the intensity of the Hadley circulation also increases and the width of it decreases. Vertical distribution of the diabatic heating changes. Peak of the dominant heating level lowers. The 30-60-day oscillations tend to be dominant with the decrease of the major heating level, depicting high sensitivity of them to the vertical profile of the heating.

1. INTRODUCTION

Diabatic heating in the atmosphere is composed of solar radiational heating, terrestrial radiational heating, heating due to phase changes of water and turbulent heat flux convergence. Figure 1 shows annual mean heating rate of the model and its major components in the meridional cross section. Characteristic feature of the diabatic heating is determined by the condensational heating. In the MRI·GCM, precipitation occurs through three processes. One is the large-scale condensation, which occurs when the mixing ratio of water vapour exceeds the saturation value at a grid point. Other is the penetrative cumulus precipitation based on the Arakawa-Schubert theory (1974), which deals with cumuli with their roots within the planetary boundary layer (PBL). The other is the precipitation through middle-level convections, which are moist convections within the free atmosphere. Figure 1 (e),(f),(g) are contributions of these three processes to the heating. They show that the condensational heating in low latitudes is almost due to penetrative cumulus convections.

We performed a series of numerical experiments with the MRI·GCM to study relations between a penetrative cumulus parameterization and low latitude circulations. The experiments show now a small change in precipitation amount, its horizontal distribution pattern, time mean circulation fields and slowly varying (or so-called intraseasonal time scale) fields, implying that the precipitation measurement in low

ISBN 0-937194-14-4

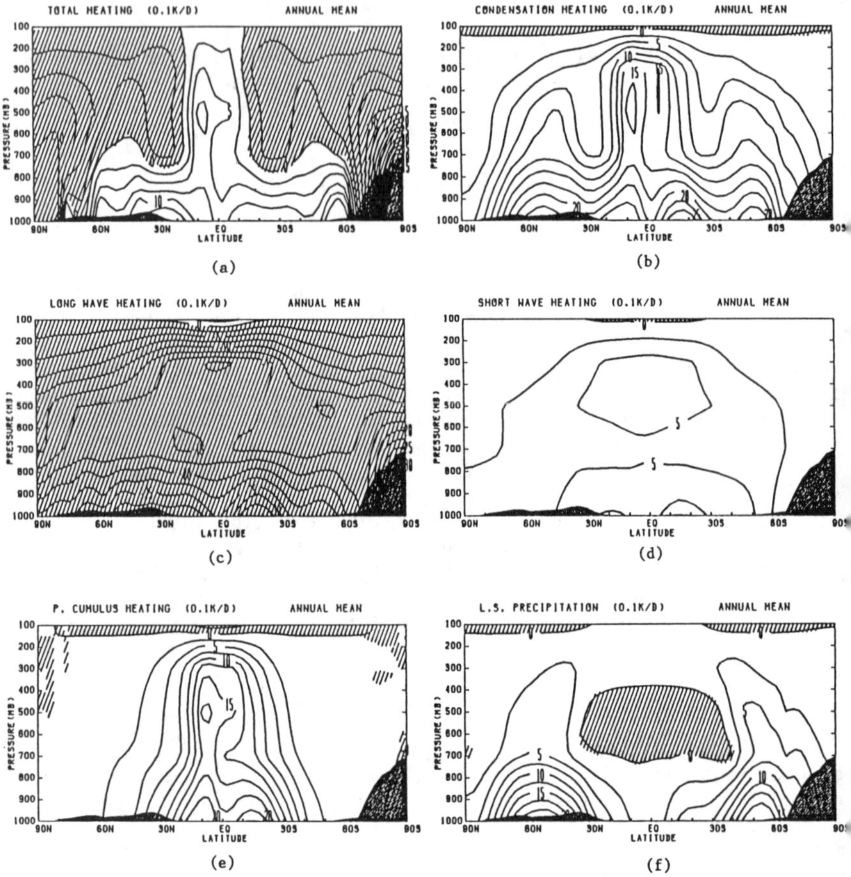

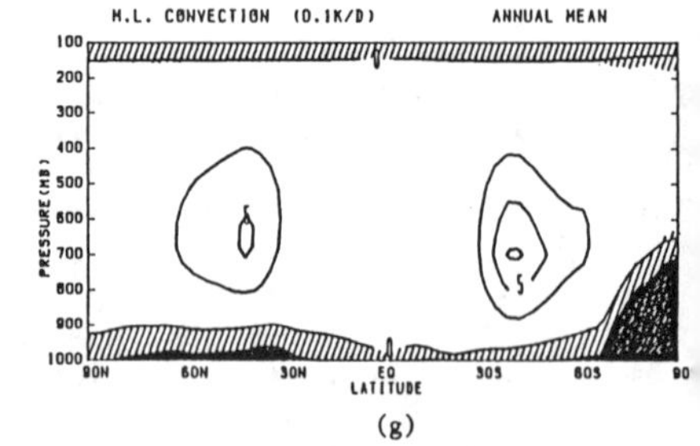

Fig.1 Annual mean heating rate of the model and its major conponents in the meridional cross section. Units are 0.1K/day.

(a) total heating
(b) condensation heating
(c) long wave heating
(d) short wave heating
(e) penetrative cumulus heating
(f) large-scale precipitation
(g) middle-level convection

latitudes is highly required to verify atmospheric models and to deepen our understanding of general circulations.

2. MODEL AND EXPERIMENTS

The model has its top at 100mb and the atmosphere is divided into five layers. The horizontal resolution is 5° and 4° in the longitudinal and latitudinal directions, respectively. The model includes most of physical processes. The Arakawa-Schubert theory (1974) is adopted to parameterize penetrative cumulus convections. For further details of the model, see Tokioka et al. (1984) or Tokioka et al. (1985). The entire earth's surface is replaced by ocean with the zonally uniform sea surface temperature to simplify experimental situations. The sun is fixed to the equinox position.

In Arakawa- Schubert theory, mass flux of each type of penetrative cumulus is determined by the quasi-equilibrium assumption. We adopt this mass flux in the control run. In other experiments, we introduce an additional constraint in determining it. We define a nondimensional constant ξ as

$$R_{max} = \xi D, \qquad (1)$$

where R and D are the radius of cumulus at its cloud base and the PBL depth, respectively. R_{max} is the maximum value of R at a grid point. Simpson (1971) shows that the entrainment rate of environment into cumulus, μ, is related to R as

$$\mu = 0.2/R. \qquad (2)$$

The Arakawa-Schubert theory determines μ for each type of penetrative cumulus through a non-buoyancy condition at a cloud top. D is a prognostic variable in the MRI GCM. If the relationship (1) holds approximately, there must be a minimum value of μ,

$$\mu_{min} = \alpha/D, \text{ where } \alpha = 0.2/\xi. \qquad (3)$$

We repeated experiments by changing α from 0 to ∞. $\alpha = 0$ is nothing but the control case, and $\alpha = \infty$ is the case of no penetrative cumuli at all.

3. TIME MEAN FIELDS

The zonally averaged precipitations of the cases $\alpha = 0$ and ∞ are shown in Fig.2, as they are two extreme cases in the present series of experiments. There are large differences in the precipitation between them. Precipitation at the equator of the case $\alpha = \infty$ almost doubles the value of the case $\alpha = 0$. Strong concentration of precipitation in the equatorial latitudes and overall increase of it in low latitudes are also noted in the case $\alpha = \infty$. Secondary maximum in the middle latitudes shifts to the equatorial side. These changes are consistent with the intensification and the latitudinal shrink of the Hadley circulation in the case $\alpha = \infty$. The lack of an efficient vertical energy transport agent in the case $\alpha = \infty$ keeps the vertical stratifica-

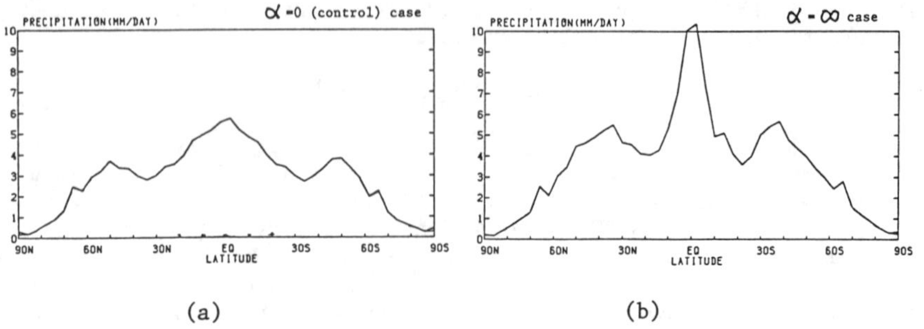

Fig.2 Zonally averaged precipitations of the aqua-planet model (a) $\alpha=0$ and (b) $\alpha=\infty$.

tion more unstable than in the control case despite the large increase of precipitation. In other words, the unstable stratification enhances precipitation in low latitudes.

4. INTRASEASONAL OSCILLATIONS IN LOW LATITUDES

Intraseasonal oscillations from 30- to 60-day periods are observationally confirmed in low latitudes. They have a dominant zonal wavenumber 1 structure. However, they are missing in the control case. A dominant quasi-10-day eastward propagating mode exists instead (see Fig.3(a)) in the zonal wavenumber 1. We found that the characteristics of intraseasonal oscillations change with the increase of α. When α = 0.1, model produces a dominant quasi-30-day eastward propagating mode in the zonal wavenumber 1 (Fig.3(b)). When $\alpha = \infty$, a dominant quasi-45-day eastward propagating mode is found in the zonal wavenumber 1. The vertical structures of the quasi-30-to-45-day oscillations are close to the observed ones (Murakami and Nakazawa, 1986). Figure 4 shows the vertical distribution of diabatic heating rate at the equator. The major heating level exists at above 500 mb level when α = 0. The major heating level lowers with the increase of α. The change of the vertical structure (especially in the zonal wind field) of the intraseasonal oscillations occurs with the decrease of the major heating level, which is produced by the increase of α through the control of the activity of deep penetrative cumuli. Intraseasonal oscillations are actually observed in the atmosphere for the period from 30 to 60 days. For the model to simulate them, present experiments suggest the use of non-zero α rather than 0.

5. EXPERIMENTS UNDER REALISTIC SITUATIONS

Experiments were repeated under the realistic surface conditions, where the realistic land-sea distribution and the observed sea surface temperature were assigned. Two time integrations were performed from the observed analysis of April 1,'84 until August 31. One is the case α = 0 and the other α = 0.05. Figure 5 shows the mean precipitation averaged from June to August for α = 0 and the deviations of α = 0.05 from that of α = 0.

We observe large changes over the North Pacific. The precipitation belt shifts toward the equatorial latitudes when $\alpha = 0.05$, and the precipitation pattern is much improved, although quantitative verification is limited at present because of the lack of reliable climatological data in that area. Simulations of the Pacific high and the zonally averaged temperature distribution are also much improved when $\alpha = 0.05$ compared with those of $\alpha = 0$. Intraseasonal oscillations are found in the $\alpha = 0.05$ case with an appropriate amplitude and period.

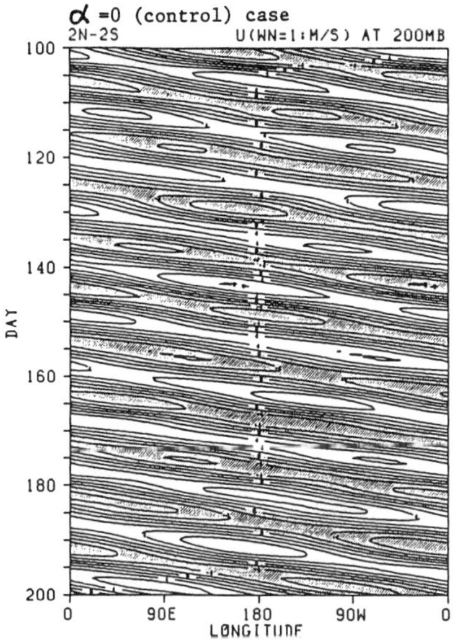

(a)

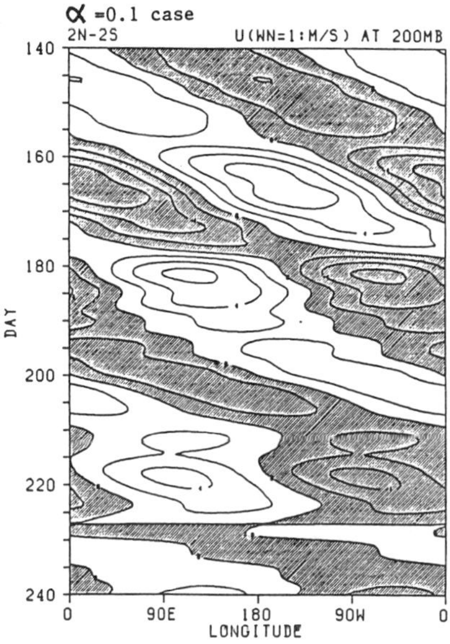

(b)

Fig.3 Longitude-time section of the wave number 1 components of equatorial 200mb zonal wind.
(a) α =0 (control) case
(b) α =0.1 case

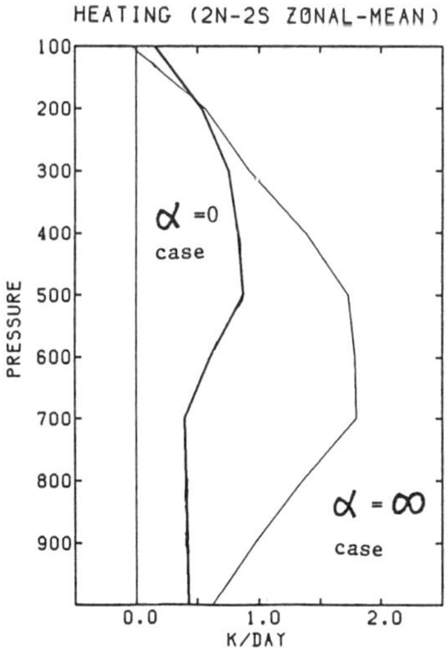

Fig.4 Vertical distribution of diabatic heating rate at the equator in the case α=0 and $\alpha = \infty$.

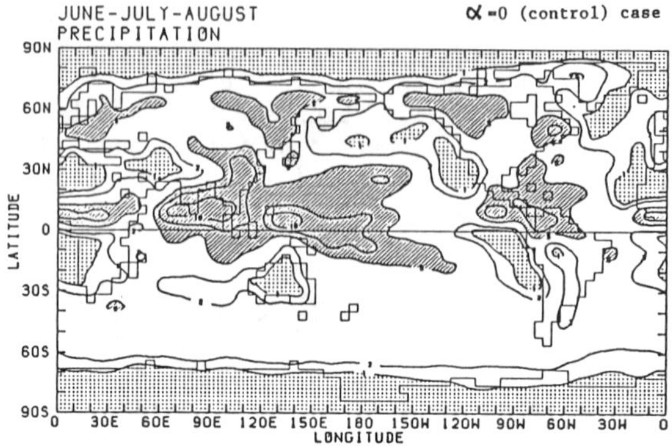

(a)

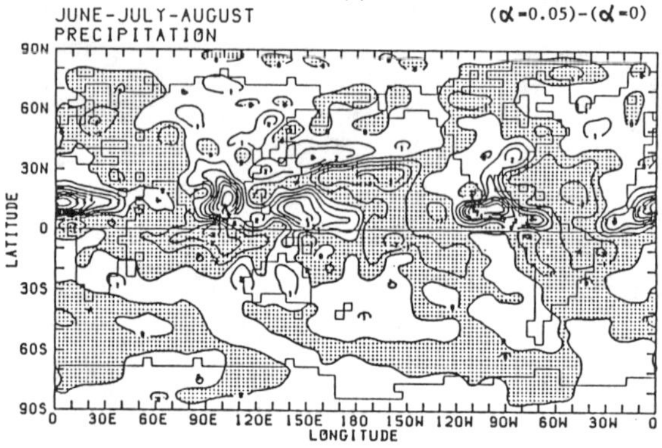

(b)

Fig.5
Mean precipitation averaged from June to August.
Units are mm/day.

(a) α =0 (control) case
(b) deviations of α=0.05 case from that of α=0 case

REFERENCES

Arakawa, A and W.H.Schubert, 1974: Interactions of a cumulus cloud ensemble with the large-scale environment, Part I. J.Atmos.Sci., 31, 674-701.

Murakami,T. and T.Nakazawa, 1985: Tropical 45 day oscillations during 1979 northern hemisphere summer. J.Atmos.Sci., 42, 1107-1122.

Simpson, J., 1971: On cumulus entrainment and one-dimensional models. J.Atmos.Sci., 28, 449-455.

Tokioka, T., K.Yamazaki, I.Yagai and A.Kitoh, 1984: A description of the MRI atmospheric circulation model (MRI·GCM-I). Tech. Rept. of the Meteorological Research Inst., No.13, 249pp.

———, A.Kitoh, I.Yagai and K.Yamazaki, 1985: A simulation of the tropospheric general circulation with the MRI atmospheric general circulation model, Part I. J.Meteor. Soc. Japan, 63, 749-778.

VARIABILITY OF RAINFALL OVER TROPICAL OCEANS: SCIENTIFIC BASIS AND JUSTIFICATION FOR TRMM

J. Shukla
University of Maryland
College Park, Maryland 20742, USA

ABSTRACT

We have investigated the following questions in this paper:

1. How good are the existing observed rainfall climatologies over the tropical oceans?

2. What are the similarities and differences between the rainfall climatologies produced by different global general circulation models (GCM)?

3. What is the magnitude and structure of the interannual variability of monthly mean rainfall over the tropical oceans as simulated by GCMs, and how does it compare with the interannual variability of the outgoing long wave radiation (OLR) anomalies?

4. What is the magnitude and structure of rainfall anomalies associated with the warm episodes of the El Nino-Southern Oscillation (ENSO) phenomenon?

We hope that an investigation of these questions will help establish the need and scientific justification for TRMM, and interalia, will help evaluate whether the expected accuracy of the proposed measurement mission is adequate.

1. INTRODUCTION

Interannual changes of monthly and seasonal mean rainfall in the tropics, and the associated changes in the tropical and the extratropical atmospheric and oceanic circulations represent one of the most important components of the global climate variability. These changes can occur either due to interactions between the atmosphere, ocean and biosphere or due to the internal dynamical processes in the atmosphere itself. In either case an uncertainty in the estimate of rainfall over the tropical oceans has remained a stumbling block in our understanding of the global hydrological cycle and in verifying global dynammical models for which changes in rainfall represent changes in the vertically integrated diabatic forcing due to latent heat of condensation.

Recognizing the need for more reliable measurements of tropical rainfall, an earth-observing system, known as the Tropical Rain Measurement Mission (TRMM), has been proposed.

ISBN 0-937194-14-4

The main scientific goals of this mission are:

1. To advance the earth system science objective of understanding the global energy and water cycle by means of providing distributions of rainfall and heating over the global tropics.

2. To understand the mechanisms through which changes in tropical rainfall influence global circulation, and to improve our ability to model these processes in order to predict global circulation and rainfall variability at monthly and longer time scales.

It is hoped that TRMM will help answer the following important questions:

1. What is the 4-dimensional structure of latent heating in the tropical atmosphere? How does it vary diurnally, intraseasonally, seasonally and annually?

2. What is the role of latent heat released in the tropics in both tropical and extra-tropical circulations?

3. What is the monthly average rainfall over tropical ocean areas of about 10^5 km^2 and how does this rain and its variability affect the structure and circulation of the tropical oceans?

4. What is the relationship between precipitation and changes in the boundary conditions at the earth's surface (e.g., sea surface temperatures, soil properties, vegetation)?

5. What is the diurnal cycle of tropical rainfall and how does it vary in space?

6. What is the relative contribution of convective and stratiform precipitation and how does the ratio vary in different parts of the tropics and in different seasons?

7. How can improved documentation of rainfall improve understanding of the hydrological cycle in the tropics?

In the present paper we have described the seasonal cycle of OLR and observed rainfall to determine the reliability of existing observed rainfall climatologies over the tropical oceans. We discuss the results of rainfall simulation by global general circulation models. In particular, we have examined the similarities and differences between the rainfall climatologies and their variability for two different GCMs. We also describe the magnitude and structure of rainfall anomalies associated with the warm episodes of the ENSO phenomenon.

2. SUMMARY AND CONCLUDING REMARKS

We have first examined the seasonal cycle of OLR and observed rainfall. Since we have only long-term monthly mean fields of observed rainfall (Jaeger, 1976), we have examined the departures of January and July mean rainfall from the annual mean. We find that the magnitude of the interannual variations of observed rainfall over the tropical oceans is far less than that over the tropical land masses. We have also compared the interannual changes of OLR over the tropical oceans with Jaeger's observed rainfall data and find that the magnitude of the interannual changes for OLR is considerably larger than that for the observed rainfall and also has larger spatial coherence. We also find rather unexpectedly that the structure of the interannual changes of observed rainfall and OLR are so similar over the tropical land masses that the OLR anomalies could be converted to the rainfall anomalies (5.7 watts m^{-2} = 1 mm day^{-1}; Arkin, personal communication). This leads us to suggest that the interannual changes of rainfall over the tropical oceans based on OLR changes can also be considered to be more realistic than those derived from scanty rainfall observations over the oceans.

We have compared the monthly mean rainfall fields simulated by the two different global general circulation models of GFDL (Geophysical Fluid Dynamics Laboratory) and GLA (Goddard Laboratory for Atmospheric Sciences) and it is found that the rainfall rates for the GLA model, particularly over the western Pacific, are larger than those for the GFDL model. Considering the uncertainty of rainfall data over the oceans we find it difficult to comment on the relative accuracy of the two model simulations. Since both the models have been successful in addressing various aspects of climate variability and predictability, such large differences in the rainfall rates simulated by the two GCMS point to the need for a quantitative measurement of rainfall over tropical oceans.

We have also examined the interannual variability of monthly mean rainfall simulated by the GFDL and GLA models and compared these with the interannual variability of the outgoing long wave radiation (OLR). There is a remarkable agreement between the variability of the model-simulated rainfall by both models and OLR. (See Figs. 1, 2, and 3). All three estimates of rainfall variability indicate that the western Pacific-eastern Indian Ocean is the region of highest rainfall variability over the globe. The value of the standard deviation ranges from 2-4 mm/day^{-1} over areas of several million square kilometers.

Finally, we have examined some results from previous studies on the sensitivity of GCM-simulated circulation and rainfall to observed SST anomalies in the tropical Pacific. We find that the magnitude of the monthly mean rainfall anomalies in the western Pacific that occur in association with the El Nino episodes are in the range of 4-8 mm/day for a spatial scale of 1000-2000 kilometers.

Based on these results of observed and model simulated rainfall variability we suggest that the TRMM observing system which is expected to have the required accuracy to resolve monthly mean rainfall anomalies of 1 mm/day^{-1} averaged over areas of about 500 km x 500 km, should be adequate to describe the interannual variability of monthly rainfall over the tropical oceans.

3. ACKNOWLEDGEMENTS

The author is extremely grateful to Mr. M. J. Fennessy for his help in conducting this study. Discussions with Mr. M. J. Fennessy and Dr. J. L. Kinter were quite useful in writing this paper. Thanks to Ms. Marlene Schlichtig for typing the manuscript and preparing the figures. The author would also like to thank the members of the science steering group of TRMM for the benefit of various stimulating discussions on this and related topics. This research was supported by NASA's Climate Program (NAGW-557) and NSF grant ATM-8414660.

REFERENCE

Jaeger, L., 1976: Monatskarten des Niederschlags fur die ganze Erde. Berichte des Deutschen Wetterdienstes, 18, No. 139. Im Selbsverlag des Deutschen Welterdienstes, Offenbach, W. Germany.

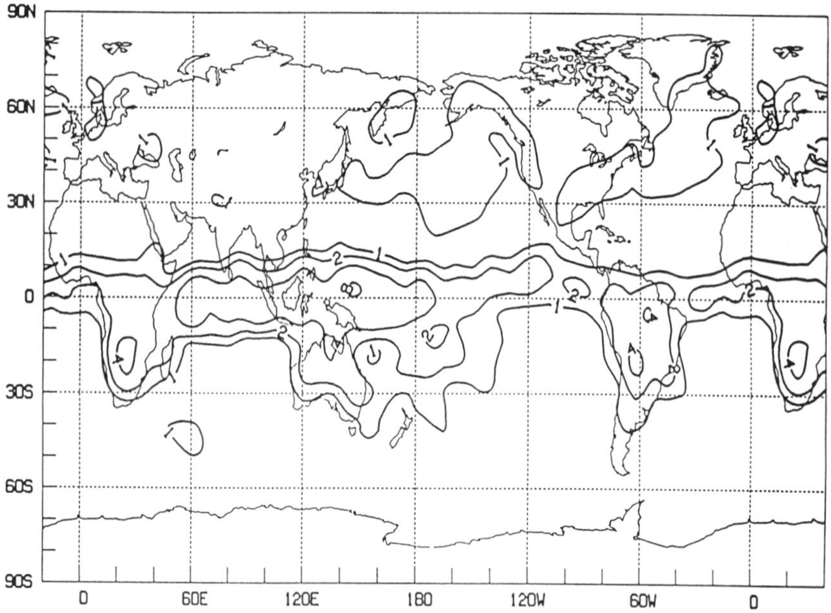

Figure 1. Standard deviation of January mean rainfall (mm/day) as simulated by the GLAS climate model.

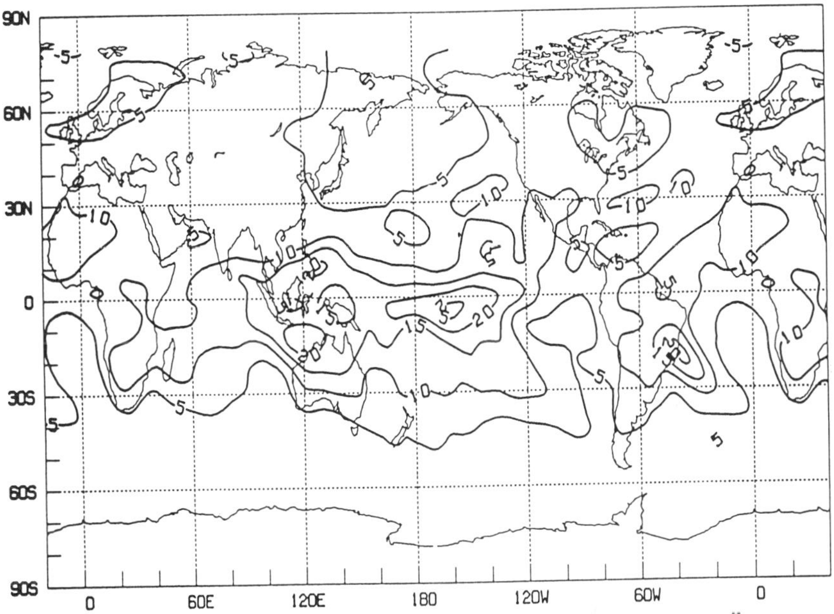

Figure 2. Standard deviation of January mean OLR (watts/m^2) based on observed data, 1974-1986.

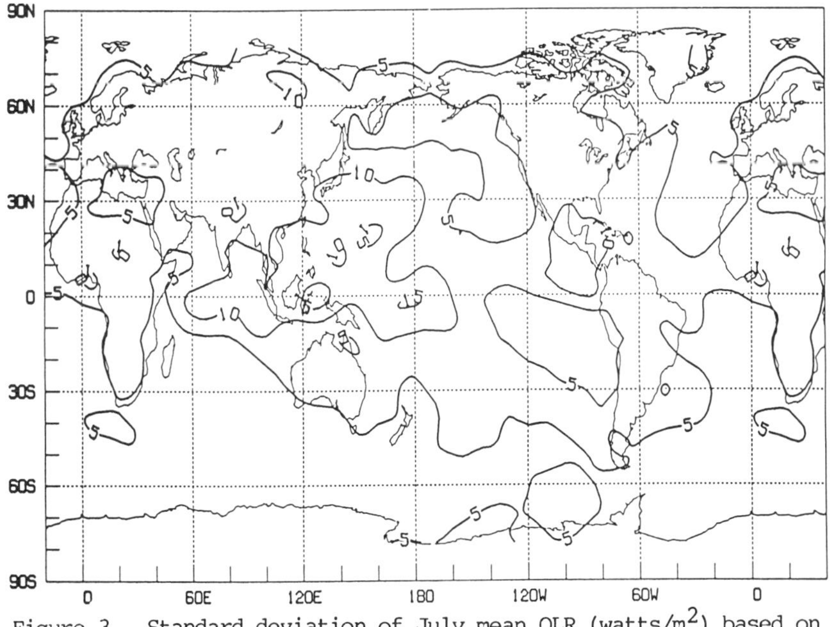

Figure 3. Standard deviation of July mean OLR (watts/m^2) based on observed data, 1974-1986.

AN APPLICATION OF SATELLITE-DERIVED PRECIPITATION IN DATA ASSIMILATION FOR NUMERICAL WEATHER PREDICTION

M. Ueno
Japan Meteorological Agency
Tokyo 100, Japan

ABSTRACT

A data assimilation method using satellite-derived precipitation is developed to improve model initial fields. In this method, a meso-scale model is integrated for an early period of numerical prediction with forced diabatic heating corresponding to the satellite-derived precipitation at each model grid point. The model with this assimilation technique improves the forecast of precipitation at an early stage and the surface pressure pattern as compared with no assimilation model.

The satellite-based precipitation is estimated using TBB (satellite-observed equivalent black-body cloud-top temperature) data of 1° lat. x 1° long. mesh and mean column relative humidity (MRH) of the lat/long grid obtained from operational objective analysis through a previously prepared statistical relation which consists of sample means of observed rainfall amount for several categories classified by TBB and MRH.

1. INTRODUCTION

In general, numerical weather prediction models have a tendency to underestimate the precipitation amount at an early stage of the integration, even if heavy precipitation occurs in the actual atmosphere at the initial time of the model. In such a case, the subsequent evolution of the precipitation system in the model may be significantly affected by the slow spin-up nature. In order to reduce the spin-up time and to incorporate meso-scale details to initial fields of a meso-scale model, Ueno et al. (1987) used a dynamic assimilation (DA) method in which diabatic heating in the model was defined from observed rainfall rate during an early period of prediction. This study follows fundamentally their work except for the next three points.

1) In order to apply the DA method to the model with a larger forecast domain, satellite-derived rain is used for determination of the imposed diabatic heating.

2) To determine the top of forced heating, TBB (satellite observed equivalent black-body cloud-top temperature) is used.

3) Evaluation of the model results is performed using some objective measures.

Satellite-derived rainfall data are used successfully by Danard (1985) for data assimilation in a limited-area model and by Krishnamurti et al. (1987) for the reduction of spin-up time in a global model.

ISBN 0-937194-14-4

2. RAINFALL ESTIMATION

Relationships between surface rainfall measurements and cloud parameters have been established by many researchers (e.g., Krishnamurti et al., 1983; DelBeato and Barrell, 1984). In this section a simple technique is presented for rainfall estimation from satellite data. The statistical relation is developed by correlating hourly rain gauge measurements from AMeDAS (Automated Meteorological Data Aquisition System) stations around Japan with TBB and MRH (mean column relative humidity) obtained from operational objective analysis scheme (Appendix by NPD JMA, 1986).

Spacing between AMeDAS stations is approximately 17 km. TBB data are operationally obtained at 12 hr intervals with 1° lat. X 1° long. mesh. MRH is calculated at the same lat/long grids from objective analysis data. MRH is introduced to reduce the impact of non-precipitating cirrus clouds.

The statistical regression analysis was performed using the data for four years' warm seasons (May 1983 - September 1986). The statistical relation consists of sample means of the observed one-hour rainfall amount, computed for 240 categories classified by TBB and MRH. Figure 1 shows some examples of the statistical results. It is found that the rainfall rates are well-separated by using MRH in addition to TBB.

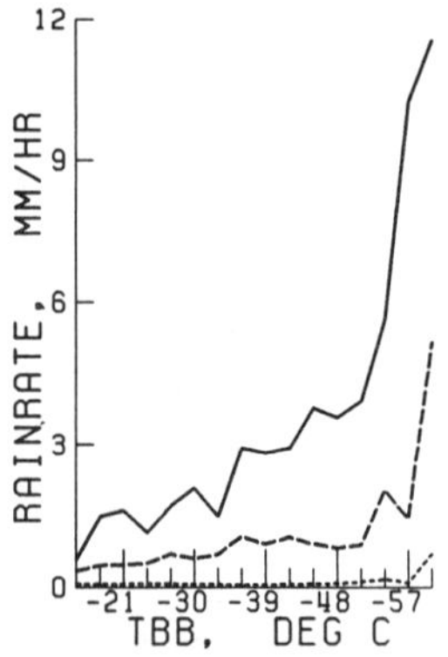

FIGURE 1. Examples of the statistical relation. Solid, dashed and dotted lines are for MRH>0.89, 0.81>MRH>0.79 and 0.63>MRH respectively.

Correlation coefficient (CC) between estimation and observation is commonly used to evaluate regression technique objectively. Table 1 gives a comparison of CC between the above-mentioned regression relation and another one which was concurrently obtained using TBB alone. The CCs are computed for independent data sets (June 1987 - August 1987) over the area where AMeDAS data are obtained with a high spatial density. CCs for the forecast of 3-hr precipitation at hr 12 of the operational very fine mesh model (Appendix by NPD JMA, 1986) are also given in the Table 1 to measure the accuracy of satellite-estimated rainfall relatively.

TABLE 1. MONTHLY-AVERAGED CORRELATION COEFFICIENTS

	TBB + MRH					TBB alone	Model
lag time (hr) =	0	+1	+2	+3	+4	0	
June	0.65	0.64	0.57	0.50	0.46	0.35	0.54
July	0.62	0.63	0.60	0.55	0.50	0.46	0.40
August	0.54	0.54	0.54	0.50	0.42	0.46	0.36

Since TBB data are obtained at 12 hr intervals, it is required to know the maximum time interval over which useful rain estimates can be obtained reliably from one-time TBB data. To do this, CCs were calculated for observed rainfall rates corresponding to times displaced from the TBB time by +1, +2, +3, +4 hr. Table 1 also gives the lag CCs. As seen in the table, the decrease in correlation with lag time is slow. Then 3 hr is chosen as the forcing period in this study.

3. DA METHOD

Details on the DA method are given in Ueno et al. (1987). The principal points of the method are outlined as follows.

At first, the satellite-derived precipitation is interpolated to each model grid. Then the diabatic heating rates converted from the precipitation rates are distributed to each time step with parabolic vertical profile in pressure. The top of the heating is determined from TBB value at each grid point and the bottom is fixed at a specified sigma level (about 955 hpa level). After the model is run with the prescribed heat forcing during the first 3 hr, the forcing is switched off and the model is run in the normal way.

4. PREDICTION MODEL

The model used is the operational very fine mesh model of the JMA (Appendix by NPD JMA, 1986). It is written in sigma coordinates, with 13 levels in the vertical and 55 X 55 grid points with a resolution of about 65 km in the horizontal. It includes physical processes such as convection and PBL processes. Other dynamic and physical specifications, initial values and lateral boundary conditions are the same as those described in Ueno et al. (1987). The forecast area is shown in Fig. 2.

5. EXPERIMENTS AND RESULTS

Two cases are selected for the experiments. Both cases include intense rainfall areas in the model domain at the initial time. one is the Baiu front (BF) case and another is Typhoon (TY) case. For each case, two experiments (i.e., with and without DA) are carried out.

Figure 2 shows the satellite-derived rainfall rates converted to forced heating rates in the model with DA. Figure 2 also shows time evolutions of rainfall rates predicted by the DA model comparing the no DA cases and observations at the selected points. The DA procedure improves the precipitation forecasts at an early stage. It is known, however, that as integration time advances, the forecasts with DA tend to become similar to the non-DA ones.

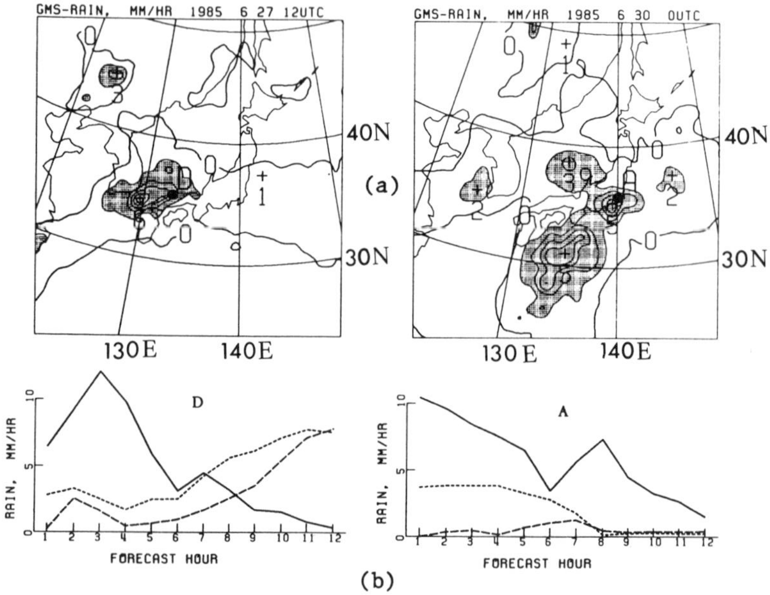

FIGURE 2. (a) Satellite-derived rainfall rates for BF cas (left) and TY case (right). The areas more than 1 mm/hr are hatched. (b) Time sequence of observed (solid) and predicted rainfall with DA (dotted) and without DA (dashed) at the point marked by solid circle in (a).

TABLE 2. OBJECTIVE SCORES

Experiment	sea-level pressure RMSE(hpa)	sea-level pressure S1-score	800 hpa height RMSE(m)	800 hpa height S1-score	rain CC
BF without DA	2.8	58.0	23.5	48.5	0.25
BF with DA	2.4	54.9	20.5	45.8	0.31
TY without DA	2.6	52.9	23.1	49.9	0.17
TY with DA	2.1	50.1	21.6	49.2	0.18

Figure 3 displays the differences in sea level pressure at 12 hr

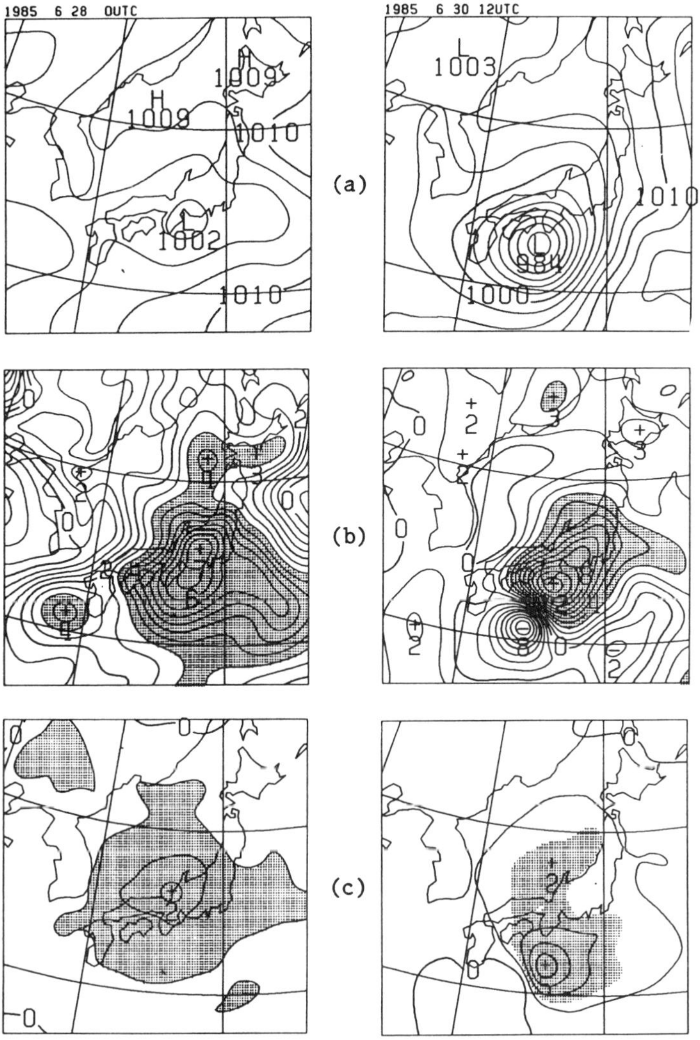

FIGURE 3. (a) Verification of the sea-level pressure (hpa) for BF case (left) and TY case (right). (b) Error map of sea level pressure (hpa) for the forecast without DA for BF case (left) and TY case (right). (c) The difference (no DA minus DA) in sea level pressure (hpa) for each case. The areas more than appropriate values are hatched for the convenience in (b) and (c).

between the two models and error maps for non-DA model. As seen in the figure, the errors for the non-DA model tend to be reduced by the inclusion of the DA procedure, although the magnitude of the reduction is relatively small. These improvements can be confirmed by some common objective scores listed in Table 2. The CCs are computed for the 6-hr rainfall between forecast hour 6 and 12.

5. SUMMARY AND DISCUSSIONS

By including the DA procedure, the forecasts of the precipitation and mass fields are improved. However, in order to make statistically significant statements concerning the impact of the DA procedure, experiments should be performed for a large number of cases.

The removal of water vapor from the model, which is equivalent to the forced heating, is not considered in the current DA procedure, for it can deplete the moisture of model atmosphere at the grid points imposed on strong heat forcing without sufficient moisture convergence to balance it. Therefore the current DA method implicitly includes moisture forcing as well as heat forcing. Since there is no ground for moisture forcing, it is necessary to scrutinize its effect on the forecast.

REFERENCES

Danard, M. B., 1985: On the Use of Satellite Estimates of Precipitation in Initial Analyses for Numerical Weather Prediction. Atmos. and Ocean, 23, 23-42.

DelBeato, R., and S. L. Barrell, 1985: Rain Estimation in Extratropical Cyclones Using GMS Imagery. Mon. Wea. Rev., 113, 747-755.

Krishnamurti, T. N., S. Cocke, R. Pasch, and S. Low-Nam, 1983: Precipitation estimates from raingauge and satellite observations Summer MONEX. FSU Report NO. 83-7, 2-5.

Krishnamurti, T. N., H. S. Bedi, W. Heckley and K. J. Ingles, 1987: On the reduction of spin up time for evaporation and precipitation in a global spectral model. FSU Report NO. 87-4, 1-22.

NPD, JMA, 1986: Outline of Operational Numerical Weather Prediction at Japan Meteorological Agency. Appendix to progress report on numerical weather prediction.

Ueno, M., R. Taira, and T. Kudo, 1987: A Dynamic Assimilation Method for a Mesoscale Model Using Observed Rainfall Rates. Special Volume of J. Met. Soc. Japan, 573-584.

LONG-TERM VARIATIONS OF CONVECTIVE ACTIVITIES IN THE TROPICAL WESTERN PACIFIC AND THEIR EFFECTS ON THE GLOBAL ATMOSPHERIC CIRCULATIONS

Tsuyoshi Nitta
Meteorological Research Institute
Tsukuba, Ibaraki 305, Japan

ABSTRACT

Interannual variations of convective activities in the tropical western Pacific and their effects on the global atmospheric circulations are investigated by using monthly mean cloud amount data for 7 years (1978-1984) derived from the Geostationary Meteorological Satellite (GMS). It is shown that there exist two dominant cloud anomaly patterns in the tropical western Pacific. One is an east-west oscillation of convection between the equatorial central Pacific and the western Pacific corresponding to the El Niño event, and the other is a north-south oscillation between northern subtropics (~20°N) and middle latitudes around the Islands of Japan corresponding to the north-south movement of convection centers. Atmospheric teleconnection patterns associated with this heat source can be seen from the tropical western Pacific to North America.

1. INTRODUCTION

It has been shown by recent observational and theoretical studies that long-term variations of tropical heat sources such as those associated with El Niño and Southern Oscillation (ENSO) play an important role in interannual variations of global-scale atmospheric circulations. The Indian Ocean and western Pacific regions are major heat source regions in the tropics. In this report interannual variations of convective activities in the tropical western Pacific and their impact on the global atmospheric circulations are investigated by using GMS high-cloud amount data (higher than 400 mb) for 7 years (1978-1984).

2. PACIFIC - JAPAN (PJ) CLOUD OSCILLATION

In order to pick up dominant oscillation patterns of convective activities, correlation coefficients of the cloud amount are computed between all pairs of grid points within the analyzed domain (90°E-170°W; 50°N-50°S). Fig. 1 shows the distribution of the maximum negative correlation. It is found that there exist two dominant cloud patterns, i.e., one is an east-west oscillation between the equatorial western Pacific and the central Pacific denoted by EN and the other is a north-south oscillation between the subtropics and middle latitudes denoted by PJ.

Fig. 2 shows time variations of the cloud amount anomaly at two centers (positive and negative) for each oscillation pattern. The EN mode increased significantly during the 1982-83 El Niño period indicating that this mode corresponds to the El Niño event. The PJ

ISBN 0-937194-14-4

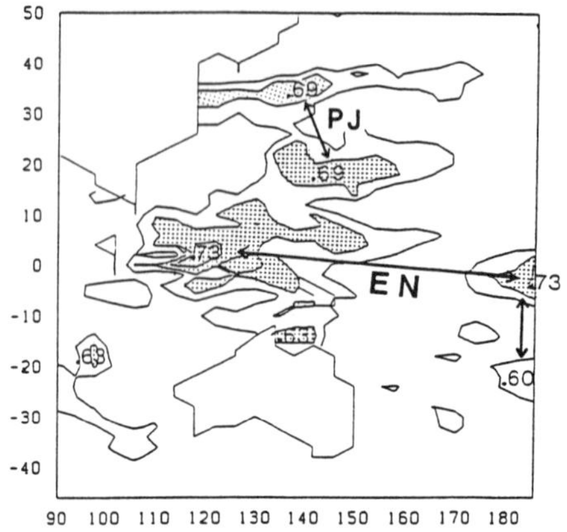

Figure 1. Distributions of the maximum negative correlation coefficient of the cloud amount for each grid point with other points. Contours larger than 0.5 are plotted with interval of 0.1. Regions larger than 0.6 are shaded.

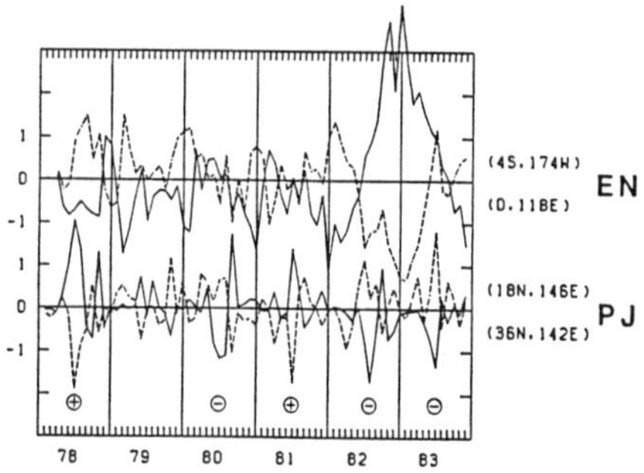

Figure 2. Time variations of the cloud amount anomaly at two centers for the EN and PJ oscillations. ⊕ (⊖) denotes the year in which the large positive (negative) PJ index (subtropics minus mid-latitudes) is observed.

mode has large amplitudes during northern summer but small amplitudes during winter.

Fig. 3 shows a one-point correlation map of the cloud amount with a reference point at 18°N,146°E corresponding to the PJ mode. Large positive correlation areas are located in the subtropical western Pacific extending zonally from the South China Sea to the dateline. Negative correlation regions are found in middle latitudes extending from East China through the Islands of Japan to near the dateline. Negative correlations are also found in the equatorial region east of New Guinea. The PJ oscillation is mainly due to the north-south movement of the convection center in the tropical western Pacific during summer.

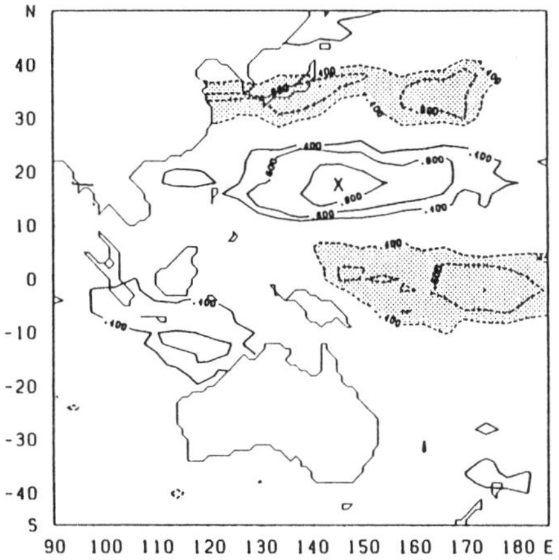

Figure 3. One-point correlation maps of the cloud amount with a reference point at 18°N, 146°E. Contours of the absolute coefficient larger than 0.4 are plotted with interval of 0.2. Regions of negative coefficients smaller than -0.4 are shaded. (After Nitta, 1987).

3. TELECONNECTION PATTERN ASSOCIATED WITH THE PJ OSCILLATION

Atmospheric teleconnection patterns associated with the PJ oscillation are obtained by using global wind data analyzed by the European Center for Medium-Range Weather Forecasting (ECMWF) during 1979-1984. The PJ index is defined as the cloud amount difference between subtropics (~18°N, 146°E) and mid-latitudes (~36°N, 138°E). Fig. 4 shows composite wind anomalies at 200 mb for 5 cases of positive extreme PJ indices (enhanced convection in the subtropics and suppressed in mid-latitudes). A cyclonic circulation is located to the northwest of the PJ heat source anomaly and a strong anticyclonic cell is found over the East Asian regions with a center around the islands of Japan. This anticyclonic cell extends far westward into the Eurasian Continent. Another anticyclonic cell is found in the North Pacific to the west of Alaska. These atmospheric anomaly circulations may be generated by the heat source in the tropical western Pacific associated with the PJ mode. These results are quite consistent with earlier results by Nitta et al.(1986) and Nitta (1987) in which cloud winds derived from GMS and global geopotential height data were analyzed.

Easterly anomalies are found along the equator and twin cyclonic cells are generated in both hemispheres over the central Pacific. These upper tropospheric wind anomalies might be due to suppressed convection in the equatorial central and eastern Pacific during the positive PJ index period. Large anomaly circulations are also detected in the southern hemisphere, but further detailed analysis may be needed to clarify the relationships between the PJ oscillation and wind anomalies in the southern hemisphere.

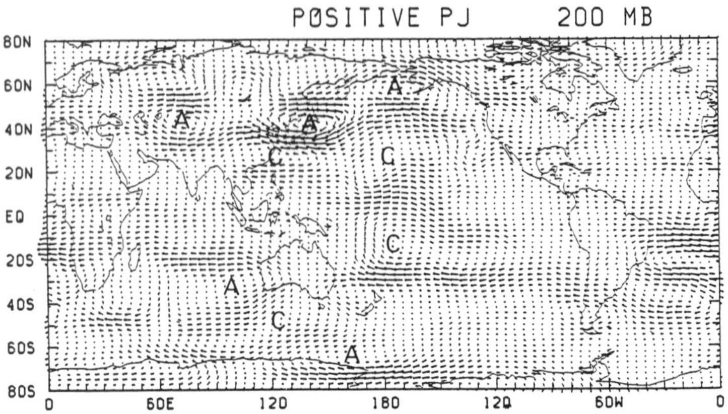

Figure 4. Composite wind anomalies at 200 mb for the positive PJ. C and A denote centers of cyclonic and anticyclonic circulations, respectively.

4. RELATIONSHIPS BETWEEN THE PJ AND SEA SURFACE TEMPERATURE

Convective activities in the tropics may be largely affected by interannual variations of the sea surface temperature(SST). From the time series of the cloud amount anomaly for the PJ mode (Fig. 3) we found that positive PJ dominates in 1978, 1981 and 1984 and negative PJ dominates in 1980, 1982 and 1983. Fig. 5 describes the distributions of SST differences in June between years during which positive PJ dominates (1978, 1981, 1984) and those during which

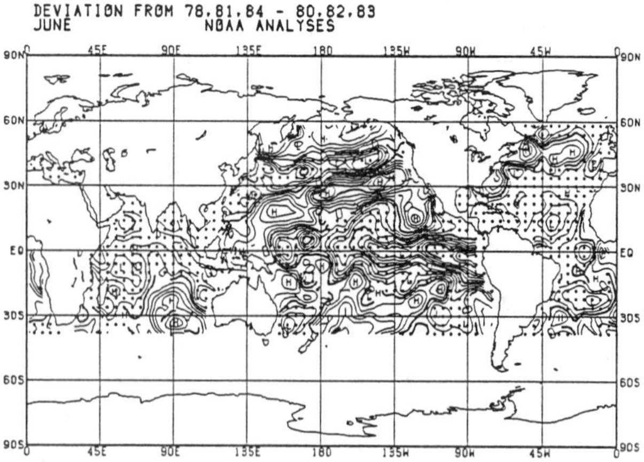

Figure 5. Distributions of the SST difference between the years of the positive PJ index (1978, 1981, 1984) and those of the negative PJ index (1980, 1982, 1983). The contour interval is 0.2°C and negative regions are shaded.

negative PJ dominates (1980, 1982, 1983). During summers in which the convection center moves northward, positive SST anomaly regions extend northeastward from the equatorial western Pacific to middle latitudes in the Northeast Pacific. SST in the western Pacific subtropics around 20°N is about 1°C warmer during the positive PJ years than during the negative PJ years. On the other hand the equatorial eastern Pacific is covered by negative SST anomalies. These SST anomaly patterns are quite similar to those for El Niño years with opposite sign indicating that the positive phase of the PJ oscillation corresponds to the non-El Niño year.

It is concluded that during summers corresponding to the non-El Niño year in which SST in the tropical western Pacific is warmer than normal, convective activities in these regions are largely enhanced and the convection center is shifted northward by about 5°-10° in latitudes. The atmosphere responds to this heat source anomaly and as a result anticyclonic circulations are generated in the East Asian regions. These regions frequently suffer from extremely hot days and severe drought under the intensification of the Pacific subtropical high during these summers. The effects of the heating associated with PJ seem to extend in higher latitudes in the northern hemisphere and also in the southern hemisphere.

REFERENCES

Nitta, Ts., 1986: Long-term variations of cloud amount in the western Pacific region. J. Meteor. Soc. Japan, 64, 373-390.

Nitta, Ts., 1987: Convective activities in the tropical western Pacific and their impact on the northern hemisphere summer circulation. J. Meteor. Soc. Japan, 65, 373-390.

Nitta, Ts., T. Maruyama and T. Motoki, 1986: Long-term variations of tropospheric circulations in the western Pacific region as derived from GMS cloud winds. J. Meteor. Soc. Japan, 64, 895-911.

ON THE ROLE OF CONVECTIVE ACTIVITY IN THE WESTERN TROPICAL PACIFIC IN THE FORMATION OF THE BAROTROPIC HIGH AROUND JAPAN DURING SUMMER

T. Tsuyuki and K. Kurihara
Japan Meteorological Agency
Tokyo 100, Japan

ABSTRACT

An observational study is made of the relationship between the subtropical high around Japan and convective activity in the western tropical Pacific during summer. It is found that when the convective activity in the area becomes intense according to the intraseasonal variation, a subtropical high with little vertical tilt develops around Japan. There exist barotropic wave trains indicated in the geopotential height field emanating from the active convective area to the northeastern Pacific. These results indicate that the formation of this subtropical high is brought about by the barotropic response of the atmosphere to the heating anomaly in the western tropical Pacific.

1. INTRODUCTION

The subtropical high in the northwestern Pacific plays an important role in governing East Asian weather, especially during summer. According to its interannual and intraseasonal variation, Japan sometimes experiences large climatic anomalies such as the cool/wet summer in 1980 and the hot/dry summer in 1984 and 1985 (e.g., Kurihara and Kawahara, 1986).

In the present paper, the role of convective activity in the western tropical Pacific in the formation of the subtropical high around Japan is investigated, based on a case study of the subtropical high in August 1984 and a statistical analysis of tropical-midlatitude interactions in the Northern Hemisphere (NH) summer.

2. DATA

The JMA objective analysis of daily geopotential height north of 20°N is used, except for Fig. 1 where geopotential height is taken from the NOAA/NMC objective analysis. As indices of convective activity in the tropics, 5-day mean equivalent black-body temperature (TBB) measured by a JMA geostationary meteorological satellite and 5-day mean outgoing longwave radiation (OLR) derived by NOAA polar-orbiting satellites are used. Spatial resolution of the data is 5°latitude x 5°longitude.

ISBN 0-937194-14-4

3. THE SUBTROPICAL HIGH IN AUGUST 1984

The subtropical high in August 1984 has been analyzed in detail by Kurihara and Tsuyuki (1987). Figure 1 shows monthly mean 500 hPa geopotential height and TBB over East Asia in August 1984. The subtropical high was shifted unusually northward and persisted over Japan for about a month, resulting in anomalously hot/dry weather in the area. That high had a barotropic structure and was seen from the surface to about 200 hPa. It is in contrast to the North Pacific High, which is located over the northeastern Pacific and only seen at lower levels. Another remarkable feature in Fig. 1 was a deep trough east of Japan.

Very low values of TBB (below -20°C) over the region from the Philippines to the Indian subcontinent indicate intense convective activity. The 30-60 day oscillation in the convective activity in the western tropical Pacific was evident in summer 1984 and it was an active phase during August. The subtropical high appeared around Japan in early August almost at the same time as the intensification of the convective activity around the Philippines. It stayed north of 30°N through the month. The development of the subtropical high seems to be closely associated with the convective activity around the Philippines.

4. STATISTICAL ANALYSIS

A correlative study between 5-day mean NH geopotential height and OLR in the global tropics on an intraseasonal time scale was carried out for NH summer in 1974-1984 (excluding 1978). Japan has a rainy season ("Baiu") from early June to mid-July. Therefore, the

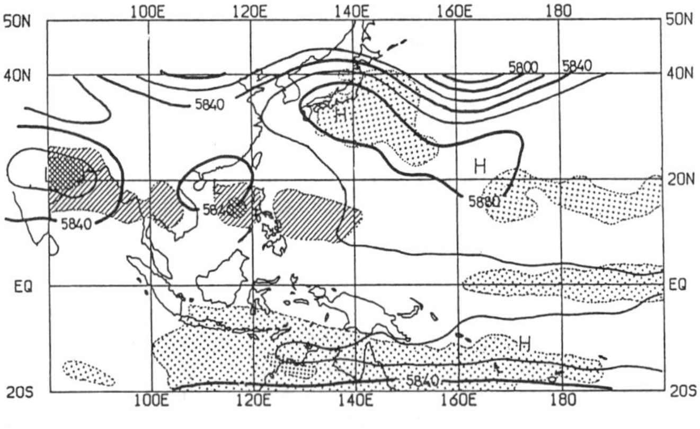

FIGURE 1. Monthly mean 500 hPa geopotential height and TBB over East Asia in August 1984. The contour interval of the geopotential height is 20 m. Light (dark) stippled areas indicate TBB above 10°C (20°C) and light (dark) hatched ones TBB below -20°C (-30°C) (After Kurihara and Tsuyuki, 1987).

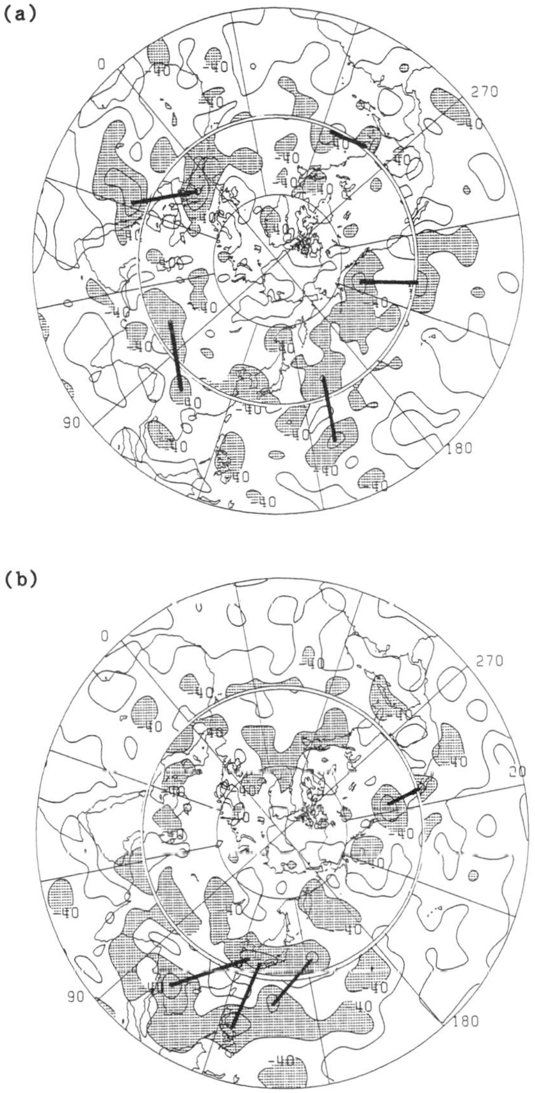

FIGURE 2. Teleconnection maps defined as the maximum negative correlation (x 100) between 5-day mean 500 hPa geopotential height north of 30°N and OLR over 0°-30°N for (a) May 26-July 14 (period I) and (b) July 15-September 2 (period II). The contour interval is 10 with areas below -40 shaded. Pairs of points with the largest correlations (correlation coefficient is less than -0.5) are connected by thick solid lines.

investigation was performed separately for May 26-July 14 and July 15-September 2. Each period consists of 50 days and, hereafter, is referred to as "period I" or "period II", respectively. The number of samples is 100 for each period. The seasonal cycle and interannual variability are removed from the data so that only intraseasonal variation remains.

4.1 TELECONNECTION

Figure 2 shows teleconection maps which show the maximum negative correlations between 500 hPa geopotential height north of 30°N and OLR over 0°-30°N. Pairs of points with the largest correlations are connected by thick solid lines. At those pairs, a positive height anomaly in the midlatitudes is associated with increased convective activity in the tropics.

For period I, large correlations are found around the Mediterranean Sea and in the Pacific. The 500 hPa geopotential height over East Asia correlates with the OLR around the Philippines, although the correlation coefficient does not exceed -0.5.

For period II, pairs of points with large correlations are located mainly over East Asia. The largest correlation exists between the geopotential height around the Korean Peninsula and the OLR around the Philippines. At each of the teleconnection pairs over East Asia, the geopotential height point is located northeast of the corresponding OLR point. Therefore, if the correlation patterns are the result of barotropic Rossby wave propagation, they must be associated with the wave propagation from the tropics to the midlatitudes, suggesting that the tropics drive the midlatitude flow.

On the other hand, the pairs of points around the Mediterranean Sea and over South Asia for period I suggest that the midlatitude flow drives the tropics. Liebmann and Hartmann (1984) found that the large correlation patterns during NH winter can be also interpreted in terms of equatorward propagation of stationary Rossby waves.

The teleconnection for the 50-day period following period II is also calculated, but correlations over East Asia are weaker than those for period I. Therefore, the influence of the heating anomaly in the tropics on the atmospheric circulation over East Asia is the most significant during period II, that is, NH midsummer through the formation of the subtropical high over the area. The following discussions are concentrated on this period.

4.2 LAG CORRELATION

Figure 3 shows lag correlation maps of 500 hPa geopotential height with the reference point of OLR at 15°N, 125°E for period II. It is found that a negative correlation area around Japan, which may be considered to represent the subtropical high, moves west-northwest from the southeast of Japan to inland China.

The intraseasonal variation of convective activity in the western tropical Pacific during NH summer exhibits west-northwestward propagation (Lau and Chan, 1986; Nitta, 1987). Lag correlation maps of OLR with the same reference point (Fig. 4) show a similar propagation. The west-northwestward propagation in Fig.3 seems to

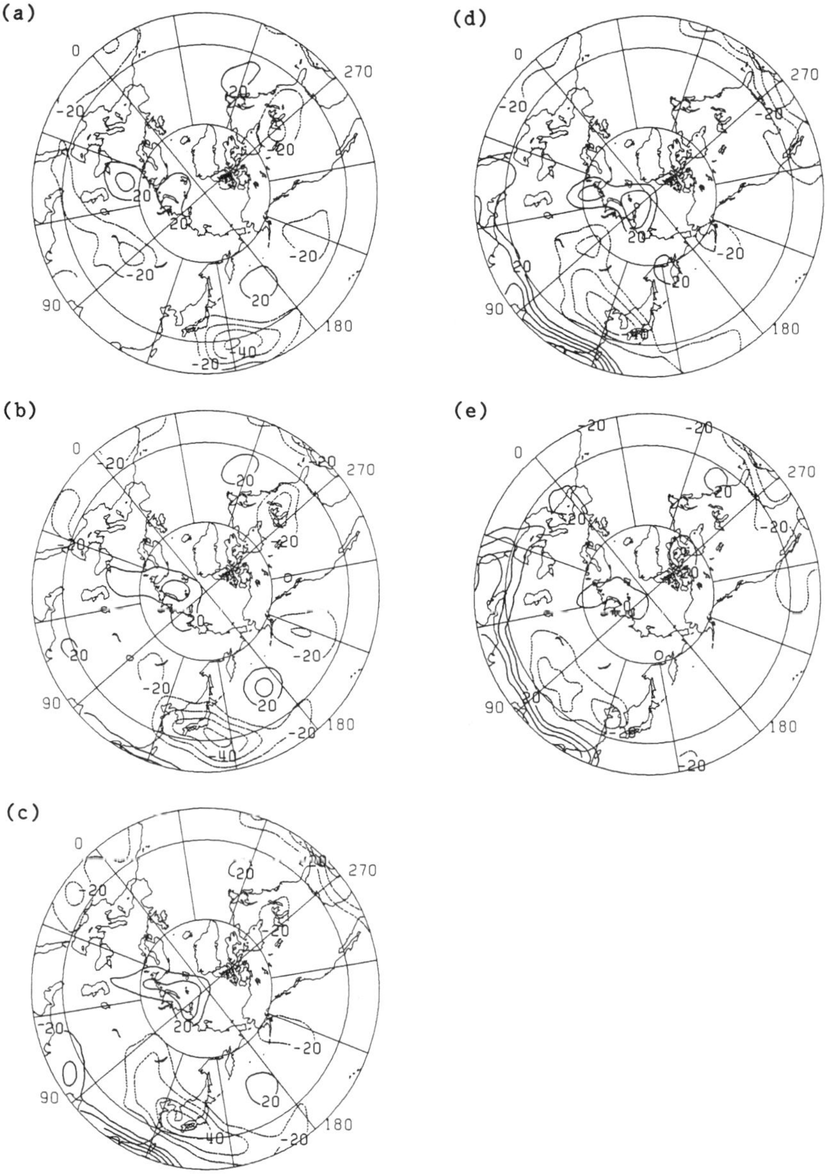

FIGURE 3. One-point correlation maps (x 100) of 5-day mean 500 hPa geopotential height with the reference point of OLR at 15°N, 125°E for July 15-September 2 with lags of (a) -5 days, (b) -2 days, (c) 0 day, (d) 2 days and (e) 5 days. The contour interval is 10 with negative contours dashed. Contours of 10, 0 and -10 are omitted for clarity.

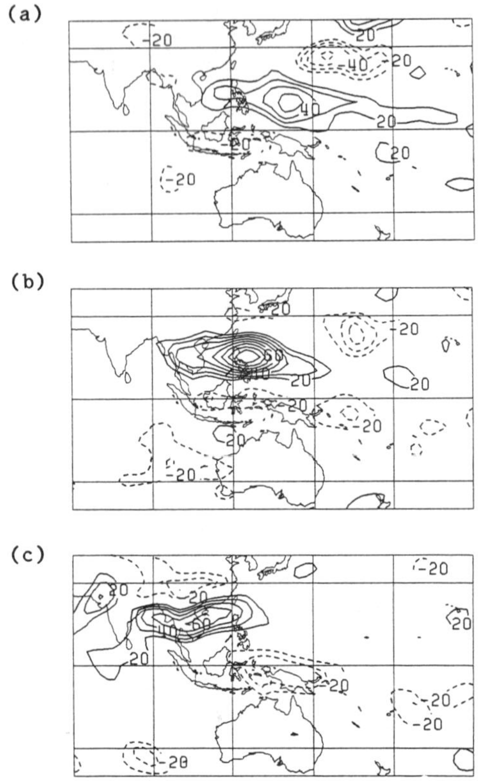

FIGURE 4. One-point correlation maps (x 100) of 5-day mean OLR with the reference point of OLR at 15°N, 125°E for July 15-September 2 with lags of (a) -5 days, (b) 0 day and (c) 5 days. The contour interval is 10 with negative contours dashed. Contours of 10, 0 and -10 are omitted for clarity.

correspond to the similar propagation of the active convective area in the northwestern Pacific.

Two wave-like patterns are seen over the Pacific in Fig.3. One of them is significant when height leads OLR by 2-3 days (Fig. 3b). It remains in the simultaneous correlation map (Fig. 3c) where the negative correlation area at the western end of the wave train moves away to Japan. The other wave-like pattern appears from the South China Sea to Alaska when height lags OLR by 2-3 days (Fig. 3d). The correlation maps for the geopotential height at other levels indicate that those wave-like patterns have a barotropic structure in middle and high latitudes.

Eastward propagations along the above wave trains are obtained from daily geopotential height data in each year (Kurihara and Tsuyuki, 1987; Nitta, 1987). The wave train in Fig. 3(d) is similar to anomaly patterns in a linear model with observed zonal mean wind in August and a vorticity source around the Philippines (Fig. 9 of Kurihara and Tsuyuki, 1987), indicating that it can be regarded as Rossby waves excited by the heating anomaly in the western tropical Pacific. The other wave train is not obtained with such a model. Further investigations for the atmospheric response under zonally non-uniform wind are desirable.

5. CONCLUSIONS

Convective activity in the western tropical Pacific has a significant influence on NH summer circulation over East Asia, especially during midsummer. When it becomes intense according to the intraseasonal variation, a subtropical high with little vertical tilt develops around Japan. The intensification of the convective activity around the Philippines results in the development of the subtropical high over Japan.

There exist barotropic wave trains indicated in the geopotential height field emanating from the active convective area to the northeastern Pacific, suggesting that the formation of the subtropical high is brought about by the barotropic response of the atmosphere to the heating anomaly in the western tropical Pacific.

ACKNOWLEDGMENT

The authors wish to express their thanks to Ms. K. Hayashi for providing and processing OLR data.

REFERENCES

Kurihara, K., and K. Kawahara, 1986: Extremes of East Asian weather during the post ENSO years of 1983/84 --- severe cold winter and hot dry summer ---, J. Meteor. Soc. Japan, 64, 493-503.

------------, and T. Tsuyuki, 1987: Development of the barotropic high around Japan and its association with Rossby wave-like propagations over the North Pacific: analysis of August 1984, J. Meteor. Soc. Japan, 65, 237-246.

Lau, K.-M., and P. H. Chan, 1986: Aspects of the 40-50 day oscillation during the northern summer as inferred from outgoing longwave radiation, Mon. Wea. Rev., 114, 1354-1367.

Liebmann, B., and D. L. Hartmann, 1984: An observational study of tropical-midlatitude interaction on intraseasonal time scales during winter, J. Atmos. Sci, 41, 3333-3350.

Nitta, T, 1987: Convective activities in the tropical western Pacific and their impact on the Northern Hemisphere summer circulation, J. Meteor. Soc. Japan, 65, 373-390.

LOW-FREQUENCY OSCILLATIONS OF TROPICAL FORCING AND GLOBAL STRUCTURES

Julia Nogues-Paegle and Beng-Chun Lee
University of Utah
Salt Lake City, Utah 84112, USA

ABSTRACT

The structure of low-frequency oscillations during the Northern Hemisphere summer is examined by projecting weekly averages of regularly gridded data into vertical and horizontal structure functions which are obtained as solutions of a primitive equation model linearized about a basic state at rest. Summer averages, weekly anomalies and intraseasonal and interannual variances of contributions by different modes to the 200 mb zonal and meridional winds are compared with those of outgoing long wave radiation and rain gauge precipitation amounts to relate these structures to the release of latent heat.

1. INTRODUCTION

Tropical latent heat release is not only the main forcing of tropical motion but also plays an important role in driving the general circulation. Several recent studies have quantified the existence of interweekly tropical variations with periods longer than one month and the impact of these oscillations in extra-tropical latitudes. Such variations are linked with tropical convection and are important contributors to tropical low-frequency oscillations. Identification of this oscillation, and its concomitant extra-tropical impact, has a potential impact on the ability to extend long range forecasts in mid-latitudes.

The present investigation studies variations of tropical forcing during the summer of 1979 using weekly averaged rain gauge data and outgoing long wave radiation (OLR) fields as a measure of latent heat release. The OLR fields are found to be most useful in this study since they provide continuous coverage around the tropical belt. This is not the case for rain gauge data, which also exhibits marked local variations due to terrain inhomogeneities. Nevertheless, OLR fields do not measure precipitation, and therefore our results have to be properly qualified due to this limitation.

The summer of 1979 exhibited a well-defined 30-60 day oscillation both in tropical forcing and in global structures (Nogues-Paegle and Mo, 1987). The present paper compares intraseasonal and interannual variations of modal structures obtained by projecting three-dimensional gridded data into the normal modes of a primitive equation model linearized about a basic flow at rest. This decomposition is particularly useful to relate observed variations of tropical forcing with patterns that have theoretical relevance.

ISBN 0-937194-14-4

The methodology used requires reliable analyses interpolated to a global grid. Such analyses have only become available after the 1979 Global Weather Experiment (GWE). Therefore, only post-GWE data are used in this study.

2. RESULTS

Figure 1 shows summer convection to maximize over India and Indonesia, with pronounced intraseasonal variations. A comprehensive set of precipitation rain gauge data is available for the 1979 year. Figure 2 compares weekly variations of OLR and daily precipitation amounts for that summer season for an area centered over India. There is a good correspondence, in general, between the two fields with low values of OLR associated with rain. The onset of the Indian monsoon is apparent in these time series, as well as two cycles of a 5- to 7-week oscillation in latent heat release. Our previous studies (Nogues-Paegle and Lee, 1987, Lee, 1987) have linked variations in tropical forcing with eastwardly propagating Kelvin and Rossby modes. During the summer season, a coherent propagation of eastwardly propagating tropical structures and the subtropical jet of the Southern Hemisphere was noted in these studies. Figures 3 and 4 show the summer average for the external (equivalent barotropic) and internal (first four height reversing structures) Rossby modes obtained using the National Meteorological Center analysis. The subtropical jet of the Southern Hemisphere has a dominant wave number one structure and projects mostly in internal modes. The weekly variability, alluded to above, projects almost equally in internal and external structures. The largest intraseasonal variations are found in the Northern hemisphere as a result of changes in the jet stream patterns that take place there as the summer season sets in. Interannual oscillations are small, with the largest signal found over South America and the Eastern Pacific.

Figures 5 and 6 show the summer structures for the tropically trapped modes. The zonal belt of tropical variability exhibits a preference for the Kelvin mode to be most variable at longitudes similar to those of maximum OLR variability. This is not the case of the mixed Rossby-gravity mode which exhibits maximum intraseasonal variability over the central Pacific Ocean. Maximum intraseasonal variances are 2 to 4 times larger than their interannual counterparts. Figure 7 shows the contribution to the Hadley cell from the internal gravity modes. The interannual variations are very weak and not plotted. The intraseasonal variance is seen to maximize at longitudes of large convective activity. These modes appear to provide a link between the tropical latent heat release and subtropical jet stream accelerations (Nogues-Paegle and Mo, 1987; Nogues-Paegle and Zhao, 1987). Such a link will be discussed more extensively in the oral presentation.

3. CONCLUSIONS

An accurate description of the modal structures discussed here requires accurate specification of the tropical forcing. This is due to the interdependence of three-dimensional data sets on the assimilating model physics and the initial guess of the objective analysis. Good rainfall estimates are essential to obtain reliable measures of tropical flow activity and its impact at higher latitudes.

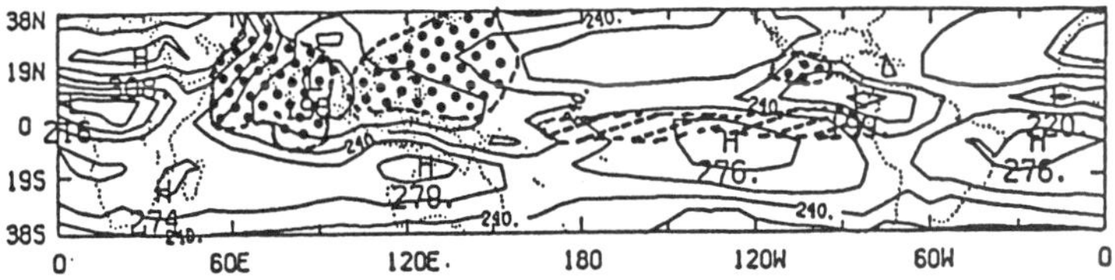

Figure 1. Outgoing long wave radiation averaged for the 1979, 1983, 1984 and 1985 summers contoured every 20 watts/m^2. Dotted and dashed regions indicate intraseasonal and interannual variances larger than 400 and 100 watts^2/m^4 respectively. Intraseasonal variances are determined from 16 consecutive weeks starting May 5. Interannual variances are determined from the 16-week averages.

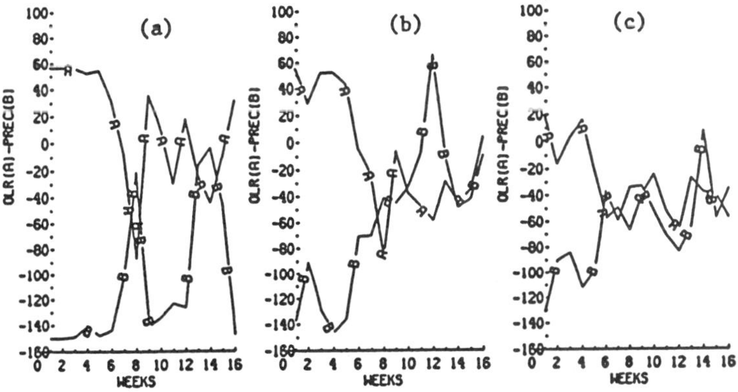

Figure 2. Weekly averages of outgoing long wave radiation subtracted from 240 watts/m^2 and of precipitation subtracted from 150 tenths of mm/day, averaged from 15° to 30°N, and from 60° to 75°E (a), 75° to 90°E (b), and 90° to 105°E (c).

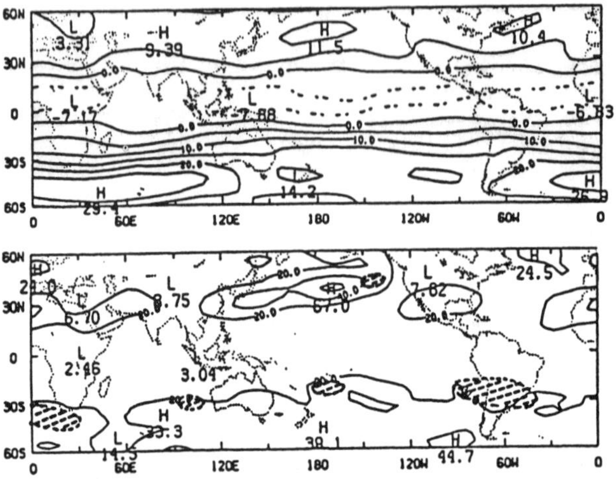

Figure 3. Contribution to the 200 mb zonal wind by external Rossby modes. Summer average (top), contoured every 5 m/s, and intraseasonal variance (bottom) contoured every 30 m^2/s^2. Dashed regions indicate interannual variances larger than 6 m^2/s^2.

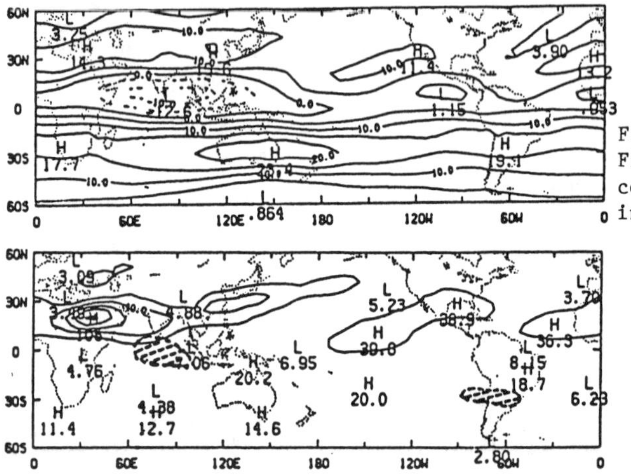

Figure 4. Same as Figure 3, but for contributions from internal Rossby modes.

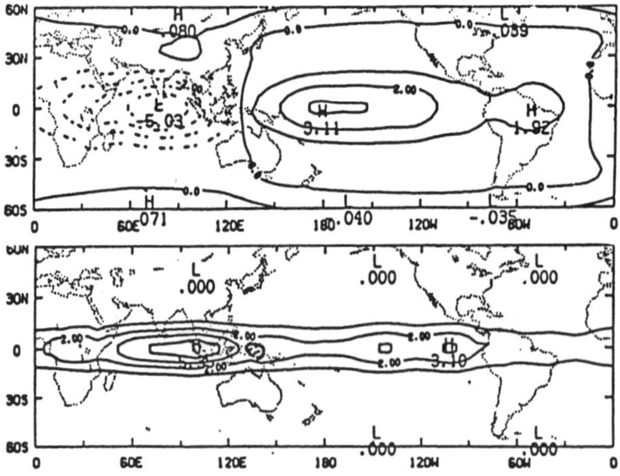

Figure 5. Same as Figure 3, but for internal Kelvin modes, contoured every m/s (top) and every m^2/s^2 (bottom). Dashed region indicate interannual variances larger than 1.5 m^2/s^2.

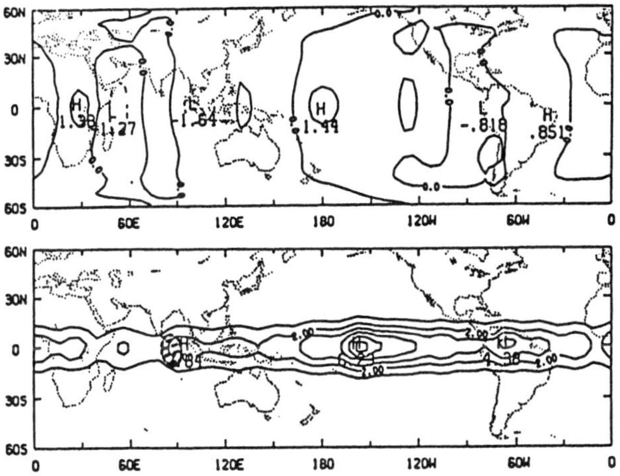

Figure 6. Same as Figure 5, but for contributions to the 200 mb meridional wind by internal mixed Rossby-gravity waves.

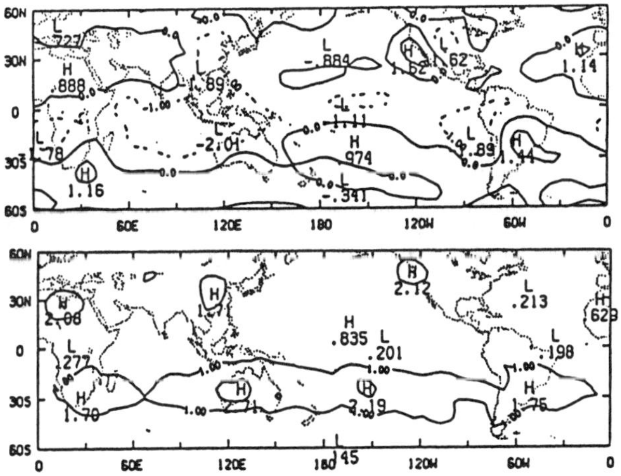

Figure 7. Same as Figure 6, but for contributions by internal inertia-gravity modes.

ACKNOWLEDGMENTS

Outgoing long wave radiation fields were obtained from V.E. Kousky (CAC/NMC/NOAA). The use of the NCAR computer facilities and the NCAR data archives is gratefully acknowledged.

REFERENCES

Lee, B.-C., 1987: Observed normal mode structure of low frequency oscillations. Ph.D. dissertation. Available from Univ. of Utah, Salt Lake City, Utah 84112.

Nogues-Paegle, J. and K.C. Mo, 1987: Spring to summer transitions of global circulations during May-July, 1979. Mon. Wea. Rev., 115, 2088-2102.

Nogues-Paegle, J. and K.C. Mo, 1988: Transient response of Southern hemisphere subtropical jet to tropical forcing. J. Atmos. Sci. In press.

Nogues-Paegle, J. and Z. Zhao, 1987: The Australian subtropical jet during the second observing period of the global weather experiment. J. Atmos. Sci., 44, 2277-2289.

QUASI-STATIONARY AND PROPAGATING FLUCTUATIONS IN LATENT HEAT RELEASE FROM TROPICAL PRECIPITATION

William K.-.M. Lau
Laboratory for Atmospheres
NASA Goddard Space Flight Center
Greenbelt, Maryland 20771, USA

ABSTRACT

Quasi-stationary and propagating fluctuations in tropical heating due to precipitation are studied using outgoing long-wave radiation and long-term station rainfall data. It was found that the most dominant large-scale tropical convection/precipitation pattern consists of (i) a transient or intraseasonal component that accounts for much of the wet and dry spells found in the near-equatorial areas and the major monsoon regions and (ii) a quasi-stationary component associated with the major dislocation of global rainfall regime. These patterns are associated with fluctuations in time scales of the 30-60-day oscillation, the quasi-biennial oscillation as well as the southern oscillation. The spatial distribution of heat sources and sinks associated with the above patterns are remarkably similar suggesting that they may be related to the same basic mode of variation in the tropical ocean-atmosphere system.

1. INTRODUCTION

Quasi-stationary and propagating fluctuations in tropical heating due to precipitation are studied using outgoing long-wave radiation (OLR) and long-term station rainfall data. This paper focuses on the variations in tropical convection associated with the 30-60-day oscillation and interannual variability such as ENSO. The present analyses differ from most previous investigations in that the data are not stratified so that the analyses can be carried out with continuous time series to include all possible time scales. As a result, intraseasonal and interannual variations are not *a priori* separated by time filtering. Empirical Orthogonal Function (EOF) analyses are performed on 12 years (1974-86) of non-overlapping 5-day mean OLR anomaly (deviations from seasonal variation) and 80 years of monthly station precipitation data.

2. TROPICAL CONVECTION (OLR)

2.1 DOMINANT SPATIAL PATTERNS

Fig. 1a shows the EOF1 (Pattern A, T1) which explains 8.3% of the the total variance. This pattern closely resembles the anomalous precipitation pattern observed during the 1982-83 ENSO. The major anomalies are found in the form of wave-like patterns along the equator stretching from the Indian Ocean, across the maritime continent to the central Pacific. There is also a secondary anomaly over Brazil and the Atlantic in agreement with the dry condition observed over these regions during the 1982-83 ENSO. Most interestingly, when the 1982-83 and the 1976-77 ENSO periods are removed, and exactly the same analysis is performed on the remaining data, a picture similar to that of Fig. 1a emerges (Fig. 1b, Pattern A, T2). All the previously noted features can be identified except that the central Pacific anomaly is less extensive and does not penetrate into the eastern Pacific. Since the ENSO periods are excluded in Fig. 1b, it can be inferred that the spatial pattern shown in Fig. 1 and Fig. 2 must also occur during non-

A. DEEPAK Publishing
ISBN 0-937194-14-4

ENSO years i.e., the ENSO anomalies may be the large-amplitude manifestation of an intrinsic anomaly pattern that frequently occurs in shorter time scales .

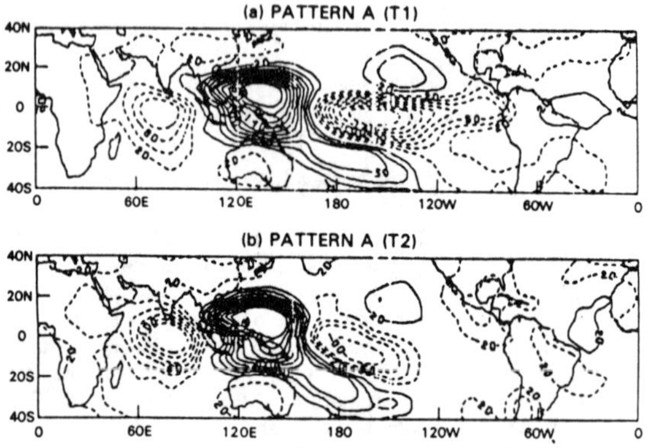

FIGURE. 1 Spatial pattern of EOF1 for tropical convection in relative units with a) the entire time series (1974-86) and b) ENSO periods excluded.

Fig. 2a shows that the EOF2 (Pattern B, T1) which explains 7% of the total variance. This pattern has an east-west dipole structure. The central Pacific anomaly has two distinct arms reaching southwest U.S. and the Indian Ocean near 90°E. The anomaly centers appear to have shifted eastward about 30-40° longitude compared with those shown in Fig. 1. With the two ENSO periods removed, the pattern shown in Fig. 2b is almost identical to that in Fig. 2a.

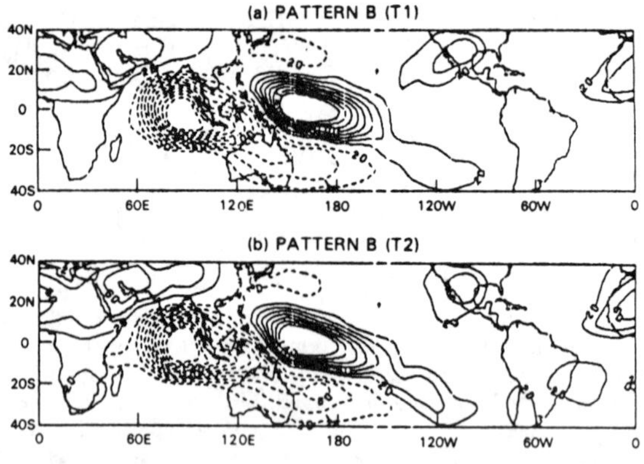

FIGURE 2. Spatial patterns of EOF2 for tropical convection with a) entire time series (1974-86) and b) ENSO periods excluded.

As is recognized from previous studies and also from the temporal information to be discussed below, the two patterns shown in Fig. 1 and 2 should not be regarded as separate modes of variation but rather as representative of two different phases of a single eastward propagating feature.

2.2 BASIC TEMPORAL VARIATIONS

The time series for EOF1 are shown in Fig. 3a. Spectral analysis (not shown) shows that the time series possesses three basic time scales; (i) a broad peak at 30-60 days, (ii) an amplitude modulation associated with the annual cycle and (iii) a strong ENSO signal. The time series for EOF1 excluding the ENSO periods are shown in Fig. 3b. It can be seen that the two ENSO periods do not significantly affect the temporal behavior of this pattern. This again supports the notion that ENSO-like anomaly can occur with substantial amplitude in the 30-60-day time scale and are modulated by the seasonal variation.

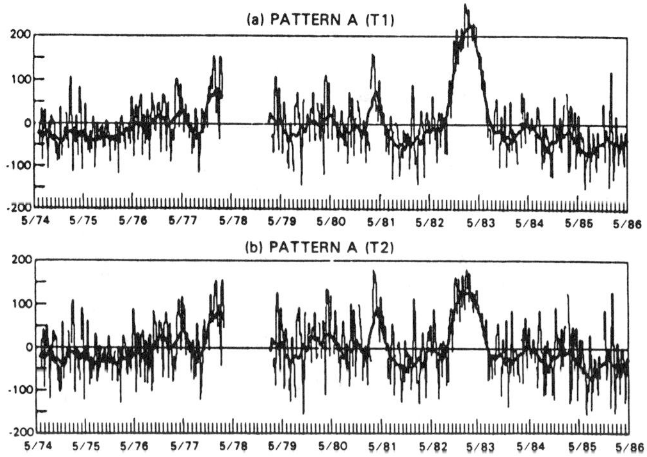

FIGURE 3. Principal component for EOF1 for a) Pattern A, T1 and b) Pattern A, T2.

2.3 THE 30-60- DAY OSCILLATION AND ENSO

Fig. 4 shows the band-passed (20-70 days) time series for EOF1 and EOF2 . It is seen that the two time series are in quadrature for most of the time especially during the enhanced periods (stippled). The temporal relationship suggests an eastward propagation of the centers of the convection anomaly . There are also periods in which the two patterns are exactly in phase (e.g. latter half of 1980 to 1985) suggesting that a stationary component may also be present. The amplitude of the oscillation which is strongest near the equator, is generally large during the northern winter. This is in agreement with previous observations of tropical convection associated with the 30-60-day oscillations. However, an exception to this modulation can be found in the large fluctuations occurring during April-June 1982. This has been noted by Lau and Chan (1986) in their OLR longitude-time sections. They suggested that these large oscillations may have provided the necessary trigger for the onset of the 1982-83 ENSO.

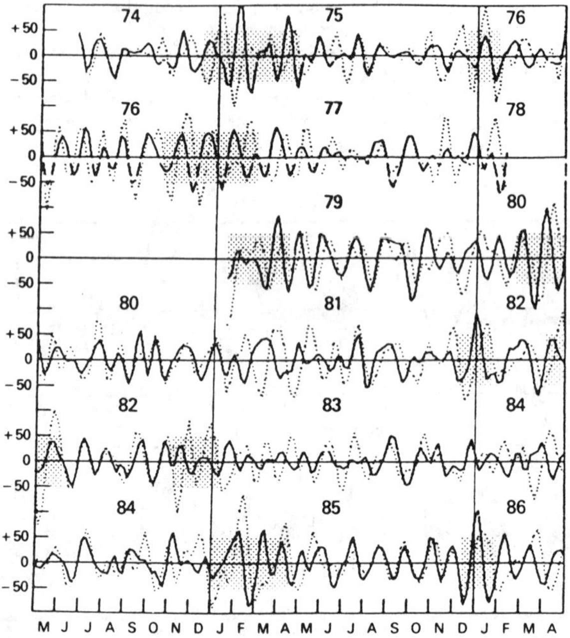

FIGURE 4. Band-passed (20-70-day) time series of the principal components of EOF1 and EOF2.

3. GLOBAL PRECIPITATION

3.1 QUASI-BIENNIAL OSCILLATON AND ENSO

Fig. 5 shows the most dominant mode of interannual variations from global monthly precipitation data (1900-1981). The key feature is the east-west dipole structure with centers over the equatorial central Pacific and the maritime continent respectively. This pattern is again similar to one of the above EOF patterns (Fig. 3).

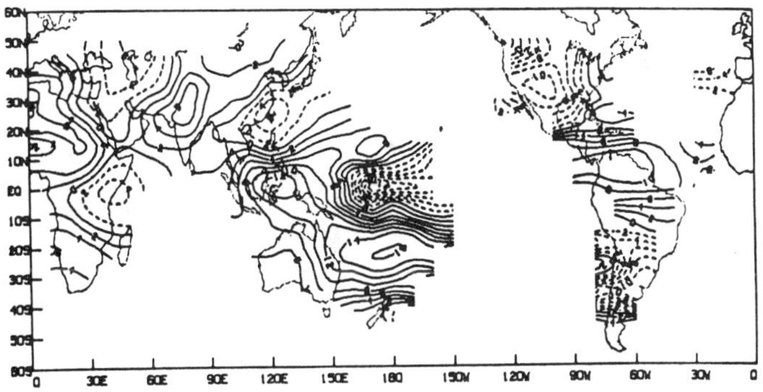

FIGURE 5. Spatial pattern of most dominant mode of global precipitation in relative units. Negative deviations are dashed.

Fig. 6 shows the power spectrum for the time variation associated with the above pattern. While it is not surprising that a large broad peak is found in the period from 4 to 6 years associated with the ENSO cycles, it is interesting to note that a well-defined peak is also found in the period around 2.5 years i.e., the quasi-biennial oscillation. The possible relationship between the QBO and the ENSO are reported elsewhere. In this paper, we only point out that the 'low-frequency oscillation" including in particular the ENSO, the QBO and the 30-60-day oscillation all have similar convection (precipitation) anomaly pattern. Are we looking at a mere coincidence of nature or is there something more fundamental that links all these 'low-frequency oscillations"? The answer to these questions are open and more work is clearly necessary in better defining these patterns and their temporal variability.

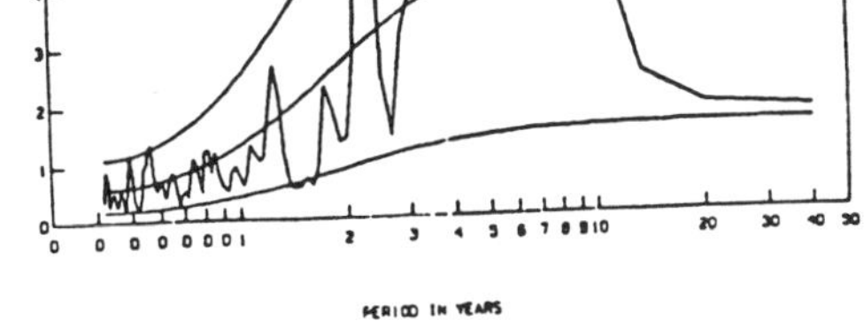

FIGURE 6. Power spectrum of time series of EOF1 in global precipitation. Abscissa in units of years. Smooth solid curve shows the background red-noise and 95% and 5% limits.

Bibliography

Lau, K.-M. and P. Chan, 1988: Intraseasonal and interannual variations of tropical convection: a possible link between the 40-50 day oscillations and ENSO. J. Atmos. Sci., 45, 506-521.

Lau, K.-M. and P. Chan, 1986: the 40-50 day oscillation and the El Nino/Southern Oscillation : A new perspective. Bull. Amer. Meteor. Soc., 67, 533-534.

Lau, K. -M., and P. J. Sheu, 1987: Annual cycle, quasi-biennial oscillation and southern oscillation in global precipitation. J. Geophys. Res. (in press).

SOME PROBLEMS ON THE INTERANNUAL VARIABILITY OF INDONESIAN MONSOON RAINFALL

Tetsuzo Yasunari and Ramasamy Suppiah
Institute of Geoscience
University of Tsukuba
Ibaraki ken 305, Japan

ABSTRACT

The vast release of latent heat over the Indonesian "maritime continent" through cumulus convection and rainfall contributes to one of the greatest heat sources of the global atmosphere.

The current study attempts to reveal the seasonal dependencies of the dominant modes of interannual variability of rainfall over the Indonesian region and their association with the atmospheric circulation by utilizing the monthly rainfall data from the Indonesian stations for a period of 46 years (1931-1976). The empirical orthogonal function (EOF) analysis is applied to the rainfall anomalies of about 18 stations to deduce the dominant modes in the interannual time scales.

It has been suggested that the quasi-biennial oscillation (QBO) is the most significant mode, particularly in the northern summer monsoon season, while the anomalies in the northern winter monsoon seem to follow those of the preceding summer monsoon season. The association of this mode with the QBO in the stratosphere and the El Nino/Southern Oscillation (ENSO) will also be discussed.

1. INTRODUCTION

The Indonesian region or the "maritime continent" releases large amounts of latent heat from the central and eastern Pacific Ocean (Ramage, 1968) and acts as a source region for the formation of a tropical east-west circulation, in particular between the western and eastern Pacific regions. This east-west circulation was named as the "Walker Circulation" by Bjerknes (1969), who studied the relationships between sea surface temperature (SST), rainfall and circulation patterns. A strong upward motion in the wind field is found over the Indonesian region and a strong downward motion in wind field is found over the central and eastern Pacific. Under normal conditions, higher SST is found over the western Pacific, centered on the Indonesian region, and thus leads to strong cumulus convection and

ISBN 0-937194-14-4

ascending motion in the wind field. By contrast, lower SST is observed over the central and eastern Pacific, where descending motion in the wind field is observed. Moreover, the negative relationship is also noticed in the seesaw pressure anomalies between the Indonesian low and the south Pacific subtropical high which was named as the Southern Oscillation by Walker (1923) and Walker and Bliss (1932). Recent observational studies on the relationships between rainfall, SST and wind anomalies (Berlage, 1966; Bjerknes, 1969; Rasmusson and Carpenter, 1982; Flohn, 1986; Behrend, 1987) and the general circulation model experiment studies (Rowntree, 1972; Julian and Chervin, 1978, etc.) proved the strong relationship between these parameters, in which the Indonesian rainfall anomalies played an important role.

In the present study, first, the spatial and temporal variations of rainfall over the Indonesian region, representing Java island, are studied by means of an empirical orthogonal function (EOF) analysis. Second, special attention has been drawn on the 30-40-month and quasi-biennial oscillation (QBO) in the time coefficients of the dominant EOF mode and their relationship to El Nino/Southern Oscillation (ENSO) phenomenon and tropospheric/stratospheric wind variations.

2. DATA

The Indonesian region itself has thousands of islands. However, for the study of interannual variations of rainfall over this region, Java island was selected for two reasons: Java island rainfall variations are strongly correlated with those of the northeast monsoon-dominated Indonesian region (see Fig. 3 of Flohn, 1984); and Java consists of many rainfall stations with fewer missing values, except for the second world war period.

Monthly rainfall values have been collected for 18 stations to study the spatial and temporal variations of rainfall of the Indonesian region. Here, three data sets have been used which are shown in Table 1.

TABLE 1. NUMBER OF STATIONS AND THE PERIOD OF ANALYSIS

Number of stations	Period of analysis	Number of years
18	1931-1945	15
6	1931-1960	30
18	1951-1976	26

3. METHODS

Normalized values were used in all calculations. For the normalization procedure, monthly anomalies were divided by the standard deviation of the specific month's time series. This procedure was applied to each month and to each station.

To deduce the dominant modes of spatial and temporal variations, an EOF analysis was applied separately for different data sets given in Table 1. This procedure helped to fill the data gap during the period from 1945 to 1950.

Spectral analysis by the Maximum Entropy Method (MEM) was used to investigate the dominant periodicities in the time coefficients of the principal modes. Band-pass filter methods were used to indicate the cycles in the time coefficients. Correlation analysis was applied to analyze the relationships between the Southern Oscillation index (SOI) and rainfall anomalies.

4. RESULTS

4.1 DOMINANT MODES AND THEIR SEASONAL DEPENDENCIES

Dominant spatial modes of rainfall anomalies for each data set explain 38, 45 and 42% of the total variance, respectively. These modes are very similar to the first EOF mode deduced by Yasunari (1981). Monthly contributions of these spatial patterns, shown in Table 2, reveal some interesting features. In particular, the contributions are higher during the dry season and summer season

TABLE 2. MONTHLY CONTRIBUTIONS (%) OF THE DOMINANT SPATIAL MODES OF DIFFERENT DATA SETS.

Eigenvalue	18 (1931-1945)	18 (1951-1976)	6 (1931-1960)
January	9.4	22.6	22.0
February	20.2	23.1	23.2
March	26.5	27.0	27.0
April	30.7	46.5	38.1
May	52.4	54.5	49.1
June	46.6	53.0	56.6
July	64.3	63.6	63.1
August	60.2	52.5	59.8
September	41.1	54.4	32.3
October	33.8	67.7	52.2
November	39.7	49.0	39.3
December	24.4	23.0	29.6

(June to September) and pre-rainy season (October-November) and lower during the wet season (December to March) which show the higher variability in rainfall during the dry and pre-rainy seasons. Special attention was given to the temporal variations of the time coefficients of the first mode because they serve as a reliable index to measure the interannual variability of Indonesian rainfall. Since the time coefficients of the overlapped period in Table 1 reveal coherent fluctuations (1931-1960 and 1951-1960) a homogeneous time series of rainfall anomalies was made by combining the time coefficients of three data sets. That is, the time coefficients in the 30-year period analysis (1931-1960) were used to fill the gap between 1945 and 1950, when more missing values were found in many stations.

To investigate the dominant periodicities in the first EOF mode, the MEM spectral analysis was applied with the maximum lag number of 60 months. Figure 1 shows the spectral densities of the time coefficients, in which the QBO and the 50-60-month oscillations are very clear.

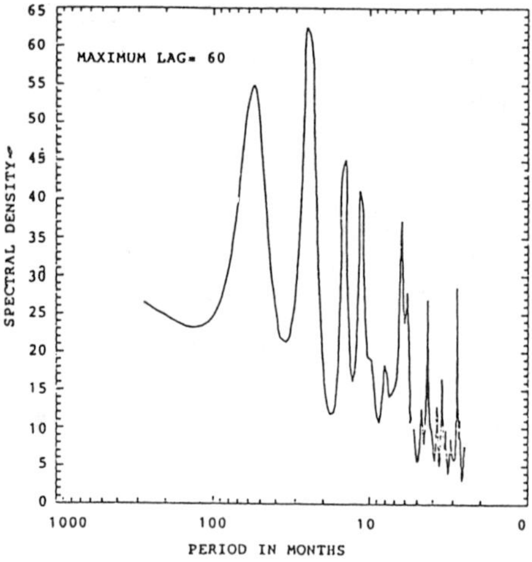

Figure 1. Spectral densities of the time coefficients of the first EOF mode.

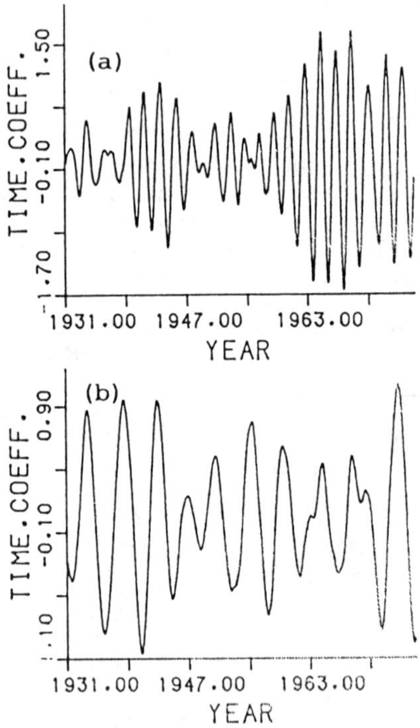

Figure 2. Band-pass filtered values showing an oscillation of 22-32 month period (a) and of 50-60 month period (b).

Figure 2a illustrates the filtered values of the time coefficients indicating the existence of strong QBO. Figure 2b indicates the filtered values of the time coefficients showing the prevalence of 50-60-month oscillation, a dominant fluctuation in the ENSO phenomenon. It should be noted that although the main rainy season comes from December through March, the interannual variabilities of rainfall of the Indonesian region are mainly regulated by the variabilities of the dry season as shown in Fig. 3. It is also noteworthy that the QBO mode is particularly related to the dry season rainfall anomalies as indicated in Fig. 3(c).

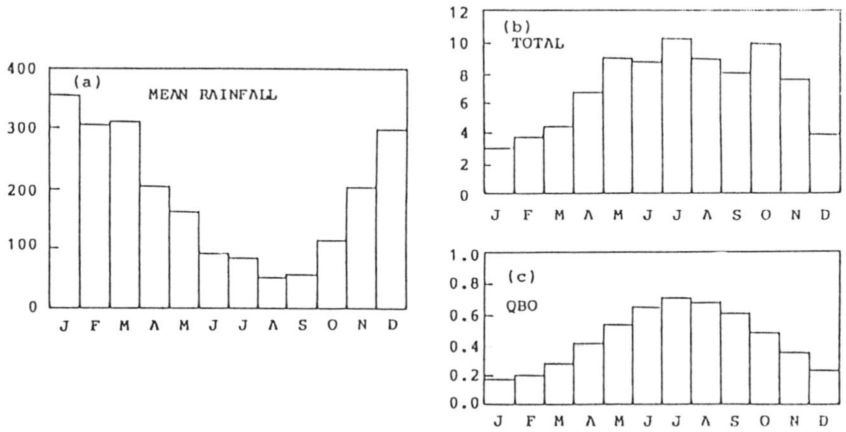

Figure 3. Mean monthly rainfall in mm (a), mean monthly contributions of the first EOF mode (b) and monthly contributions of the QBO mode of the first EOF mode.

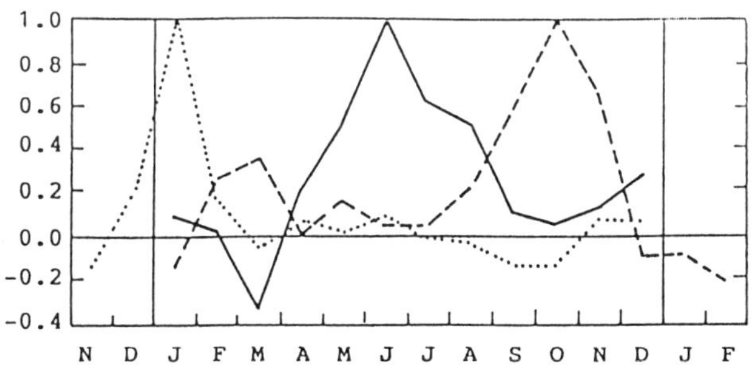

Figure 4. Autocorrelations for the rainfall of the months of January (dotted line), June (solid line) and October (dashed line).

To further examine the seasonal dependencies of the interannual variabilities, auto correlations of rainfall anomalies were computed for each month of the year. Fig. 4 shows the results for January, June and October, which reveal the fact that the variabilities during the main rainy season are not correlated with those in preceding or subsequent seasons. Also, the month-to-month persistence is very low, even within the rainy season (see the autocorrelation for January). On the other hand, the dry and the pre-rainy seasons show a relatively high persistence, at least within the seasons, but the correlation between the dry and the pre-rainy season seem to be low. Indeed, the QBO mode is more closely related to the variability during the dry season as shown Fig. 3(c). On the other hand, the pre-rainy season anomalies particularly represent the ENSO mode (50-60- month period), which is shown in the lag correlation between the Southern Oscillation Index (SOI) and the rainfall anomalies of each month (Fig. 5). It is interesting to note in Fig. 5 that the SOI in the dry season is not correlated with the rainfall in the concurrent season, but it is strongly correlated with the subsequent seasons.

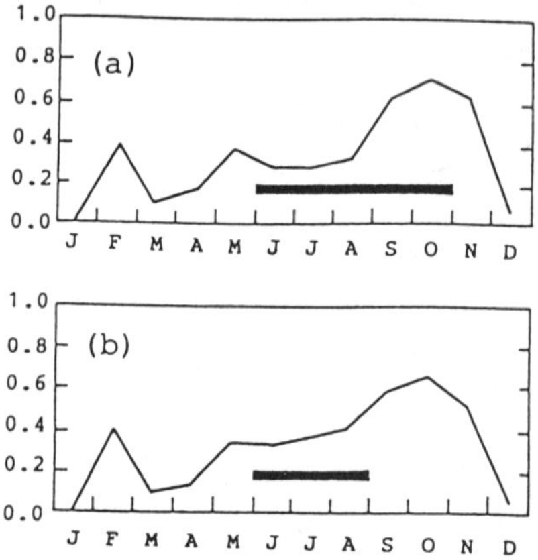

Figure 5. Lag-relationships between monthly rainfall and the SOI of June through October (a) and of June through August (b).

4.2 FURTHER COMMENTS ON QBO

The QBO is widely observed in tropical rainfall anomalies (Fleer, 1981), tropical stratospheric wind anomalies (Ebdon, 1971), SST anomalies (Rasmusson and Carpenter, 1982) and pressure anomalies in the seesaw relationship between Tahiti and Darwin (Chen, 1983). To further analyze the phase relationship between Indonesian rainfall anomalies and the QBOs in SST anomalies and in tropospheric wind anomalies along the equatorial belt, 22-32-month band-pass filter was applied to the time series of the parameters. Fig. 6 shows the longitude-time section of SST anomalies for 15 years from 1964 through 1978. The phases of westerly (easterly) maxima of the filtered zonal

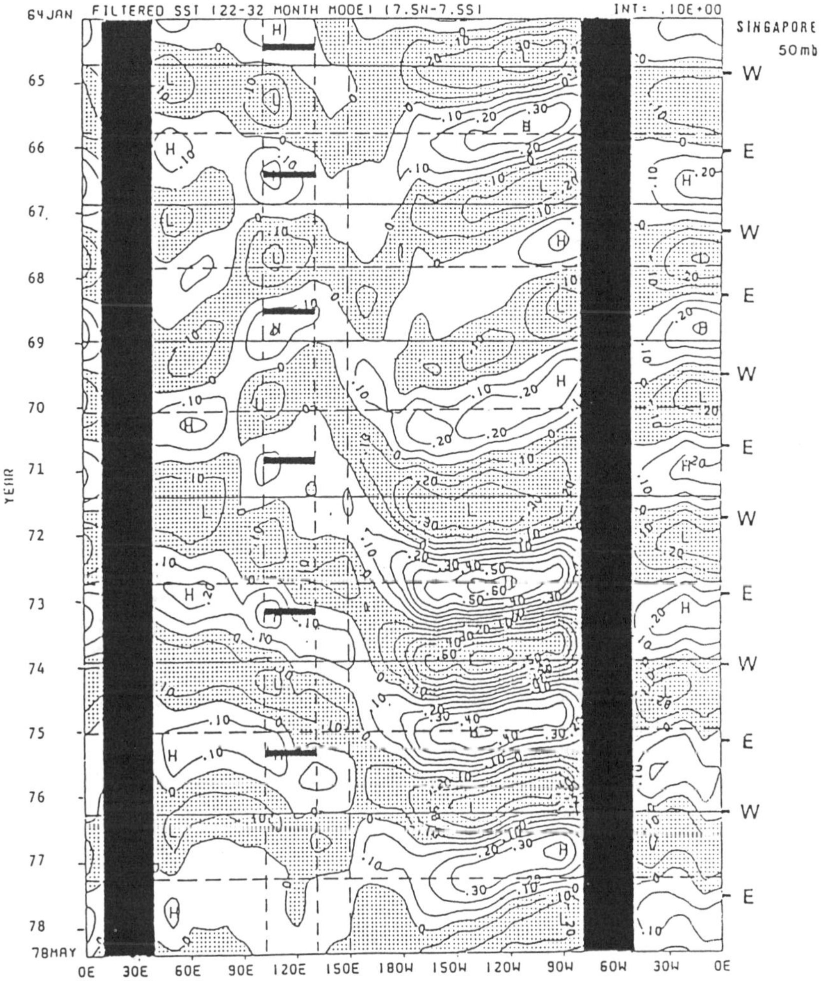

Figure 6. Longitude-time section of filtered SST anomalies (QBO mode) along the equatorial belt (7.5°N-7.5°S). Units are 0.1°C and negative anomalies are shaded. Phases of westerly (easterly) maxima of filtered zonal wind (QBO mode) at 850 mb over Singapore are shown with solid (dashed) lines. Phases of westerly (easterly) maxima of zonal wind at 50 mb are also plotted at the right-hand side of the diagram with capital letters of (W) (E). Phases of maximum rainfall over Indonesia are shown as thick bars.

wind at the 850 mb level over Singapore are shown with solid (dashed) lines. Phases of the maximum rainfall over Indonesia are also indicated by black thick bars over the Indonesian longitudes. This figure shows that the maximum phases in rainfall of the Indonesian region clearly coincide with those maximum phases of the QBOs in SST. Moreover the QBO in the zonal wind anomalies over Singapore is also consistently coupled with that in SST anomalies and also the QBOs in the stratosphere and troposphere seem to be linked with each other as indicated at the right-hand side of the diagram. Further discussions on the link of QBOs among stratosphere, troposphere and SST in the tropics have been made recently by Yasunari (1987). The above results, in any case, suggest that the QBO in rainfall anomalies over the Indonesian region is apparently related to the QBOs in SST and the troposphere and even posssibly to the stratosphere.

5. CONCLUSIONS

By utilizing the homogeneous rainfall index for the Java island, which was deduced from the time coefficients of the first dominant mode of EOF analysis, some characteristics of the interannual variabilities of the Indonesian region were investigated. The variability is largest in the dry season (northern summer monsoon) and in the pre-rainy season. These two seasons show contrasting patterns of interannual variabilities. In particular, the dry season shows a dominant QBO while the pre-rainy season indicates an oscillation around a 50-60-month period, which is associated with the ENSO cycle. The QBO in the dry season has been suggested to be linked with those oscillations of zonal wind in troposphere, stratosphere and SST anomalies. One of the remaining important problems may be why the ENSO cycle and the QBO show strong seasonal dependencies in the rainfall fluctuations of the Indonesian region. We earnestly hope that the Tropical Rainfall Measuring Mission (TRMM) will provide an answer with adequate rainfall data over the whole maritime continent including the ocean area where the rainfall data are not measured at present.

REFERENCES

Behrend, H., 1987: Teleconnections of rainfall anomalies and of the southern oscillation over the entire tropics and their seasonal dependence. Tellus, 39a,138-151.

Berlage, H. P., 1966: The southern oscillation and world weather. Mededelin en Verhandelingen Koninklijk Nederlands Meteorologisch Instituut, Debilt, No. 88,1-152.

Bjerknes, J., 1969: Atmospheric teleconnections from the equatorial Pacific. Mon. Wea. Rev., 97, 163-172.

Chen, W.Y., 1982: Assessment of southern oscillation sea-level pressure indices. Mon. Wea. Rev., 110, 800-807.

Ebdon, R. A., 1971: Periodic fluctuations in equatorial stratospheric temperatures and winds. Met. Mag., 100, 84-90.

Fleer, H., 1981: Large-scale tropical rainfall anomalies. Bonner Met. Abhand., No. 26, 1-114.

Flohn, H. (ed), 1984: Tropical rainfall anomalies and climatic change. Bonner. Met. Abhand., No. 31, 1-135.

Flohn, H., 1986: Indonesian droughts and their teleconnections. Berliner Geogr. Stud., Band 20, 251- 265.

Julian, P.R. and R. M. Chervin, 1978: A study of the southern oscillation and Walker circulation phenomenon. Mon. Wea. Rev., 106, 1433-1451.

Ramage, C. S., 1968: Role of a tropical "maritime continent" in the atmospheric circulation. Mon. Wea. Rev., 96, 365-370.

Rasmusson, E. M. and T. H. Carpenter, 1982: Variations in tropical sea surface temperature and surface wind fields associated with the southern oscillation/El nino. Mon. Wea. Rev., 110, 354-384.

Rowntree, P. R., 1972: The influence of tropical east Pacific ocean temperature on the atmosphere. Quart. J. Roy. Met. Soc., 98, 290-321.

Walker, G.T. Correlation in seasonal variations of weather, VIII: A preliminary study of world weather (world weather I), Mem.India. Met. Dept., 24, 275-332.

Walker, G. T. and E. W. Bliss, 1932: World weather V. Mem. Roy. Met. Soc., 4, 53-84.

Yasunari, T., 1981: Temporal and spatial variations of monthly rainfall in Java, Indonesia. Southeast Asian Studies, 19, 170-186.

Yasunari, T., 1987: A possible link of QBOs among stratosphere, troposphere and sea surface temperature in the tropics. submitted to J. Atmos. Sci.

GLOBAL SIMULTANEITY OF THE ABRUPT SEASONAL CHANGES IN PRECIPITATION DURING MAY AND JUNE OF 1979

J. Matsumoto[1], *M. Shinoda*[1], *T. Eguchi*[1], *K. Masuda*[1], *T. Mikami*[2], *K. Iwasaki*[3], *and H. Kitajima*[4]
[1]University of Tokyo, Tokyo 113, Japan
[2]Tokyo Metropolitan University, Tokyo 158, Japan
[3]Hokkaido University, Sapporo 060, Japan
[4]University of Shinshu, Matsumoto 390, Japan

ABSTRACT

Seasonal changes in global continental precipitation distribution are examined to determine the regional differences in the sequence of the seasonal precipitation cycle, utilizing FGGE level II-c precipitation and snow data sets. The distribution of the Outgoing Longwave Radiation (OLR) from TIROS-N satellite observations is investigated including oceanic regions.

Time-latitude sections of the 10-day precipitation totals along the coasts of continents and several meridians, and of the 10-day 30° sector means of OLR reveal that significant changes in precipitation take place in early May and early June almost simultaneously over the globe. We can identify four regions based on the time when marked change occurs. These changes are thought to be related to an important process in seasonal changes of atmospheric circulation, and precise descriptions of these phenomena are required in order to study regional climates, examine global heat budgets, and provide basic materials for the general circulation models. Some problems related to the precipitation data are pointed out and the requirements for rainfall observation are stressed.

1. INTRODUCTION

The seasonal march of weather is not always gradual. It sometimes experiences significant changes which take place simultaneously over a wide region. For example, the onset of the Indian summer monsoon in early June is almost concurrent with the beginning of the rainy season over Central China (Dao, 1958), over the Pacific Coast of Japan (Suda and Asakura, 1955), and with the northward shift of the rain-belt along the eastern coast of Africa (Shinoda, 1985). These facts imply that global precipitation patterns change abruptly with the onset of the Indian summer monsoon.

These large-scale changes of precipitation distribution have influence on the distribution of tropical heat sources and global atmospheric circulation. In the preceding study, however, the seasonal changes including precipitation have generally studied in a particular region, such as over India, China, Japan and so on, and then correlated each other. Therefore, seasonal changes are still analyzed inadequately on a global scale.

The FGGE provided us one year precipitation data sets with the

ISBN 0-937194-14-4

best data coverage ever achieved over land areas. We can also utilize the OLR data from a satellite including oceanic regions. In the present paper, the authors present results that examine the global simultaneity of seasonal changes in precipitation and OLR focusing on May and June of 1979. Furthermore, they also point out some problems related to the precipitation and OLR data in the present study.

2. DATA AND ANALYSIS PROCEDURES

We utilized the FGGE level II-c precipitation and snow data set which was compiled at the National Climatic Data Center in the U.S.A. Approximately 6000 stations are selected to obtain daily and 10-day precipitation totals. The 10-day precipitation maps during the FGGE period are already published in Matsumoto et al. (1985), and the extent of the non-precipitation area in winter and summer is discussed in Eguchi et al. (1986). The seasonal cycle of convective activity in the tropics is examined in Masuda et al. (1985).

We made time-latitude sections of 10-day precipitation along coast lines of continents and several meridians in order to identify the seasonal shift of heavy precipitation areas (above 50 mm / 10-day) and non-precipitation areas over the continents. Moreover, we constructed time-latitude sections of 10-day averaged OLR in 30° sectors to examine the state over the ocean using the daily OLR data compiled on $2.5^\circ \times 2.5^\circ$ grid by NESDIS/NOAA from TIROS-N satellite observations. The results of the precipitation analysis have already been shown in Shinoda et al. (1986). Therefore, we will briefly report the regional differences in the seasonal cycles of precipitation and present our results using the OLR data.

In this paper, the terms "early", "middle", and "late" of each month represent the periods from 1st to 10th, 11th to 20th, and 21st to the last day of the month, respectively.

3. ABRUPT SEASONAL CHANGES IN GLOBAL PRECIPITATION AND OLR

3.1 ANALYSIS OF PRECIPITATION

According to the time-latitude sections of 10-day precipitation maps, we can identify two pronounced stages of abrupt seasonal change during May and June. We can also recognize four sectors according to the period of abrupt changes (Fig. 1).

The first stage takes place in early May. In Southeast Asia, the heavy precipitation area disappears over Indonesia, in transition from early to middle May, accompanied by the appearance of heavy precipitation areas over the Philippines, Burma and the east coast of India (Sector IV in Fig. 1). Therefore, this stage is the onset phase of the monsoon rainy season over the Indochina Peninsula. At the same time, the non-precipitation area extends over a large part of mid-latitude North America, and the mid-western part of Eurasia (Sector II). In Sector III, precipitation over equatorial East Africa increases at this time, whereas it increases over equatorial West Africa one 10-day period earlier. At the same time, a northward shift of the non-precipitation area around the

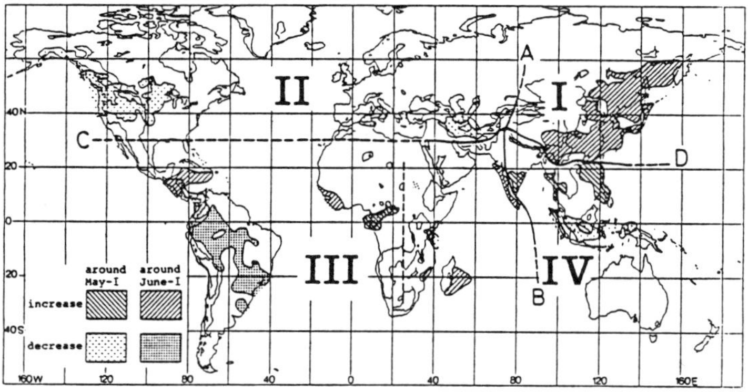

FIGURE 1. Areas which undergo marked seasonal change in 10-day precipitation around early May and June of 1979. Areas with changes over 50mm/ 10-day in the tropics and over 10mm/10-day in mid-latitudes are shaded or hatched (after Shinoda et al., 1985).

(a)90 E - 120 E

(b)120 E - 150 E

(c)150 W - 180

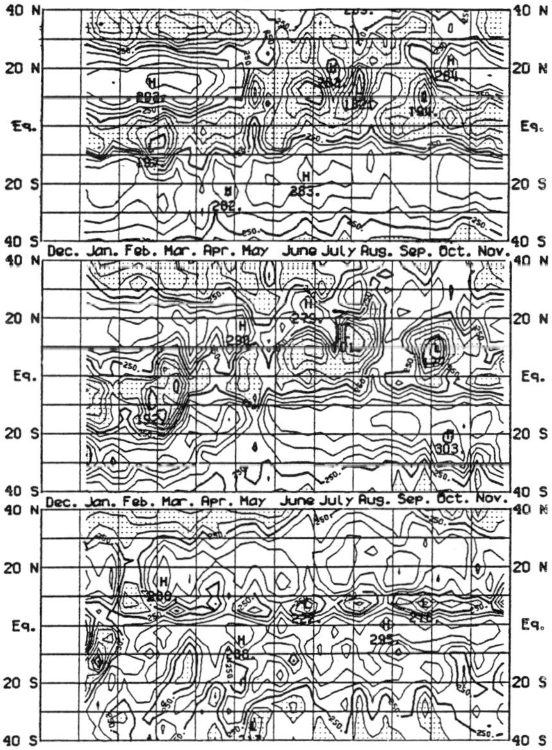

FIGURE 2. Time-latitude section of 10-day 30° sector averaged OLR. Units are W/m^{-2}. Contour intervals are 10 W/m^{-2}. Areas less than 240 W/m^{-2} are lightly shaded.

west coast of equatorial South America. It seems that these regional changes are not only simultaneous, but correlated with each other and are accompanied by global scale changes.

The second major stage is around early June. It corresponds to the onset phase of the Indian monsoon. The west coast of India is included in the eastern part of Sector III in Fig. 1. The amount of 10-day precipitation abruptly increases between early June and the middle of June over the west coast of India and the east coast of Madagascar, while it decreases in equatorial East Africa.

Prior to the onset of the Indian monsoon, between late May and early June, precipitation increases in a wide region over East and Northeast Asia (Sector I). It also increases over the Caribbean Sea and West Africa around 10° N (western part of sector III), accompanied by decreases over the Amazon Basin and the equatorial west Africa.

3.2 ANALYSIS OF THE OLR

Time-latitude sections of 10-day averaged OLR in several 30° sectors are presented in Fig. 2. The northward shift of the low OLR region (below 250 W/m^{-2}) is recognized in 10-20° N zone over the land areas around early May in 90-120° E (Fig. 2a) and around early June in 120-150° E (Fig. 2b). These features coincide with the occurrence of precipitation. On the contrary, a marked northward shift of the low OLR region is not seen over the Pacific Ocean (Fig. 2c). Therefore, the eastern boundaries of Sectors I and IV in Fig. 1 are not coincident with the western boundaries of Sectors II and III. There seemed to be a zone without pronounced seasonal change in precipitation over the Pacific Ocean. However, the precise extent of this zone is not revealed due to the lack of precipitation data.

4. SOME PROBLEMS RELATED TO THE PRECIPITATION AND OLR DATA

There are two main problems related to the continental precipitation data. One is the non-uniform observational station distribution. Although the FGGE data seem to be the best data coverage available, observation data are so sparse that it is difficult to draw meaningful isohyets in some regions. The data sparse regions are; over Sahara, Arabian Peninsula, western and northern part of the Tibetan Plateau, eastern part of Indochina Peninsula, central part of the southern Africa, California Peninsula, in the Andes, and the western coast of South America.

The other problem is the non uniformity in the times at which measurements are made to record daily amounts of precipitation. It is different among nations. In the present study, we redefine the date so that the longest part of 0-24GMT overlaps the observation "day". This problem prevents us from minute examination of the precipitation distribution in a small time scale beyond the national boundaries.

The OLR data also have some problems. They cannot express the non-precipitation areas. The precipitation estimated from the value of OLR is not very reliable.

In order to conquer these problems, adequate distribution of rain gauge stations, standardization of the observing time, and better remote estimation of precipitation amounts are required.

5. CONCLUDING REMARKS

The abrupt seasonal changes in precipitation and OLR distribution are shown to occur in early May and early June in transition from northern spring to summer. These changes are global in extent except over the Pacific Ocean.

Such world-wide changes have not been fully noticed yet. If they occur every year, the seasonal march over the globe can be viewed quite systematically. Thus, year-long precise measurement of precipitation over the globe is necessary.

REFERENCES

Dao, S.-Y., 1958: The relationship between May-yu in Far East and the behavior of circulation over Asia, Acta Meteor. Sinica, 29, 119-134. (in Chinese with English abstract)

Eguchi, T., J. Matsumoto, H. Kitajima, K. Iwasaki, M. Shinoda, T. Mikami, and K. Masuda, 1986: Global climatic classification based on seasonal distribution of non-precipitation areas. Geogr. Rev. Japan, 59, Ser. B, 43-54.

Masuda, K., M. Shinoda, T. Eguchi, and J. Matsumoto, 1985: Seasonal march of tropical heat sources and precipitation. In Proceedings International Conference on Monsoons in the Far East, Ocean Research Institute, University of Tokyo, 167-173.

Matsumoto, J., K. Iwasaki, H. Kitajima, T. Eguchi and M. Shinoda, 1985: Global 10-day precipitation distribution using FGGE daily precipitation data. In Y. Sakaguchi (ed.), Environmental Changes During Post Last-Glacial Period, Faculty of Science, University of Tokyo, 81-98. (in Japanese, available from Dept. of Geography, Univ. of Tokyo)

Shinoda, M., 1985: Seasonal migration of rainbelt along the southeast coastal region of Africa. Geogr. Rev. Japan, 58, Ser.A, 135-154. (in Japanese with English abstract)

----, T. Mikami, K. Iwasaki, H. Kitajima, T. Eguchi, J. Matsumoto, and K. Masuda, 1986: Global simultaneity of the abrupt seasonal changes in precipitation during May and June of 1979. J. Meteor. Soc. Japan, 64, 531-546.

Suda, K., and T. Asakura, 1955: A study on the unusual "Baiu" season in 1954 by means of Northern Hemisphere upper mean charts. J. Meteor. Soc. Japan, 33, 233-244.

AN OVERVIEW OF THE BAIU FRONT IN EAST ASIA AS A COUPLED SYSTEM OF THE TROPICS AND MIDDLE LATITUDES

Hiroshi Tanaka, Kuranoshin Kato, Sadamu Saitoh, Guosheng Liu, Yutaka Ishizaka, and Takao Takeda
Water Research Institute
Nagoya University
Chikusa-ku, Nagoya 464, Japan

ABSTRACT

The Baiu front in East Asia in early summer is the most significant subtropical front in the world and is one of the important climate systems in the atmospheric general circulations. In this review the large-scale characteristics of the Baiu front in East Asia from spring to summer are illustrated. In addition, the features of the meso-scale disturbances on the Baiu front are briefly shown.

1. INTRODUCTION

In East Asia the quasi-stationary heavy precipitation zone called the Baiu front appears May to July along the northern boundary of the maritime airmass. The thermal and kinematic structures are rather different through their several stages together with the two rainfall types: convective or continuous.

The Baiu front is the most significant subtropical front and coupled climate system of the circulations in both the tropics and middle latitudes. In other words, it is greatly affected by the global monsoon, baroclinicity and long waves. Thus the studies of the Baiu front are very important in order to understand the general circulation system, to validate the quality of a GCM, and to verify whether the principal systems such as subtropical high in the western Pacific area and longwave trough are properly simulated.

In the present review the large-scale features of the Baiu front in East Asia from spring to summer are illustrated by using the figures in Kato (1988), one of the present authors, which is submitted to J. Meteor. Soc. Japan. In addition to the overall structure of the Baiu front, this review includes several types of mesoscale disturbances, all of which play important roles in producing latent heat due to the precipitation process and in transporting it poleward. These mesoscale disturbances differ significantly from those seen in the usual baroclinic disturbances.

TROPICAL RAINFALL MEASUREMENTS
John S. Theon and Nobuyoshi Fugono (Eds.)

ISBN 0-937194-14-4

2. CHARACTERISTICS OF THE BAIU FRONT

The features of the Baiu front in East Asia and the large-scale circulations associated with it show great variability in place and time (Akiyama, 1973; Kato, 1985, 1987; Murakami and Huang, 1984; Ninomiya et al., 1981; Ninomiya, 1984; Ninomiya and Muraki, 1986).

First of all we show the features of the Baiu front in the peak stage of the Baiu season in Japan and central China. Figure 1 shows the distribution of the outgoing longwave radiation (OLR) for the period from 20 to 29 June (after Kato, 1988). An active convective area (OLR $\leqq$ 220 W/m^{-2}) extends from the Arabian Sea to the western Pacific Ocean, corresponding to the global monsoon trough. A huge cloud belt extends over Japan starting from the Tibetan Plateau. This northern branch of cloud coverage is the Baiu front. Following Ninomiya (1984), the Baiu front is typical as a subtropical front with the weak potential temperature gradient and the strong equivalent potential temperature gradient across it. The frontal zone associated with the typical extratropical baroclinic wave is quite opposite in a sense of its strong baroclinicity and weak moisture gradient. The Baiu front is controlled by both mid-latitudinal and tropical systems.

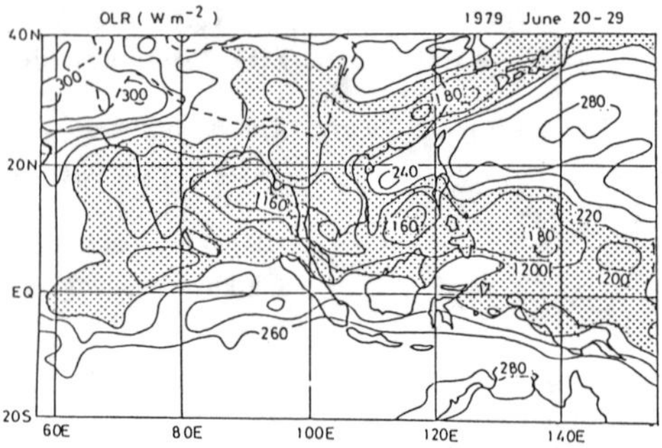

FIGURE 1. Distribution of the mean OLR for the period from 20 to 29 June in 1979. in W/m^{-2}. Areas where OLR$\leqq$ 220 W/m^{-2} are shaded (Kato, 1988).

Figures 2a, 2b, and 2c show the patterns of the large scale circulation around the Baiu front in the three stages (A), (B), and (C), respectively (after Kato (1988)). The strong low-level southerly flow transports an enormous amount of moisture and produces convective instability sustaining heavy rainfalls in the Baiu frontal zone (Ninomiya, 1984). The moisture is transported not from the Indian monsoon region but from the area of the subtropical high (Pacific High) in the western Pacific. It should be noted that the subtropical high with the strong low-level southerly wind plays an essential role in sustaining the Baiu front.

The Baiu front in Stage (A) is a baroclinic front associated with the southern branch of the middle-level westerly jet around the Tibetan Plateau (cf. Yoshino, 1965; Murakami, 1981; Murakami and Huang, 1984; Kato, 1988; Ninomiya and Muraki, 1986). The Baiu front is widely

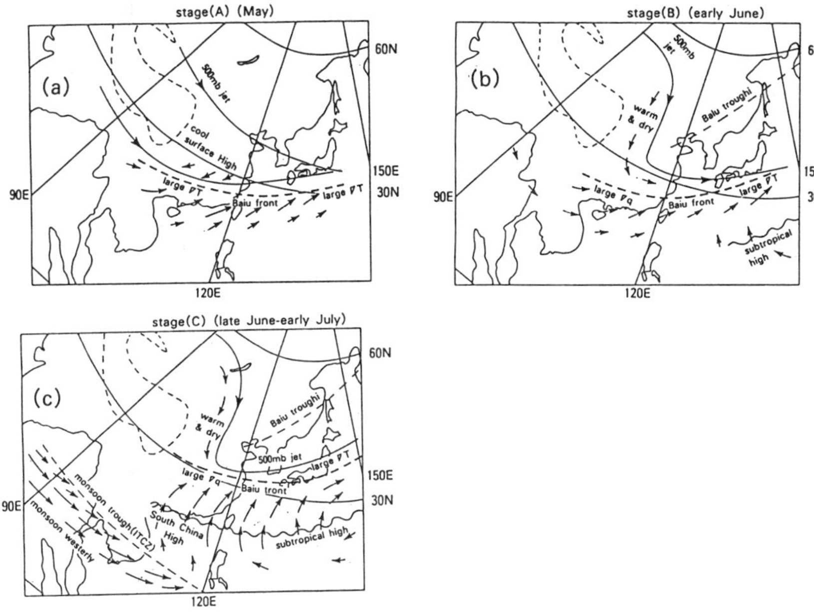

FIGURES 2. Typical patterns of the large-scale circulation on the Baiu front for stages (A), (B) and (C). (Kato, 1988).

covered by stratiform clouds. The frontogenetic field and the vertical circulation across the Baiu front with upward motion in south China (~25°N) and subsiding motion in central China (~35°N) (Kato, 1988) are suggested in Stage (A). Eventually the Baiu front in Stage (A) is primarily driven by the effects of the Tibetan Plateau.

In late May the Baiu front is still located in south China. However, the cross-frontal temperature gradient is abruptly decreases over the continental part of China. As a result, the Baiu front in China is characterized by the moisture gradient only. The southern branch of the middle-level westerly jet around the Tibetan Plateau disappears in that period and the Baiu trough is formed in the polar airmass (Ninomiya and Muraki, 1986). This is the beginning of Stage (B).

The final stage (C) starts in the middle of June. An anticyclonic circulation called the South China High produces strong low-level southerly wind (see Figure 2c). This anticyclone is enhanced by the intensification of the local Hadley cell associated with ITCZ in the South China Sea area after onset of the global summer monsoon (see

Kato, 1988). Also in this stage, the northerly wind associated with the quasi-stationary Baiu trough advects the cold air toward the eastern part of the Baiu front. The shift from stage (B) to (C) is also rapid.

Besides the characteristics of the Baiu front stated above, we should note that remarkable intraseasonal and interannual variations are also seen in the rainfall activity associated with the Baiu front (Kato and Kurihara, 1988; Ninomiya and Mizuno, 1987).

3. DISTURBANCES IN THE BAIU FRONTAL ZONE

Rainfall in the Baiu frontal zone is often organized in mesoscale structures. The atmosphere is almost saturated in the lower troposphere; thus convective instability tends to be released without strong upward forcing.

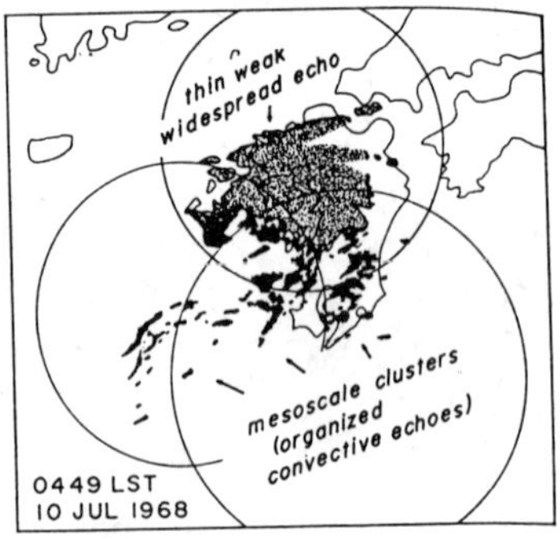

FIGURE 3. Radar echo pattern associated with the medium-scale disturbance (Akiyama, 1978).

Reduction of the equivalent Richardson number down to unity results in the generation of medium-scale cyclones which are called meso-α scale disturbances (Stone, 1966; Gambo, 1970a,b; Tokioka, 1970, 1973). The rainfall area associated with the medium-scale cyclones extends over a hundred kilometers and is divided into two patterns; weak rainfall spread widely and convective rainfall organized on bands (Akiyama, 1978, 1979; see Figure 3).

The Baiu front is a show window of mesoscale disturbances. Meso-α, β, γ scale disturbances interact with each other in a complicated way (e.g., Takeda and Seko, 1986 for convective cloud bands; Saitoh and Tanaka, 1987a,b for meso-β disturbances; Thorpe, 1987 for meso-α scale disturbances). The mesoscale disturbances also produce several types of low-level jet nearby due to the combined effects of baroclinicity and water vapor supply.

4. CONCLUDING REMARKS

The Baiu front should be considered as a part of the global atmospheric circulation. There are a number of factors that are responsible for formation and maintenance of the Baiu front and its associated rainfall activity: dynamical and thermal effects of the Tibetan Plateau, behavior of the long waves in high latitudes, low-level wind and moisture fields associated with the subtropical high, effects of tropical heat sources on variation of the subtropical high, and several types of mesoscale disturbances generated there.

Rainfall observation is a key factor for the research of the Baiu front. For this purpose, high-accuracy, long-range TRMM observations of rainfall amounts, as well as instantaneous snapshots of rainfall patterns, for the entire region of the Baiu front (20° to 37° N) are required.

REFERENCES

Akiyama,T., 1973: The large-scale aspects of the characteristic features of the Baiu front. Pap. Meteor. Geophys., 24. 157-188.

Akiyama,T., 1978: Mesoscale pulsation of convective rain in medium-scale disturbances developed in the Baiu front. J. Meteor. Soc. Japan, 56, 267-283.

Akiyama,T., 1979: Thermal stratification in Baiu frontal medium-scale disturbances with heavy rainfalls. J. Meteor. Soc. Japan, 57. 587-598.

Gambo,K., 1970a: The characteristic feature of medium-scale disturbances in the atmosphere (I). J. Meteor. Soc. Japan, 48, 173-184.

Gambo,K., 1970b: The characteristic feature of medium-scale disturbances in the atmosphere (II). J. Meteor. Soc. Japan, 48, 315-330.

Kato,K., 1985: On the abrupt change in the structure of the Baiu front over the China Continent in late May of 1979. J. Meteor. Soc. Japan. 63, 20-36.

Kato,K., 1987: Airmass transformation over the semiarid region around North China and abrupt change in the structure of the Baiu front in early summer. J. Meteor. Soc. Japan, 65, 737-750.

Kato,K., 1988: Change in the large-scale circulations around the Baiu front in China at the onset of the northern summer monsoon in 1979. Submitted to J. Meteor. Soc. Japan.

Kato,K. and T.Kurihara, 1988: Structure of the subtropical high and the characteristics of the Baiu front duing a short-break period of the Baiu in Kanto District (A case study for the second decade in June 1979). Submitted to Tenki (in Japanese).

Murakami,T., 1981: Orographic influence of the Tibetan Plateau on the Asiatic winter monsoon circulation. Part I: Large-scale aspects. J. Meteor. Soc. Japan, 59, 40-65.

Murakami,T. and W.G.Huang, 1984: Orographic effects of the Tibetan Plateau on the rainfall variations over the Central China during the 1979 summer. J. Meteor. Soc. Japan, 62, 895-909.

Ninomiya,K., 1984: Characteristics of the Baiu front as a predominant subtropical front in the summer northern hemisphere. J. Meteor. Soc. Japan, 62, 880-894.

Ninomiya,K., M.Ikawa and T.Akiyama, 1981: Long-lived

medium-scale cumulonimbus cluster in Asian subtropical humid region. J. Meteor. Soc. Japan, 59, 564-577.

Ninomiya,K. and H.Muraki, 1986: Large-scale circulations over East Asia during Baiu period of 1979. J. Meteor. Soc. Japan, 64, 409-429.

Ninomiya,K. and H.Mizuno, 1987: Variations of Baiu precipitation over Japan in 1951-1980 and large-scale characteristics of wet and dry Baiu. J. Meteor. Soc. Japan, 65. 115-127.

Saitoh,S. and H. Tanaka, 1987: Numerical experiments of conditional symmetric baroclinic instabilty as a possible cause for frontal rainband formation. Part I A basic experiment. J. Meteor. Soc. Japan, 65, 675-708.

Saitoh,S. and H.Tanaka, 1987: Ibid. Part II Effects of water vapor suply. Submitted to J. Meteor. Soc. Japan.

Stone,P.H., 1966: On non-geostrophic baroclinic stability. J. Atmos. Sci., 29, 419-426.

Takeda,T. and K.Seko, 1986: Formation and maintenance of band-shaped convective radar echoes. J. Meteor. Soc. Japan, 64, 941-955.

Thorpe,A.J., 1987: Evolution of mid-latitude cyclones and fronts using a semi-geostrophic model with parameterized moist processes. Extended Abstracts of the Third Conference on Mesoscale Processes, 177-178p.

Tokioka,T., 1970: Non-geostrophic and non-hydrostatic stability of a baroclinic fluid. J. Meteor. Soc. Japan, 48, 503-520.

Tokioka,T., 1973: A stability study of medium-scale disturbances with inclusion of convective effects. J. Meteor. Soc. Japan, 51, 1-10.

Yoshino,M.M., 1965: Four stages of rainy season in early summer over East Asia (part I). J. Meteor. Soc. Japan, 43, 321-345.

INFLUENCE OF TROPICAL FORCING ON THE ATMOSPHERIC CIRCULATION OF MIDDLE AND HIGH LATITUDES

Huang Ronghui
Institute of Atmospheric Physics
Academia Sinica
Beijing, PRC

ABSTRACT

Many observed facts show that the onset of El Nino may cause severe cold surges in northeastern China and the Hokkaido region of Japan.

The propagating laws of quasi-stationary planetary waves, forced by a forcing source at low latitudes in a realistic current, are discussed theoretically and numerically. It is pointed out that the quasi-stationary planetary waves can horizontally propagate from the troposphere at low latitudes to the upper troposphere at middle and high latitudes. It is explained that the anomaly of heat source at low latitudes due to EL Nino events can cause the atmospheric circulation anomalies at middle and high latitudes.

1 INTRODUCTION

Recently, there have been many observational studies on the anomalous warming of sea surface temperature (SST) in the equatorial Pacific. Rasmusson and Carpenter (1982) have summarized the characteristics of the sea surface temperature anomaly (SSTA) in the equatorial Pacific. They have found that SSTA in winter is the largest in the eastern Pacific between 100°W-160°W, and the maximum anomaly may attain about +2°C. In the winter of 1982, the maximum SSTA attained about +4°C in that area. Because of the increase of SST, the sensible heat transfer from the ocean to the atmosphere will largely increase. Besides, according to the Doberitiz (1968) and Bjerknes (1969) investigations, with the SSTA in the equatorial Pacific, the region of deep cumulus convection will shift eastward with the warm water. Moreover, it has been verified by satellite data. This deep cumulus convection corresponds to about 15 mm day^{-1} of precipitation anomaly. Thus, the anomaly warming in the sea surface will cause the heat source anomaly over low latitudes, i.e., the heat source will be enhanced. This heat source anomaly will cause the circulation anomalies over middle and high latitudes. As a consequence, it is necessary to investigate, theoretically and numerically, the circulation anomalies over middle and high latitudes due to the heat source anomaly over low latitudes.

ISBN 0-937194-14-4

2 RELATION BETWEEN THE CIRCULATION ANOMALY OVER MIDDLE LATITUDES AND SSTA IN THE EASTERN EQUATORIAL PACIFIC

Many observed facts show that the onset of El Niño events may cause severe cold surges in northeast China and the Hokkaido region of Japan. For example, the continued anomalies of the atmospheric circulation occurred over wide areas of middle and high latitudes during Dec. 1976-Feb. 1977. The surface temperature in the northeast China and Hokkaido was 4°C below normal. The same phenomena also occurred during the winters of other El Niño events, except for the winter of 1972. Fig.1 shows the anomaly distribution of 500hPa observed disturbance height field over the Northern Hemisphere during Dec. 1976-Feb. 1977, by synthesizing wave components of k=1-3. From Fig. 1 we can see that a positive anomaly of disturbance height field is formed over the tropical Pacific and a negative anomaly over the North Pacific. This negative anomaly shifts northeastward to Hokkaido, Japan and to the northeast of China. Another positive anomaly region may be found in Alaska and the northern region of North America. Another negative anomaly region may be found in the central and eastern parts of the U.S. This pattern is the so-called PNA pattern of atmospheric circulation anomaly.

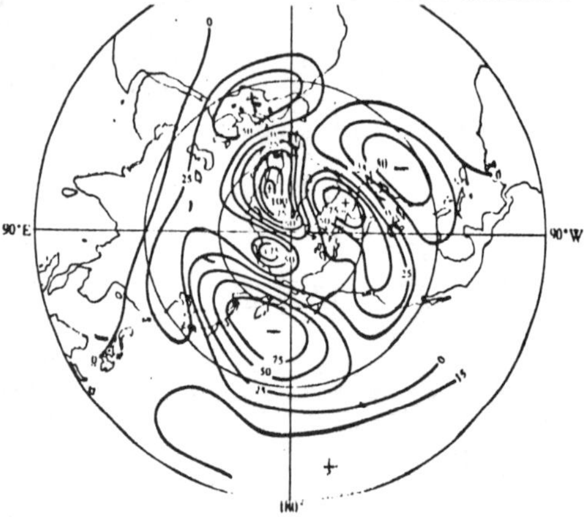

Fig.1 Anomaly distribution of 500hPa disturbance height field for Dec. 1976-Feb. 1977 (units in m).

As is well-known, the SST in the eastern equatorial Pacific was above normal during the winter of 1976-77, i.e., the onset of El Niño event. The same phenomena also occurred in the winters of 1956-57, 1960-61, 1964-65, 1968-69, 1972-73, 1982-83, and 1986-87.

3 PROPAGATING CHARACTERISTICS OF QUASI-STATIONARY PLANETARY WAVES FORCED BY A FORCING SOURCE AT LOW LATITUDES

Because of the vertical and meridional shears in the realistic basic flow, and the sphericity of the earth, the propagating rays of quasi-stationary planetary waves will be refracted during the propagation.Assuming the angle between a propagating ray of wave and the horizontal is $\hat{\alpha}$, Huang Ronghui and Gambo (1984) have obtained the following relation using the kinematic relation between wavenumber and frequency of waves.

$$\frac{d_g \hat{\alpha}}{dt}=\frac{1}{2Q_k}\vec{i}\,\vec{C}'_g \times \nabla Q_0 \tag{1}$$

where $\vec{i}$ is a unit vector in the λ direction, $\vec{C}'_g$ is the projection of group velocity onto the meridional plane and we define

$$\frac{d_g}{dt}=\frac{\partial}{\partial t}+\vec{C}'_g\cdot\nabla, \tag{2}$$

as the advective derivate along with a propagating ray. The variations of a propagating ray of waves are determined by Q_k and the gradient of Q_0.Obviously, the propagating rays of waves are always towards the direction of the gradient of Q_0. Hunag Ronghui and Gambo (1984) have computed the distribution of Q_0 from the winter current. The minimum of $\bar{Q}_0$ is located in the upper stratosphere at high latitudes and another minmum of $\bar{Q}_0$ is located in the lower stratosphere at middle latitudes. As analysed in the paper of Huang Ronghui and Gambo, there should be two wave guides in the propagations of quasi-stationary planetary waves. Waves can propagate from the troposphere into the stratosphere through the so-called polar wave guide. They can also propagate from the lower troposphere at middle and high latitudes into the upper troposphere at low latitudes through another guide which is located between the region of low latitudes and that of middle latitudes.

However, as shown in Fig.2, for planetary waves forced by a forcing source at low latitudes, it is not easy for the waves to be directly refracted into the polar wave guide at high latitudes. Thus, the waves cannot be propagated directly toward the stratosphere and quasi-horizontally propagate toward the upper troposphere at high latitudes. So, the propagating laws of quasi-stationary planetary waves, forced by a forcing source at low latitudes are in a sphere. Using WKBJ method, the variations of the propagating ray path of waves, the time propagating from a forcing source to turning point and the local distribution of wave amplitude during the propagation may be obtained.

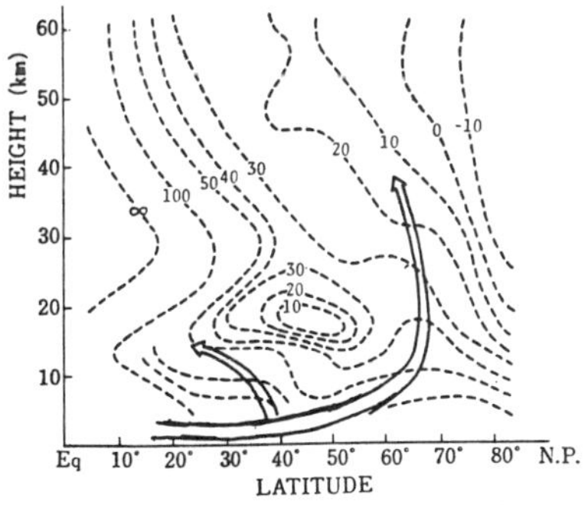

Fig.2 Schematic picture of wave guides of the planetary wave forced by a forcing source at low latitudes. Dashed curves show the refractive index squared for wave number 1.

4 NUMERICAL SIMULATION OF THE INFLUENCE OF TROPICAL FORCING ON ATMOSPHERIC CIRCULATION OF MIDDLE AND HIGH LATITUDES

In order to investigate the influence of tropical forcing on the atmospheric circulation over middle and high latitudes during the Northern Hemisphere winter,a quasi-geostrophic, 4-level spherical coordinate model, with Rayleigh fraction, Newtonian cooling and the horizontal eddy thermal diffusivity (see Huang Ronghui and Gambo 1982), is used to compute the anomalies of disturbance pattern due to an idealized heat source anomaly. Fig.3 shows the anomaly distribution of 500hPa disturbance height field caused by the heat source anomaly over the eastern tropical Pacific.

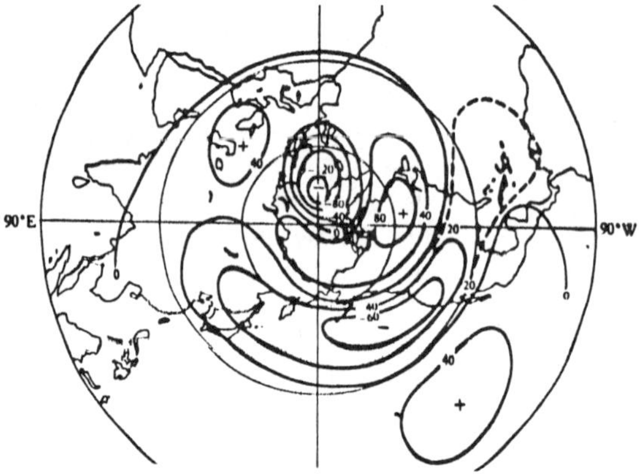

Fig.3 The anomaly distribution of 500hPa disturbance height computed by the model (units in m).

We can clearly see that when a heat source anomaly is caused over the eastern tropical Pacific, a positive anomaly of disturbed height field is found over the eastern subtropical Pacific and a negative anomaly may be found over the North Pacific. This negative region shifts to the northeast of China and Hokkaido region of Japan. Another positive anomaly is formed over the northern part of North America, while another negative anomaly is formed over the central and eastern part of U. S. Thus, the anomaly distribution of the disturbance height field, computed by the model, is in good agreement with the observed results.

5 CONCLUSIONS

In this paper, it is explained,with the help of theoretical and numerical arguments, that the anomaly of heat source at low latitudes due to El Nino events can cause the atmospheric circulation anomalies at middle and high latitudes and can be the cause of severe cold surges in the northeastern regions of China in the onset period of El Nino events.

REFERENCES

Bjerknes, J., 1969: Atmospheric teleconnections from the equatorial Pacific. Mon. Wea, Rev., 97, 162-172.

Doberitz, R., 1968: Cross spectrum analysis of rainfall and sea temperature at the equatorial Pacific ocean. Bonn. Meteor. Abh., 8, 61.

Huang Ronghui and K. Gambo, 1982: The response of a hemispheric multi-level model atmosphere to forcing by topography and stationary heat sources. Part I, II, I Meteor Soc Japan 60 78-108.

Huang Ronghui and K. Gambo, 1984: On other wave guide in stationary planetary wave propagations in winter Northern Hemisphere. Scientia Sinica (Series B) 27, 610-624.

Rasmusson, E., and T. Carpenter, 1982: Variations in tropical sea surface temperature and surface wind fields associated with the Southern Oscillation/El Nino. Mon. Wea. Rev., 110, 354-384.

ON THE FEEDBACK ROLE OF TROPICAL CONVECTION

Li Chong-yin
Institute of Atmospheric Physics
Academia Sinica
Beijing, PRC

ABSTRACT

Based on the CISK theory, the results of studies on the feedback role of tropical convection are presented in this paper. The results show that: a) The vertical distribution of convective condensation heating is very important. It not only affects the growth rate but also the structure of the disturbance. b) The convection heating is able to cause the CISK-oscillatory unstable mode. This mode could drive the mixed Rossby-gravity wave and Kelvin wave in the tropical atmosphere and what kind of wave can be excited depends on the vertical distribution of the convection heating. c) The feedback of the convection heating can also excite a moving CISK mode and drive 30-50-day monsoon oscillation; d) For studies of tropical precipitation, the measurement of the vertical distribution of convective condensation is also important.

1. INTRODUCTION

The tropical atmosphere is often in a moist conditional instability situation. The cumulus convection is very important. Therefore, the feedback role of the cumulus convection on the large-scale atmospheric motion has consistently held meteorologists' attention The conditional instability of the second kind (CISK) indicated a positive feedback process between the tropical depression and the cumulus convection. The wave-CISK expanded the feedback role of cumulus convection into more tropical systems. Thus, many phenomena of the atmospheric motion in the tropics should associate with cumulus convection.

The discovery of the mixed Rossby-gravity waves and Kelvin waves in the lower stratosphere and upper troposphere in the tropics is looked upon as one of important advances in the tropical meteorology (Matsuno, 1966; Yanai and Maruyama, 1966). But their formation mechanisms have not yet been found satisfactorily. Some meteorologists suggested these waves may be caused by cumulus convection heating and are related to the cumulus convection. Holton (1972) got the disturbances, which are similar to the mixed Rossby-gravity waves and Kelvin waves in the equatorial stratosphere, by using numerical diagnostic model with a periodic condensation heating in the limited region. The question remaining is whether the cumulus convection can cause oscillatory unstable mode without periodic heating. Thirty to fifty day low-frequency oscillation in the tropical atmosphere has been extensively studied (Krishnamurti, et al., 1982; Murakami, et al., 1984; Wallace, et al., 1983; Lau, 1985). What is the relationship between the formation of 30-50-day oscillation and the cumulus convection?

In this paper, we have studied the feedback role of tropical convection based on the CISK theory and indicated the important effect of the cumulus convection heating, which causes tropical waves and 30-50-day low-frequency oscillation, and the importance of vertical distribution of the convection heating.

TROPICAL RAINFALL MEASUREMENTS
John S. Theon and Nobuyoshi Fugono (Eds.)

ISBN 0-937194-14-4

2. NUMERICAL MODEL

In a cyclindrical model including Ekman pumping and the cumulus friction, the perturbation governing equations can be written as follows:

$$\frac{\partial v}{\partial t} = -fu - \frac{\partial}{\partial p}[M(v - v_B)] \tag{1}$$

$$\frac{\partial \phi}{\partial r} = fv \tag{2}$$

$$\frac{\partial \phi}{\partial p} = -\frac{RT}{p} \tag{3}$$

$$\frac{\partial (ru)}{\partial r} + r\frac{\partial \omega}{\partial p} = 0 \tag{4}$$

$$\frac{\partial T}{\partial t} - \frac{pS}{R}\omega = \frac{Q}{c_p} \tag{5}$$

where u, v and ω, respectively, refer to the radial, azimuthal and p-vertical velocity components. $S \equiv -\frac{R}{p}\left(\frac{\partial \overline{T}}{\partial p} - \frac{R\overline{T}}{pc_p}\right)$ is the static stability. M is the cloud mass flux. V_C is the velocity component of the cloud air. All other notations have common meanings.

Both Ekman pumping and cumulus friction are included in the numerical model. These two processes can cause the secondary circulation and lead to positive feedback between the cumulus convection and large-scale disturbance. The convection heating should include two parts and can be written as follows:

$$Q = c_p \varepsilon \eta(p)\left[M_* \frac{1}{r}\frac{\partial r v_*}{\partial r} + N\frac{1}{r}\frac{\partial r v_B}{\partial r}\right] \tag{6}$$

where η (p) is the convection heating function; $N = \frac{Kp_B}{2f}$, K is the frictional coefficient. The footnote star and B represent the reference level and the bottom boundary respectively.

For convenience, a new variable ξ and streamfunction ψ are introduced. The model solution for ξ and ψ is in the form of

$$\begin{aligned} \xi(r,p,t) &= \varphi(r)A(p)\mathrm{Exp}(\sigma t) \\ \psi(r,p,t) &= \varphi(r)E(p)\mathrm{Exp}(\sigma t) \end{aligned} \tag{7}$$

We finally obtain the following governing equations for E(p) and A(p)

$$\sigma A = -f\frac{dE}{dp} - \frac{d}{dp}[M(A - A_B)], \tag{8}$$

$$\sigma f\frac{dA}{dp} = -\lambda^2 SE + \frac{\lambda^2 H}{f}[M_* A_* + NA_B]. \tag{9}$$

where $H \equiv \frac{\varepsilon R f \eta}{p}$ is a heating parameter, λ^2 is a constant.

3. THE IMPORTANCE OF VERTICAL HEATING PROFILE

The influence of condensation heating profile on the unstable mode is investigated by using numerical solution of Eqs. (8) and (9) with different vertical distribution of the convection heating.

The results are as follows:

(1) The growth rate of unstable mode is closely related to the heating profile. The greater growth rate might be obtained when the maximum convection heating is in the middle-lower troposphere.

(2) For different profile of the convection heating, the vertical structure of the unstable mode is different.

(3) The vertical distribution of the convection heating also affects the production and transformation of the energy. The higher the maximum heating layer is, the higher the maximum production layer of available potential energy will be. It is especially interesting that the efficiency of available potential energy transforming into the disturbance kinetic energy is higher if the maximum heating layer is higher (Fig.1).

Therefore, the vertical distribution of the convection heating is very important for the growth rate and the structure of the unstable mode.

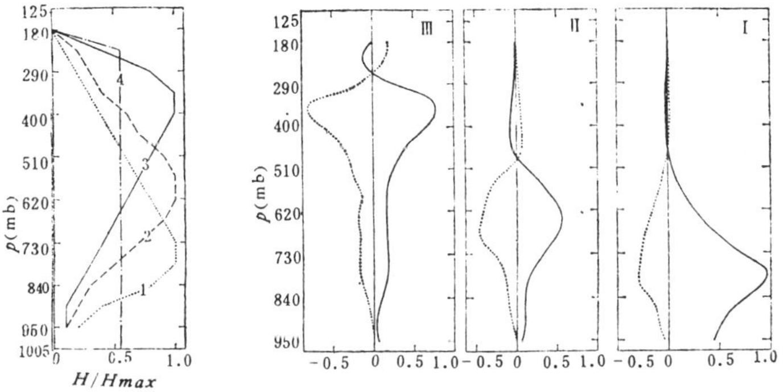

Figure 1 The vertical distribution of the heating parameter (left) and the production (solid line) and conversion (dotted line) of available potential energy (right).

4. TROPICAL WAVES AND CONVECTION HEATING

The results from the above numerical model show that through the CISK mechanism, convective condensational heating is able to produce, not only the stationary unstable mode, but also an oscillatory unstable mode with periodic variation (called CISK-oscillatory unstable mode). This CISK-oscillatory unstable mode is a planetary scale disturbance. Its vertical structure is shown in Fig.2. In addition, the model results also show that it is advantageous to produce the CISK-oscillatory unstable mode if there is a maximum convection heating in the middle-lower troposphere.

Through the CISK mechansim, what kind of the periodic wave can be excited by the convection heating and the cumulus friction? This depends, to a great extent, on the vertical distribution of the convection heating. As the heating is maximum in the middle-lower troposphere, for the normal convection, it is favorable to produce the oscillatory unstable mode corresponding to the Kelvin waves. However it is favorable to produce the oscillatory unstable mode corresponding to the mixed

Rossby-gravity waves, if the heating is maximum at the middle-upper troposphere, for deep cumulus.

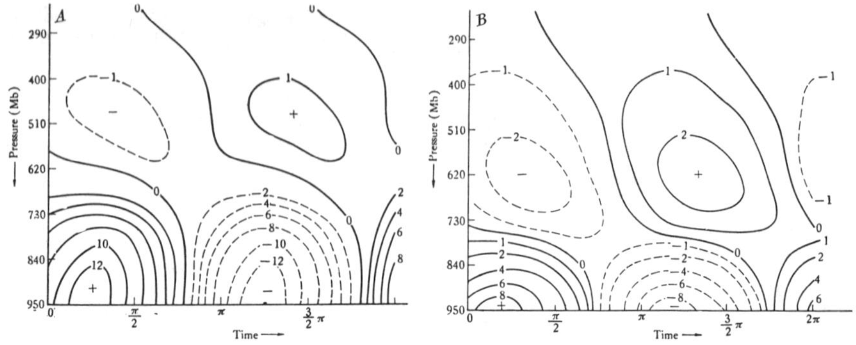

Figure 2 The temperature section (a) and tangential velocity section (b) for CISK oscillatory unstable mode.

5. 30-50-DAY OSCILLATION AND CONVECTION HEATING

Based on the analysis of the observation data during the summer monsoon season in 1979, a 30-50 day periodic oscillation of the summer monsoon in South Asia is found. The studies show that the 30-50-day oscillation of the monsoon in South Asia has a phase velocity northward. The dispersion of the energy of tropical cyclones and monsoon depressions associated with 30-50-day oscillation is eastward. How is this low-frequency oscillation excited?

We introduce vertical shearing of basic flow, which is the most obvious in South Asia during the summer monsoon season, into a CISK model. A moving CISK mode is obtained (Li, 1985).

As we know, the active monsoon and break monsoon are two major phenomena of the summer monsoon. In fact, the active and break monsoons are associated with the actions of the monsoon troughs (ridges). The monsoon trough is an active convection band, including the monsoon depression or tropical cyclone. The action of the monsoon troughs (ridges) should represent the large-scale character of the summer monsoon.

In order to study the action of the monsoon trough (active convection band), we use a CISK model with the shearing basic flow in a plane-symmetrical coordinate. For the two-level model, the phase speed C and the group velocity C_g of the moving CISK mode can be obtained as follows:

$$C_i \approx \frac{f\sigma_r}{S_2\Delta[(1-\mu H)l^2+\eta]}\frac{d\bar{v}_i}{dp} \qquad (i=x, y)$$

$$C_{gi} \approx C_i\left[1-\frac{2l^2(1-\mu H)}{(1-\mu H)l^2+\eta}\right].$$

where σ_r is the growth rate of unstable mode, μ is a heating parameter, η and H are the condition parameters of tropical atmosphere, l is horizontal wave-number, $\frac{d\bar{v}_i}{dp}$ is the vertical shearing of basic flow.

For the atmospheric basic state during the summer monsoon in South Asia, the moving CISK mode propagates northward with a speed of 0.6 degree latitude per day. It has about a 34-day oscillation period and a group velocity southeastward. These are similar to the results from analyzed MONEX data. Therefore, it can be preliminarily concluded that 30-50-day oscillation of the summer monsoon is driven by a moving CISK mode which is caused by the convection heating feedback.

We can further suggest that the convection heating feedback is an important mechanism to produce a 30-50-day oscillation in tropical atmosphere because the convection heating is a major energy source of the atmospheric motion in the tropics.

6. CONCLUSIONS

According to our work, the conclusions can be summarized as follows:

Convective condensation heating has an important feedback role for the atmospheric motion in the tropics. The vertical distribution of the convection heating is also important. It can affect the growth rate and the structure of the disturbance.

The convection heating is able to cause the CISK-oscillatory unstable mode. This unstable mode could drive the mixed Rossby-gravity wave and Kelvin wave in the tropical atmosphere. The normal convection is favorable to produce the oscillatory unstable mode corresponding to the Kelvin waves. The deep convection is favorable to produce the mode corresponding to the Rossby-gravity waves.

The feedback of the convection heating can also excite a moving CISK mode. This moving CISK mode will cause a 30-50-day low-frequency oscillation of monsoon.

It is necessary to measure the vertical distribution of convective condensation in the tropical atmosphere.

REFERENCES

Holton, J.R., 1972: Waves in the equatorial stratosphere generated by troposphere heat sources, J. Atmos. Sci., 29, 368-375.

Krishnamurti, T.N., et al., 1982: The 30-50 day mode at 850 mb during MONEX, J. Atmos. Sci., 39, 2088-2095.

Lau, K.M., et al., 1985: Aspects of the 40-50 day oscillation during northern winter as inferred from outgoing longwave radiation, Mon. Wea. Rev., 113, 1889-1909.

Li Chong-yin, 1985: Actions of summer monsoon troughs (ridges) and tropical cyclones over South Asia and the moving CISK mode, Science in China, 28, 1197-1207.

Matsuno, T., 1966: Quasi-geostrophic motions in the equatorial area, J. Met. Soc. Japan, 44, 25-43.

Murakami, T., et al., 1984: On the 40-50 day oscillations during the 1979 Northern Hemisphere Summer. Part I: phase propagation, J. Met. Soc. Japan, 62, 440-468.

Wallace, J.M., et al., 1983: Observation of low frequency atmospheric variability, Large-Scale Dynamical Processes in the Atmosphere, Academic Press, London, 55-94.

Yanai, M. and T. Maruyama, 1966: Stratospheric wave disturbances propagating over the euqatorial Pacific, J. Met. Soc. Japan, 44, 291-294.

PRECIPITATION IN NORTHERN PERU DURING THE 1982–83 EL NINO: VARIATIONS IN DAILY PRECIPITATION PATTERNS

T. Mikami
Tokyo Metropolitan University
Tokyo 158, Japan

ABSTRACT

Temporal variations in daily precipitation patterns in Northern Peru during the 1982-83 El Nino are analyzed using principal component analysis for the data set of 74 stations. The results suggest that the eigenvector pattern for component 2 which is characterized by the opposite signs between coastal plains and Andes Mountains should be important to explain the causes of El Nino precipitation in this region.

1. INTRODUCTION

Northern Peru experienced unusual heavy rainfalls and severe floods during the 1982-83 El Niño. In 1985, the Japanese Ministry of Education, Science and Culture funded two projects to collect meteorological data on the 1982-83 El Niño in northern Peru. A brief

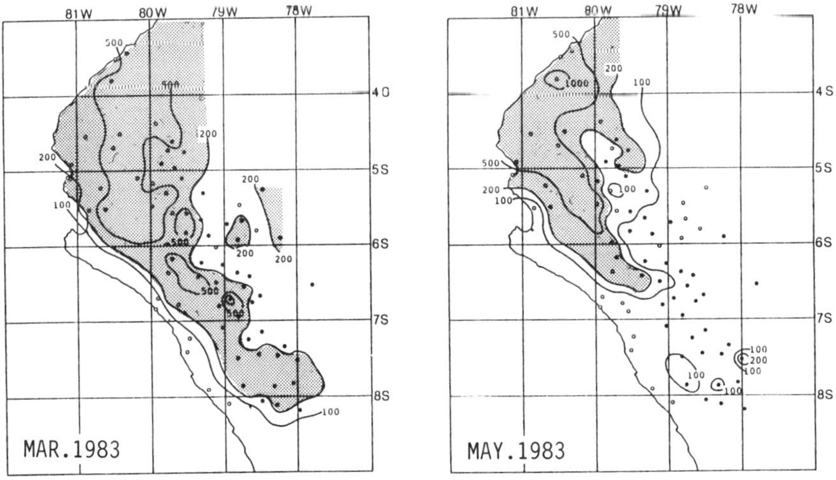

FIGURE 1. Monthly precipitation distribution in March 1983 (left) and May 1983 (right). Precipitation areas over 200 mm are shaded.

ISBN 0-937194-14-4

report including collected meteorological data has already been published (Nogami ed.,1987). On the other hand, spatial and temporal variation of clouds and rainfall along the coast of Northern Peru during 1983 was analyzed in detail by Horel et al.(1986).

However, the overall feature of the variations in daily precipitation patterns in Northern Peru during the 1982-83 El Nino is unclear. The purpose of this paper is to analyze a few basic daily precipitation patterns and their temporal variation quantitatively.

2. DATA AND METHODS

The precipitation data were obtained from SENAMHI (Servicio Nacional de Meteorologia y Hidrologia) and CORPAC (Cooperacion Peruana de Aviacion Comercial) in Peru. Considering the spatial density of collected station data, 74 stations were selected for the present study. The duration of precipitation data covers the period from December 1982 to June 1983. As a first step to make an overview of the variations in precipitation patterns during this period, we drew monthly precipitation distribution maps. In December 1982, the area of more than 200 mm precipitation is restricted to a narrow band over the Andes Mountains and no precipitation can be found along the coastal plains in northern Peru. From January to April 1983, heavy precipitation area extends over the coastal plain and Andes Mountains from northwest to southeast. Then the reverse trend can be seen from May to June 1983. Fig.1 shows typical examples of monthly precipitation distribution maps.

In order to analyze the temporal variation in spatial precipitation patterns more in detail and quantitatively, a principal component analysis was applied to the 74 daily precipitation data sets (212 days) in northern Peru.

3. RESULTS AND DISCUSSION

3.1 SPATIAL AND TEMPORAL VARIATIONS IN EIGENVECTORS

The analysis made it clear that 53.8 percent of the cumulative variance could be explained by the first 10 principal components. In particular, spatial patterns and temporal variations of the principal components 1 and 2 were considered to be important (Fig.2).

The principal component 1 explains 21.5 percent for the total variance. Positive eigenvectors appear all over the study area, among which higher values extend along the coastal plain near the western slope of the Andes Mountains. A time series of the principal component scores suggests that the peak of rainfalls over the whole area appeared from March to April in 1983 with a large day-to-day variability.

The variance of the principal component 2 is 7.4 percent. The eigenvectors show positive signs over the coastal plains where the altitude is below 400 m and negative signs over the Andes Mountains above 1000 m. Amplitude of the principal component scores is rather small as compared with that of the component 1. It should be noted that positive scores appeared continuously from April to the middle of June indicating active coastal precipitation in this period.

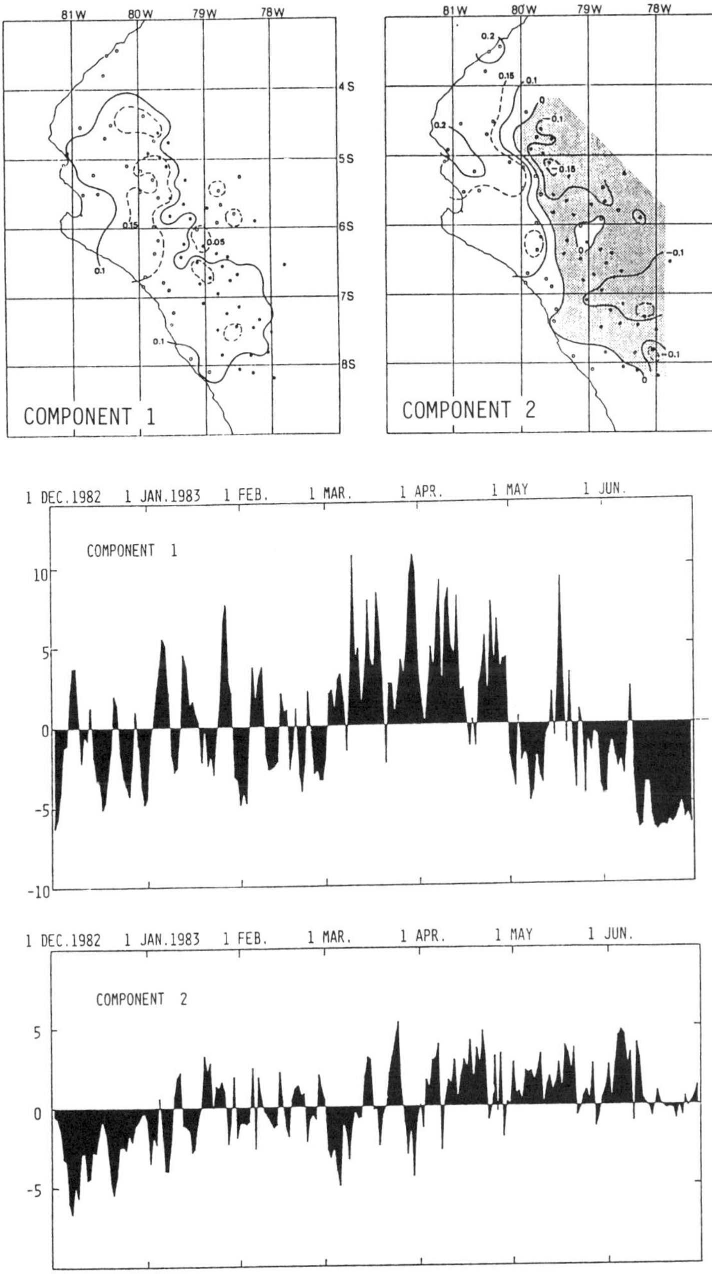

FIGURE 2. Eigenvector patterns (upper panels) and time series of scores (lower panels) for the principal components.

3.2 EFFECT OF VARIATIONS IN SST

A heavy rainfall along the northwestern coastal plain which appeared frequently from April to June is considered to have been affected by the high sea surface temperature during this period. Zuta et al.(1984) pointed out that the sea surface temperature at Paita (around 5°S-81°W) was near 30°C from April to mid-June that was approximately 2°C higher than that in the previous two months. It is apparent that convective activity along the northwestern coastal plain was intensified by the unusual high SST in this region. These facts suggest that there existed some relationships between the principal component 2 and the sea surface temperature variations near the northwestern coast.

REFERENCES

Horel, J.D. and A.G.Cornejo-Garrido, 1986: Convection along the Coast of Northern Peru during 1983:Spatial and Temporal Variation of Clouds and Rainfall, Monthly Weather Review, 114, 2091-2105.

Nogami,M.(Ed.),1987: Rainfall and River Discharge during 1983 El Nino Event and Geomorphological and Geological Significance of the Past El Nino Phenomena in the Northern Peru, Department of Geography,Tokyo Metropolitan University, Tokyo 158 Japan, 308 pp.

Zuta,S., M.Farfan and O.Moron, 1984: Caracteristicas Principales del Mar Frente a Peru durante la Evolucion del Evento El Nino 1982-1983, Revista de la Comision Permanente del Pacifico Sur, No.15, 141-178.

ESTIMATING CLIMATIC SCALE TROPICAL PRECIPITATION FROM SATELLITE OBSERVATIONS

Phillip A. Arkin
Climate Analysis Center
National Oceanic and Atmospheric Administration
Washington, DC 20233, USA

ABSTRACT

Measurements of large-scale precipitation are critical to the monitoring, modeling and prediction of the earth-atmosphere climate system. Satellite observations provide the only practical source of information from which such measurements may be derived over many parts of the globe, including in particular the tropical oceans. Analyses of data from the GARP Atlantic Tropical Experiment have shown that the time-mean fraction of large areas covered by cold cloud is highly correlated with accumulated rainfall over the same region. Following these findings, histograms of equivalent black body temperature for areas of 2.5° latitude/longitude were compiled from GOES imagery for each month during the period December 1981-November 1986 and used to estimate large-scale precipitation over the Americas and the eastern Pacific. The 5-year mean seasonal estimated precipitation contains features grossly consistent with published descriptions of climatological rainfall. The interannual variability associated with such climatic phenomena as the El Niño/Southern Oscillation phenomenon appears to be accurately depicted. The validation of these estimates is extremely difficult; comparisons with station observations over the United States during the warm season show good agreement in temporal variability but large errors in magnitude. These errors represent a satellite overestimate which appears to be related to distance from the coast. These results have led the World Climate Research Programme to institute a Global Precipitation Climatology Project to use estimates based on geostationary satellite data, along with conventional observations and passive microwave estimates, to produce monthly maps of near-global rainfall for use in the Tropical Ocean-Global Atmosphere Programme.

1. INTRODUCTION

The monitoring of variability in large-scale rainfall over the tropical oceans is a critical component of the mission of the Climate Analysis Center (CAC) of the U.S. National Oceanic and Atmospheric Administration. Such variability is a key aspect of many global climate phenomena such as the El Niño/Southern Oscillation (ENSO) (Rasmusson and Wallace, 1983).

Since conventional observations provide very little information on the distribution of large-scale rainfall and convection over the tropical oceans, the use of satellite imagery for this purpose has

A. DEEPAK Publishing
ISBN 0-937194-14-4

been attempted for many years. Arkin (1979) and Richards and Arkin (1981) compared the temporal variations in the fractional coverage of clouds with equivalent black body temperatures (EBBT) less than various thresholds with those of radar and rain gage-based estimates of rainfall during the GARP Atlantic Tropical Experiment. They found that correlations between time series of rainfall and cloudiness were strongly positive for areal averages on scales of 1.5° and 2.5° latitude/longitude and for temporal averaging scales ranging from 1-24 hours. Correlations were >0.8 for a range of thresholds, with highest values found for a threshold of 235 K.

These results led to an effort to collect data which could be used to calculate a rainfall estimate from the U.S. geostationary satellites (GOES). The data, along with some aspects of the data processing and quality control, are described in Arkin and Meisner (1987). In brief, the data consist of 16-class histograms of EBBT for each pixel in 2.5° latitude/longitude areas from 175°E-25°W and from 50°N-50°S. In this paper, results for the 5-year period December 1981-November 1986 will be discussed.

2. ANNUAL CYCLE

The difficulties in applying an algorithm developed from data taken during one short period of one year in a single relatively small area in such a general fashion are obvious. Nevertheless, we will present the results from such an attempt. Estimated rainfall is shown only equatorward of 30° latitude. The 5-year mean seasonal rainfall (Figs. 1-4) is somewhat distorted due to periods when data from only a single GOES was available. This is most noticeable during March-May (MAM) when the heavy rainfall associated with the 1982-83 ENSO warm episode was detected only east of 130°W.

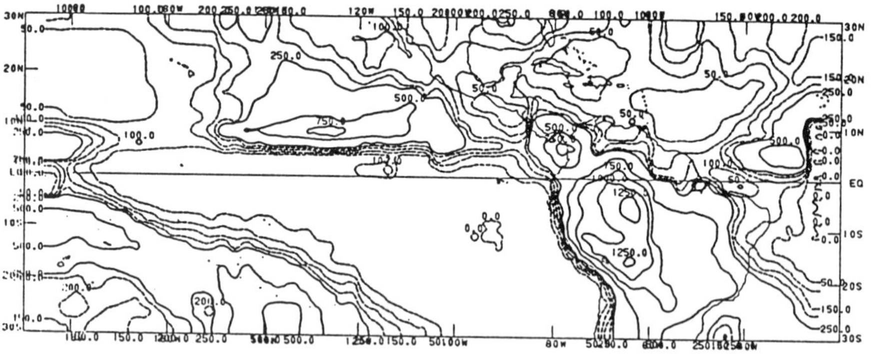

FIGURE 1. 5-year mean estimated rainfall for the December-February season. Contour interval 250 mm with dashed contours at 50 mm intervals below 250 mm.

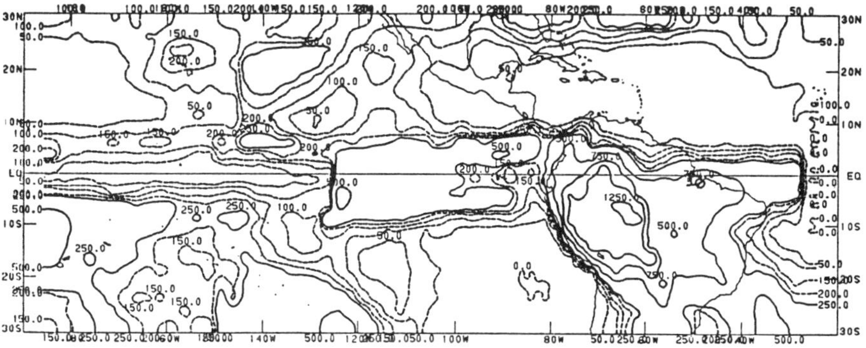

FIGURE 2. 5-year mean estimated rainfall for the March-May season. Contour interval 25Ø mm with dashed contours at 5Ø mm intervals below 25Ø mm.

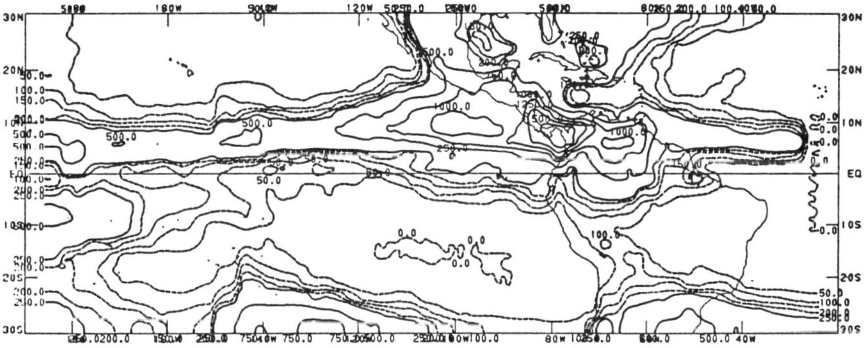

FIGURE 3. 5-year mean estimated rainfall for the June-August season. Contour interval 25Ø mm with dashed contours at 5Ø mm intervals below 25Ø mm.

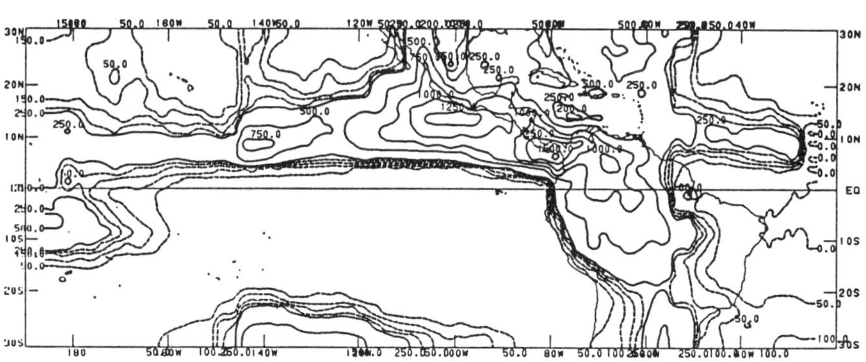

FIGURE 4. 5-year mean estimated rainfall for the September-November season. Contour interval 25Ø mm with dashed contours at 5Ø mm intervals below 25Ø mm.

The expected spatial features are shown clearly during December-February (DJF), June-August (JJA) and September-November (SON). A continuous band of rainfall >250 mm, presumably associated with the Intertropical Convergence Zone, is found in the Pacific north of the equator during JJA and SON, with an area south of the equator extending eastward from the date line. Dorman and Bourke (1979) present maps of long-term mean Pacific Ocean rainfall with similar features. Areas of low rainfall are found in the eastern Pacific, south of the equator, and farther to the west, north of the equator, in most seasons.

The validity of specific characteristics of these mean fields, let alone the amounts, is very difficult to verify. The most critical need for improving the estimation of climatic-scale variations in rainfall is for the availability of more direct measurements of actual rainfall against which the indirect estimates can be validated.

3. INTERANNUAL VARIABILITY

The 1982-83 ENSO warm episode has already been the subject of extensive descriptive and modeling studies. The extraordinary magnitude of this event was associated with rainfall anomalies of exceptional proportions in parts of South America (Horel and Cornejo-Garrido, 1986) and, by inference from proxy data, with equally strong anomalies in the equatorial eastern Pacific (Arkin et al., 1983).

Although the loss of the western GOES in December 1982 prevented the estimation of rainfall in the central Pacific during the event, departures from the 5-year mean for DJF 1982-83 and MAM 1983 for the area visible to GOES-EAST (Figs. 5 and 6) illustrate this situation. In MAM, throughout much of the eastern equatorial Pacific, very little rainfall was estimated other than during this season. One can also see from these figures that the very large rainfall anomalies observed at coastal stations in Peru and Ecuador are not well estimated by this technique.

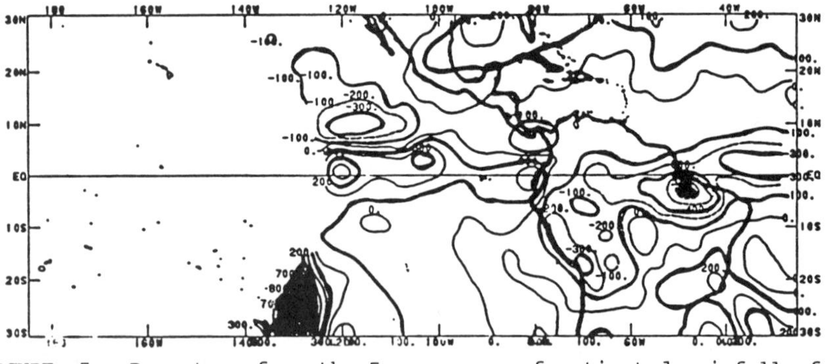

FIGURE 5. Departure from the 5-year mean of estimated rainfall for DJF 1982-83. Contour interval 100 mm with negative contours dashed; anomalies with absolute magnitude greater than 500 mm are shaded.

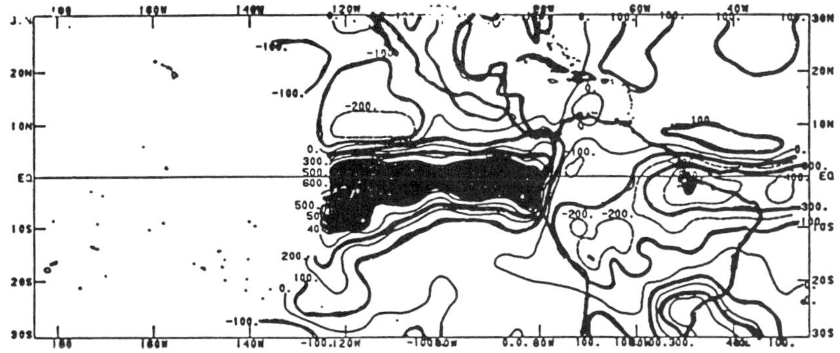

FIGURE 6. Departure from the 5-year mean of estimated rainfall for MAM 1983. Contour interval 100 mm with negative contours dashed; anomalies with absolute magnitude greater than 500 mm are shaded.

4. VERIFICATION

Quantitative verification of estimates, such as those presented here, is only possible in general for restricted areas and times. In fact, no good set of independent measurements of oceanic rainfall exists (to the author's knowledge) for the spatial and temporal domain of this study. In an attempt to use station observations to at least begin the validation process, Arkin and Meisner (1987) compared these estimates for the first three years to rainfall data from the U.S. Climate Divisions (CDs). An extension of their study to include all 5 years confirms their finding that the temporal variability in rainfall is well-defined by these estimates, but that a substantial, and apparently systematic, bias exists.

The temporal correlation between warm season (April-September) CD rainfall and estimated rainfall (Fig. 7) is positive and large across most of the eastern part of the U.S. However, in a quantitative sense, the satellite technique overestimates the rainfall by an amount which appears to be a function of the distance from the coast (Fig. 8). While these results cannot be considered directly applicable to the problem of validating oceanic estimates, they do suggest that a simple threshold estimate may be valid in regions other than those for which it was derived.

5. CONCLUSIONS

The application of a simple threshold scheme to the estimation

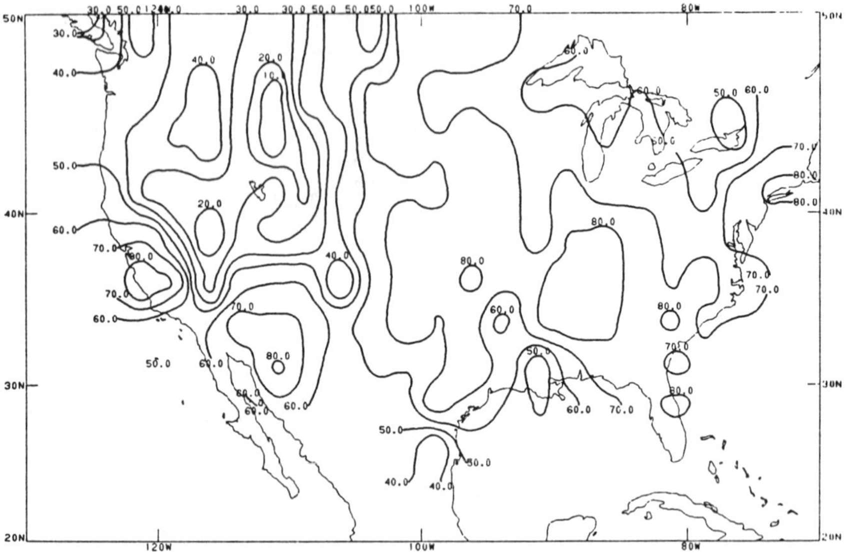

FIGURE 7. The temporal correlation (multiplied by 100) between monthly CD rainfall and estimated rainfall over the warm seasons (April-September) of 1982-1986. Contour interval is 10, with negative contours dashed.

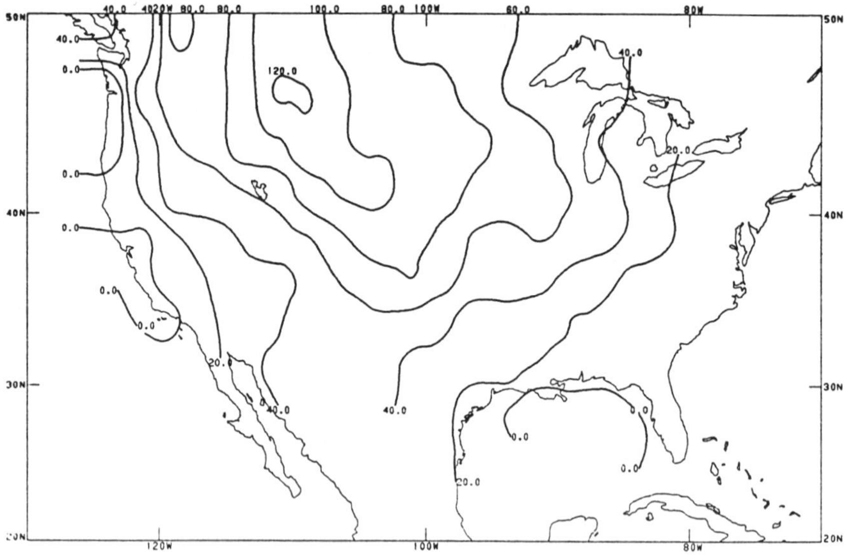

FIGURE 8. Monthly mean bias of estimated rainfall relative to CD rainfall for the warm seasons of 1982-1986. Contour interval is 20%.

of climatic scale precipitation over the tropical oceans has yielded sufficiently encouraging results to lead the World Climate Research Programme to sponsor a Global Precipitation Climatology Project (GPCP). The GPCP will use such a technique, supplemented by conventional and other satellite observations, to provide near-global rainfall data for the Tropical Ocean/Global Atmosphere (TOGA) Programme.

However, some serious flaws are also evident. This technique, as with any technique based on observations of clouds rather than actual rainfall, requires calibration and validation. Over the tropical oceans this is an extremely difficult problem. Conventional observations cannot, in general, supply the needed data over the oceans. One candidate, which is becoming increasingly viable, is the development of a dedicated satellite which will use multiple sensors, including radar, to measure not only the rain intensity but its vertical distribution. Such a satellite, in combination with other available observations, would go far toward solving the problem of oceanic rainfall measurements.

ACKNOWLEDGMENTS

The author wishes to thank Mrs. Kathleen Stevenson for typing the manuscript and John Kopman for helping to prepare the figures.

REFERENCES

Arkin, P. A., 1979: The relationship between fractional coverage of high cloud and rainfall accumulations during GATE over the B-scale array. Mon. Wea. Rev., 107, 1382-1387.

Arkin, P. A., J. D. Kopman and R. W. Reynolds, 1983: 1982-1983 El Niño/Southern Oscillation Quick Look Atlas. Climate Analysis Center, National Weather Service, Washington, DC, 95 pp.

Arkin, P. A. and B. N. Meisner, 1987: The relationship between large-scale convective rainfall and cold cloud over the Western Hemisphere during 1982-84. Mon. Wea. Rev., 115, 51-74.

Dorman, C., and R. H. Bourke, 1979: Precipitation over the Pacific Ocean, 30°S to 60°N. Mon. Wea. Rev., 107, 896-910.

Horel, J. D. and A. G. Cornejo-Garrido, 1986: Convection along the coast of northern Peru during 1983: spatial and temporal variation of clouds and rainfall. Mon. Wea. Rev., 114, 2091-2105.

Rasmusson, E. M. and J. M. Wallace, 1983: Meteorological aspects of the El Niño/Southern Oscillation. Science, 222, 1195-1202.

Richards, F. and P. Arkin, 1981: On the relationship between satellite-observed cloud cover and precipitation. Mon. Wea. Rev., 109, 1081-1093.

THE EVOLUTION OF MESOSCALE CONVECTIVE SYSTEMS: EVIDENCE FROM RADAR AND SATELLITE OBSERVATIONS

Edward J. Zipser
National Center for Atmospheric Research*
Boulder, Colorado 80307, USA

ABSTRACT

The structure of mesoscale convective systems (MCSs) at maturity has received more attention than their evolution. Most MCSs go through their life cycle in less than one day, so it is important to learn how they differ during their growth and decay phases. Examples of quantitative infrared satellite data combined with radar data are shown. Typically, the area of cloud tops colder than a specified threshold temperature grows and then decays without attaining a steady state. When radar data are used to determine the areas of convective and stratiform rain, the coldest cloud tops are observed to shift away from the convective and toward the stratiform region with time. A conceptual model of the evolution of MCSs is proposed which is consistent with the preliminary data.

1. INTRODUCTION

A large fraction of tropical rainfall and mid-latitude warm season rainfall comes from mesoscale convective systems. The structure of mature MCSs is now well-documented in the literature, and owes much to studies of GATE data (review by Houze and Hobbs, 1982). But there is increased acceptance of the proposition that many mid-latitude systems are similar (Ogura and Liou, 1980; numerous others).

The evolution of MCSs is difficult to deal with observationally because quantitative data are needed over the lifetime of a system, covering the area over which it moves and its environment. Nevertheless there has been progress. Leary and Houze (1979) proposed a radar schematic for the life cycle of MCSs based on GATE data, but it has validity for some mid-latitude systems (cf. Leary and Rappaport, 1987). The Oklahoma-Kansas PRE-STORM (1985) program was designed to study MCS evolution (Cunning, 1986), and this paper draws on that data set.

There is agreement that clouds and precipitation in MCSs follow a predictable pattern. First, deep convective clouds and convective rain increase, then the coexistence of convective and stratiform rain, then convective and finally stratiform rain dies out. While only the well-studied squall line system of 10-11 June 1985 is illustrated in this paper, several PRE-STORM case studies are under way which seem to have some important similarities.

This paper illustrates how the structure of the wind, cloud, and precipitation fields changes as the system ages, and how the relationship between radar and satellite data changes systematically, pointing to the possible need to account for these changes in the design of future satellite-based observing systems such as TRMM.

---

* The National Center for Atmospheric Research is sponsored by the National Science Foundation.

ISBN 0-937194-14-4

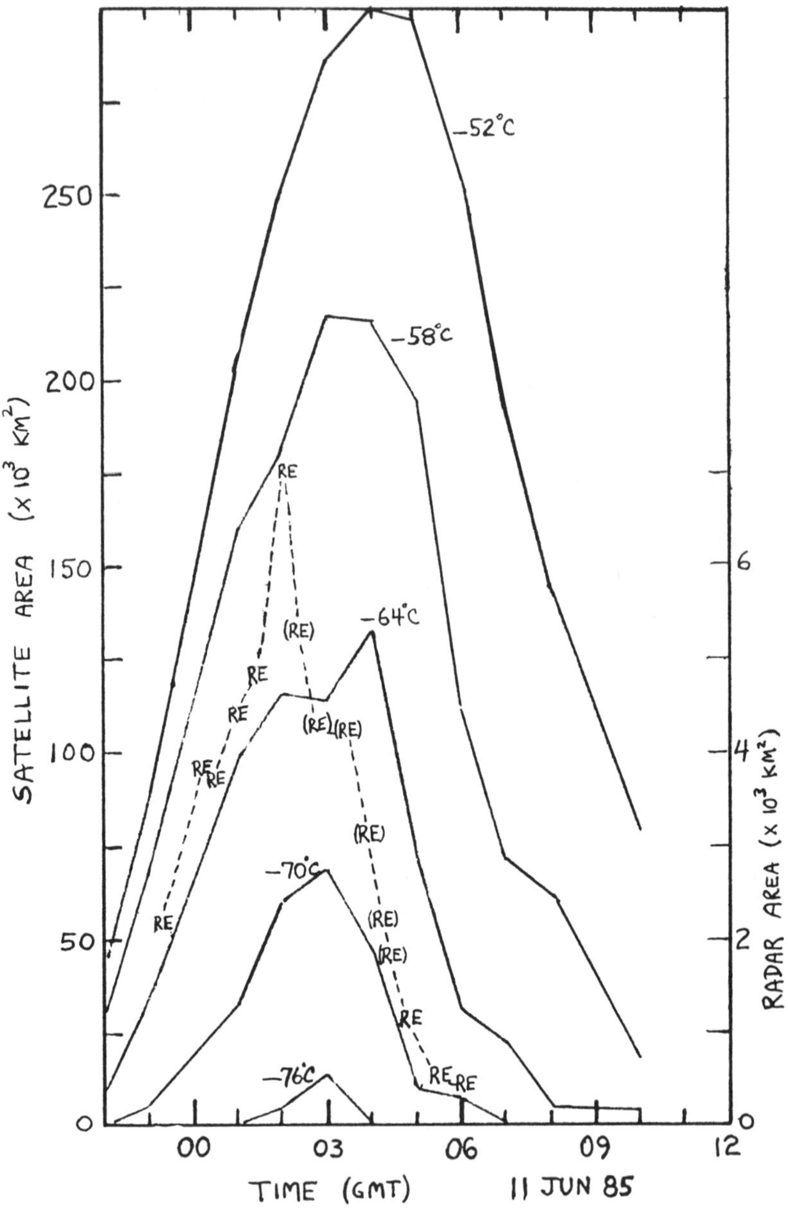

Figure 1. Time variation of area covered by infrared temperatues colder than indicated values in a large squall line system over Kansas and Oklahoma during PRE-STORM. Area covered by intense radar echoes (above 50 dBZ) indicated by data points labelled RE; values in parentheses partially estimated.

2. RELATIONSHIPS BETWEEN COLD CLOUDS AND RADAR ECHOES

Maddox et al. (1986) proposed the working hypothesis that long-lived canopies of cold clouds signal persistent mesoscale lifting in the upper half of the troposphere. They note, significantly, that observed relationships between satellite-observed cloud top temperatures and the convective structure beneath the anvil vary dramatically during the life cycle of a convective complex. For example, there is a much higher coverage of precipitation beneath a given isotherm later in its lifetime. Their implication is that one should not apply techniques of rain estimation that are static and ignore evolution. While not the issue here, it is significant that Maddox et al. find that severe weather is strongly associated with the organizing stage of a convective complex and is much less common when the area covered by cold clouds approaches its maximum.

The 10-11 June squall line system had a rather typical evolution and passed through the center of the PRE-STORM network. The area of cold tops grew rapidly, and the maximum area covered by intense radar echoes was attained at 0200 GMT while the cold cloud area was expanding rapidly (Fig. 1). Hail and tornado occurrences preceded this time. During several hours centered at 0400 GMT, when area covered by cold clouds was at maximum, there was heavy rainfall, and strong winds along the gust-front, but no other severe weather.

The radar echo distribution (Fig. 2) is typical of squall line systems, with trailing stratiform precipitation despite westerly shear in the environment. The cloud top temperature pattern is asymmetric (Fig. 3), with coldest tops north of the squall-line convection and becoming progressively less cold with increasing distance. This is not surprising in itself, although the fact that the coldest tops are *not*, in general, coincident with the strongest convective echoes might be. Only one hour later, at 0300 GMT, the centroid of coldest tops is displacd more than 50 km from the nearest convective echoes (Fig. 4). The observed winds at 200 mb are shown relative to the system, and they support a displacement distance close to 100 km along a relative trajectory.

The time evolution of the relationship between the convection and the cold tops is quite revealing. By 0400 GMT, just two hours after Figs. 2 and 3, most of the convective radar echoes are totally separated from the coldest tops and are some 200 km away from the centroid of the area colder than -70C along a relative trajectory (Fig. 5). (If one were to use the area enclosed by the -52C isotherm to represent the satellite-observed intensity of the system, as is done by Maddox for MCCs, 0400 would be the peak time for this system.)

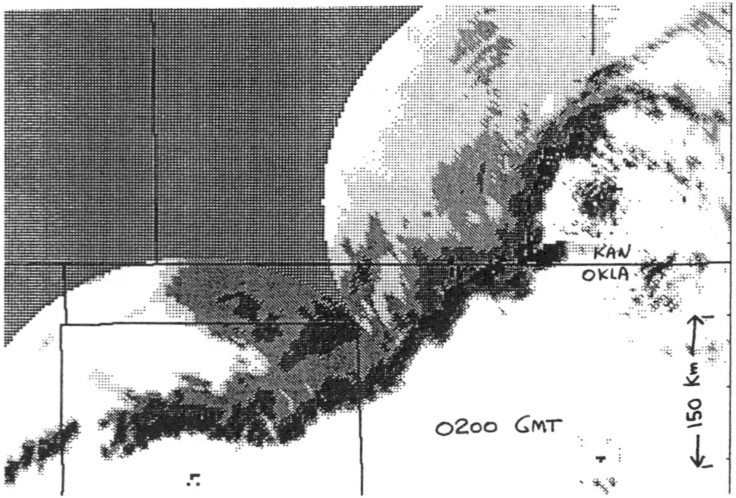

Figure 2. Low–level pattern of radar reflectivity at 0200 GMT 11 June 1985, corresponding to time of maximum coverage of intense echoes in Fig. 1. Contours are for 15, 25, 35, and 50 dBZ. The WSR-57 radars at Amarillo, Wichita, and Oklahoma City were used in the composite. Total dimensions of the figure are 700 x 700 km.

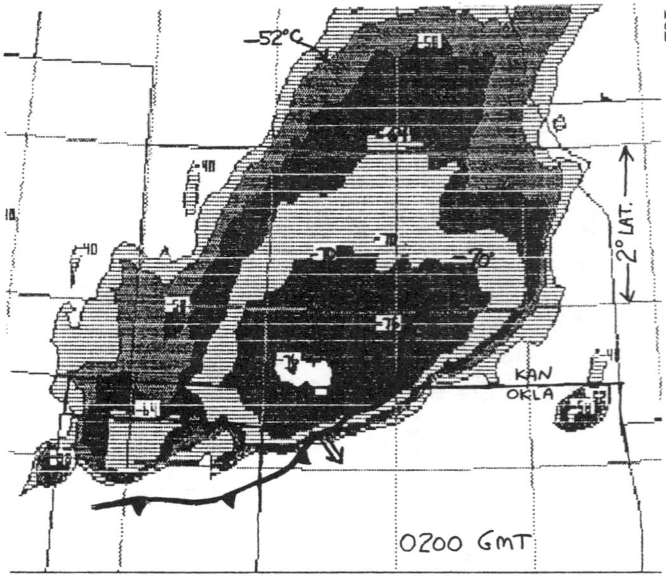

Figure 3. Infrared satellite picture at 0200 GMT 11 June 1985. Contours are at -40, -52, -58, -64, -70, and -76C. The leading edge of the convective line (see Fig. 2) is superimposed; the line is about 25 km wide.

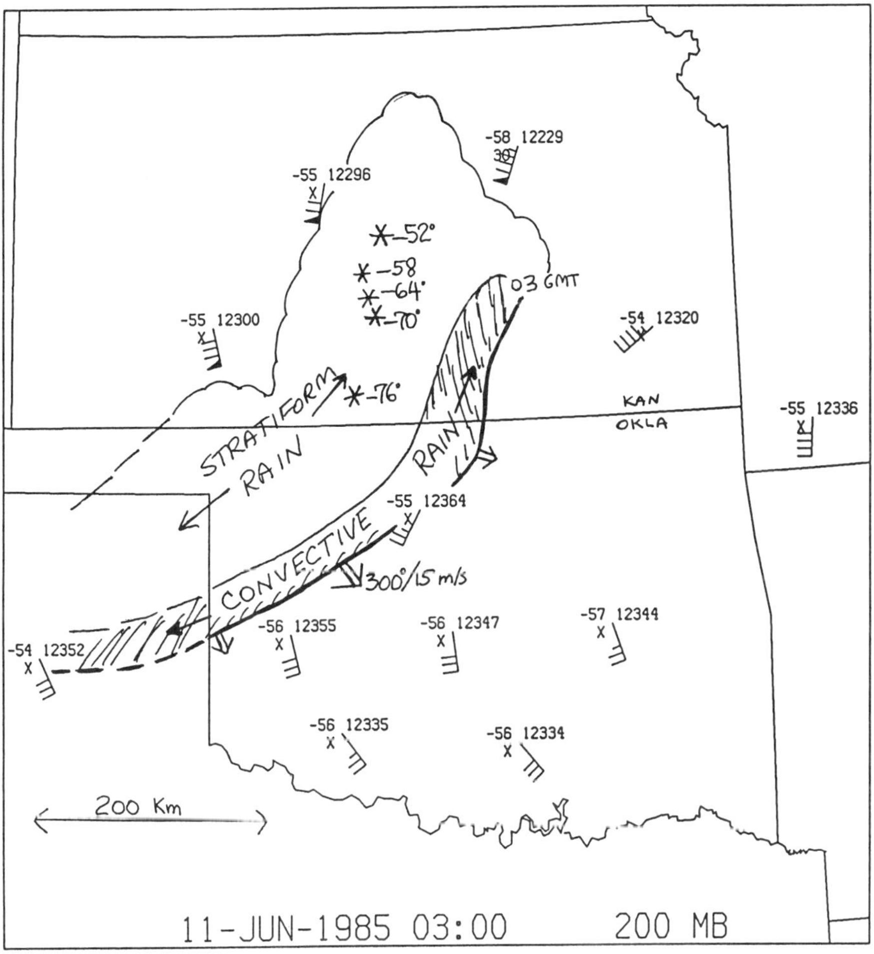

Figure 4. Wind plot relative to the squall line at 0300; full barbs are 5 m/s, flags 25 m/s. The convective band is superimposed, with the area of stratiform rain reaching the ground indicated. The centroid of areas with satellite infrared temperatures colder than specified values are located by asterisks (*).

3. DISCUSSION

For this and many other systems, the upper cloud shield is becoming displaced from the convection with time. Obviously, if we imagine the very early stage of the system when it consists of only deep cumulonimbus clouds, the coldest tops would be coincident with the convection. What is changing, and why?

Tentatively, two answers can be proposed, although we have much to learn by additional studies. First, the slope of the system, steep at first, becomes less steep with time. Mesoscale circulations develop over a period of several hours which increases the horizontal scale of the overall wind system. They include the front-to-rear airstream which transports air, momentum, and hydrometeors from the convective to the stratiform region, and the rear inflow jet just below the base of the stratiform cloud shield (Augustine and Zipser, 1987). The sloping interface between these two airstreams is proposed to become more horizontal with time (Fig. 6), which is consistent with the increasing displacement of the leading edge from the cold cloud centroid with time. A second explanation is that upward motion is observed within the stratiform cloud shield in the upper troposphere whch could have the effect of lowering the temperature of its satellite-observed top. Neither of these observations is new, or unique to this case, but the relationship between radar and satellite-observed fields has received less attention.

The implications of this "typical" behavior of MCSs for satellite sensing strategies should be examined. If rainfall from any single MCS is to be estimated accurately, these effects are of first-order importance. Convective rain accounts for some 60–70% of the total but occupies a small area, which may not have the coldest tops during the second half of the system's lifetime. The stratiform cloud shield occupies a huge area, may have very cold tops for many hours, and yet produce only modest rainfall, and none under a large fraction of its area. During future ground truth experiments it may be important to sample these cloud shields with instruments being tested for future spacecraft; aircraft overflights would seem to be an important component of such experiments.

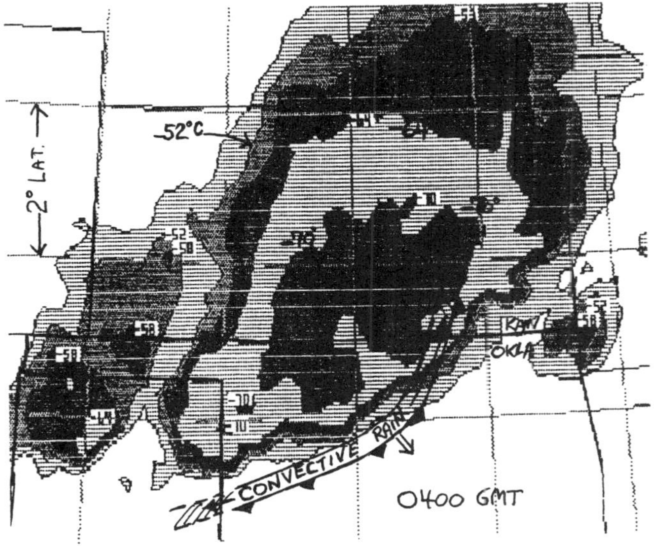

Figure 5. Infrared satellite picture at 0400 GMT 11 June 1985. Contours as in Fig. 3. Both the leading and trailing edge of the convective line are superimposed.

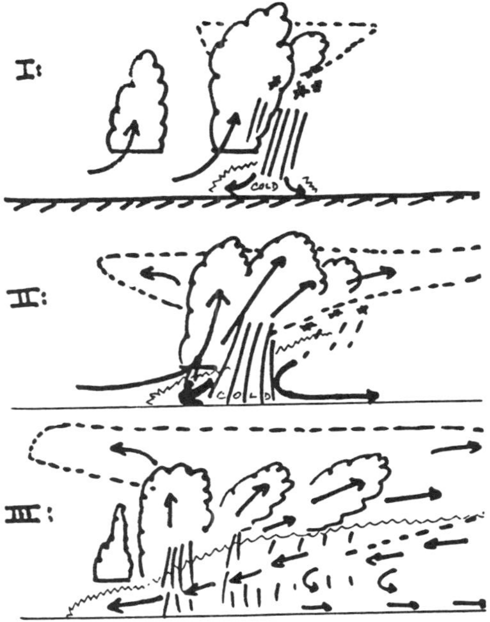

Figure 6. The schematic of the evolving structure of a mesoscale convective system in vertical cross-section, during initiation (I), organizational (II), and mature (III) stages. The interface separates upper tropospheric outflow from (often) mid-level inflow, and becomes more horizontal with time. See text.

ACKNOWLEDGMENTS

The cooperation of the Weather Research Program, ERL/NOAA, for data and use of facilities is gratefully acknowledged, especially the work done by John Augustine, David Jorgensen, José Meitin, and Irv Watson in producing analysis products. Discussions with Tom Matejka and Dhiren Sikdar, as well as access to their analyses, were of great help in preparing this paper.

REFERENCES

Augustine, J. A., and E. J. Zipser, 1987: The use of wind profilers in a mesoscale experiment. *Bull. Amer. Meteor. Soc.*, **68**, 4–17.

Cunning, J. B. 1986: The Oklahoma-Kansas Preliminary Regional Experiment for STORM-Central. *Bull. Amer. Meteor. Soc.*, **67**, 1478–1486.

Houze, R. A. Jr., and P. V. Hobbs, 1982: Organization and structure of precipitating cloud systems. *Adv. in Geophysics*, **24**, 225–315.

Leary, C. A., and R. A. Houze Jr., 1979: The structure and evolution of convection in a tropical cloud cluster. *J. Atmos. Sci.*, **36**, 553–574.

Leary, C.A., and E.N. Rappaport, 1987: The life cycle and internal structure of a mesoscale convective complex. *Mon. Wea. Rev.*, **115**, 1503–1527.

Maddox, R. A., K. W. Howard, D. L. Bartels and D. M. Rodgers, 1986: Mesoscale convective complexes in middle latitudes. In *Mesoscale Meteorology and Forecasting*, P. Ray, Ed., Amer. Meteor. Soc., 390–413.

Ogura, Y., and M. T. Liou, 1980: The structure of a mid-latitude squall line: a case study. *J. Atmos. Sci.*, **37**, 553–567.

METHODOLOGY FOR USING TRMM-CONTEMPORARY REMOTE DATA SETS IN SUPPORT OF TRMM OBJECTIVES

James C. Dodge
National Aeronautics and Space Administration
Washington, DC 20546, USA

ABSTRACT

The fundamental data set to be obtained from the TRMM program is a 3-year sequence of monthly-averaged, tropical, rainfall-rate probabilities averaged over geographical grid squares 500 km on a side. These gridded monthly averages are then to be used to characterize appropriate diabatic initialization in weather forecasting models.

During the time period of the 1990's, five other satellite systems offer the promise of collecting data sets which can be useful to TRMM in several ways. This paper suggests ways in which the observed data might be used and urges the collection, calibration, and processing of the full complement of available, relevant satellite observations into a geographically referenced data base which would be made available to TRMM researchers.

1. INTRODUCTION

1.1 OVERVIEW

The scientific planning for the Tropical Rainfall Measuring Mission (TRMM) program has resulted in a data analysis strategy, i.e., monthly averages over 500 x 500 km areas. This can be used to great advantage with other contemporary satellite system remote sensor data in order to develop a markedly improved understanding of oceanic hydrologic processes, as well as to create tropical data sets of a wide range of geophysical variables. These data sets may be used to establish global climatologies to study interannual fluctuations even in the traditionally data-sparse tropical regions. Continued satellite monitoring and accumulation into sufficiently large-area and long-time averages could well provide the information

A. DEEPAK Publishing
ISBN 0-937194-14-4

to detect early indications of global change through detection of regional deviations from the established tropical zone climatologies of key geophysical remote observables and through estimates of the magnitude of hydrologic processes based upon remote observables.

This paper is intended to summarize the potential wealth of geophysical information related to oceanic hydrologic processes which can be obtained if wide-area and long-term averages are accumulated from all available satellite sensors. Also, synergistic observations from sensors on the same observing platform can be used conjunctively to estimate processes which are not observable directly. Certain advantages can be obtained by using data from polar-orbiting and geostationary satellites together when they are observing the same regions. Geostationary satellites can fill in the details of the diurnal variations of certain geophysical parameters such as cloudiness and convective rainfall rate.

This capability for establishing global climatologies, and providing a database that will allow routine searching for deviations from regional and seasonal climatological means, can become a key element in the international effort to measure and understand global change. Recently, many world and national organizations have begun to advocate studying the bio-geochemical and physical environments of the Earth as a single interacting system, in order to begin to understand the complex interrelationships among processes which were studied traditionally within isolated scientific disciplines. Of vital concern in the tropical regions, where the maximum inputs of heat and moisture to the atmosphere from the oceans occur, is the hydrologic cycle. This paper demonstrates how TRMM observations analyzed in conjunction with other contemporary satellite sensors can yield insight into the major tropical hydrologic processes.

1.2 THE HYDROLOGIC CYCLE

The global hydrologic cycle, if studied in its entirety, and in terms of the continuity equation for water, would have to include; (1) Measurement of the water equivalent contents of all short- and long-term storage reservoirs, including the oceans, lakes, rivers, groundwater, aquifers, snow, ice, clouds, and atmospheric water vapor, and (2) The fluxes of water between the different reservoirs. Remote sensing can measure certain, but not all, aspects of this cycle.

Remote sensing is most effective in determining the areal extents of water, snow, and ice storage reservoirs with land as a background. Over the oceans, the principal hydrologic parameters are: (1)The remotely observable moisture amounts stored in gaseous and liquid forms in the atmosphere, which are measured as precipitable water and liquid water content, respectively; (2)evaporation from the ocean surface; and (3) precipitation back into the oceans. The horizontal transport within the atmosphere is not likely to be measurable on large-area and climatic scales because of the absence of the continuous remote sensing of both wind and water vapor as a function of height within the atmosphere. Rainfall into the ocean does change the salinity of the upper layer; however, its consequences are also difficult to assess from remote measurements. Thus, the most easily observed or estimated hydrologic parameters for the oceanic regions are evaporation, precipitation, and the amount of water being stored in the atmosphere. By limiting our study to the oceanic regions of the tropics, we can avoid the more difficult-to-measure aspects of transpiration from plants, remote determination of plant types and coverage, soil moisture, and the many aspects of sub-surface ice and snow characteristics which make their water equivalents difficult to estimate.

Evaporation from the oceans represents a water volume of 517,000 km^3 or a global sea level reduction of 1 m per year. Direct precipitation back into the oceans is about 78% of this total, with the remainder of the sea level replenishment coming from runoff from the land. The evaporation from the ocean surface occurs because turbulent transfer from the saturated layer of air next to the sea surface, to the sub-saturated layer above it, reduces the moisture content of the lowest layer to less than saturation and allows direct evaporation from the sea surface into the boundary layer of air. The whole process constitutes a net transfer of heat and moisture to the middle troposphere. One can calculate the amount of exaporation and the latent heart flux if we have simultaneous measurements of the sea surface temperature, the near-surface wind speed, and the surface-level humidity,(Liu, 1984). Current satellite remote sensing provides adequate sea surface temperatures and winds. However, the boundary-layer humidity only can be estimated from large-scale monthly averages of satellite-derived precipitable water,(Liu, and Niiler, 1984).

Future microwave sounders promise to make direct meaurements of humidity in the lower atmospheric layer. This should make possible short-term estimates of regional evaporation and latent heat flux from a satellite platform which simultaneously measures sea surface temperature, sea surface wind speed, and boundary layer humidity.

2. REGIONAL-SCALE CLIMATOLOGIES OF REMOTE OBSERVABLES

TRMM plans to be able to produce, as a data product, monthly means of precipitation rate over 500 x 500 km squares. Therefore, it is important to consider what other remote observables could also be collected into similar regional-scale averages to allow comparisons of physically related trends that might provide a better understanding of the bulk properties of the air-sea interaction.

The satellite observables useful for studying oceanic hydrologic processes and for developing regional climatologies for climatic change studies related to air-sea interactions are listed in Table 1. The assumptions made in developing the table are that (1) the Advanced Microwave Sounding Unit (AMSU) will be included on NOAA-I,J,and K operational polar orbiters, (2) the Special Sensor Microwave Imager (SSM/I) will be included on all subsequent DMSP satellites and that its data will be routinely available, and (3) current plans for the baseline Earth Observing System (Eos) polar platform series will be implemented. In preparing the table, choices have been made to ensure highest available resolution at the surface and long-term continuity for climatological averages. It should be noted that while any individual sun-synchronous polar orbiting satellite aliases the measurement of any variable which has substantial diurnal variation, there are to be two each of the NOAA, DMSP, and EOS polar platforms, each with two overpasses daily. For the NOAA and EOS satellite systems, the overpasses including both satellites are spaced moderately uniformly throughout the day. Thus, several forms of analysis can either remove or determine the diurnal variations for appropriate monthly averages over moderate-sized regions. For the two DMSP satellites, their morning overpasses are only about four hours apart, thus the four daily observations are not so evenly spaced throughout the day. The diurnal cycle should be sampled more evenly in the EOS era with morning and afternoon platforms each having a full complement of operational and research instruments.

Since some of the sensors in Table 1 may be unfamiliar to the reader, as many are still in the planning stage, a summary of the measurement bands is provided in Table 2. More information may be found in Butler (1984).

TABLE 1.

SOURCES OF SATELLITE OBSERVATIONS
SUITABLE FOR TROPICAL OCEAN
CLIMATOLOGIES OF HYDROLOGIC PROCESSES & INTERANNUAL VARIATIONS

OBSERVABLE	SOURCE		
	1987-1991	1991-1996	1996-2006
Sea Surface Temperature	NOAA:AVHRR/2	NOAA:AVHRR/3	EOS:MODIS-N EOS:AMSR EOS:AMRIR TRMM:AVHRR
Sea Surface Wind Speed	DMSP:SSM/1	DMSP:SSM/I ADEOS:SCATT	DMSP:SSM/I EOS:AMSR EOS:SCATT TRMM:SSM/I
Atmospheric Layer Temp.	NOAA:HIRS/2	NOAA:HIRS/3 NOAA:AMSU	EOS:AMRIR EOS:AMSU
Precipitable Water	NOAA:HIRS/2 DMSP:SSM/I	NOAA:HIRS/3 NOAA:AMSU DMSP:SSM/I	EOS:AMRIR EOS:AMSU EOS:AMSR TRMM:SSM/I DMSP:SSM/I
Humidity in Atmospheric Layers	NOAA:HIRS/2	NOAA:HIRS/3 NOAA:AMSU	EOS:AMRIR EOS:AMSU
Cloud Cover (low & high)	NOAA:AVHRR/2 GMS:VISSR	NOAA:AVHRR/3 GMS:VISSR NOAA:AMSU(ice)	EOS:AMRIR GMS:VISSR EOS:AMSU (ice) EOS:MODIS-N

Table 1 (Cont'd)

SOURCES OF SATELLITE OBSERVATIONS SUITABLE FOR TROPICAL OCEAN CLIMATOLOGIES OF HYDROLOGIC PROCESSES & INTERANNUAL VARIATIONS

OBSERVABLE	SOURCE		
	1987-1991	1991-1996	1996-2006
Precipitation Rate(Light rain:direct)		NOAA:AMSU	EOS:AMSU EOS:AMSR TRMM:Radar
Precipitation Rate(Heavy rain inferred from IR or HF microwave)	DMSP:SSM/I GMS:VISSR NOAA:AVHRR/2	DMSP:SSM/I GMS:VISSR NOAA:AMSU NOAA:AVHRR/3	DMSP:SSM/I GMS:VISSR EOS:AMSU EOS:AMRIR EOS:AMSR TRMM:SSM/I TRMM:AVHRR TRMM:Radar
Evaporation (direct)			EOS:AMRIR (for SST) EOS:AMRIR & AMSU (surf. humid.) EOS:SCATT (surf. wind)

Table 2

MEASUREMENT BANDS OF SENSORS
SUGGESTED FOR STUDYING
HYDROLOGIC PROCESSES
OVER THE OCEANS
(Several bands are usually within each range)

SENSOR	MEASUREMENT BANDS						
	VISIBLE	NEAR IR	THERMAL IR	5-37 6HZ	50-60 6HZ	85-90 6HZ	183 6HZ
AMRIR	X	X	X				
AMSR				X			
AMSU					X	X	X
AVHRR	X	X	X				
HIRS	X	X	X				
MODIS-N	X	X	X				
SCATT				X			
SSM/1				X		X	
TRMM Radar				X			
VISSR	X		X				

3. INTERPRETATION OF THE TABLES

Tables 1 and 2 show that if plans for NOAA-I, J, K, EOS, and TRRM are carried out with their current baseline sensors included, then useful mean monthly regional climatologies of sea surface temperature, sea surface wind speed, atmospheric layer humidity, and cloud cover could begin now with available operational sensors.

A substantial improvement will be achieved in the mid-1990's when NOAA-I and ADEOS are in operation. Then, better sea surface temperatures and winds will be possible. Sea surface evaporation can be calculated indirectly from the surface wind speed, sea surface temperature and surface humidity derived from monthly averages of precipitable water.

An even more impressive increase in the data available for air-sea interaction regional climatologies and studies of hydrologic processes will begin in approximately 1996 when EOS and TRMM will make possible better estimates of all the observables. At that time, direct observations of important hydrologic processes, such as evaporation and precipitation, can begin. Multiple sensors will allow improved measurement accuracies, while multiple satellite systems should remove diurnal biases by allowing measurements several times during each day.

It is well to note that as early as 1991, climatology of the direct observation of light rain can begin if the data from AMSU-A are collected and the low-altitude moisture-sensitive channels are interpreted.

Table 2 also shows an interesting fact. That is, with the exception of the 85.5 GHz channel on the SSM/I, the only remote sensor baselined with frequencies in excess of 37 GHz is the AMSU, which will become the all-weather operational sounder for NOAA in 1991. These higher microwave frequencies of AMSU-B also can detect the regions of high-altitude ice crystals above intense convection, and thus can be used to estimate the rate of that convective precipitation. Since the AMSU-B surface resolution will be an impressive 15 km, this sensor should be used to collect full-global data sets of morning and afternoon orbits. Used in this way, the AMSU-B will yield an excellent database from which to determine the geographic, seasonal, diurnal, and interannual variations of ice clouds and inferred convective precipitation. Ironically, the same instrument, using the lower sounder frequencies of AMSU-A near 50 GHz, should be able to detect the very low rainfall rates of thin stratiform precipitation through the absorption mechanism.

4. SUMMARY

It has been shown that, if one considers all of the potential satellite sensors on likely satellite systems over the next 20 years, good global climatologies can be developed of the key observables required to understand air-sea interactions over the open oceans. In the last half of the period, we can begin to collect climatologies of both oceanic evaporation and rainfall back into the oceans. The ability first to estimate and then to observe evaporation directly, using a combination of sensors on the same spacecraft, is of great importance. This is because evaporation is a continuous process over the vast oceans and its measurement is not bothered by the sampling problems of small size, brief duration and non-linear development evolutions related to precipitation with various sensors is still very important. Even while evaporation is occurring over vast ocean areas, and the associated latent heat to be eventuallly deposited can be calculated from the evaporation; the direct observation of precipitation, and the cloud systems which produce it in varying amounts, will permit researchers to know exactly where the latent heat is being released. This information can be used as inputs to numerical weather and climate models to determine higher latitude effects and possible interannual changes.

5. ACRONYMS

ADEOS	-	Advanced Earth Observing System (JAPAN)
AMRIR	-	Advanced Medium Resolution Imaging Radiometer
AMSR	-	Advanced Microwave Scanning Radiometer (Japan)
AMSU	-	Advanced Microwave Sounding Unit
AVHRR	-	Advanced Very High Resolution Radiometer
DMSP	-	Defense Military Satellite Program
EOS	-	Earth Observing System
GMS	-	Geostationary Meteorological Satellite (Japan)
HIRS	-	High-Resolution Infrared Radiometer Sounder
MODIS-N	-	Moderate Resolution Imaging Spectrometer-Nadir
NOAA	-	NOAA Operational Polar-Orbit in Satellites (TIROS-N Derivative)
SCATT	-	Scatterometer
SSM/I	-	Sensor System Microwave Imager
TRMM	-	Tropical Rainfall Measuring Mission
VISSR	-	Visible and Infrared Spin-Scan Radiometer

REFERENCES

Butler, D.M. 1984: Science and Mission Requirements Working Group of the Earth Observing System, NASA Technical Memorandum No. 86129, VOl. 1., 21-24.

Liu, W.T., 1984: Estimation of latent heat flux with SEASAT SMMR, a case study in North Atlantic. Large-Scale Oceangraphic Experiments and Satellites, C. Gautier and M. Fieux, Eds., Reidel, Dordrecht, 205-221

Liu, W.T., and P. P. Niiler, 1984: Determination of Mean Monthly Humidity in the Atmospheric Surface Layer Over Oceans from Satellite Data, J. Phys. Oceanogr., 14. 1451-1457.

BIBLIOGRAPHY

Murphy, R. et al, 1987: High-Resolution Multifrequency Microwave Radiometer, Instrument Panel Report, Vol. IIe, Earth Observing System Reports, NASA.

Thiele, O.W., 1987: On Requirements for a Satellite Mission to Measure Tropical Rainfall, NASA Reference Publication No. 1183.

THE USES OF THE HIGHLY REFLECTIVE CLOUD DATA SET IN LARGE-SCALE TROPICAL RAINFALL ESTIMATION

Oswaldo Garcia
NOAA Environmental Research Laboratories
Boulder, Colorado 80303, USA

ABSTRACT

This paper discusses the use of the Highly Reflective Cloud (HRC) data sets in large-scale tropical rainfall estimation over the global tropics. The method used is a linear regression linking monthly rainfall measured at several hundred stations to the number of days per month when HRC were recorded over the corresponding station locations.

1. INTRODUCTION

The HRC data set consists of a daily record over the global tropics of areas of organized convection which generally appear as bright clouds in satellite pictures. Areas of HRC are interpreted subjectively by an analyst using both visual rànge and infrared picture mosaics from the daytime passes of the NOAA polar-orbiting satellites. In the process of identifying the HRC areas, the analyst uses his knowledge of satellite meteorology to exclude from consideration non-convective areas that also appear bright in the satellite picture mosaics, such as the extensive cirrostratus associated with jetstream regions.

HRC data are archived at one degree of latitude by longitude resolution, between 25N and 25S, in a continuous daily record from January 1971 to the present. Monthly maps of the number of days with highly reflective clouds over the entire tropics are used in monitoring shifts in large-scale patterns of tropical convection. An Atlas of the first thirteen years of the data set can be obtained from the author (Garcia, 1985).

Assuming that a large proportion of tropical rainfall occurs in areas of organized convection , one can use HRC data to estimate large-scale patterns of monthly rainfall by means of a regression equation linking raingauge-measured monthly rainfall amounts at different stations to the number of days with HRC at corresponding locations. That was the approach taken by Kilonsky and Ramage (1976). They used rainfall data from forty Pacific atoll stations and the corresponding HRC values to estimate open-ocean rainfall. Here we present a more extensive analysis of the relationship between the number of days with HRC and monthly rainfall at several hundred stations throughout the global tropics for the period 1971 to 1985.

A. DEEPAK Publishing
ISBN 0-937194-14-4

2. DATA USED

2.1 HRC DATA

Monthly total values of HRC for the period 1971 to 1985 were used in this study. In order to effect a meaningful comparison with other data sets, we reduced the resolution of the HRC data from one to 2.5 degrees.

2.2 RAINFALL DATA

WMO monthly rainfall data for tropical stations were obtained for the period 1971 to 1985. The number of stations available in this data set varied significantly from month to month. Figure 1 shows the distribution of rainfall data for the month of July 1975. Note the relative absence of data from west central Africa and Indonesia, a situation encountered for a significant portion of this fifteen year period. Rainfall data for all stations were consolidated into grid squares measuring 2.5 degrees of latitude by longitude. Where more than one station fell within a grid square we averaged the monthly rainfall values for the stations. In the example shown in Figure 1, a total of 439 grid squares had rainfall data that could be used for comparison with satellite data, out of a total of 2,736 2.5 degree grid squares located between 25N and 25S.

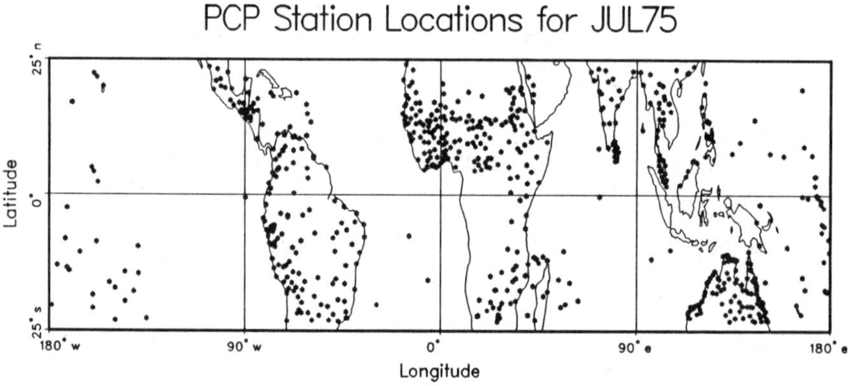

FIGURE 1. Location of stations for which rainfall data were available for July 1975.

3. ANALYSIS PROCEDURE

Following the method used by Kilonsky and Ramage (1976), we assumed a linear relationship between monthly rainfall measured at raingauge stations and the number of days with HRC recorded during the month at the corresponding grid squares. With the number of days with HRC as the independent variable and rainfall as the dependent variable, the theory of least squares was used to derive a regression equation and correlation coefficient for each month of the fifteen year period examined.

To study regional differences in the relationship between the two variables, the Pacific Ocean, Africa and South America were considered separately, in addition to the entire tropical belt. In brief, for each month in each of the four regions, the following items were produced:

- A locator map of each of the stations with rainfall data available for the month.
- A scattergram for the pairs of values for each grid square used for the month.
- The correlation coefficient.
- The slope and y-intercept of the linear regression equation linking the two variables.

4. RESULTS

4.1 THE SCATTERGRAMS

Figure 2 illustrates a sample scattergram for July 1975. The regression line and four parallel lines located 0.25 and 1 standard deviations on either side of it are shown. There is large scatter about the regression line, with some extreme outliers. For this month the correlation between the two variables was 0.66, while the regression line had a slope of 50.84 and a y-intercept of 44.19. A total of 439 grid squares contained rainfall data.

A study of the outliers in these scatter diagrams reveals interesting facts about the relationship between HRC and rainfall. The point labelled "A" in Figure 2 represents Mangalore, India. During July 1975 this station recorded 1,556 mm of rain, nearly three times the 536 mm predicted by the regression equation. Mangalore is on the west coast of India and is subject to copious orographic rainfall triggered by the action of the Ghat mountains on the moist airflow of the southwest monsoon. Although a significant amount of HRC occurred in the area during this month, apparently the bulk of rain was a result of relatively shallow orographic clouds. On the other hand, Salina Cruz, Mexico (point "B") on the Pacific coast, had 153 mm of rain, instead of the 418 mm predicted. In this case the station was in an area of strong HRC gradients, with much higher HRC values offshore. One can see that, by systematically eliminating from consideration stations where non-convective rainfall predominates, the accuracy of the regression equations could be significantly improved.

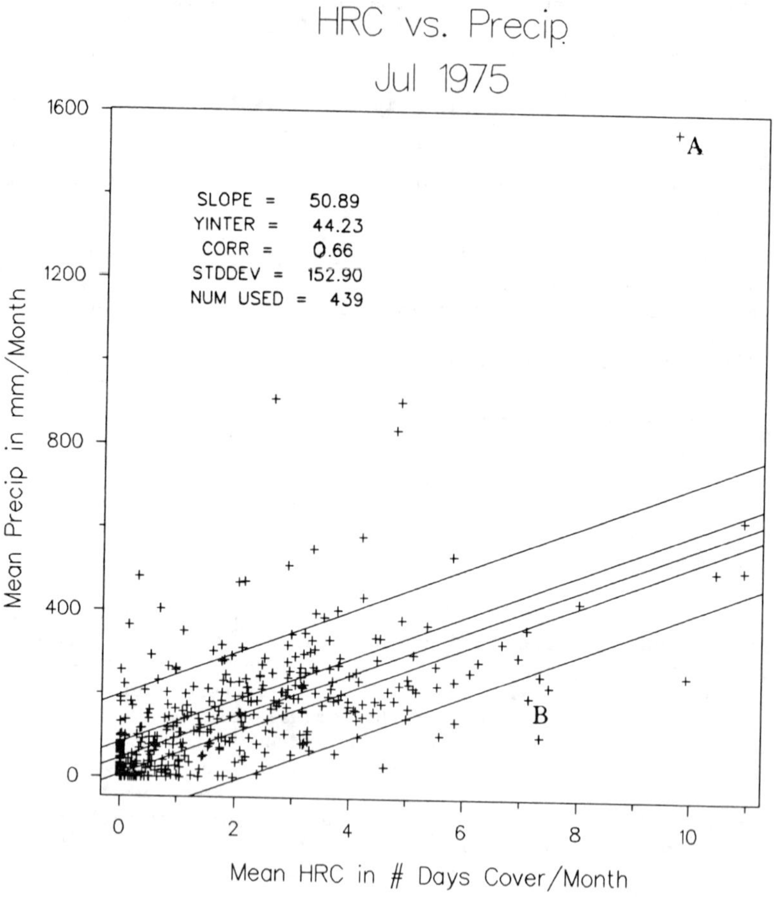

FIGURE 2. Scattergram of the number of days with HRC and rainfall recorded at the stations shown in Fig. 1

4.2 THE CORRELATION COEFFICIENTS

In order to illustrate the variability of correlation coefficients that can be found, the values for January and July for the fifteen years studied are shown on Table 1. There is remarkably little change in the worldwide correlation coefficients from year to year during the month of January, with the range extending from .60 to 0.71 and an average value of 0.66. By contrast, the July values ranged from 0.51 to 0.66 and averaged at 0.58. There are no large differences between the regions, with Africa having the largest mean correlation coefficient values during January and the Pacific during July. In general, the smaller the region the more likely it is to have large year-to-year changes in the correlation coefficient. Overall, there is surprisingly little difference in correlation coefficients obtained during the early years of the data set, when the daytime pass of the polar-orbiting satellites occured around 9AM local time, and the recent years with an afternoon satellite pass time.

TABLE 1. CORRELATION OF HRC WITH PRECIPITATION

Year	All Data		Pacific		Africa		S. America	
	Jan.	July	Jan.	July	Jan.	July	Jan.	July
1971	0.71	0.64	0.54	0.74	0.76	0.63	0.48	0.41
1972	0.67	0.55	0.61	0.54	0.76	0.55	0.65	0.52
1973	0.70	0.54	0.73	0.45	0.71	0.37	0.47	0.60
1974	0.64	0.57	0.69	0.76	0.65	0.56	0.22	0.56
1975	0.68	0.66	0.48	0.64	0.78	0.73	0.59	0.53
1976	0.63	0.65	0.77	0.68	0.67	0.49	0.46	0.68
1977	0.68	0.51	0.67	0.49	0.63	0.52	0.46	0.40
1978	0.66	0.57	0.52	0.67	0.71	0.50	0.51	0.21
1979	0.65	0.56	0.35	0.45	0.45	0.52	0.63	0.39
1980	0.66	0.59	0.56	0.53	0.78	0.41	0.56	0.76
1981	0.63	0.64	0.59	0.56	0.62	0.80	0.56	0.69
1982	0.65	0.59	0.71	0.48	0.83	0.47	0.62	0.60
1983	0.61	0.63	0.82	0.84	0.87	0.64	0.43	0.72
1984	0.60	0.51	0.51	....	0.74	0.59	0.38	0.64
1985	0.66	0.52	0.68	0.70	0.74	0.71	0.70	0.56
Mean	0.66	0.58	0.62	0.61	0.71	0.57	0.51	0.55

4.3 THE SLOPE OF THE REGRESSION EQUATIONS

Table 2 lists the slopes of the monthly regression equations relating HRC days to rainfall for January and July. The slopes can be interpreted as the increase in estimated rain (in mm) from a one day increase in HRC at any given location. The magnitude of the slope varies significantly from year to year. This is probably a consequence of the subjective interpretation of HRC data, which can result in large differences in the overall density of HRC around the tropics. Note that the correlation coefficients vary much less than the slopes of the regression equations, indicating a more consistent relationship between the monthly patterns of HRC and rainfall than their respective magnitudes.

When all tropical stations are considered, there is no large seasonal difference in the slopes between January (50.7) and July (47.3). Over Africa, more rainfall is observed with an additional HRC day in January (48.5) than in July (42.3). This may be due to the moister environment in Southern Africa during the height of the rainy season in January than the corresponding July situation over the semi-arid Sahel. The reverse situation occurs over South America, where in the mean, more rainfall occurs associated with HRC in July than in January. The mean relationship between HRC and rainfall appears more steady as a function of season over the Pacific.

TABLE 2. SLOPE OF THE REGRESSION EQUATIONS OF HRC VS RAINFALL

Year	All Data		Pacific		Africa		S. America	
	Jan.	July	Jan.	July	Jan.	July	Jan.	July
1971	53.9	37.3	53.2	55.8	49.8	30.5	37.7	26.3
1972	37.6	49.9	40.5	75.1	31.3	32.6	29.9	45.4
1973	61.3	64.9	68.6	39.3	63.0	72.3	33.9	62.8
1974	54.3	45.6	50.4	48.1	46.8	27.0	22.1	49.6
1975	43.4	50.9	32.5	37.4	38.8	41.4	25.2	37.0
1976	60.4	44.8	68.7	35.7	51.4	24.7	47.2	43.5
1977	56.8	57.2	58.6	38.0	40.8	49.8	31.5	40.3
1978	44.5	50.0	30.6	70.0	44.3	44.0	41.2	29.1
1979	74.6	54.0	44.5	38.0	58.5	48.1	52.2	89.4
1980	47.2	58.9	31.4	47.4	62.0	51.2	42.2	41.3
1981	62.5	45.7	45.4	33.1	38.1	50.1	43.7	46.5
1982	60.1	50.7	61.1	48.5	63.1	50.1	40.9	53.0
1983	40.5	43.2	67.0	53.3	39.2	40.4	31.4	45.7
1984	29.5	28.1	32.2	48.3	35.8	29.7	14.9	46.4
1985	34.0	28.1	42.2	43.1	23.6	43.3	36.1	38.7
Mean	50.7	47.3	45.8	47.4	48.5	42.3	36.0	46.3

5. CONCLUSIONS

When compared with other tropical rainfall estimation methods discussed in this volume, the HRC technique only provides crude estimates of monthly rainfall over the global tropics. This technique has the advantage of being a record of many years duration. The regional differences in the slopes of the regression equations indicate that care must be taken in the interpretation of HRC data for rainfall estimation.

REFERENCES

Garcia, O., 1985. Atlas of Highly Reflective Clouds over the Global Tropics. U.S. Dept. of Commerce, Wash. D.C.

Kilonsky, B. J. and C. S. Ramage, 1976. A technique for estimating tropical open-ocean rainfall from satellite observations, J. Appl. Meteor., 15, 972-975.

MICROWAVE OBSERVATIONS AND MODELING OF DEEP CONVECTION

Robert F. Adler[1], *N. Prasad*[2], *Wei-Kuo Tao*[2], *H.-Y. Yeh*[3],
Robert A. Mack[2], *and Joanne Simpson*[1]
[1]Laboratory for Atmospheres, NASA Goddard Space Flight Center
Greenbelt, Maryland 20771, USA
[2]General Science Corporation, Laurel, Maryland 20707, USA
[3]CAELUM Research Corporation, Silver Spring, Maryland 20906, USA

ABSTRACT

Aircraft microwave observations at 18, 37, 92, and 183 GHz over deep convection are presented and shown to compare favorably to radar- and cloud-model-based simulations. A three-dimensional, multi-cloud model is then used to simulate fields of microwave radiances which can be area-averaged to simulate satellite microwave observations.

1. INTRODUCTION

In order to understand the relations among convective cloud structure, rainfall, and remote passive microwave observations, an effort has been initiated to combine aircraft overflight radiometer observations, calculations from cloud dynamical/microphysical models, and radiative transfer models. Results should lead to simulations of past, present, and future satellite passive microwave observations with cloud model/radiative model procedures which have been checked against the aircraft observations. In this paper, some results from the above-described strategy will be given.

2. EXAMPLE OF AIRCRAFT MICROWAVE OBSERVATIONS OVER DEEP CONVECTION

During the summer of 1986, a series of flights by NASA's ER-2 aircraft (20-km operating altitude) was made as part of the Cooperative Huntsville Meteorological Experiment (COHMEX). The aircraft carried a number of remote-sensing instruments. They include (1) the Microwave Precipitation Radiometer (MPR), having dual-polarized channels at 18 and 37 GHz with an along-track view at 45° from nadir toward the front of the aircraft, and (2) the Advanced Microwave Moisture Sounder (AMMS), which scans across the aircraft track ±45° from nadir at 92 GHz and at 183±2, ±5, and ±9 GHz. The instrument scientist for both instruments is Thomas Wilheit of NASA Goddard Space Flight Center. On 29 June 1986, a pair of overpasses was made of an area of intense convection in southeastern Virginia. All of the convection cores were over land, although a portion of the anvil material extended off the coast.

A. DEEPAK Publishing
ISBN 0-937194-14-4

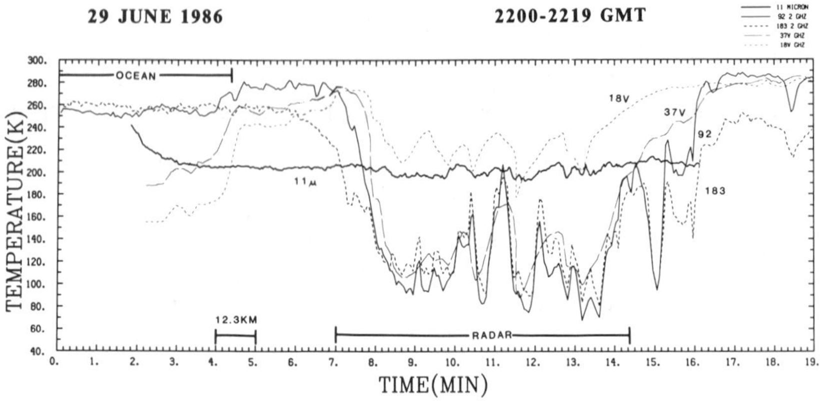

Fig. 1. Observed brightness temperature (T_B) at 11 microns and 18, 37, 92, and 183 GHz over a line of thunderstorms on 29 June 1986 from a high-altitude (20 km) aircraft.

Figure 1 shows a time profile along the aircraft track of four microwave frequencies (from 18 GHz [V] to 183 GHz) and one wavelength at the infrared window (11 micron) during one aircraft pass over a line of thunderstorm cores and associated cirrus anvil material. For the first 4 minutes, the aircraft was over the ocean, so that the 18, 37, and 90 GHz T_B values are relatively low due to the low surface emissivity of the ocean surface. As the aircraft crosses the coastline from ocean to land, the T_B's at 18, 37, and 92 GHz increase due to the higher land surface emissivity. During this period (0 to 6 minutes on the figure's time scale), the 11-micron IR observation is showing evidence of a thickening cirrus anvil downwind of the convective area. This is confirmed from aircraft photography and limited radar data, which indicates an anvil depth of ~5 km at time mark 7 minutes. The anvil top is at the tropopause and the IR T_B of ~205 K indicates an IR cloud emissivity of ~1.0. The anvil cloud is nearly transparent at the microwave frequencies of 18, 37, and 92 GHz, and even the 183 GHz T_B shows no significant decrease until minute 6. As the aircraft overflies the series of thunderstorm cores (minutes 8 to 16), small negative perturbations (10 to 15 K) in the 11-micron data are noted due to the overshooting convective tops. At the microwave frequencies, the intense thunderstorm cores produce T_B's as low as 70 K (at 92 GHz). Even at 18 GHz, a minimum of 180 K is reached. These low T_B's are caused by scattering by the large concentrations and relative large sizes of ice particles in the middle and upper parts of the thunderstorm cores.

The T_B's in the thunderstorm cores of Fig. 1 have been simulated using radar- and cloud-model-based hydrometeor profiles. The results are shown in Table 1 as compared to observed T_B's in an intense core. The radar profile is generated by converting the reflectivity to liquid or ice water content with an assumed transition from water to ice. The resulting profile is used as input into a radiative transfer

TABLE 1. COMPARISON OF AIRCRAFT-OBSERVED WITH MODEL-PRODUCED BRIGHTNESS TEMPERATURES FOR A STRONG CONVECTIVE TOWER

SOURCE OF HYDROMETEOR PROFILE	T_B FOR FREQUENCY (GHz) 18 H/V	37 H/V	92
AIRCRAFT OBSERVED	179/180 K	78/90 K	72 K
RADAR CONSTRUCTED	193/195 K	80/80 K	74 K
ONE-DIMENSIONAL (CHENG, 1981)	168/169 K	62/62 K	52 K
TWO-DIMENSIONAL TAO (RUTLEDGE AND HOBBS, 1984)	125/129 K	50/50 K	51 K
TWO-DIMENSIONAL TAO (LIN ET AL., 1983)	184/188 K	77/76 K	61 K

model described by Szejwach et al. (1986), which is a modification of that described by Wilheit et al. (1982). The results compare favorably with the observed values, except for the 18-GHz values being slightly too high. The cloud model-based profiles were generated by producing model clouds similar in cloud top height, radar structure, etc., as observed in the 29 June case. Hydrometeor profiles, in a typical core in a mature model storm, are then selected and input into the radiative transfer model. The results again compare favorably with the observations.

3. SIMULATED MICROWAVE OBSERVATIONS BASED ON A THREE-DIMENSIONAL CLOUD MODEL

To help understand the complex relations among the vertical and horizontal structures of raining clouds, rain rate at the surface, and the upwelling microwave radiance, the output from a three-dimensional, multi-cloud model (Tao and Soong, 1986) is used to calculate the upwelling radiance on a horizontal grid of 1.5-km spacing. By calculating area averages of the calculated T_B's and rain rates, simulations of satellite-observed T_B values and associated area-averaged rain rates can be made.

Figure 2 shows rain rate results from a three-dimensional Tao cloud model run with the Rutledge and Hobbs (1984) three-class ice parameterization, using the same environment observed for fast-moving squall lines in the GATE. So far, only the outlined 18 x 18 km box (12 x 12 grid points) has been selected for the purpose of linkage with the radiative model.

Calculations have been made at various frequencies between 10 and 183 GHz. Here, 18 GHz will be used as as example. Figure 3 shows the scatter plot of T_B versus rain rate for 18 GHz (horizontal polarization). Figure 3a shows the relationships of both Wu and Weinman (1984) and Wilheit et al. (1977) superimposed on the model

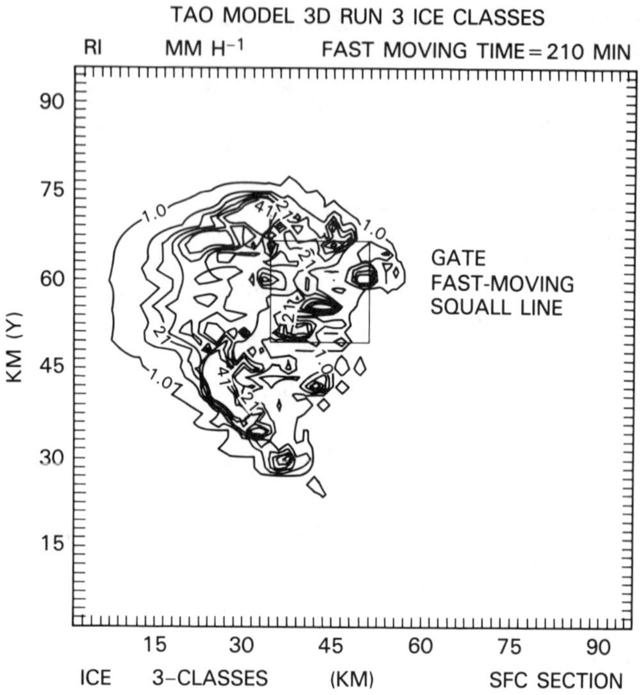

Fig. 2. Runs from a three-dimensional cumulus population model by Tao containing three classes of ice. The environmental conditions are those of a fast-moving oceanic tropical squall line in GATE. Surface section of rain rates in mm/h^{-1}. Horizontal resolution of the model is 1.5 x 1.5 km. The y axis is parallel to the squall line, while the x axis is perpendicular to it. Rain contours are for 10 mm/h^{-1}, beginning at 1 mm/h^{-1}, so that maximum values exceed 31 mm/h^{-1}. The small black-outlined box is the 18 x 18 km region where the radiative model was applied.

plot for all 144 profiles in the grid, while Fig. 3b shows results when the variables are averaged over 9 x 9 km boxes. Both the Wu and Weinman and Wilheit et al. curves now significantly underestimate the rain rate. This result brings home the impact of the beam-filling problem. The plane-parallel model curves are constructed for horizontally uniform hydrometeor profiles. When horizontal variability occurs within the averaging area, the rain rate derived by the plane-parallel model curve is biased, the more so the greater the departure from linearity. If, in Fig. 3b, a T_B of 235 K is selected, the Wu and Weinman curve underestimates the rain rate by about 43%. This bias is undoubtedly due to the fact that many of the footprints were not totally filled or had large horizontal variability of rain rate and cloud structure within the averaging area. Similar results,

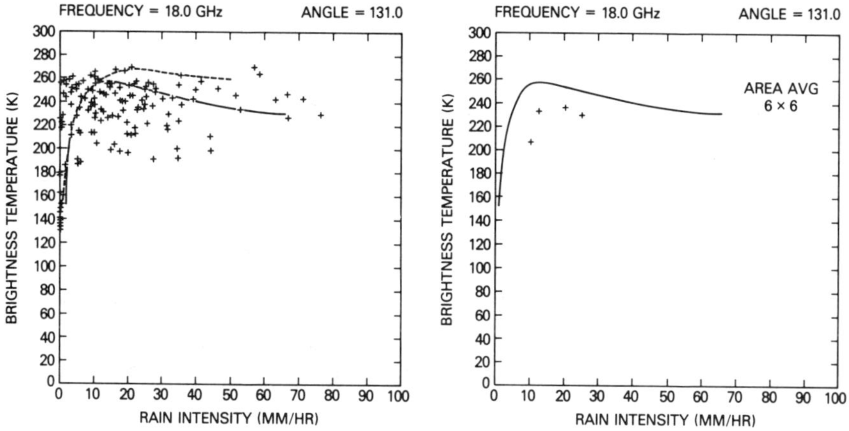

Fig. 3. Coupled radiative model with Tao model results for 18 GHz (horizontal polarization): (a) T_B plotted against rain rate showing all 144 profile points with Wu and Weinman (1984) curve the superimposed solid line and the Wilheit et al. (1977) curve dashed; (b) T_B plotted against rain rate for 18 GHz when both variables are averaged over 9 x 9 km squares, with the Wu and Weinman curve the superimposed solid line.

but with a smaller bias, were found at 10 GHz, due to the more linear relation at that frequency. Many more tests need to be made to establish these relationships and to really understand the effects of different hydrometeor structures and different averaging areas for varying types of convective systems.

From Figs. 3a and 3b, it is not clear what the "real" relation is between 18-GHz T_B (horizontal polarization) and surface rain rate. When the calculations at full model resolution (1.5 km in the horizontal direction) are examined (Fig. 3a), the large scatter suggests that the complexity of the convection, when viewed on a small spatial scale, renders the T_B-rain rate relations almost meaningless on that scale. When spatial averaging is carried out, the scatter is reduced and a curve shape like that of the theoretical curve emerges. However, the bias related to the non-linear response also appears. The simple, smooth theoretical curve does not exist in the real world at the small scale because of the complexity of the convection and does not exist in the spatially averaged data because of the non-linear response.

A similar analysis with calculations at 90 GHz (figures not shown here) indicates that at 1.5-km resolution there is a poor negative correlation between T_B and rain rate. However, when the results are averaged over 6 x 6 km boxes, the scatter decreases dramatically, with the resulting slope being somewhat steeper than determined by Wu and Weinman. The lower T_B's at the same rain rate (as compared to Wu and Weinman) are due to significantly more ice in the upper parts of the

Tao model clouds as compared to the schematic cloud profiles of Wu and Weinman. The nearly linear T_B-rain rate relation at 90 GHz indicates that "beam-filling" problems will be muted at this wavelength. The strong correlation of T_B (90 GHz) with area-averaged rain rate clearly indicates the usefulness of this frequency in estimating rain rate and agrees with findings by Hakkarinen and Adler (1987), who compared 92-GHz aircraft data with ground-based radar data.

4. CONCLUSIONS

This paper has shown the utility of combining aircraft microwave observations with results from cloud models coupled with radiative transfer models. One- and two-dimensional cloud models and ground-based radar data have served as the basis for simulating the observed T_B's at 18, 37, and 92 GHz over intense thunderstorm cores. A three-dimensional cloud model has been used to produce realistic T_B-rain relations and the effect of spatial averaging on these relations. Results show the complexity of the rain/cloud structure/T_B interrelations for active convection and are the first step in simulating satellite microwave observations starting from a realistic three-dimensional structure of cloud systems.

REFERENCES

Cheng, C. P., 1981: Numerical simulation of the dynamics, cloud microphysics and radar echo structure of tropical and mid-latitude convection. Ph.D. Dissertation, University of Washington, 217 pp.

Hakkarinen, I. M., and R. F. Adler, 1988: Observations of convective precipitation at 92 and 183 GHz: Aircraft results. Meteor. and Atmos. Phys. (accepted).

Lin, Y.-L., R. D. Farley, and H. D. Orville, 1983: Bulk parameterization of the snow field in a cloud model. J. Clim. Appl. Meteor., 22, 1065-1092.

Rutledge, S. A., and P. V. Hobbs, 1984: The mesoscale and microscale structure and organization of clouds and precipitation in midlatitude clouds. XII: A diagnostic modeling study of precipitation development in narrow cold frontal bands. J. Atmos. Sci., 41, 2949-2972.

Szejwach, G., R. F. Adler, I. Jobard, and R. A. Mack, 1986: A cloud model-radiative model combination for determining microwave T_B-rain rate relations. Preprints, 2nd Conference on Satellite Meteorology/Remote Sensing and Applications, Amer. Meteor. Soc., Williamsburg, VA, USA, 444-449.

Tao, W.-K., and S.-T. Soong, 1986: The study of the response of deep tropical clouds to mesoscale processes: Three-dimensional numerical experiments. J. Atmos. Sci., 43, 2653-2676.

Wilheit, T. T., A. T. C. Chang, M. S. V. Rao, E. B. Rodgers, and J. S. Theon, 1977: A satellite technique for quantitatively mapping rainfall rates over the oceans. J. Appl. Meteor., 16, 551-560.

Wilheit, T. T., J. L. King, E. B. Rodgers, R. A. Nieman, B. M. Krupp, A. S. Milman, J. S. Stratigos, and H. Siddalingaiah, 1982: Microwave radiometric observations near 19.35, 35, 92 and 183 GHz of precipitation in tropical storm Cora. J. Appl. Meteor., 21, 1137-1145.

Wu, R., and J. A. Weinman, 1984: Microwave radiances from precipitating clouds containing aspherical ice, combined phase, and liquid hydrometeors. J. Geophys. Res., 89, 7170-7178.

A TROPICAL RAIN SYSTEM OBSERVED BY THE NIMBUS-7 SCANNING MULTICHANNEL MICROWAVE RADIOMETER[1]

Roy W. Spencer
NASA Marshall Space Flight Center
Huntsville, Alabama 35812, USA

ABSTRACT

Measurements of a tropical rain system at 37, 18, and 10.7 GHz from the Nimbus-7 Scanning Multichannel Microwave Radiometer (SMMR) are presented. The effects of emission and scattering at these different frequencies are discussed in the context of their impact on passive microwave precipitation retrieval at high spatial resolution, as is proposed for TRMM.

1. INTRODUCTION

The Nimbus-7 Scanning Multichannel Microwave Radiometer (SMMR) has provided useful insights into, and confirmation of, the effects that precipitation (both liquid and frozen) have on radiation upwelling from land and ocean surfaces. These insights should be taken into account in our attempts to retrieve rain rates from any passive microwave instrumentation flown on TRMM. For rain rate retrieval to be at all possible, contrast between a rain system's brightness temperature (T_B) and the background is absolutely necessary. Thus, over the radiometrically cold oceans, brightness temperature increases due to absorption and re-emission is usually the desired effect. This restricts the frequency range to below about 20 GHz, where absorption is the dominant extinction mechanism for all but the largest hydrometeors (e.g. Wilheit et al., 1977). Over the radiometrically warm land background, ice-dominated volume scattering-induced T_B decreases are necessary, restricting the frequency range to above about 30 GHz (e.g. Spencer et al., 1983). Additionally, two approaches allow scattering measurements over the ocean to be made. Frequencies above the 60 GHz oxygen absorption complex maximize the scattering effect on T_B, while minimizing the cooling effect of the ocean background on T_B through increase water vapor absorption (e.g. at 92 GHz - Wilheit et al., 1982). Secondly, frequencies from about 30 GHz to just below the oxygen complex can be used if polarization information is available. This information allows separation between the T_B depressing effects of the radiometrically cold

---

[1] This work was supported by the NASA Mesoscale Atmospheric Processes Research Program

ISBN 0-937194-14-4

ocean surface (highly polarized) and volume scattering by precipitation (mostly unpolarized - see Spencer, 1986). Because of the relative insensitivity of frequencies below 20 GHz to scattering by ice, low-frequency oceanic emission-based retrievals are more directly related to how much precipitation arrives at the earth's surface than the ice-scattering-based retrievals. Therefore, the emission-based methods should be given priority for rain rate estimates over the ocean.

Here we will give examples of observed effects from a tropical oceanic rain system at the SMMR frequencies of 37, 18, and 10.7 GHz. It will be shown that multiple microwave channels not only allow us to isolate the most basic microphysical state of a rain cloud both above and below the freezing level, but will also provide complementary information allowing the most confidence in a retrieved rain rate.

2. MIXED PHASE MODEL CALCULATIONS

Before presenting the SMMR observations, it is helpful to review some radiative transfer calculations at the frequencies of interest. The Wu and Weinman (1984) model is used for this purpose. The model assumes liquid rain up to about 4 km, linearly mixed liquid and ice up to about 9 km, then all ice above 9 km. The vertical profiles of water content are the result of averaging the results of radar studies. The hydrometeor size distribution is assumed to be Marshall-Palmer, and T_B were calculated for a 50° look angle at 37, 18, 10.7, and 6.6 GHz (Fig. 1). This model is relatively simple. Its most questionable assumptions are a Marshall-Palmer size distribution for all of the ice, no cloud water, and some of the precipitation is allowed to exist in the liquid phase at very high levels. However, our SMMR case study will support the major features of the model.

It can be seen from Fig. 1 that at all but the lowest oceanic rain rates, the lower frequency T_B increases with increasing rain rate, whereas the higher frequency T_B decreases with rain rate. The 37-GHz channel saturates quickly at the approximate thermometric temperature of the rain, then its T_B decreases due to scattering by precipitation-size ice associated with the higher rain rates. As frequencies decrease, scattering by ice becomes a small effect, and the raindrops' absorption cross section decreases as well. These effects combine to produce a gradual flattening of the curves, offering lesser sensitivity to light rain, but greater sensitivity to emission by heavy rain.

The contribution of different rain rates to total annual rainfall at Miami, Florida is also shown in Fig. 1. Note that one-half of the total rainfall is made up of rates below 20 mm/h (this might be unrepresentatively high for all tropical rainfall).

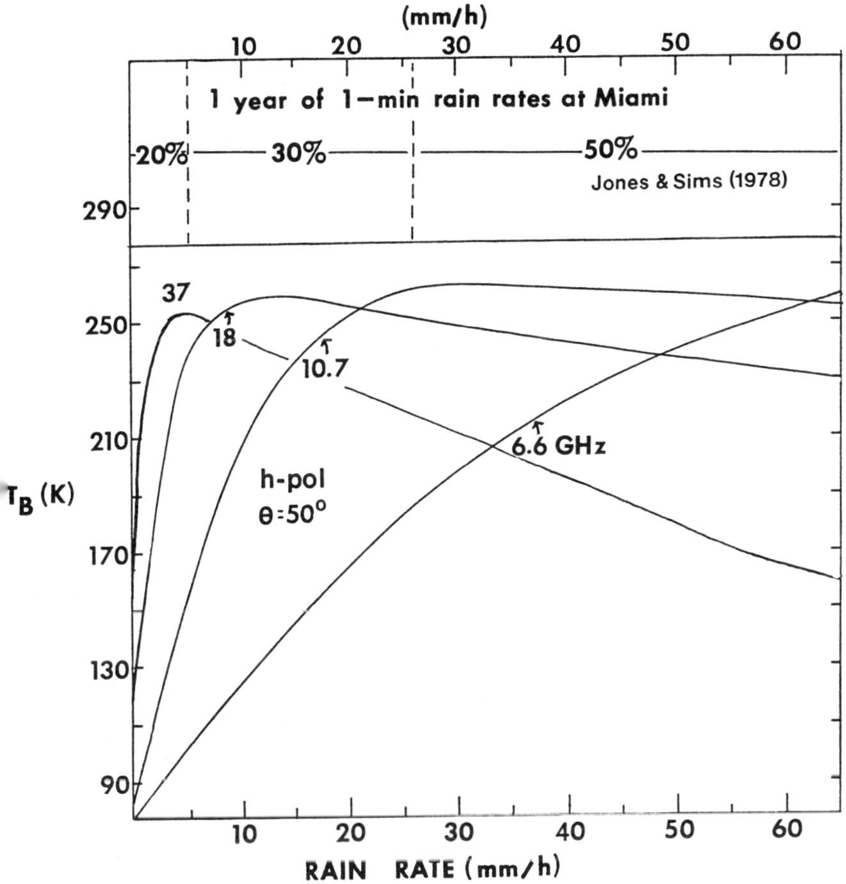

Fig. 1 Oceanic brightness temperature - rain rate relationships at 37, 18, 10.7, and 6.6 GHz for a vertical distribution of liquid and ice Marshall-Palmer hydrometeors (horizontal polarization, adapted from Wu and Weinman, 1984). Also shown is the contribution of different rain rates to the total rainfall during one year at Miami, Florida.

Next we will show an example of these effects from SMMR observations of a tropical oceanic thunderstorm.

3. AN INTENSE TROPICAL PACIFIC THUNDERSTORM

An unusually large thunderstorm was viewed by the Nimbus-7 SMMR and GOES-West visible and infrared sensors on January 22, 1982 over the central tropical Pacific (Fig. 2). The storm's large size makes it easier to study with the large SMMR footprints. The edge of the storm's cirrus anvil has been traced out with a black line in all channels presented. Minimum IR temperatures of 193 K located about 40 km from the western (left) edge of the anvil (Fig. 2a) correspond to convective turrets resulting from intense convective activity. These turrets are enhanced in the image with a folded-over black-to-white intensity scale.

Of particular importance to interpretation of the microwave channels responses are the many low clouds extending into the area surrounding the storm anvil in the corresponding GOES visible image (Fig. 2b).

These low clouds appear to be heavily influencing the response of the SMMR 37-GHz channel (Fig. 2c), which is reaching saturation with T_B above 250 K. Much of this low cloud area has IR-derived tops in the 3- to 4- km range. Assuming an average cloud water content of 0.75 g/kg, these T_B can be explained by cloud water alone. Alternatively, it is possible that these clouds contain a mixture of rain and cloud water, although with the low cloud tops their rain would be initiated without an ice phase. At the same location as the IR cold spot (convective turrets) a 40° to 50° relative T_B depression exists, with T_B as low as 206 K in the small dark "hole". Because it is coincident with the turrets and because it is only slightly polarized, this feature is most likely due to volume scattering by ice. Ice having a size being a significant fraction of the 37- GHz wavelength (8 mm) has virtually all of its extinction in the form of scattering. A large number of these hydrometeors will scatter back downward much of the radiation upwelling from the rain below. If the model in Fig. 1 were correct for this storm, 206 K would correspond to a rain rate of 28 mm/h (for the 30-km diameter footprint of the 37 GHz channel). If we spatially average to 50 km (for later comparisons) a 217 K T_B results, corresponding to 22 mm/h.

At 18 GHz (Fig. 2d) scattering is not unambiguously seen as it was at 37 GHz. The low clouds surrounding the storm have much less influence than at 37 GHz. Saturation of the T_B signal appears to not yet be reached in the storm core. Maximum T_B are 235 K, 115° warmer than the background, and just 25° short of the maximum pos-

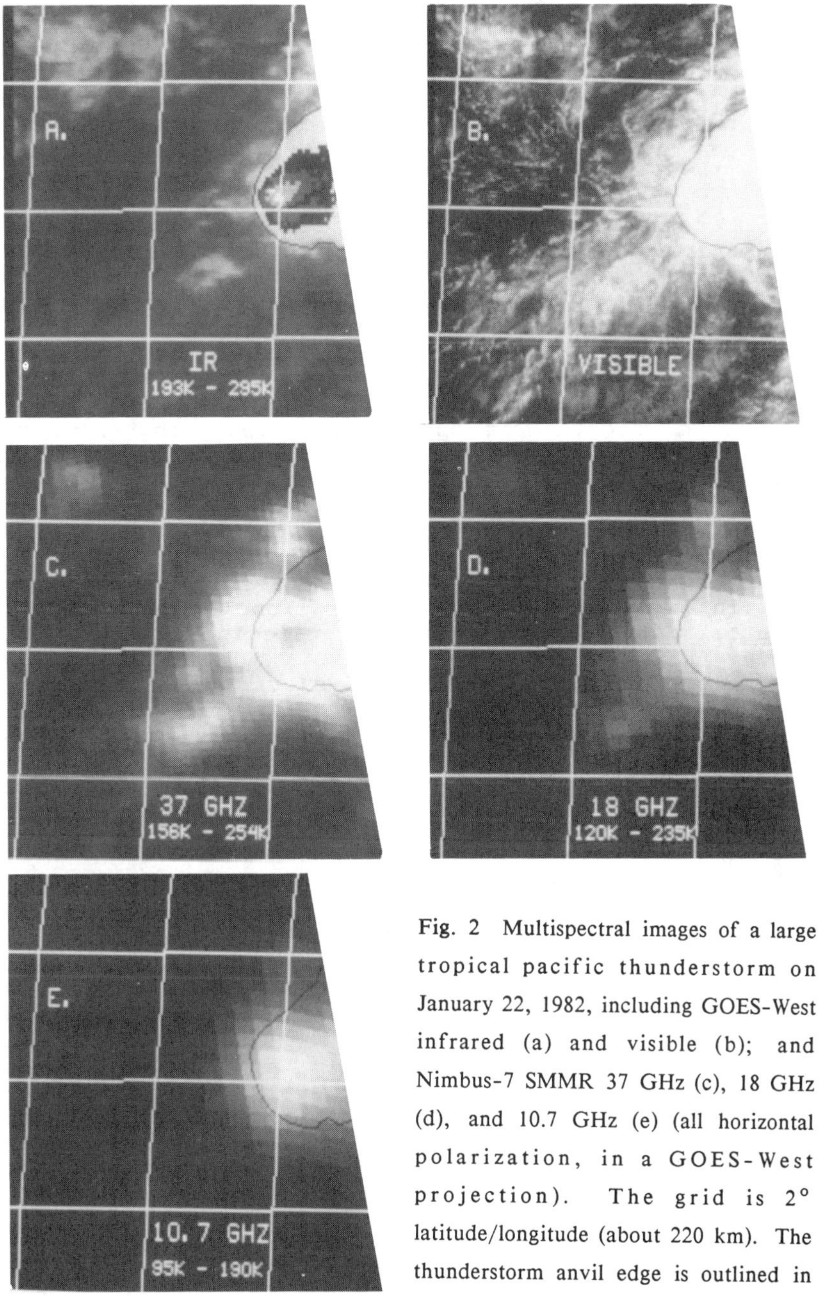

Fig. 2 Multispectral images of a large tropical pacific thunderstorm on January 22, 1982, including GOES-West infrared (a) and visible (b); and Nimbus-7 SMMR 37 GHz (c), 18 GHz (d), and 10.7 GHz (e) (all horizontal polarization, in a GOES-West projection). The grid is 2° latitude/longitude (about 220 km). The thunderstorm anvil edge is outlined in black, and the range of T_B for each image is indicated.

sible model T_B of about 260 K. Because of the longer wavelength at 18 GHz, it is not clear how much of the ocean is shining through the rain (left of the "saturation hump" in Fig. 1), or how much scattering is influencing the measurement (to the right of the saturation hump). Therefore, the T_B of 235 K could be interpreted as either 5 mm/h or 50 mm/h, assuming either a partially transparent rain layer, or an opaque rain and ice layer, respectively. It is likely that a mixture of both processes are operating, rendering an accurate rain rate estimate difficult. Smaller footprints (TRMM is attempting to achieve 8 km at 19 GHz) will mostly alleviate this ambiguity by providing fields of view small enough to assume much more uniform conditions within.

The 10.7-GHz response (Fig. 2e) has been deconvolved to the 18-GHz resolution. This was acheived by a simple along-track and cross-track weighted differencing procedure. The weights were optimized by analyzing the warming produced by the Galapagos Islands in the eastern tropical Pacific. Assuming that the Islands by themselves would have the same T_B at both 18 and 10.7-GHz, the 10.7-GHz spatial distribution of T_B response was sharpened until the 10.7-GHz warming matched that of the 18-GHz warming for several SMMR passes. The different footprint weights were then applied to the present 10.7-GHz storm data, resulting in the image in Fig. 2e. The maximum T_B in the storm core is 190 K, 95° above the 95 K background, but still over 70° short of the theoretically expected maximum of 264 K (Fig. 1). This would correspond to a model rain rate of about 10 mm/h. Support is seen for the general model trend of the lower frequencies providing less sensitivity to light precipitation, but greater sensitivity to heavier precipitation.

Thus we have estimates of 22 mm/h at 37 GHz, between 5 and 50 mm/h at 18 GHz, and 10 mm/h at 10.7 GHz. Which of these is closer to the truth? The model-assumed ice contents are a large source of uncertainty, making the detection of ice at 37 GHz not a good measure of surface rain rate. Being double valued, the 18 GHz curve presents too great a problem in data interpretation for such an intense storm. Because the 10.7-GHz channel is relatively unaffected by scattering, we would be inclined to trust its estimate first. However, because the T_B - rain rate relationship at 10.7 GHz is nonlinear, any contributions to the measured T_B from heavy rain (far on the right of the T_B curve) will lead to a low bias in the estimate of rain rate. Because this is the only single-valued curve, we therefore have confidence in assigning a lower bound at 10 mm/h. Taking into account the aforementioned effects, the most likely storm core rain rate on the 50-km scale is in excess of 10 mm/h, with a value approaching 20 mm/h not out of the question. Unfortunately, the 6.6-GHz observations from the SMMR have such large footprints (about 150 km), it is difficult to apply the quasi-linear 6.6-GHz curve in Fig. 1 to the SMMR observations with any hope of deriving a good rain estimate on the 50-km spatial scale.

4. CONCLUSIONS AND RECOMMENDATIONS

Firstly, 37-GHz measurements should not be used in the "emission" mode to relate brightness temperature increases to rain rate over the ocean because of (1) rapid saturation of the signal at very low rain rates or high cloud water contents, and (2) the moderate to strong scattering that often accompanies the emission when precipitation-size ice is present. Although the ambiguity between volume scattering by ice and rain transparency can be relieved with polarization information (Spencer, 1986), the resulting microphysical insight can not compete with a good low-frequency emission estimate for surface rain rate retrieval. Recent work at 92 GHz has shown that this frequency's high sensitivity to ice phase precipitation will provide valuable additional microphysical information on storm conditions above the freezing level.

The 18-GHz brightness temperatures will reach saturation more often than 10.7-GHz observations at high spatial resolution. Thus, the sensitivity to heavy rain afforded by a 6 or 10-GHz channel would make a nice complement to the higher resolution provided by a frequency near 18 GHz. Even though the 6.6-GHz frequency provides only 5° contrast per 1 mm/h in rain rate, this is still much more than has been successfully utilized at 37 GHz over land (about 1° per 1 mm/h). Therefore, if a TRMM orbit near 300 km is achieved, an antenna size at or above 0.8 m could provide 50-km spatial resolution at 6.6 GHz, which would be very useful for providing a constraint on the retrievals. A full complement of four or five frequencies from 6 to 90 GHz is therefore preferred for tropical rain rate retrievals and cloud physics surveys.

REFERENCES

Jones, D.M.A., and A.L. Sims, 1978: Climatology of instantaneous rainfall rates. J. Appl. Meteor., 17, 1135-1140.

Spencer, R.W., W.S. Olson, W. Rongzhang, D.W. Martin, J.A. Weinman, and D.A. Santek, 1983: Heavy thunderstorms observed over land by the Nimbus 7 Scanning Multichannel Microwave Radiometer. J. Climate Appl. Meteor., 22, 1041-1046.

Spencer, R.W., 1986: A satellite passive 37 GHz scattering based method for measuring oceanic rain rates. J. Climate Appl. Meteor., 25, 754-766.

Wilheit, T.T., A.T.C. Chang, M.S.V. Rao, E.B. Rodgers, and J.S. Theon, 1977: A satellite technique for quantitatively mapping rainfall rates over the oceans. J. Appl. Meteor., 16, 551-560.

Wilheit, T.T., A.T.C. Chang, J.L. King, E.B. Rodgers, R.A. Nieman, B.M. Krupp, A.S. Milman, J.S. Stratigos, and H. Siddalingaiah, 1982: Microwave radiometric observations near 19.35, 92, and 183 GHz of precipitation in Tropical Storm Cora. J. Appl. Meteor., 21, 1137-1145.

Wu, R., and J.A. Weinman, 1984: Microwave radiances from precipitating clouds containing aspherical ice, combined phase, and liquid hydrometeors. J. Geophys. Res., 89, 7170-7178.

A STATISTICAL-PHYSICAL INTERPRETATION OF ESMR-5 BRIGHTNESS TEMPERATURES OVER THE GATE AREA

David A. Short
NASA Goddard Space Flight Center
Greenbelt, Maryland 20771, USA

ABSTRACT

Brightness temperatures (T) from the Nimbus-5 Electrically Scanning Microwave Radiometer, reprocessed to correct location errors and identify calibration problems, have been re-examined along with radar estimates of rainrate (R) from Phase I of the GARP Atlantic Tropical Experiment. An idealized statistical model of the R to T transformation, incorporating field-of-view inhomogeneities in R, has been developed. The model clearly demonstrates that R, when inferred directly from T, will under-estimate R(true) by a factor which depends on the mean and variance of the R distribution (for GATE rainrate parameters the under-estimation factor is about 0.50). The model also suggests that moments of T may be useful for inferring parameters of the underlying rainrate distribution. Coincident ESMR-5 and GATE radar rainrate observations are being used to develop and test this method.

1. INTRODUCTION

Rainrate remote sensing over the oceans with a space-borne single channel microwave radiometer is faced with the fundamental problems of a non-linear rainrate/brightness temperature (R/T) response and field-of-view (fov) inhomogeneities in R. These factors combine to produce large random errors in satellite-inferred rainrates. The errors are also biased, systematically under-estimating the true rainrate (Smith and Kidder, 1978). Accurate rainrate estimates could be obtained for averages over a large collection of observations if the appropriate bias correction factor was known. However, conditions under which the bias depends on the rainrate itself are easily demonstrated with idealized mathematical models. This may be the general case in nature. It is the purpose of this paper to introduce a model to address the fundamental problems outlined above and to examine alternative approaches to rainfall retrieval suggested by the model.

A. DEEPAK Publishing
ISBN 0-937194-14-4

2. BASIC MODEL

Microwave radiative transfer theory and calculations suggest an R/T relationship of the form

$$T = A - B \exp(-C R) \tag{1}$$

near 20 GHz (Olsen et al., 1978; Atlas and Ulbrich, 1974). We have adopted this simple form in order to examine consequences of the non-linearity.

The brightness temperature resulting from an integration of Eq. (1) over a gamma distribution of rainrates is directly calculable as

$$E[T] = \int_0^\infty [A - B \exp(-C R)] \frac{R^{\alpha - 1}}{\Gamma(\alpha)\beta^\alpha} \exp(-R/\beta) dR$$

$$= A - B(1 + \beta C)^{-\alpha} \tag{2}$$

because the exponential form of Eq. (1) gives a form of the gamma moment generating function for Eq. (2). The gamma distribution also allows representation of a "long tail" of high rainrates, an important feature of rainrate distributions We refer to this formulation as the "ensemble field-of-view" model because it has been assumed that the entire distribution of rainrates exists within the fov.

The rainrate inferred from the brightness temperature of this "ensemble field-of-view" is given as

$$\tilde{R}\{E[T]\} = \frac{\alpha}{C} \log_e(1 + \beta C) \tag{3}$$

by equating (1) and (2) and solving for R. Thus, the difference between the true rainrate and the inferred rainrate, expressed as a fraction of the true rainrate, is given by

$$Bias = \frac{[R(true) - \tilde{R}\{E[T]\}]}{R(true)} = \frac{\alpha\beta - \frac{\alpha}{C}\log_e(1 + \beta C)}{\alpha\beta}$$

$$= 1 - \frac{1}{\beta C}\log_e(1 + \beta C) \tag{4}$$

For this idealized model the bias is seen to depend on the rain column height (through C) and the scale parameter of the rainrate distribution (beta). Figure 1 shows the bias as a function of beta for rain column heights of 5 and 3 km. A bias of about 0.50 is indicated for GATE rainrate conditions (10 < beta < 15 and a 5-km rain column height).

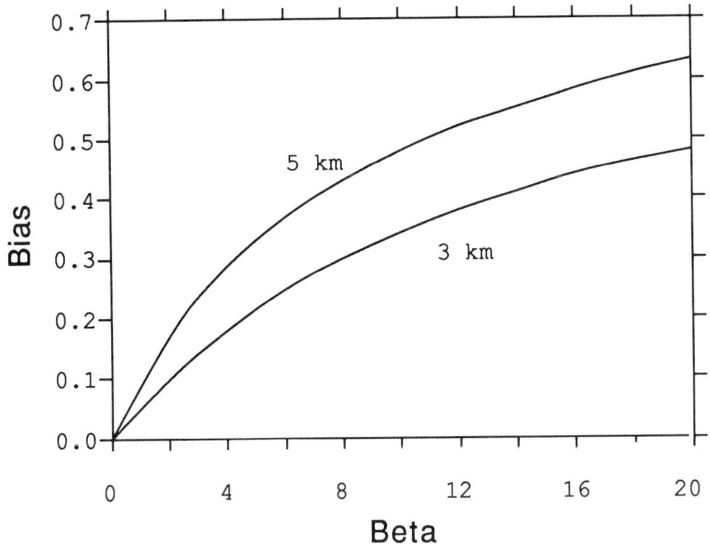

Figure 1. Bias versus the scale parameter beta.

The basic model can be extended to include a portion of the fov having zero rainrate. The bias then increases to approximately 60% for GATE rainrate parameters. While this formulation gives us some insight into the bias problem a physically more realistic model is required to interpret the 25-km resolution ESMR-5 observations.

2.1 N ELEMENT FOV MODEL

We first consider a two-element fov with random samples of rainrate from identical, independent gamma distributions in each element. T for this model can be written as

$$T = \frac{1}{2}\left[A - B\,\exp\left(-C\,R_1\right)^{x}(S)^{1-x}\right] + \frac{1}{2}\left[A - B\,\exp\left(-C\,R_2\right)^{y}(S)^{1-y}\right] + noise \tag{5}$$

where x and y are Bernoulli random variables equal to one with probability P. This formulation includes the mixed distribution properties of rainrate. P represents the probability of rain. S is the brightness temperature in the absence of rain. The noise is normally distributed with mean zero. It is independent of rainrate and rain occurrence. The mean and variance of T for this model are

$$E[T] = (1-P)\,S + P\left[A - B(1+\beta C)^{-\alpha}\right] \tag{6}$$

and

$$VAR\,(T) = \tfrac{1}{2}\left[(1-P)S^2 + P\left(A^2 + B^2 Z_2 - 2ABZ_1\right) - (E[T])^2\right] + VAR\,(noise) \tag{7}$$

where

$$Z_n = (1 + n\,\beta C)^{-\alpha} \tag{8}$$

We can now use the method of moments to find parameters of the gamma distributed rainrate. Figure 2 shows contours of $\alpha\beta$ (the average rainrate when raining) as a function of the mean and standard deviation of the brightness temperature.

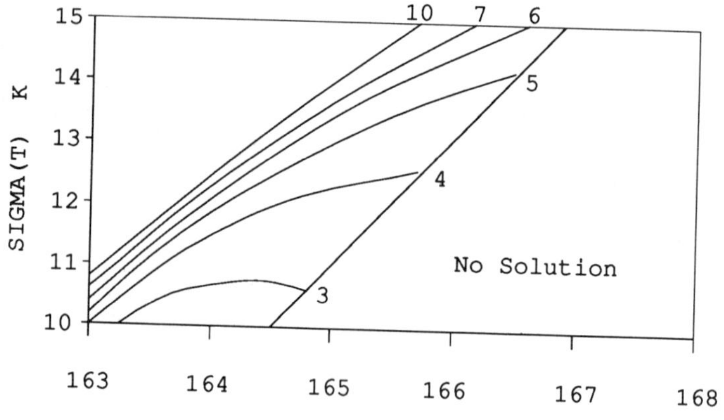

Figure 2. Contours of mean rainrate (when raining) versus the standard deviation and mean of the brightness temperature for Eqs. (6 and 7) with P = 0.10.

The constants of Eqs. (6 and 7) used for Fig. 2 were A = 270 K, B = 100 K, C = 0.16, VAR (noise) = 9 K^2 and S = 160 K. These values have been estimated from Wilheit (1986) and the GATE radar rainrate data (Hudlow and Patterson, 1979). They are intended to be representative of conditions over the GATE area.

The model generalizes to an N element fov by changing the factor of 1/2 in Eq. (7) to 1/N. The solution is most sensitive to the number of independent elements.

During Phase I of GATE 30 useful overflights by ESMR-5 occurred. Figure 3 shows the histogram of brightness temperatures from the ESMR-5 data.

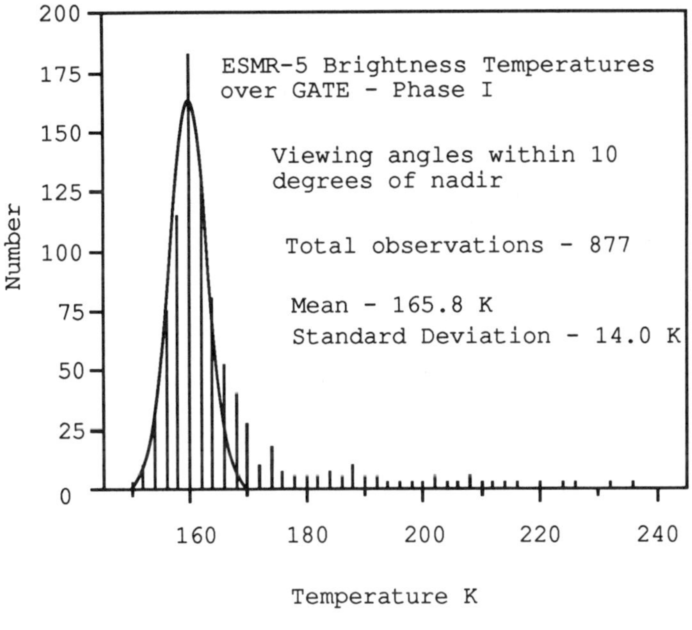

Figure 3. ESMR-5 brightness temperatures observed within 200 km of the center of the GATE area during Phase I. Bin size is 2 K. The smooth curve represents a normal distribution with mean 160 K and standard deviation 3 K.

Using Fig. 2 we can infer a rainrate of 5.5 mm/hr from the mean and standard deviation of the ESMR-5 data. The average rainrate (where raining) for the corresponding 30 radar scenes was 5 mm/hr. In principle the model can be solved for P in terms of higher moments of T. The level of agreement is encouraging but depends critically on the assumption of the number of independent elements.

Further development of the model, including correlation structure in the rainrate fields, effects of uncertainties in the basic assumptions and comparison with the GATE data is underway and will be reported at a later date.

References

Atlas, D., and C. W. Ulbrich, 1974: The physical basis for attenuation-rainfall relationships and the measurement of rainfall parameters by combined attenuation and radar methods. J. Rech. Atmos., 8, 275-298.

Hudlow, M. D., and V. L. Patterson, 1979: GATE radar rainfall atlas. NOAA Special Report, Center for Environmental Assessment Services, NOAA, 155 pp. [Govt. Printing Office Stock No. 003-019-00046-2].

Olsen R. L., D. V. Rogers and D. B. Hodge, 1978: The aR^b relation in the calculation of rain attenuation. I.E.E.E. Trans. AP-26, 318-329.

Smith, E. A., and S. Q. Kidder, 1978: A multispectral satellite approach to rainfall estimates. Unpublished manuscript. Colorado State University, Fort Collins, Colorado, 80523, 26 pp. plus tables and figures. See Fig. (9.7) in Barrett, E. C., and D. W. Martin, 1981. The Use of Satellite Data in Rainfall Monitoring, Academic Press, London, page 160.

Wilheit, T. T., 1986: Some comments on passive microwave measurement of rain. Bull. Amer. Meteor. Soc., 67, 1226-1232.

SUMMARY OF RAIN MEASUREMENTS WITH AN AIRBORNE MICROWAVE RAIN SCATTEROMETER/RADIOMETER

Masaharu Fujita
Kashima Space Research Center
Radio Research Laboratory
Kashima, Ibaraki 314, Japan

and

Ken'ichi Okamoto, Shin Yoshikado, Harunobu Masuko, and Kenji Nakamura
Radio Research Laboratory
Koganei, Tokyo 184, Japan

ABSTRACT

The results of rain measurements with an airborne microwave rain scatterometer/radiometer system are summarized in the present article. A dual-wavelength algorithm was newly developed and was used for the data analysis of the present airborne scatterometer. The rain rate estimates were compared with rain rates derived from a ground-based radar to demonstrate the validity of the estimates from an aircraft. Data from the airborne radiometer were analyzed using the rain rate estimates by the scatterometer. It is shown that the microwave antenna temperature correlated very well with the path-integrated rain rate derived from the scatterometer.

1. INTRODUCTION

Radio Research Laboratory (RRL) developed an airborne microwave rain scatterometer/radiometer system (Okamoto et al., 1982) for a feasibility study of rain measurements from the topside of clouds. The system consists of a dual wavelength scatterometer and a dual-wavelength radiometer both operating in 10 GHz and 35 GHz bands. In each frequency band, the same antenna is used for the scatterometer and the radiometer alternately with a period of approximately 1 or 2 milliseconds depending on the pulse repetition frequency of the scatterometer of 440 Hz or 220 Hz, respectively. Almost the same volume of rain can, therefore, be measured both actively and passively. Flight experiments of about 100 hours were carried out since 1980 in various rain conditions over ocean and land backgrounds. Some experiments were conducted in cooperation with a ground-based radar for quantitative comparison of the rain rate estimates from the aircraft and the ground. In addition to the experiments, a dual-wavelength algorithm was examined for efficient analysis of the airborne scatterometer data (Fujita, 1983).

In the present article, results of analysis are summarized for the scatterometer and radiometer measurements of rain from an aircraft altitude for possible application to future satellite data. The present article is based on two papers published elsewhere (Fujita et al., 1985a, b).

ISBN 0-937194-14-4

2. EXPERIMENT

An airborne experiment was carried out on 15 June 1981 in cooperation with a ground-based C band rain radar located in Kashima Space Research Center, RRL. Figure 1 shows a part of the flight trajectory on that day derived from the LORAN-C data. The altitude of the aircraft was around 6 km above the surface. Beginning at 16:15 the aircraft flew straight toward Kashima Space Research Center, while the ground-based radar was operated to obtain range height indicator (RHI) data in which rain rate distributions were measured within a vertical plane in a specific direction. Since some disturbances might cause fluctuations of the flight direction and the aircraft attitude, the azimuth direction for RHI measurement was changed by 1.5° every minute between 84.3° and 90.3° so as to increase the possibility that the antenna beams of the airborne and the ground-based instruments intersected each other. The airborne antennas were pointed downward from the aircraft and were scanned perpendicularly to the flight direction up to $\pm$23.2° from nadir. The data from the scatterometer and the radiometer were recorded on a computer compatible tape on the aircraft together with ancillary data such as aircraft altitude and attitude.

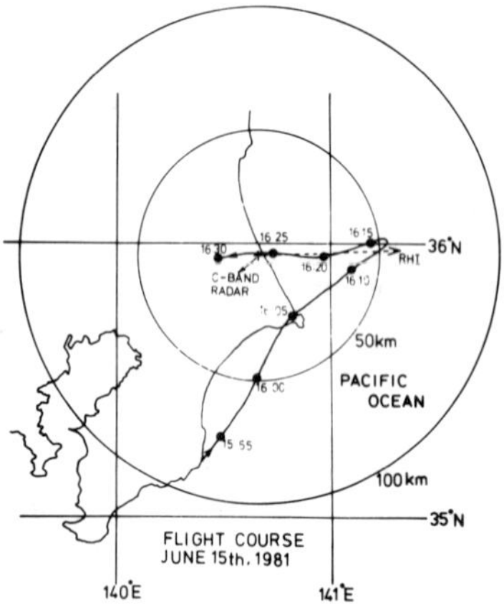

FIGURE 1. Flight trajectory on 15 June, 1981 (after Fujita et al., 1985a). © American Geophysical Union

3. ANALYSIS

Since the frequencies of the scatterometer were above 10 GHz, it was necessary to compensate for rain attenuations of echoes due to intervening rain. For rain rate estimation by an attenuating radar, Hitschfeld and Bordan (1954) proposed an algorithm and Meneghini (1978) extended the algorithm to increase its estimation accuracy. Since their algorithms were developed for a single-wavelength radar, radar calibration was indispensable for relating echo powers to rain rates. Thus, we developed a new algorithm for rain rate profiling by a dual-attenuating-wavelength radar for analyzing the data from the airborne scatterometer (Fujita, 1983; Fujita et al., 1985a). The algorithm estimates a rain rate profile by minimizing the squared difference between the quantities derived from the data and those calculated theoretically from a radar equation on the assumption of attenuation coefficients of the radar waves. As in the other dual-wavelength algorithms (e.g., Eccles and Mueller, 1971; Goldhirsh and Katz, 1974), the present algorithm is independent of the radar calibration.

Figure 2 shows a comparison of the rain rate estimates by the airborne scatterometer (a, upper panel) and the ground-based radar (b, lower panel). The data time is from 16:18 to 16:24. In the data processing of the ground-based radar, $Z=200R^{1.6}$, where Z and R stand for radar reflectivity in mm^6m^{-3} and rain rate in mm/h, respectively, was assumed to estimate rain rate. As can be seen in Figure 2, the estimated rain rate distributions are similar to each other; however, they do not necessarily correspond to each other especially in their detailed structures. The main reason for this difference might be attributed to differences in the beam locations where the measurements by two instruments were carried out. An error included in a calibration constant for the ground-based radar might be another possible cause of the above differences. For a quantitative comparison, the rain rate estimates from the aircraft and the ground were scatter-plotted as shown in Figure 3. Although there was a difference in their absolute values as discussed above, the rain rate estimates correlate well with each other. The results clearly demonstrate the feasibility of rain rate measurements from topside of clouds by the dual-attenuating-wavelength radar.

Several attempts were made to estimate rain rate by a microwave radiometer from space (e.g., Spencer et al., 1983; Weinman and Guetter, 1977; Wilheit et al., 1977). They compared the satellite microwave data to radar rain rates from the ground. Although the results indicated a strong correlation between the

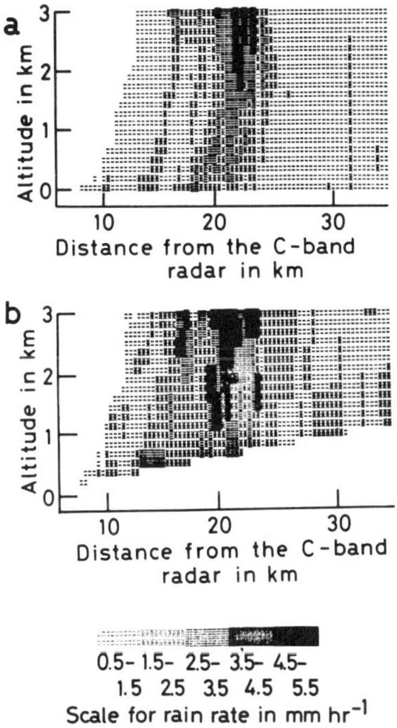

FIGURE 2. Comparison of rain rate estimates from the aircraft and the ground (after Fujita et al., 1985a).

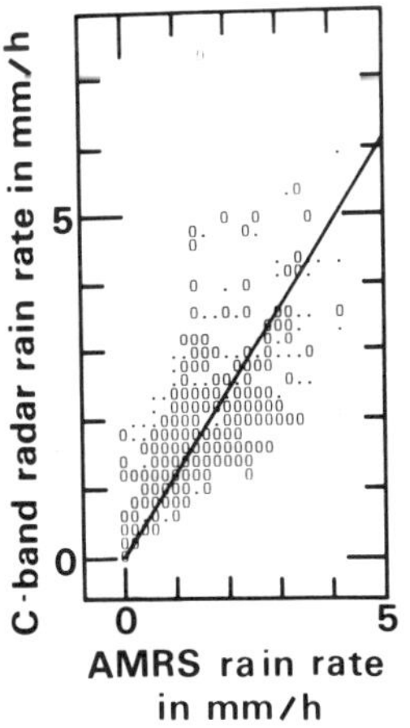

FIGURE 3. Scattergram of rain rate estimates from the aircraft and the ground (after Fujita et al., 1985a).

satellite microwave data and the ground-based radar rain rates, differences between the resolution areas by the satellite antenna and the ground antenna might cause somewhat large scattering of the data points. The difficulty in the data registration of the satellite and ground data might be the other cause of the data scattering.

The present system used the same antenna for the scatterometer and the radiometer alternately within a very short period of time so that it was easy to compare the active and passive measurement data for the same area of rain. Figure 4 shows a scattergram of the measured and calculated excess values of the brightness temperature relative to those for the no rain condition in the 10-GHz band. The theoretical calculation was made by the equations of transfer of polarized radiation (Weinman and Guetter, 1977) and the rain rate profiles were estimated by the scatterometer as described previously. Microwave scattering due to raindrops was neglected in the theoretical calculation. The measured and calculated excess brightness temperatures agreed well with each other as shown in Figure 4. In Figure 5 is shown the relation between the excess brightness temperature and the path-integrated rain rate in km mm/h which was obtained by integrating a rain rate profile estimated from the scatterometer data. From these figures, the usefulness of 10-GHz microwave radiometer for quantitative measurements of rain over the ocean was proven experimentally.

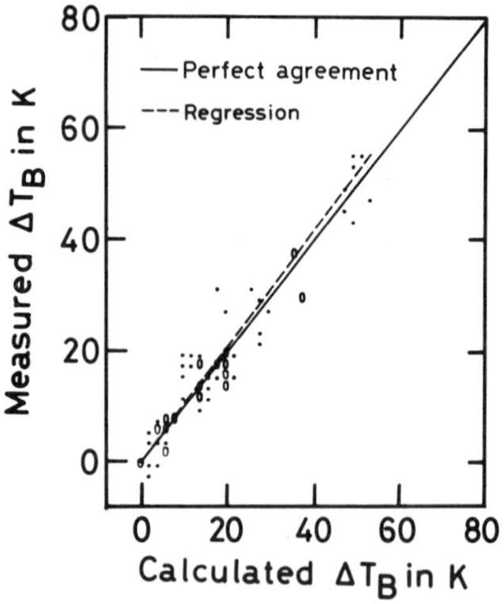

FIGURE 4. Scattergram of measured and calculated excess brightness temperature in 10 GHz band over the ocean (after Fujita et al., 1985b).
© American Geophysical Union

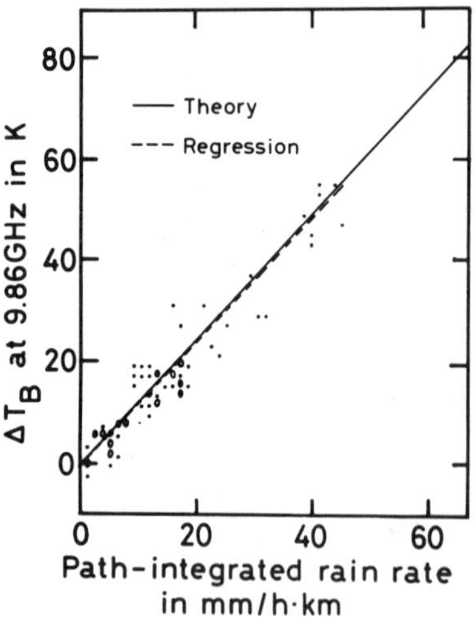

FIGURE 5. Scattergram of measured excess brightness temperature and path-integrated rain rate estimated from the scatterometer data (after Fujita et al., 1985b).
© American Geophysical Union

4. CONCLUDING REMARKS

In this article, we have reviewed the results of rain measurements from the topside of clouds with an airborne microwave rain scatterometer/radiometer system. The airborne data were analyzed by means of a newly developed algorithm for a dual-attenuating-wavelength radar. The rain rate estimates from the aircraft and the ground agreed well with each other, demonstrating the validity of the aircraft estimates of rain rate. The 10-GHz brightness temperature correlated very well with the path-integrated rain rate obtained from the scatterometer data. The present results indicate the promising role for a dual-attenuating-wavelength radar in a future satellite-borne system.

REFERENCES

Eccles, P.J. and E.A. Mueller, 1971: X-band attenuation and liquid water content estimation by a dual-wavelength radar. J. Appl. Meteor., 10, 1252-1259.

Fujita, M., 1983: An algorithm for estimating rain rate by a dual-frequency radar. Radio Sci., 18, 697-708.

Fujita, M., K. Okamoto, S. Yoshikado and K. Nakamura, 1985a: Inference of rain rate profile and path-integrated rain rate by an airborne microwave rain scatterometer. Radio Sci., 20, 631-642.

Fujita, M., K. Okamoto, H. Masuko, T. Ojima and N. Fugono, 1985b: Quantitative measurements of path-integrated rain rate by an airborne microwave radiometer over the ocean. J. Atmos. Ocean. Tech., 2, 285-292.

Goldhirsh, J. and I. Katz, 1974: Estimation of raindrop size distribution using multiple wavelength radar system. Radio Sci., 9, 439-446.

Hitschfeld, W. and J. Bordan, 1954: Errors inherent in the radar measurement of rainfall at attenuating wavelength. J. Meteorol., 11, 58-67.

Meneghini, R., 1978: Rain-rate estimates for an attenuating radar. Radio Sci., 13, 459-470.

Okamoto, K., S. Yoshikado, H. Masuko, T. Ojima, N. Fugono, K. Nakamura, J. Awaka and H. Inomata, 1982: Airborne microwave rain-scatterometer/radiometer. Int. J. Remote Sens., 3, 277-294.

Spencer, R.W., B.B. Hinton and W.S. Olson, 1983: Nimbus-7 37 GHz radiances correlated with radar rain rates over the Gulf of Mexico. J. Climate Appl. Meteor., 22, 2095-2099.

Weinman, J.A. and P.J. Guetter, 1977: Determination of rainfall distributions from microwave radiation measured by the Nimbus 6 ESMR. J. Appl. Meteor., 16, 437-442.

Wilheit, T.T., A.T.C. Chang, M.S.V. Rao, E.B. Rodgers and J.S. Theon, 1977: A satellite technique for quantitatively mapping rainfall rates over the ocean. J. Appl. Meteor., 16, 551-560.

TROPICAL RAINFALL MEASURING MISSION RAIN RADAR

K. Okamoto, T. Kozu, K. Nakamura, and T. Ihara
Radio Research Laboratory
Ministry of Posts and Telecommunications,
Koganei-shi, Tokyo 184, Japan

ABSTRACT

Conceptual designs of Tropical Rainfall Measuring Mission (TRMM) rain radar have been studied by Radio Research Laboratory (RRL) in cooperation with three major electric companies of Japan. The mission requirements proposed by NASA were used as the guideline to design the TRMM rain radar. Three cases of dual-frequency radar systems at 13.8 GHz and 24.15 GHz were studied. Case 1 and case 2 adopt two matched beam planar arrays with Travelling-Wave Tube Amplifiers (TWTAs) and Solid State Power Amplifiers (SSPAs), respectively. Case 3 uses TWTAs and a cylindrical parabolic antenna reflector, which is commonly used by both frequencies. All cases adopt the electric scanning phased array antenna and the pulse compression technique. Radar system parameters of these three cases are shown with power consumption and weight values. Performances of these three cases satisfy the mission requirements established by the Science Steering Group. The critical items for the development of TRMM rain radar are also described.

1. INTRODUCTION

Tropical rain is one of the key parameters that affects the global climate change, the global heat balance, and the global water cycle. However, the quantitative measurement data of tropical rain, especially over the ocean, have been very poor until now.

The one-year feasibility study of the TRMM project is being performed under the joint efforts of the United States and Japan. TRMM has three mission instruments (AVHRR, microwave radiometers, and dual-frequency rain radars), which aim to measure rain from space. Radars can directly measure the rainfall rate quantitatively and only radar can measure vertical profiles of rain. Although radar is a powerful means to observe rain and is widely used on the ground, a satellite-borne rain radar has not yet been realized. TRMM rain radar will be the first spaceborne radar. Following is the tentative report of the conceptual design study of TRMM rain radar made by RRL in a one-year feasibility study.

2. CONCEPTUAL DESIGN STUDY OF TRMM RAIN RADAR

The mission requirements for the TRMM rain radar by NASA are summarized and are shown in Table 1. These requirements were used as the guideline for the design of TRMM rain radar. As the satellite

ISBN 0-937194-14-4

TABLE 1. MISSION REQUIREMENTS

Frequencies (two):	13.8 GHz and 24-35 GHz
Satellite altitude:	300 km
Scan angle:	± 20 deg
Contiguous mapping over the swath:	Swath width = 220 km
Range resolution:	≤ 250 m
Horizontal resolution	≤ 4 km(at nadir)
Minimum measurable rain rate:	0.5 mm/h at storm top for both frequencies
Observation range:	20 km/cos θ m θ m : Maximum scan angle =20°
Beamwidth:	Matched beamwidth or synthesized matched beam by data processing
Number of independent samples	≥ 100
Antenna sidelobe level:	Low enough to suppress the effect of ground clutter

altitude is about 300 km, the deployment of a large solar paddle is not practical because of the large air drag effects. Therefore, the available electric power for dual-frequency radar is limited to 300-400 W. The system design of the radar must be performed under this limitation of power allocation. Rainfall conditions of 5-km rain height, a uniform rain rate, and a bright band of 0.5-km thickness just above the rain are assumed for the rainfall model.

Figure 1 shows three cases studied in the conceptual design study of TRMM rain radar at 13.8 GHz and 24.15 GHz. Case 1 and case 2 adopt two matched beam planar arrays with TWTAs (passive array) and SSPAs (active array), respectively. Case 3 uses TWTAs and a cylindrical parabolic antenna reflector, which is commonly used by both frequencies. All cases adopt the electric scanning phased array antenna and pulse compression technique, which is introduced to attain high peak power, good range resolution, and a large number of independent samples in the range direction.

Radar system parameters of case 1 and case 3 are shown in Table 2. Although case 3 uses a cylindrical parabolic antenna that is common to both frequency bands, other parameters are the same as those of case 1. In both cases, the total system is a passive array radar, which means it divides high output power from TWTAs and feeds the power into each array element and then combines received echoes by each array element and forwards them to a low noise amplifier (LNA). Radar system parameters of case 2 are shown in Table 3. Case 2 is the active array radar. In this case, there exists a separate transmitter and receiver unit just behind each array element; i.e., the system is composed of 130 small transmitters and receivers. The estimations of power consumption and weights in all three cases are shown in Table 4.

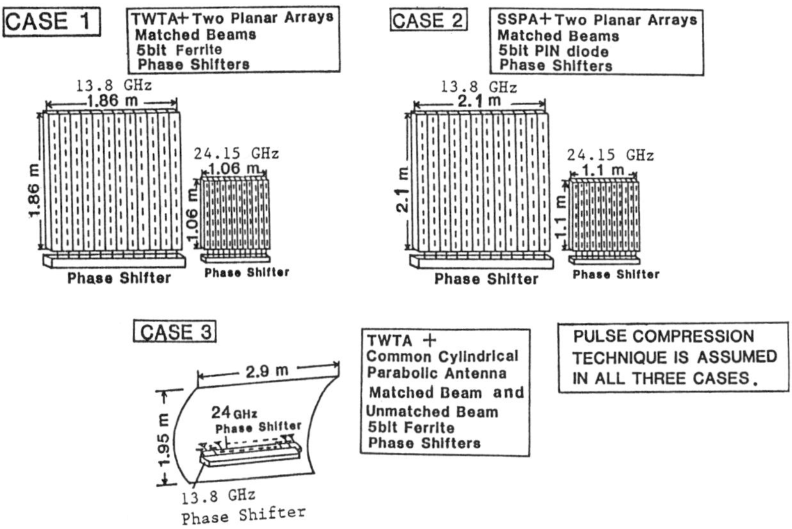

FIGURE 1. TRMM Rain Radar Conceptual Design Study. Three cases studied at 13.8 GHz and 24.15 GHz.

Figure 2(a), (b) shows calculated results of signal-to-noise ratio of rain echoes (both rain top and rain bottom echo) and sea surface reference echo at nadir in case 1 as a function of rainfall rates for 13.8 GHz, (a) and 24.15 GHz, (b), respectively. Table 5 summarizes radar performances. In all three cases, the mission requirements are nearly satisfied.

3. CONCLUSIONS

Trade-off studies of antenna type and transmitter type will be continued. The adaptive scan strategies, which are introduced to reduce power consumption and increase independent sample numbers by concentrating radar power onto the rain area, must also be carefully studied. Key technology items that require further and more immediate detailed studies are: the development of low sidelobe antennas, phase shifters and driver with low power consumptions; development of high-efficiency TWTAs and SSPAs and drivers; reduction of pulse compression range sidelobe levels; and development of a LNA High Electron Mobility Transistor (HEMT).

TABLE 2. RADAR SYSTEM PARAMETERS (CASES 1 AND 3)

	13.8 GHz	24.15 GHz
Antenna		
Type	Planar Array	Planar Array
	(*1: Cylindrical Parabolic Antenna Common to 13.8 GHz and 24.15 GHz)	
Polarization	HH	HH
Gain	46.3 dB	46.3 dB
Beamwidth	0.76° x 0.76°	0.76° x 0.76°
Aperture	1.86 m x 1.86 m	1.06 m x 1.06 m
	(*1: 2.9 m x 1.95 m, Common to 13.8 GHz and 24.15 GHz)	
Sidelobe	less than -30 dB	less than -30 dB
Phase shifter(*2)	Ferrite (5 bits)	Ferrite (5 bits)
Number	116 x 2	116 x 2
Loss(1)	1.2 dB	1.5 dB
Scan angle	± 20 deg.	± 20 deg.
Scan period	0.54 s	0.54 s
Feeder loss(2)	2.8 dB	3.6 dB
Transmitter		
Type	TWTA	TWTA
Peak power	150 W	150 W
Efficiency(*3)	44%	24%
Pulse width	50μs	50μs
Pulse Repetition Frequency	2245 Hz	2245 Hz
Duty	11.2 %	11.2 %
Transmission line loss(3)	0.7 dB	0.9 dB
Receiver		
Noise figure	2.5 dB	3.5 db
IF frequency	160 MHz	160 MHz
Bandwidth	3.1 MHz	3.1 MHz
Compressed pulse width	0.417 μs	0.417 μs
Compressed ratio	120	120
Dynamic range	More than 70 dB	More than 70 dB
Linearity	± 0.5 db	± 0.5 dB
Transmission line loss(4)	0.7 dB	0.9 dB
Total system loss ((1)x2+(2)+(3)+(4))	6.6 dB	8.4 dB
Resolution	62.5 m	62.5 m
Independent sample number	92	92
Data rate	170 kbps	-

(*1) Applicable only to case 3.
(*2) Separate phase shifters are used for both transmitters and receivers.
(*3) Including the power supply.

TABLE 3. RADAR SYSTEM PARAMETERS (CASE 2)

	13.8 GHz	24.15 GHz
Antenna		
Type	Planar array	Planar array
Polarization	HH	HH
Gain	47.5 dB	47.5 dB
Beamwidth	0.76° x 0.76°	0.76° x 0.76°
Aperture	2.1 m x 2.1 m	1.1 m x 1.1 m
Sidelobe	less than -30 dB	less than -30 dB
Phase shifter(*1)	PIN-diode(5 bit)	PIN-diode(5 bit)
Number	130 x 2	130 x 2
Loss	-	
Scan angle	± 20 deg.	± 20 deg.
Scan period	0.54 s	0.54 s
Transmitter		
Type	SSPA(x130)	SSPA(x130)
Peak power	48.5 W	32 W
Efficiency(*2)	20 %	12 %
Pulse width	50 μs	50 μs
PRF	2247 Hz	2247 Hz
Duty(*3)	13 %	13 %
Receiver		
Noise figure	2.5 dB	4.0 dB
Bandwidth	3.89 MHz	3.89 MHz
Compressed pulse width	0.333 μs	0.333 μs
Compression ratio	150	150
Total system loss	2.0 dB	2.5 dB
Resolution	50 m	50 m
Independent sample number	115	115
Data rate	170 Kbps	

(*1) Separate phase shifters are use for transmitter and receiver.
(*2) Including power supply.
(*3) Including margin of 16 %.

TABLE 4. ESTIMATIONS OF POWER CONSUMPTIONS AND WEIGHTS IN ALL THREE CASES

	Power Consumption(W)		Weights(kg)	
CASE 1	13.8 GHz	24.15 GHz	13.8 GHz	24.15 GHz
Antenna	24	24	106	72
Transmitter	86	90	36	36
Receiver	15	15	(*1)	(*1)
Signal processor	20	(*2)	10	(*2)
Interface unit	10	(*2)	8	(*2)
Others			15	(*2)
Subtotal	155	129	175	108
Margin	71			
Total	355		283	
CASE 2	13.8 GHz	24.15 GHz	13.8 GHz	24.15 GHz
Antenna	(*3)	(*3)	60	30
Transmitter	67.3	71.2	106	106
Receiver	80.2	80.2	63	63
Signal processor	10	10	5	5
Others	0.5	0.6	9	9
Subtotal	158	162	243	213
Margin	78			
Total	398		456	
CASE 3	13.8 GHz	24.15 GHz	13.8 GHz	24.15 GHz
Antenna	24	24	152	(*2)
Transmitter	86	90	36	36
Receiver	15	15	(*1)	(*1)
Signal processor	20	(*2)	10	(*2)
Interface unit	10	(*2)	8	(*2)
Others			15	(*2)
Subtotal	155	129	221	36
Margin	71			
Total	355		257	

(*1): Included in the transmitter.
(*2): Common to 13.8 GHz.
(*3): Included in the transmitter and receiver.

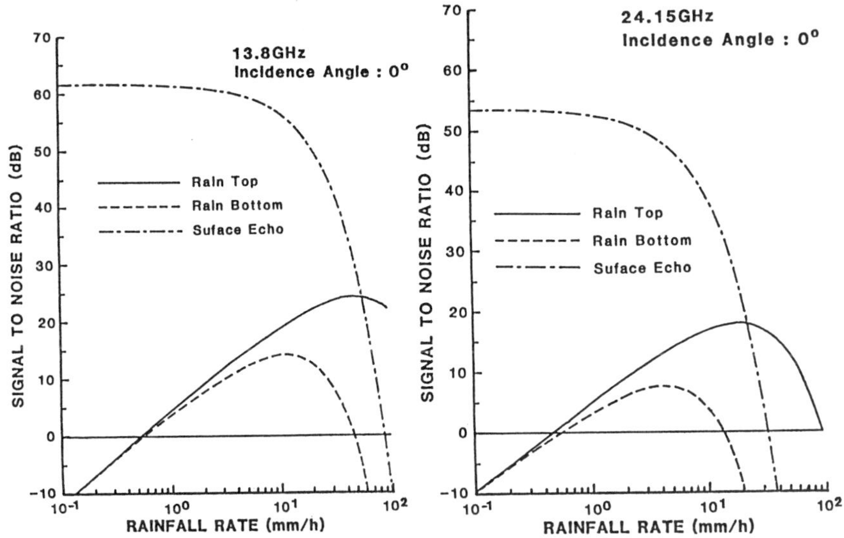

(a) 13.8 GHz (b) 24.15 GHz

FIGURE 2. Signal-to-noise ratios of rain echoes (rain top and rain bottom) and sea surface reference echo at nadir in the case 1 as a function of rainfall rate.

TABLE 5. SUMMARY OF RADAR PERFORMANCE

	13.8 GHz	24.15 GHz
Sensitivity		
Rain top (θ = 20°)	0.5 - > 100 mm/h	0.5 - 90 mm/h
Rain bottom (θ = 20°)	0.5 - 45 mm/h	0.7 - 12 mm/h
Sea surface reference (nadir)	0 - 85 mm/h	0 - 32 mm/h
Horizontal resolution	4 km	4 km
Range resolution	250 m	250 m
Swath width	220 km	220 km
Maximum scan angle	± 20 deg.	± 20 deg.
Number of independent samples	92(case 1,3) 115(case 2)	92(case 1,3) 115(case 2)

THE ESTIMATION OF PRECIPITATION INTENSITY IN THE PRESENCE OF SURFACE BACKSCATTER AND THE CONVERSE USING A SPACEBORNE RADAR

K. Eastwood Im
Jet Propulsion Laboratory
California Institute of Technology
Pasadena, California 91109, USA
and
David Atlas
7420 Westlake Terrace, Bethesda, Maryland 20817, USA

Abstract

The 'rain-modified' surface backscattering coefficient and the rain intensity can be estimated simultaneously by analyzing the return echo waveforms obtained by a non-nadir pointing spaceborne radar. Preliminary simulation results, using system parameters for the TRMM rain mapping radar, indicate that path-averaged rain attenuation coefficient can be estimated to 20% by such technique.

1. Introduction

The estimation of the path-averaged rain intensity (r) from space requires the *apriori* knowledge of the 'rain-modified' backscattering coefficient ($\tilde{\sigma}^o$) of the reference ocean or land surfaces. In the conventional approach (Eckerman et al., 1978; Meneghini et al., 1983), $\tilde{\sigma}^o$ is approximated by the 'dry' σ^o obtained either in the vicinity of the raining areas or from prior observations. One drawback of this approach is that the biases and uncertainties associated with the approximation may cause large errors in rainrate estimation. More recently, Meneghini and Atlas (1986) have proposed a technique to simultaneously estimate both $\tilde{\sigma}^o$ and r over the ocean by using a spaceborne radar. However, such a technique is only applicable for near-nadir incidence.

In this paper we present a study of a technique to estimate both r and $\tilde{\sigma}^o$ simultaneously from surface backscatter measurements obtained by a non-nadir pointing, narrow beam, spaceborne radar. This technique is based upon the assumptions that the surface echoes of the ocean and uniform land are almost symmetric relative to the radar boresight; and the rain echo is almost asymmetric with respect to the boresight for uniform rain across the radar beam. With these assumptions, the observed echo waveform asymmetry is contributed by precipitation and can be used to infer both r and $\tilde{\sigma}^o$. The results obtained by this technique will be useful not only in the precipitation-related scientific applications, but also in other disciplines such as ocean scatterometry.

2. Basic Derivation

The geometry used in our model in deriving the backscatter powers is illustrated schematically in Fig. 1. Here it is assumed that the radar pulses are transmitted off-nadir from a spaceborne antenna of circular aperture along the beam axis to the earth surface and back to the radar receiver. It is also assumed that the rain cell completely fills the radar beam along the propagation direction at altitudes below the height of the rain cell. As the radar pulses propagate towards the earth surface, the radar beam widths will encompass both the rain and the surface simultaneously. Therefore, within the same pulse volume the average echo power will consist of backscattered power contributions from both the rain and the surface, and the averaged radar received power can be expressed as

$$P(t_i) = P_r(t_i) + P_s(t_i); \quad i = -n, \ldots, n \tag{1}$$

ISBN 0-937194-14-4

where

$$t_{\mp n} \simeq 2(R_0 \mp r_b \tan\varphi_0)/c$$

defines the time interval where the pulse is intercepted by the surface, R_0 is the range distance to the center of the surface footprint (boresight), $r_b = R_0 \tan\theta_b$ is the radius of the beam circle at R_0 formed by the half-width of the antenna beam measured at the half-power point (θ_b), φ_0 is the angle of incidence at the surface footprint center, c is the speed of light, and the index i corresponds to the range bin number.

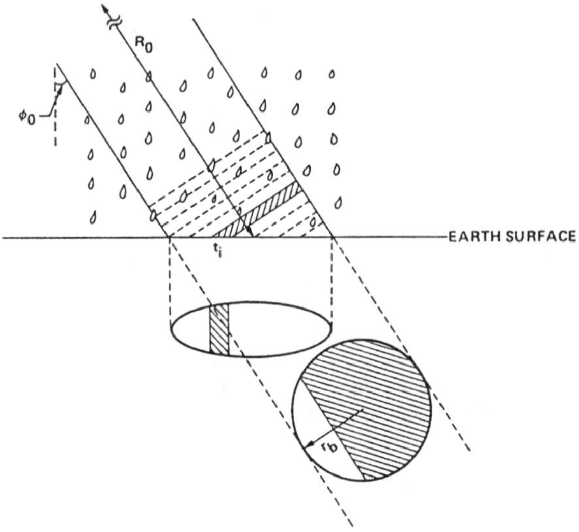

Figure 1. Geometry of the rain/surface interface.

For a rectangular transmit pulse with a peak transmitted power P_T and of duration τ , P_r and P_s can be expressed as

$$P_r(t_i) = \frac{\lambda^2 G_0^2}{(4\pi)^3} \int_{\Delta\phi_i} d\phi \int_{\Delta\theta_i} d\theta\, g^2(\theta,\phi) \sin\theta \int_{\Delta R_i} dr_1 \frac{P_T(t_i - 2r_1/c)\eta_r(r_1,\theta,\phi)}{r_1^2} \times 10^{-0.2\int_0^{R_i} dr_2\, k(r_2,\theta,\phi)} \quad (2)$$

and

$$P_s(t_i) = \frac{\lambda^2 G_0^2}{(4\pi)^3} \int_{\Delta S_i} d\bar{\rho} \frac{P_T(t_i - 2R_i(\bar{\rho})/c)\, g^2(\bar{\rho})\tilde{\sigma}^\circ(\bar{\rho})}{R_i^4(\bar{\rho})} 10^{-0.2\int_0^{R_i(\rho)} dr\, k(r)} \quad (3)$$

where λ is the radar wavelength, G_0 is the antenna gain at boresight, g is the normalized antenna radiation pattern, η_r (1/m) is the rain reflectivity, k (dB/m) is the atmospheric attenuation coefficient along the propagation direction to and from the scattering elements, which includes attenuations by both raindrops and other atmospheric constituents, $\tilde{\sigma}^\circ$ is the 'rain-modified' surface backscattering coefficient, and R_i is the slant range to the scattering element at $t_i/2$. Notice that in Eq. (2) the spherical coordinate system has been adopted so that $\Delta V_i = (\Delta R_i, \Delta\theta_i, \Delta\phi_i)$ corresponds to the portion of the transmitted pulse volume which is intercepted by the rain at $t_i/2$. In Eq. (3) the cylindrical coordinate system is adopted and $\Delta S_i = (\Delta\rho_i, \Delta\phi_i)$ corresponds to the area of the surface which intercepts the transmit pulse at $t_i/2$. Because of the complexity involved, numerical technique is usually required to evaluate these two equations for a given set of radar geometry, rain and surface scattering models. In our attempt to derive an algorithm for estimating the backscattering and extinction parameters of both the precipitation and the earth surface, we have used this following set of assumptions :

(i) The radar slant range is taken to be much greater than the spatial extent of the surface footprint so that R_i is essentially constant within the interval of integration.

(ii) The signal attenuation due to other atmospheric constituents relative to that due to precipitation is negligible.

(iii) The rain cell is uniform in height and intensity. With this assumption the rain reflectivity is constant throughout the entire rain cell, and the rain attenuation is only a linear function of the portion of the pathlength that intervenes the rain cell.

(iv) The 'rain-modified' earth surface is smooth, homogeneous and isotropic such that $\tilde{\sigma}^o$ is constant over the entire surface footprint area.

With this set of assumptions, Eqs.(2) and (3) can be simplified to

$$P_r(t_i) \simeq C_r \eta_r \left(\frac{c\tau}{2}\right) r_b^2\, g_{r_i}^2 \left[\cos^{-1}(f_i) - f_i(1-f_i^2)^{1/2}\right] \times 10^{-0.2k_r\left[\Delta R_{r_i} - \frac{r_b}{2}(1-f_i)\tan\varphi_0\right]} \tag{4}$$

and

$$P_s(t_i) \simeq C_r \tilde{\sigma}^o \left(\frac{c\tau}{\sin\varphi_0}\right) r_b\, g_{s_i}^2\, (1-f_i^2)^{1/2}\, 10^{-0.2k_r\,\Delta R_{r_i}} \tag{5}$$

where

$$f_i = i\,/\,n; \tag{6}$$

$$C_r = \frac{P_T G_0^2 \lambda^2}{(4\pi)^3 R_0^4}\,; \tag{7}$$

$$g_{r_i} = \frac{\int_{\Delta\phi_i} d\phi \int_{\Delta\theta_i} d\theta\, g(\theta,\phi)}{\int_{\Delta\phi_i} d\phi \int_{\Delta\theta_i} d\theta}\,; \tag{8}$$

$$g_{s_i} = \int_{\Delta S_i} d\bar{\rho}\, g(\bar{\rho})\,/\,\Delta S_i\,; \tag{9}$$

$$\Delta R_{r_i} \simeq \Delta R_{r_0} = \Delta h_r / \cos\varphi_0\,; \tag{10}$$

k_r is the attenuation coefficient of rain; ΔR_{r_i} is the pathlength measured from top of the rain cell to the center of the scattering elements, and it is nearly constant for all values of i. Δh_r is the altitude of the rain cell; C_r is the radar constant factor; g_{r_i} is the 'averaged antenna pattern' for the rain echo; and g_{s_i} is the 'averaged antenna pattern' for the surface echo. It should be noted that g_{s_i} is symmetric with respect to the boresight. That is, $g_{s_{-i}} = g_{s_i}$. From Eqs. (4) and (5) we observe that the P_r is a decreasing function of time, while P_s is symmetric relative to the surface return at the radar boresight, $P_s(t_0)$. Consequently, the peak value of the received echo power waveform $P(t)$ always occurs at $t_{peak} < t_0$ in the presence of precipitation. Specifically, in the heavy rain situations t_{peak} approaches t_{-n}, and in light rain or rain-free situations t_{peak} approaches t_0 and the received echo waveform appears symmetric. Therefore, the relative position of the peak echo power can be used to indicate, in the qualitative sense, the presence/absence of precipitation.

3. Parameter Estimations

3.1 Rain Attenuation Coefficient

Because of the symmetry of $P_s(t_i)$, the difference between $P(t_{-i})$ and $P(t_i)$ completely removes the backscattering power contributions from the surface echo. Furthermore, $P(t_{-n})$ consists of only the backscattering power from the precipitation. Since these two quantities are directly proportional to the rain reflectivity η_r for uniform rain, by ratioing them we can obtain $\hat{k}_r$, the estimate of the rain attenuation coefficient. Using Eqs. (4) and (5), we obtain

$$\pi \frac{P(t_{-i}) - P(t_i)}{P(t_{-n})} = \left(\frac{g_{r_{-i}}}{g_{r_{-n}}}\right)^2 \left[\cos^{-1}(f_{-i}) - f_{-i}\,(1 - f_{-i}^2)^{1/2}\right] 10^{-0.1\hat{k}_r r_b (1+f_{-i}) \tan\varphi_0}$$
$$- \left(\frac{g_{r_i}}{g_{r_{-n}}}\right)^2 \left[\cos^{-1}(f_i) - f_i\,(1 - f_i^2)^{1/2}\right] 10^{-0.1\hat{k}_r r_b (1+f_i) \tan\varphi_0} \quad (11)$$

Since g_{r_i}, r_b, φ_0 depend on the antenna parameters and geometry and are deterministic, $\hat{k}_r$ can be computed using Eq. (11) and the measured echo waveform. Notice that Eq. (11) is simply the difference between two differential rain attenuation measurements except that the subtraction must be done in the appropriate order to eliminate the surface echo contributions. Notice also that only two samples plus one reference sample at the first range bin are required to compute $\hat{k}_r$, so if the echo consists of $2m+1$ samples, in principle $(m-1)$ independent estimates of k_r can be obtained. In the presence of noise, however, only a portion of the samples will be useful in estimating $\hat{k}_r$.

3.2 Rain Reflectivity and Rain Rate

The rain reflectivity η_r can be estimated from the computed $\hat{k}_r$ and the power difference

$$P(t_{-i}) - P(t_i) = P_r(t_{-i}) - P_r(t_i) \quad (12)$$

where $P_r(t_i)$ was given in Eq. (4). It should be noted that a separate measure of the rain top altitude is required in order to compute $\hat{\eta}_r$.

The associated rain rate and raindrop size distribution can now be estimated from the computed $\hat{k}_r$ and $\hat{\eta}_r$ via existing algorithms (e.g. Meneghini et al., 1983).

3.3 'Rain-Modified' Surface Backscattering Coefficient

In order to estimate $\tilde{\sigma}^o$, one needs to first compute the $\hat{P}_r$ using Eq. (4), $\hat{\eta}_r$, $\hat{k}_r$, and the radar parameters. After subtracting $\hat{P}_r$ from P, $\tilde{\sigma}^o$ can be estimated via Eq. (5).

4. Preliminary Results

We shall show the characteristics of the received power and the algorithm by means of an example. The algorithm will also be tested for some non-ideal cases. The radar parameters assumed here follow those of the rain mapping radar for the Tropical Rainfall Measuring Mission (TRMM) (Li et al., 1987). These parameters, together with the assumed precipitation and surface parameters, are given in Table 1. For the radar operating at 13.8 GHz, Li et al. showed that the echoes backscattered from rain volumes near the earth surface are highly detectable (with SNR $\geq$ 5 dB) for rain rates between 3 and 50 mm/hr and for incidence angles $\leq 20°$ We shall therefore choose $r = 40$ mm/hr and $\varphi_0 = 20°$ in this example.

Table 1. Radar, precipitation, and geometric parameters.

Frequency (GHz)	13.8
Antenna Aperture (m)	2 × 2
Transmit Peak Power (W)	125
Scan Angle (deg)	± 20
Expanded Pulse Width (μs)	80
Compressed Pulse Width (μs)	0.28
Orbit Altitude (km)	300
Altitude of Rain Cell (km)	5
Rain Reflectivity (m^{-1})	$3.02 \times 10^{-7}\, r^{1.57}$
Rain Attenuation Coefficient (dB/km)	$2.37 \times 10^{-2}\, r^{1.17}$

For the spacecraft altitude of 300 km and the antenna size of 2 m, the 'averaged antenna patterns' for both the rain and surface echoes, g_{r_i} and g_{s_i} are computed numerically in Fig.

2. In the calculation we have assumed the antenna radiation pattern to be Gaussian. The corresponding averaged powers, $P_r(t_i), P_s(t_i)$, and $P(t_i)$, based on the various assumptions, are calculated as shown in Fig. 3. In reality, however, the echo waveforms will be randomly distorted by various noises and inhomogeneities of both rain and surface. It is our purpose here to show some of these effects through simulation.

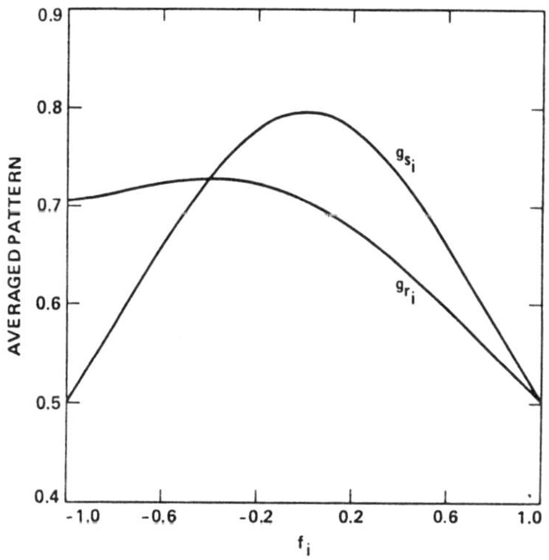

Figure 2. Antenna patterns normalized by the radar-illuminated rain volumes (g_{r_i}) and surface cross-sections (g_{s_i}) at different range bins.

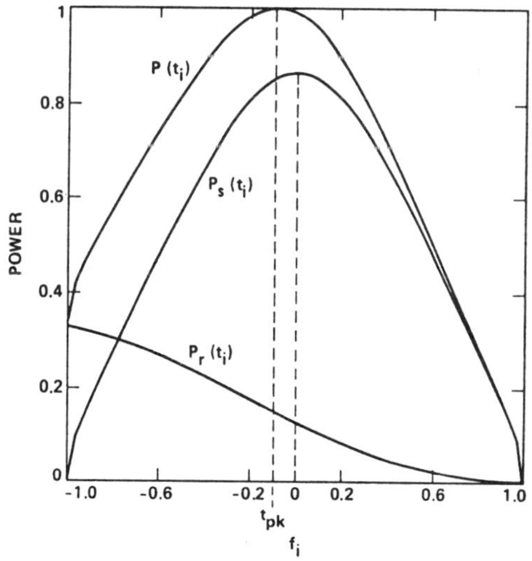

Figure 3. The mean backscattered powers of the rain and the surface; and the total mean power at 20° incidence.

In our present simulation the rain cell is assumed to be uniform. However, the surface scatterers are assumed to be uniformly distributed over the entire radar footprint. Sufficient number of independent signal-plus-noise and noise-only measurements is assumed for noise averaging purposes so that the noise variations can be modelled as Gaussian. As for the signal processing, the mean noise floor is first removed from the signal by subtracting the pulse-averaged noise measurements from those of the signal-plus-noise measurements. However, the random noise residues remain within the signal. For the results that we present here the echo waveforms are fitted, in the least-square sense, by a sixth-order polynomial. It should be noted that this method is by no means the best processing approach. In fact, other available approaches such as the Maximum-Likelihood estimation will probably perform better. However, this is beyond the scope of this paper. The algorithm was tested in cases of varying amounts of thermal noise and also with varying number of independent measurements. The $\tilde{\sigma}^o$ values we assumed in this simulation have a mean of 0 dB with a 3-dB fluctuation. The two rain parameters are assumed to have values of $\eta_r = 9.81 \times 10^{-5}/\text{m}$ and $k_r = 1.77$ dB/km at $r = 40$ mm/hr. The simulation results, which include the fractional biases and standard deviations of the estimated η_r, k_r and $\tilde{\sigma}^o$, are summaried in Table 2. (In the case of $\tilde{\sigma}^o$, the bias and standard deviation are being normalized by the mean of 0 dB). We observed from these limited simulated results that the algorithm in general underestimates the parameters of interest. The algorithm seems to be most effective in estimating the rain attenuation coefficients. In this particular example, it can estimate k_r to $\leq 20\%$ using 500 independent measurments with SNR $\geq$ 5 dB. On the other hand, in order to estimate η_r and $\tilde{\sigma}^o$ with reasonable accuracies, a larger number of measurements ($\geq$1000) with higher signal strength (SNR of 10 dB) is needed.

TABLE 2. THE FRACTIONAL BIASES AND STANDARD DEVIATIONS OF THE ESTIMATED RAIN REFLECTIVITY, RAIN ATTENUATION COEFFICIENT, AND SURFACE BACKSCATTERING CROSS-SECTION.

	$\hat{\eta}$			$\hat{K}$			$\hat{\sigma}^o$			
N \ SNR (dB)	2	5	10	2	5	10	2	5	10	
200	-4.29 (6.22)	-2.20 (3.10)	-1.21 (1.82)	-0.25 (0.35)	-0.15 (0.26)	-0.07 (0.22)	-10.31 (17.12)	-4.52 (6.77)	-2.58 (3.90)	BIAS (STD)
500	-1.63 (2.55)	-0.56 (1.33)	-0.26 (0.91)	-0.10 (0.25)	-0.01 (0.21)	0.05 (0.19)	-3.19 (5.64)	-1.06 (2.43)	-0.32 (1.68)	BIAS (STD)
1000	-1.22 (1.22)	-0.65 (0.81)	-0.26 (0.62)	-0.10 (0.21)	-0.04 (0.19)	-0.14 (0.16)	-1.96 (2.00)	-1.06 (1.29)	0.19 (0.32)	BIAS (STD)

It should be noted that this set of preliminary results is intended to be used to illustrate the characteristics of the algorithm. A more rigorous error analysis is underway for the evaluation of the algorithm.

5. DISCUSSION

The technique proposed here for remote sensing of the rain intensities with the surface background is motivated by two factors. First, the existing techniques for estimating the path-averaged rain intensity from space required information on the surface backscattering properties in the presence of rain. Second, the next-generation spaceborne earth-observing or rain mapping radars would likely employ wide-swath scanning antennas for sampling and coverage purposes, and rain/surface interferences at off-nadir incidence would introduce biases when measurements obtained by such instruments are used. The development of this algorithm is based on the uniform rain and surface assumptions so that, on the average, the power contribution from the surface is symmetrical relative to the echo power at the middle range gate and can be removed from the total received echo power. The ratio of the power difference to the echo power at the first range gate is a measure of the differential power attenuations by rain absorption, and thus, can be used to estimate the rain attenuation coefficient. Using this estimated parameter, the rain reflectivity and the 'rain-modified' surface backscattering coefficient can be determined accordingly.

To make an initial estimate of the sensitivity of this algorithm from the noise background, we have simulated the noisy return echo waveforms from the random surface in the presence of uniform rain. The results indicate that the algorithm is most effective in estimating the rain attenuation coefficient. The other two parameters can also be accurately estimated, provided that a large number of independent measurements ($\geq$ 1000) are used for noise averaging and that the signal-to-thermal-noise ratios for such measurements are sufficiently large ($\sim$10 dB).

There are other error sources in addition to the ones mentioned in the paper. For instance, at a large angle of incidence, part of the radar beam may be outside the rain cell, and thus, may introduce errors if the scattering properties of the surface outside the rain cell are significantly different from those inside the cell. Antenna sidelobes may also introduce additional errors since in this case the received power is likely to be higher.

Depending upon the rain intensity, surface backscattering cross-section, radar wavelength, and the echo's signal strength, we show that it is possible to detect rain in the clutter background and the converse, thereby, allowing the use of a wide swath (scanning) spaceborne radar for precipitation and/or surface remote sensing when the radar beam width encompasses both the rain and the surface background.

ACKNOWLEDGMENTS

The research described in this paper by K.E. Im was performed at the Jet Propulsion Laboratory, California Institute of Technology, under contract with the National Aeronautics and Space Administration.

REFERENCES

Eckerman, J., Meneghini, R., and Atlas, D., 1978; Average rainfall determination from a scanning beam spaceborne meteorological radar, NASA Tech. Memo. 79664

Meneghini, R., and Atlas, D., 1986; Simultaneous ocean cross section and rainfall measurements from space with a nadir-looking radar, J. Atmos. Oceanic Tech., 3, p.400

Meneghini, R., Eckerman, J., and Atlas, D., 1983; Determination of rain rate from a spaceborne radar using measurements of total attenuation, IEEE Trans. GeoSci. Remote Sensing, GE-21, p.34

Li, F.K., Im, K.E., Wilson, J.K., and Elachi, C., 1987; On the design issues for a spaceborne rain mapping radar, Proceedings of International Symposium on Tropical Precipitation Measurements, Tokyo, Japan, 1987

AIRBORNE RAIN MAPPING RADAR

W.J. Wilson, G.S. Parks, F.K. Li, K.E. Im, and R.J. Howard
Jet Propulsion Laboratory
California Institute of Technology
Pasadena, California 91109, USA

ABSTRACT

An airborne scanning radar system, operating at 13.8 and 24.1 GHz, is being developed. This system will demonstrate the capability for accurate remote rain mapping from an aircraft and verify the system design concept for the spaceborne Tropical Rainfall Measuring Mission (TRMM) radar. This radar system will also gather *in situ* data for TRMM and be a test bed for future spaceborne rain mapping radars.

1. INTRODUCTION

An Airborne Rain Mapping Radar (ARMAR) is being developed by NASA/JPL to demonstrate accurate, large-scale remote rain mapping. A number of airborne rain radars have been previously developed, e.g. Okamoto et. al., 1982. This radar system is being developed to support the development of the Tropical Rainfall Measuring Mission's (TRMM) radar system, and follows the recommendations from the 1985 TRMM workshop (Thiele, 1987). ARMAR will confirm the TRMM system design concepts, such as the frequency selection and the pulse compression techniques. It will also be used in a coordinated experiment program for rain rate retrieval algorithm development. Following the TRMM launch, ARMAR will be used to collect *in situ* data to verify the system performance of the TRMM rain radar. Finally, ARMAR will serve as a test bed to develop advanced techniques and technology for future rain mapping missions.

2. SYSTEM CHARACTERISTICS

The baseline design for ARMAR consists of two radar frequency channels at 13.8 and 24.1 GHz. These frequencies were selected based on an analysis of the signal-to-noise ratio of various frequencies by Li, et. al, 1987. It will have the capability to scan its antenna beam over ±20 degrees from the antenna boresight to give a typical swath width of 7 km. Figure 1 shows a sketch of the experiment configuration. The horizontal spatial resolution at nadir will be ~500 m and will have a range (rain height) resolution of 120 m. ARMAR is designed to provide data usable to retrieve rainfall rates between 0.1 to 60 mm/hr at the Earth's surface. This radar will also measure linear polarization signatures and the raindrop's fall velocity. ARMAR will fly on the NASA AMES DC-8 and the high altitude ER-2 aircraft.

ISBN 0-937194-14-4

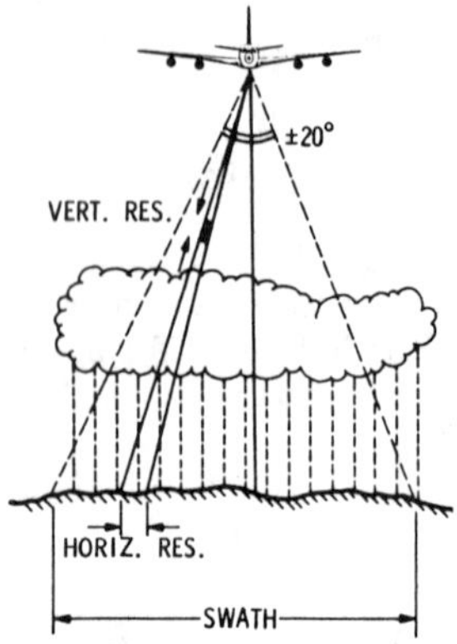

Figure 1. ARMAR Flight Configuration

Table 1 gives the proposed system characteristics, and Figure 2 shows the system block diagram for one frequency channel. A digitally controlled frequency synthesizer generates the expanded transmitted pulse. The amplitude and phase of the transmitted pulse is set so the received compressed pulse will have very low range sidelobes. Frequency diversity is used to obtain more independent samples. This signal is upconverted and then amplified by a Traveling Wave Tube Amplifier (TWTA). (The TWTA operates in the non-saturated mode for control of the pulse shape.) The amplified signal is connected to the antenna feed and the polarization is selected by a switch.

TABLE 1. ARMAR SYSTEM CHARACTERISTICS

Operating Frequency (GHz)	13.8	24.1
Polarization	HH/VV	HH/VV
Antenna Size (m)	0.5	0.5
Antenna Beamwidth (deg)	2.9	1.7
Antenna Gain (dB)	35	40
Antenna Scan Angle (deg)	±20	±20
Antenna Boresight (deg)	±25	±25
Transmitter Power (W)	250	20
Transmitter Pulse Width (μsec)	10-80	10-80
PRF (kHz)	1-4	1-4
Frequency Diversity	1-4	1-4
Bandwidth (MHz)	4	4
Dynamic Range (dB)*	70	60
System Noise Temperature (K)	700	900
Independent Samples**	>500	>500

* Prior to pulse compression
** The two channels operate simultaneously

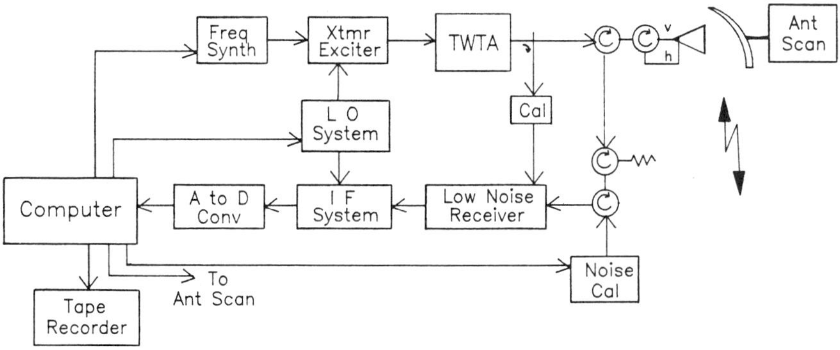

Figure 2. ARMAR System Block Diagram

Figure 3 shows a sketch of the mechanically scanned offset parabolic antenna system as it will be installed in the ER-2 wing pod. A dual frequency, dual linearly polarized scalar feed horn illuminates the offset parabolic reflector. The offset parabola has an F/D ratio of 0.18 and scans over a 40° angle around a boresight which can be varied $\pm 25^{\circ}$ from the nadir. With this offset parabolic antenna, the main beam efficiency will be ≈90%, the sidelobes <30 dB below the main lobe and the cross-polarization ≤-20 dB.

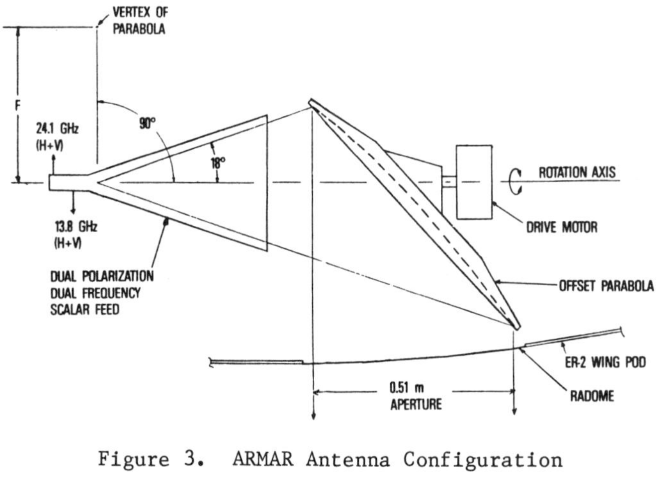

Figure 3. ARMAR Antenna Configuration

A low noise amplifier amplifies the received signal. This signal is then down-converted to 100 MHz, further amplified and then mixed down to the baseband frequency. The signal is digitized, formatted with other system data and recorded on a high-speed digital tape recorder. A microprocessor controls the data sampling, formatting, and the tape recorder. The microprocessor also controls the transmitter synthesizer, polarization switch, antenna scanner, and the receiver gain and synthesizer. The operator system interface is also through the microprocessor.

For calibration, the transmitted signal power is measured and the receiver calibrated with a noise source. In addition, the transmitted signal is attenuated and sent through the receiver to measure the system amplitude and phase distortions. These data are used for calibration in the ground processing. An external ground transponder will also be used for an end-to-end system calibration.

ARMAR will operate on the NASA DC-8 and the ER-2 aircraft. In the DC-8, the system will mount in the cargo bay looking down through a radome. In the ER-2, the system will mount in the rear of a wing pod, and look down through a radome. The advantages of the DC-8 are the larger working areas and having operators along during the measurements. However, the maximum operating altitude is only 12 km. The NASA ER-2 can fly at altitudes of 16 km or higher and can be used for many high altitude storm measurements.

3. SYSTEM MEASUREMENTS

During the experiments, ARMAR measures the rain and surface reflectivity at 13.8 and 24.1 GHz. Figure 4 shows the calculated signal-to-noise ratios (SNR) for rain returns at a flight altitude of 8 km. This shows that the SNR is ≥ 1 from rain rates of 0.1 to 75 mm/hr throughout the rain cloud. These measurements will be used to test the many backscatter and attenuation rain retrieval algorithms.

ARMAR can also measure the dual linearly polarized returns and the Doppler frequency at nadir. Algorithms for rain retrieval using the differential polarization and the mean fall velocity will also be tested. In coordinated experiments, combined radar and radiometer data will also be evaluated for rain measurements. Techniques for measurement of the bright-band attenuation and location, Doppler related wind, and removal of the surface clutter can also be tested with this data.

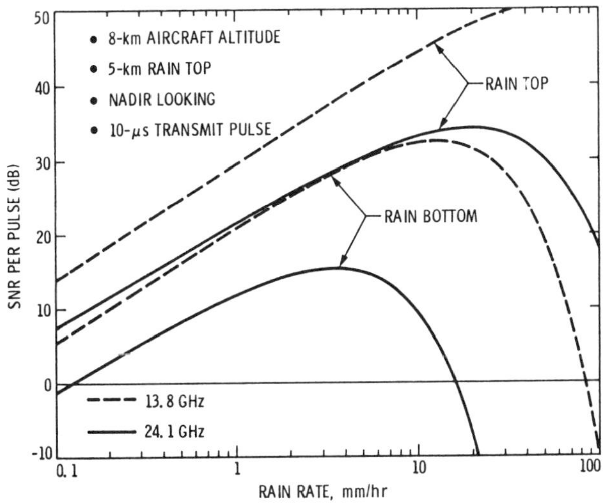

Figure 4. ARMAR Signal-to-Noise Ratio per Pulse

ACKNOWLEDGMENT

This research was performed by the Jet Propulsion Laboratory, California Institute of Technology, under contract with the National Aeronautics and Space Administration.

REFERENCES

Li, F. K., Im, K. E., Wilson, W. J., Elachi, C., 1987: On the design issues for a spaceborne rain mapping radar, Proc. of Int. Symp. on Tropical Precipitation Measurements, Tokyo, Japan, 1987.

Okamoto, K., Yoshikado, S., Masuko, H., Ojima, T., Fugono, N., 1982: Airborne microwave rain-scatterometer/radiometer, Int. J. Remote Sensing, 3, 277-294.

Thiele, O. W., Ed., 1987: On requirements for a satellite mission to measure tropical rainfall, NASA Ref. Pub. 1183, 33-36.

SOME CHARACTERISTICS OF THE MIRROR-IMAGE RETURN IN RAIN

R. Meneghini
NASA Goddard Space Flight Center
Greenbelt, Maryland 20771, USA
and
K. Nakamura
Radio Research Laboratory
Tokyo 184, Japan

ABSTRACT

Observations of the radar return from an aircraft viewing the rain at near-nadir angles show three distinct types of return: (1) The direct backscattered power from the hydrometeors, (2) The echo from the surface, and finally (3) A return which corresponds to a number of multiple scattering mechanisms. It has been hypothesized that among these multiple scattering contributions, the dominant component arises from the scattering of the incident pulse from the surface out to the rain, a fraction of which is returned to the surface where it undergoes a second scattering from the surface back to the radar. To distinguish this scattering mechanism from others, the term "mirror-image" (MI) return has been used.

In this paper, an attempt is made to understand some of the characteristics of this return and its relationship both to the direct rain return and the scattering properties of the surface. A numerical evaluation of the MI power provides a way to test a simplified expression for this quantity as well as to identify the volume of rain that contributes to the return. The results suggest that the MI power can be expressed as an integral of the direct rain return and a function which represents the spatial broadening of the MI rain volume. The paper concludes with a discussion of algorithms for the estimation of path attenuation.

1. NUMERICAL INTEGRATION OF THE MI RETURN

In Meneghini and Atlas (1986), a closed-form expression for the MI return was derived. Because of the assumptions which were used, the theory is provisional. As an initial test, results of a numerical integration are compared to those predicted by the closed-form expression for the MI return. Since the details of the numerical integration are somewhat involved, the procedure is discussed qualitatively. For convenience, we assume a rotationally symmetric gaussian radar beamwidth, nadir incidence, and a delta function transmitted pulse. To compute the MI return at a particular time requires the evaluation of a seven-fold integral: four representing the integration over the surface on the first and second scatterings and three over the rain volume. Because of the symmetry of the scattering process, the volume integral can be reduced to an

A. DEEPAK Publishing
ISBN 0-937194-14-4

area integral over any half-plane. The contribution to the MI return from any such "rain" area element is found by summing all possible scattering paths which include the selected area element.

Each path consists of four segments: (1) A ray directed from the radar to an area element on the surface, (2) from the surface element to the selected "rain" area, (3) from the rain element traced back to a second and generally different surface element, and (4) the final ray from the second surface element back to the radar. If the length of each of the four rays comprising a particular scattering path is denoted by r1 through r4, then a requirement that the path contribute to the MI return at a time t is: r1 + r2 + r3 +r4 = ct, where c is the speed of light. Use of the numerical integration eliminates several assumptions in the previous derivation. For example, no limitation is needed either on the height or beamwidth of the radar; moreover, the Fresnel reflectivity of the surface is allowed to vary with the local incidence angle. Most importantly, no approximations are needed in evaluating the integrals. On the other hand, several assumptions remain: the Kirchhoff scalar approximation (Ishimaru, 1984) is used to describe the bistatic scattering from the surface; in addition, the scattering from the hydrometeors is assumed to be equal to the backscattering cross-section.

To compare the results of the numerical integration with those of the analytic expression, it is convenient to normalize the MI return, P_m, by the direct rain return, P_d. This ratio can be written (Meneghini and Atlas, 1986)

$$\frac{P_m\,(t)}{P_d\,(t')} = \frac{(H - h_j)^2\, R^4\, \sigma^0\, 10^{-0.4A_j}}{\sigma^0 H^2 + 11\, R^2\, h_j^2/\theta_B^2} \tag{1}$$

where $\sigma°$ is the differential backscattering cross section of the surface at nadir, H is the height of the radar, θ is the 3-dB beamwidth, R^2 is the Fresnel reflectivity of the surface, h_j is the distance from the surface to the MI volume and A_j is the path attenuation between the surface and h_j. For $P_m(t)$ and $P_d(t')$ to be viewing approximately the same volume of rain, the times t and t′ must be related by $t = t' + 4h_j/c$.

In Table 1, the ratio of P_m/P_d, from (1), (shown by the left-most entries in the table) are compared to the corresponding results from the numerical integration for three different surface roughness scales ($\sigma°$ = 5, 10, 15 dB) at several heights h_j. In the left hand portion of the table, a TRMM-type geometry is assumed, H = 300 km and θ = 0.6°, while in the right hand table a geometry typical of the RRL/NASA airborne rain experiment is chosen: H = 5 km, and θ = 8°. Rain attenuation is not included in Table 1. To include it, the results should be modified by the factor $\exp(-0.92kh_j)$ where k is the attenuation coefficient in dB/km. An inspection of the results shows that the differences between the analytic and numerical results are typically less than 1 dB, with a maximum difference of 1.6 dB. Although the analytic expression predicts that the P_m/P_d ratio should be basically constant near the surface (h_j = 100 m), the numerical results are generally larger and vary somewhat with changes in sigma-zero. Apart from these differences, the analytic and numerical results predict similar behavior. In particular, the amount of decrease in P_m/P_d with height is strongly dependent on sigma-zero, with the decrease being more rapid for rough surfaces ($\sigma°$ = 5 dB) than for smooth surfaces ($\sigma°$ = 15 dB). Moreover, for a fixed surface roughness, the decrease in P_m/P_d with height is greater for the broad-beam airborne radar than for the narrow-beam spaceborne radar. This implies that for comparable signal-to-noise ratios in the direct rain return, a greater fraction of the MI return should be detectable from satellite altitudes than from aircraft altitudes.

TABLE 1. COMPARISONS OF ANALYTIC (LEFT-HAND ENTRIES) AND NUMERICAL RESULTS FOR Pm/Pd (dB); THE MINUS SIGN FOR ALL ENTRIES IS SUPPRESSED

	H = 300 KM,	θ = 0.6°			H = 5 KM,	θ = 8°	
		$\sigma°$ (dB)				$\sigma°$ (dB)	
hj(km)	5	10	15	hj(km)	5	10	15
0.1	4.1/2.9	4.1/3.6	4.1/4.2	0.1	4.5/3.6	4.3/4.1	4.3/4.7
1	5.1/4.3	4.4/3.9	4.2/4.3	0.6	9.3/8.5	6.9/7.2	5.8/6.9
5	12.6/11.2	8.8/8.4	6.3/6.5	1.5	17.5/17.5	13.3/13.7	10.1/11.7
10	18.3/17.2	13.6/13.3	9.6/9.9	3	28.1/28.2	23.3/23.1	18.9/20.1

As noted previously, the numerical integration provides a means of identifying the volume of rain that contributes to the MI return at a particular time. In Fig. 1, the -3 dB, -6 dB and -9 dB contours of the MI return are shown for three observation times. In all cases, the -3 dB is the innermost contour. The levels have been defined relative to the maximum value: e.g., the -3 dB

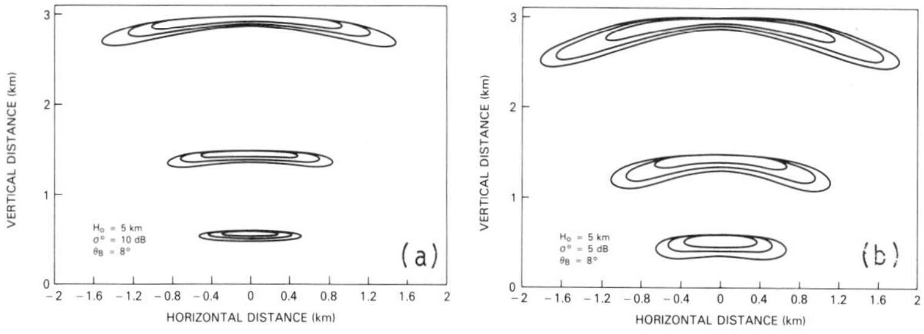

Figure 1. Power Level Contours of -3, -6 and -9 dB for the MI return at three altitudes for (a) $\sigma°$ = 10 dB and(b) $\sigma°$ = 5 dB.

contour is defined as the set of elements whose power is down 3 dB from the element that contributes the maximum MI power. For both figures H = 5 km, θ = 8°, and the observation times are selected to correspond to heights of the MI volume equal to 0.6, 1.5 and 3 km. In Fig. 1a, $\sigma°$ = 10 dB while in Fig. 1b $\sigma°$ = 5 dB. The illustrations show that the volume of rain that contributes to the MI return increases both with height and surface roughness so that the effective MI rain volume at distances far from the surface can be many times greater than the direct rain volume.

This broadening of the MI volume implies that the rain reflectivities appearing in the expressions for P_m and P_d are generally not equal as they refer to quantities averaged over different volumes of rain. Algorithms which assume their equivalence are therefore subject to an error. Moreover, relative to the direct return, the MI return should appear smoothed because of the greater spatial averaging. Finally, it should be noted that because of the much smaller beamwidths used for spaceborne rain sensing, the MI volume exhibits less broadening than from a typical airborne radar. Errors arising from mismatched rain volumes are therefore less severe.

2. GENERALIZED EXPRESSION FOR THE MI POWER

The results of Fig. 1 suggest that for a vertically stratified storm, the MI return can be expressed as an integral of the direct rain return and a function which represents the broadening of the MI return. As a first step, we replace the rain reflectivity, η, which appears in Pm with a spatially weighted version of the same quantity; that is, we replace η (hj) in

$$P_m(h_j) = \frac{C\, R^4\, \sigma^0\, 10^{-0.2(A_n+A_j)}\eta(h_j)}{\sigma^0 H^2 \theta_B^2 + 11 R^2 h_j^2} \tag{2}$$

(where C is a radar constant), with

$$N(h_j) \int_{h=0}^{h_j} \eta(h)\, f(h_j, h)\, dh \tag{3}$$

where

$$N(h_j) = \left[\int_{h=0}^{h_j} f(h_j, h)\, dh \right]^{-1}$$

Next, the rain reflectivity which appears in Eq. (3) is expressed in terms of the directly measured rain return, P_d, by

$$\eta(h) = P_d\,(H\text{-}h)\, \theta_B^2\, (H\text{-}h)^2\, 10^{0.2(A_n - A_j)}/C \tag{4}$$

Note that the direct rain return is measured from the radar, whereas the reflectivity in Eq. (2) and Eq. (3) is, by convention, assumed to be measured from the surface. Substituting Eq. (4) into Eq. (3) and then into Eq. (2) yields

$$P_m(h_j) = \frac{R^4\, \sigma^0\, 10^{-0.4A_j}\, N(h_j)}{\sigma^0 H^2 + 11 R^2 h_j^2/\theta_B^2} \int_{h=0}^{h_j} f(h_j, h)\, P_d\,(H\text{-}h)\,(H\text{-}h)^2\, dh \tag{5}$$

Eq. (5) is a generalization of Eq. (2) which accounts for the spatial broadening of the MI volume. One way to determine the distribution function f is by means of the numerical integration described previously. Since this has been found to be very time consuming, however, we employ an approximation based on simple scattering arguments. For the results to be presented, f(h,h') is given by the integral of $f'(h_j,h;\rho)$ over all positive values of ρ. The expression for f' is quite long: we quote the result only for the special case of $\rho = 0$.

$$f'(h_j, h; \varrho=0) = \frac{h_j^2-h^2}{(h_j+h)^4} \exp\{-2[(h_j^2-h^2)/s^2(h_j+h)^2 + 2.76\theta_B^{-2}(h_j^2-h^2)/H^2]\}$$

where s, the rms surface slope, is related to $\sigma°$ by $s^2 = R^2/\sigma°$. Several features of f' are worth noting. The function is zero for $h > h_j$; as $\sigma°$ increases the main contribution from f' comes from the region where h is nearly equal to h_j ; f' is not expressible as a function of the difference $h_j - h$ so that it is not range invariant (i.e., the width of the distribution changes as a function of the maximum height of the MI volume, h_j).

To test Eq. (5) we use data obtained during the RRL/NASA rain-measuring experiment. The solid lines in Fig. 2 represent the range-normalized radar return powers as a function of the range. Both examples were measured at nadir incidence over the ocean at an altitude of about 4.5 km. Only the non-attenuating (3 cm) return is shown. The three components of the return are clearly distinguishable since the large, sharply peaked surface return at about 4.5 km provides a demarcation between the direct return (left) and the MI return (right). The three discontinuous lines in each figure represent the estimated MI return based on Eq. (5) for values of $\sigma°$ equal to 5, 10, and 15 dB, assuming negligible attenuation. In Fig. 2a, good agreement is obtained between the measured and estimated MI profiles if $\sigma°$ is assumed to be between 10 and 15 dB. A more interesting example is shown in Fig. 2b. In this case, discrepancies are evident both in shape and in magnitude. One

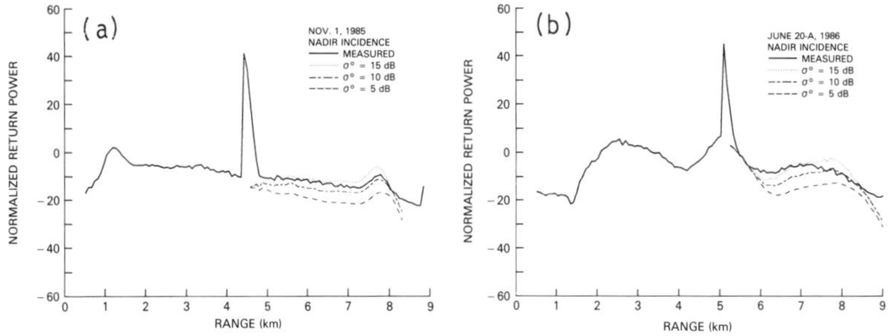

Figure 2. Comparisons of the measured radar return (3 cm) with the predicted MI return (Eq. (5)) for three values of $\sigma°$.

possible reason for the differences is that Eq. (5) is limited by the assumption of vertically stratified reflectivity. When this is not the case, the MI and the direct returns will no longer be simply related. Another reason may be that a better approximation is needed for the distribution function f.

3. ALGORITHMS FOR PATH ATTENUATION

The most useful property of the MI return is its potential for estimating rain attenuation at near-nadir incidence angles. There are two basic approaches for deriving algorithms for attenuation. In the first approach, we define the quantity Q(j,k) by

$$Q(j,\ k) = P_m(j)\ P_d(k)/P_d(j)\ P_m(k) \tag{6}$$

where the notation $P_m(j)$, $P_d(j)$ is used for the MI and direct returns at the jth range gate above the surface, corresponding to a height h_j. Using the expression for P_d and the simplified form of P_m (Eq. 2) in Eq. (6) yields

$$Q(j,\ k) = 10^{-0.4(A_j-A_k)} \left(\frac{H-h_j}{H-h_k}\right)^2 \left[\frac{(H\ \theta_B)^2\ \sigma^0 + 11\ h_k^2\ R^2}{(H\ \theta_B)^2\sigma^0 + 11\ h_j^2\ R^2}\right] \tag{7}$$

There are two cases for which the attenuation from h_k to h_j is easily solved. In the first case, the minimum distance from the surface is much greater than the field of view (FOV). Using this approximation in Eq. (7) and solving for $A_j - A_k$ gives

$$A_j - A_k \equiv \int_{h_k}^{h_j} k(s)\ ds = -\frac{5}{2} \text{Log} \left\{ Q(j,\ k) \left[\frac{h_j\ (H-h_k)}{h_k\ (H-h_j)}\right]^2 \right\} \tag{8}$$

At the other extreme, where the FOV is much greater than the maximum distance from the surface, the interval attenuation becomes:

$$A_j - A_k = -\frac{5}{2} \text{Log} \left[Q(j,\ k) \left(\frac{H-h_k}{H-h_j}\right)^2 \right] \tag{9}$$

In both Eq. (8) and Eq. (9), the estimates are independent of $\sigma°$ and of the radar constants. To obtain estimates under more general conditions we use the radar equation for surface scattering, assuming beam-limiting conditions, to express $\sigma°$ by:

$$\sigma^0 = P_s\ H^2\ 10^{0.2A_n}/C_s \tag{10}$$

where C_s is a radar constant, P_s is the measured return power from the surface, and A_n is the path attenuation through the entire rain column comprising n range gates. Substituting Eq. (10) into Eq. (7)

and letting j -> n and k -> 1 (so that $A_k = 0$) yields a cubic equation in x where $x = \exp(0.46\ A_n)$. A solution therefore provides an estimate of A_n. It may be worth noting that Eq. (8) has been derived previously while Eq. (9) and the procedure just outlined are new.

The second approach to finding attenuation algorithms begins with the ratio $T(j) = P_m(j)/P_d(j)$. By substituing the expressions for P_m and P_d into this equation, using Eq. (10) and letting j -> n, an estimate of total path attenuation is found:

$$A_n = 5 \text{ Log} \left[\frac{b}{2a} \left\{ [1 + 4ac/T(n)b^2]^{1/2} - 1 \right\} \right] \qquad (11)$$

where $a = H^4 P_s\ \theta_B^2$, $b = 11\ R^2 h_n^2 C_s$, and $c = (H - h_n)^2 R^4 H^2 P_s \theta_B^2$.

As before, simplifications result by assuming that the FOV is either much greater or much less than the height h_n of the MI return.

Fig. 3 shows two scatter plots of the rain rates estimated from the Z-R relation (using the direct rain return at 3 cm), compared with two of the above algorithms as applied to the 0.87 cm MI return. Because of the low altitude of the aircraft on this day (Oct. 2, 1985) and the good sensitivity in the 0.87 cm channel, this is the only flight in the experiment for which extensive 0.87 cm mirror-image data were acquired. The MI algorithm used in Fig. 3a is given by Eq. (11) while for Fig. 3b, the appropriate algorithm is given by Eq. (9). In both cases, the FOV is assumed to be much greater than the relevant MI height. Although good correlation coefficients are obtained between the ZR and MI methods (0.88 for Fig. 3a, 0.85 for Fig. 3b), the MI results are obviously positively biased with respect to the Z-R results. Initial tests of MI algorithms for which the opposite extreme holds (FOV << h), specifically Eq. (9), show a decrease in the magnitude of the rain rate estimates. On the other hand, the correlation decreases and the number of cases increases for which the method fails.

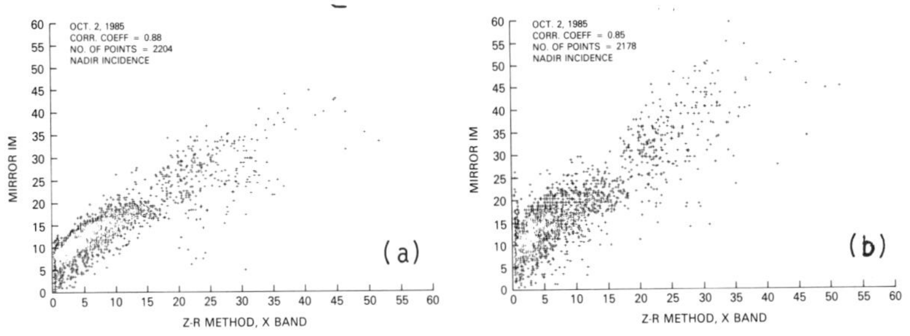

Figure 3. Comparisons of the Z-R method (3 cm) with two MI algorithms (0.87 cm) using Eq. (11) for (a) and Eq. (9) for (b).

4. CONCLUSIONS

In this paper, two tests of the mirror-image (MI) return were outlined. In the first, a comparison of the numerically integrated MI return with a previously derived closed-form expression showed fairly good agreement between the two. Because of the broadening of the MI volume, a generalized expression for the MI power is proposed. As a second test of the theory, estimated rain rates derived from the MI return were compared to those found from the Z-R method. Despite an apparent bias in the MI results, good correlations between the MI and other methods are obtained. Although these initial results are encouraging, further analyses and airborne measurements are required.

REFERENCES

Ishimaru, A., 1978: Wave Propagation and Scattering in Random Media. Vol. 2, Academic Press, 572 pp.

Meneghini, R. and D. Atlas, 1986: Simultaneous Ocean Cross Section and Rainfall Measurements from Space with a Nadir-Looking Radar, J. Atmos and Oceanic Tech, Vol 3, 400-413.

RADAR RAINFALL MEASUREMENT SYSTEM USING SAR AND CONICAL-SCAN METHODS

*Richard K. Moore and Shou-Sheng Xie**
Radar Systems and Remote Sensing Laboratory
University of Kansas Center for Research, Inc.
Lawrence, Kansas 66045-2969, USA

ABSTRACT

Vertical profiling of rain with space radars demands very narrow elevation beamwidths that require large apertures (in one dimension). For radars that scan to the side, synthetic-aperture methods allow reduction of the other dimension. Availability of new large and light space antenna structures will permit use of scanning real-aperture radars, with conical scan preferred to make the cells all the same size and the surface clutter uniform.

Here we describe the required aperture dimensions for different swath widths for the vertical dimensions. Next, we describe a 15-GHz SCANSAR that can achieve a 600-km swath with good profiling capability and a 3.5-km sq. surface footprint. We then describe a forward-scanning real-aperture 15-GHz system using a 7 x 3.37-m aperture. With an angle of incidence of 32° and scanning 45° either side of the flight line, this system covers a 441-km swath. Finally we discuss a "squint-mode" SAR example. This uses conical scan for the receiver with a wide conical transmit beam. It leaves a coverage gap along the flight line since SAR does not work in the forward direction. Swath coverage is somewhat less than for the other configurations.

These approaches should be considered for future spaceborne rain radars.

1. INTRODUCTION

Spaceborne rain radars can be built in many forms. The purpose of this paper is to present some ideas for rain radars that can profile the rain rate in elevation. The radars as presented use the normal relationship between backscatter and rain rate, although, of course, these relations take many forms and one must select the proper one. The methods, however, could be adapted to surface reference and double-reflection radars with some changes. Moreover, these radars, like others, could be combined with microwave radiometers using the same antenna system, although the microwave radiometers would have considerably poorer spatial resolution.

*on leave from Institute of Electronics, Academia Sinica, Beijing, P.R.C.

ISBN 0-937194-14-4

These radars use a fine range resolution and a narrow vertical beamwidth for the antenna. Figure 1 illustrates the general idea. A narrow vertical beam illuminates a section of rain at a scan angle of θ relative to the vertical at the ground. The fine range resolution permits distinguishing the returns from a small region of the rain (shown cross-hatched). Because the vertical beamwidth cannot be made narrow enough to obtain resolution similar to that in range, the cell is slanted as shown, resulting in a larger vertical resolution (between the top of the cylindrical cell and the bottom) at larger tilt angles. If the vertical beam is small enough the vertical resolution can be kept within reasonable limits for reasonable values of θ.

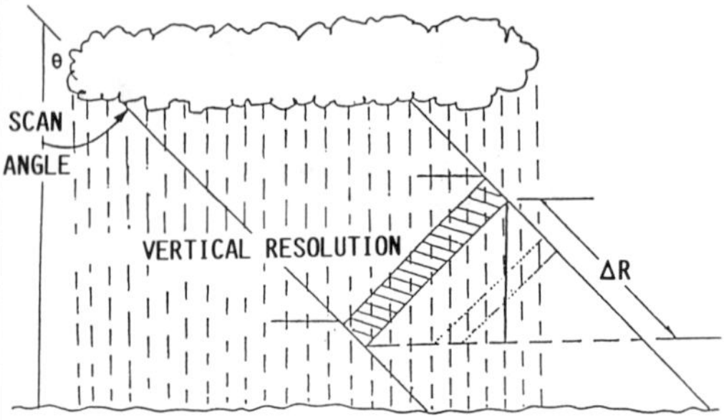

Figure 1. Concept of fine-vertical-beam rain profiling.

Figure 2 illustrates the same concept, but shows the elements of the geometry. With a vertical (elevation) beamwidth, β_E, the slanted width of the cell is shown to be $\beta_E R$. The slant range resolution is r_R. At vertical incidence ($\theta = 0$) the vertical resolution is simply the slant range resolution r_R. At any other value of θ the vertical resolution r_V is as shown in the picture. Clearly, this value is given by

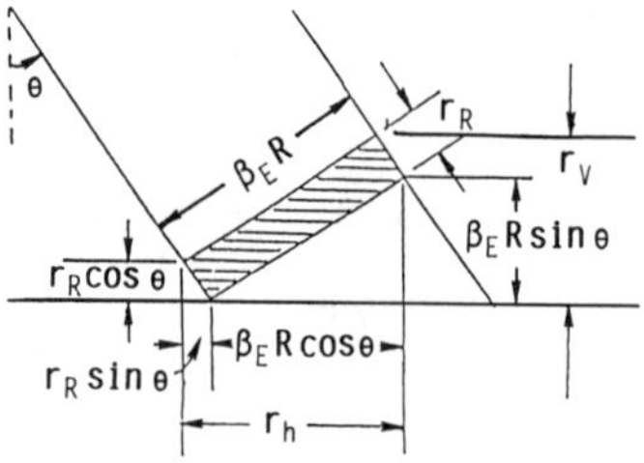

Figure 2. Vertical resolution in rain for scanning radar.

$$r_V = \beta_E R \sin\theta + r_R \cos\theta \tag{1}$$

The horizontal resolution on the ground is $\beta_E R/\cos\theta$. The horizontal projection of the resolution in the rain itself is given by

$$r_h = \beta_E R \cos\theta + r_R \sin\theta \tag{2}$$

For relatively large angles of incidence, θ, the vertical beamwidth, β_E determines the vertical resolution in the rain, so β_E must be very narrow, requiring a large aperture in this direction. The transverse horizontal resolution (normal to the paper) is determined by the transverse (horizontal) beamwidth of the antenna, β_h, and the horizontal resolution in the other direction is determined by β_E and r_R as shown in Fig. 2. The transverse beamwidth β_h may be obtained either with a real aperture or by synthesizing an aperture.

The basis for the rain measurement in the examples given is the Z-R relation.

$$Z = 372\ R_r^{\ 1.54} \tag{3}$$

If one prefers, one of the many other forms of this relation may be used (Im et al., 1987). As long as we assume that such a relation between the backscatter parameter Z and the rainfall rate R_r applies, we may write the volume scattering coefficient σ_V as

$$\sigma_V = 1.9 \times 10^{-9}\ Z \tag{4}$$

where σ_V is measured in m^2/m^3 and z is in mm^6/m^3.

This may then be used in the standard radar equation for volume scatter in which attenuation through the medium is accounted for:

$$S = \frac{P_t A_r \sigma_v r_R^2\ L_b L_r 10^{-0.2 \int_R^{R_t} \alpha(z) dR}}{2\pi R^2 kTFc} \tag{5}$$

where P_t = transmitter power (W), A = antenna area (m^2), r_R = range resolution (m), L_P = bright-band loss factor, L_r = radar loss factor (accounting for wave-guide losses, antenna inefficiencies, etc.), α - extinction coefficient (dB/km), R = slant range (m), k = Boltzmann's constant, T = temperature (K), F = receiver noise factor, and c = speed of light. R_t in the attenuation integral is the slant range to the top of the raining region.

2. SCANNING RADAR TECHNIQUES

Three different scanning modes are discussed here for the radar with narrow elevation beamwidth β_E. These are shown in Fig. 3. Figure 3a shows a side-to-side scanning synthetic aperture radar (SAR). The same approach could be used with a real-aperture radar

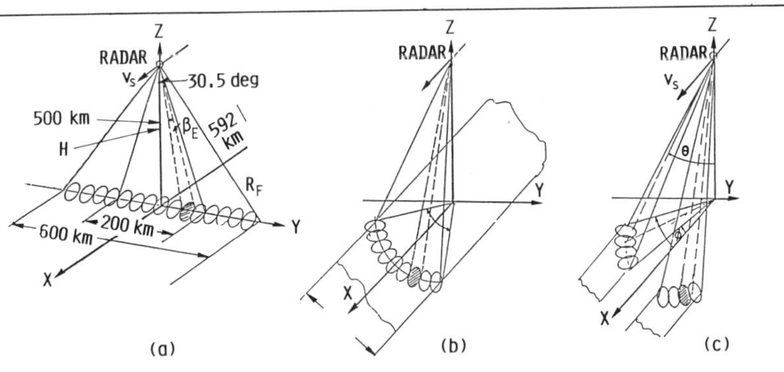

Figure 3. Three concepts for scanning radars to profile rain rates; a) Side-to-side scanning SAR, b) Forward-scanning real-aperture radar, c) Conical-scanning SAR.

(RAR), but the antenna would have to be much longer in the direction of flight for a RAR than for a SAR. In the figure, the beam is scanned successively from a maximum angle on one side of the spacecraft to smaller and smaller angles, through vertical and out to a maximum angle on the other side of the spacecraft. For the examples considered here, we treat a 500-km height H and a total scanning swath of 600 km. This requires, for a spherical Earth, that the beam be scanned at the spacecraft over a range ± 30.5°. The concept of a scanning SAR is discussed later. Figure 3b shows a different configuration, a forward-scanning RAR. As for a side-scanning RAR, the horizontal dimension of the antenna must be relatively large to achieve a reasonable resolution normal to the radiation direction. In the forward-scanning mode, synthetic-aperture techniques cannot be used. The forward scan, however, has significant advantages, if the large antenna can be accommodated, because it provides a wide swath at a constant angle of incidence. The geometry of each cell is the same as that as every other cell, which can be convenient. Moreover, the interfering backscatter from the surface is essentially constant at a given angle of incidence, although some variation can be expected because of differences in the characteristics of the surface. These variations are usually smaller than those resulting from changing the angle of incidence.

Figure 3c shows a means to use the synthetic-aperture technique for a constant-angle-of-incidence conical scan. In this case, the nature of the synthetic aperture prohibits use of the forward direction, but allows the use of directions away from the forward direction along the arc corresponding to constant angle of incidence. Of course, this leaves a gap in the center of the region of coverage, but for some purposes this may not be a big enough disadvantage to overcome the advantages due to constant angle of incidence and smaller along-track antenna size. This is a so-called "squint-mode" SAR.

The side-to-side scanning SAR is illustrated graphically in Fig. 4. This SCANSAR concept was originally proposed by Moore (Moore et al., 1981). The basic idea is to use the total time that a point is illuminated by the antenna beam to scan to different positions. The entire time could be used for a single scan position to create a synthetic aperture as long as the area on the ground is illuminated by the real aperture. The synthetic resolution in the along-track direction would then be one-half the antenna length (L/2). If we only use a portion of this time to build a synthetic aperture, dividing the time up into N pieces so that the time for the aperture is T_a/N, the resolution is

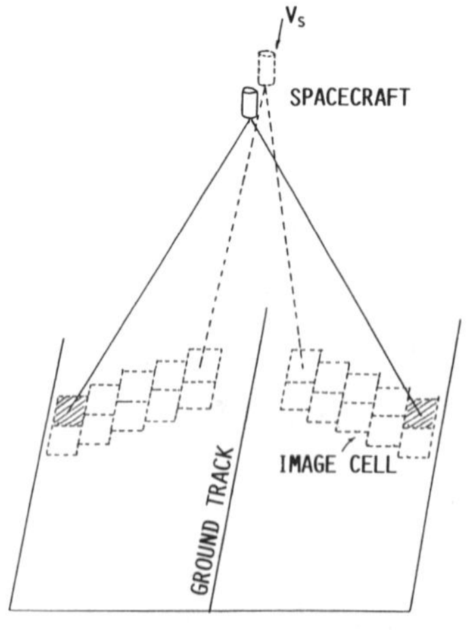

Figure 4. Pictorial concept of side-to-side scanning SAR.

poorer by a factor of N. Hence, the resolution for this situation is NL/2. For example, if we wish a 3.5-km resolution and use an antenna 2 m long for a rain radar, 3500 subapertures of duration T_a/N are possible. The time not used for a single aperture is T_a (1 - 1/N).

This extra time can be used to scan the beam to other elevation angles or to produce multiple looks at the same point to allow non-coherent averaging to obtain greater precision. In a typical spaceborne rain radar, a combination of these two techniques would be used, so that multiple beam positions and averaging to improve performance are both possible.

This situation is shown in Fig. 4. Initially, the first position closest to the spacecraft is illuminated. After the spacecraft moves a small distance, the beam jumps out to the next position. This process is repeated until the cross-hatched positon is reached, at which point the beam must return to the steeper initial angle of incidence so that the next cell closest to the spacecraft can be imaged. The number of beam positions used is simply the ratio of the total swathwidth to the ground beam footprint in the cross-track direction (determined by the elevation beamwidth).

The ability to scan to many beam positions and the modest resolution requirement mean that the rain radar can use a small enough vertical beamwidth to measure the rain rate as a function of height and still be able to average many samples to improve precision. The same concept may be applied either to a side-to-side scan or to a conical scan, although the shape of the cells for the conical scan is more like a diamond than like a rectangle.

3. SIDE-TO-SIDE SAR RAIN RADAR EXAMPLE

To illustrate the radar parameters appropriate for rain measurement from space, we have chosen an example with a radar at a frequency of 15 GHz, a swath width of 600 km, and a horizontal resolution of 3.5 km. The parameters of the radar are listed in Table 1.

TABLE 1. SIDE-TO-SIDE SAR EXAMPLE

Frequency	15	GHz
Antenna (Cross-Track)	7	m
(Along-Track)	1.5	m
Cross-Track Beam Width	0.16°	
Scan Angle	30.5°	
Swath Width	600	km
Range Resolution	30	m
Horizontal Resolution		
Cross-Track	3.5	km
Along-Track (1 Look)	280	m
(13 Looks Averaged)	3.5	km
SNR for 0.5 mm/hr to 50 mm/hr		
Rain Top	35	dB
Rain Bottom	7	dB

The range resolution is set at 30 m, but multiple samples are averaged together (5-10) to obtain better measurement precision. For the example, the cross-track antenna aperture is 7 m long, but the along-track antenna aperture is only 1.5 m. The along-track aperture could be made smaller, but more power would be required. Figure 5 illustrates the effect on the vertical resolution of choosing different cross-track antenna aperture sizes. If only 1 m were used, the minimum rain height that could be profiled (1 vertical resolution) clearly is quite large. It is 1 km at a distance of only about 60 km to the side of the surface track. However, if a 15-m aperture were used. the vertical resolution could be kept smaller than about 300 m out to the full swathwidth of 350 km. For the example chosen, we have selected 7 m, which gives reasonable results (less than 1 km out to the full 600-km swath).

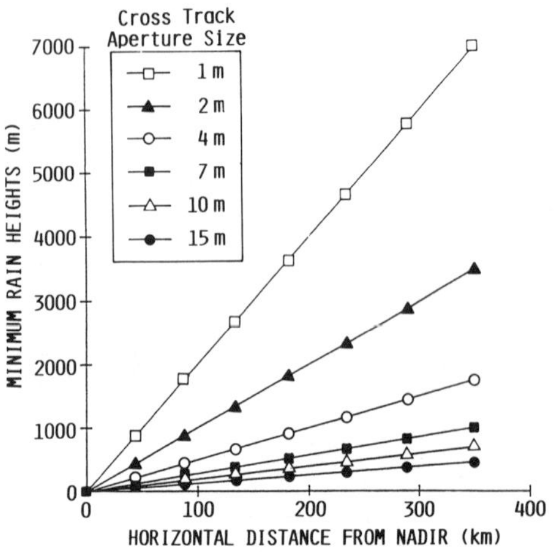

Figure 5. Performance of cross-track SCANSAR with different antenna sizes.

With a peak transmitter power of 100 watts, the signal-to-noise ratio per pulse is illustrated in Fig. 6 for both the bottom and top of a 5-km-thick rain cell at different rainfall rates. Clearly, the signal-to-noise ratio is better than 0 dB for very small rainfall rates. It is better than 10 dB for rainfall rates exceeding 0.5 mm/hr at the top of the rain, and up to 50 mm/hr at the bottom, where attenuation reduces the return. Thus a radar like this can achieve a 600-km swath with good signal-to-noise ratio for a wide range of rainfall rates. Figure 6 shows

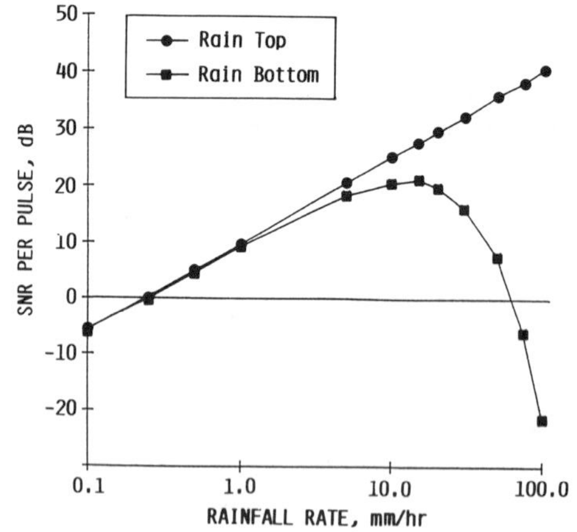

Figure 6. Performance of 100-watt SCANSAR at vertical incidence.

the situation at vertical incidence and Fig. 7 shows the same situation at 33.2° angle of incidence (at the ground), corresponding to 300 km to the side of the spacecraft track. When we perform all the possible averages, we find that the measurement error in rain echo power is as shown in Fig. 8 for 300-km and 600-km altitudes. Clearly, the higher altitude is satisfactory, although the lower altitude could be used to allow better resolution for a radiometer on the same spacecraft.

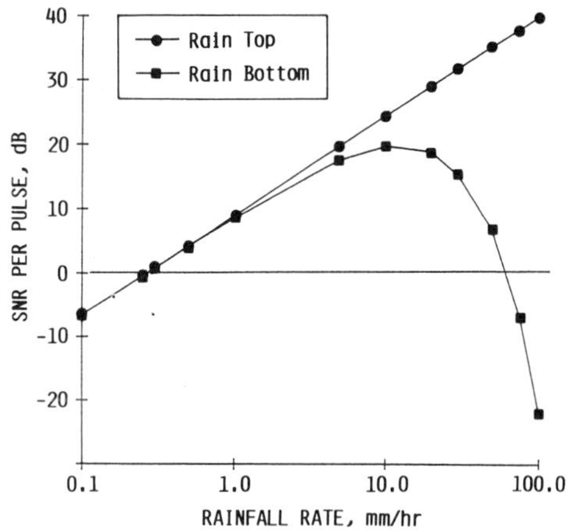

Figure 7. Performance of 100-watt SCANSAR at 33.2° angle of incidence.

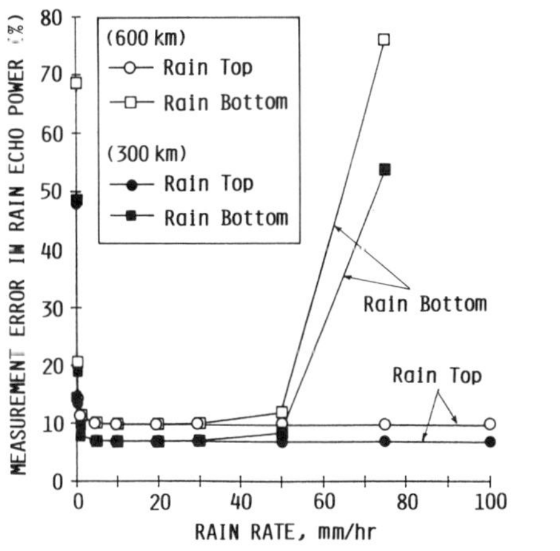

Figure 8. Measurement error for rain scatter with 100-watt SCANSAR at 32.2° angle of incidence. (Based on SNR and averaging.)

4. EXAMPLE OF FORWARD-SCAN REAL-APERTURE RAIN RADAR

The forward-scanning configuration shown in Fig. 3b has significant advantages, but requires both more power and a larger antenna. For the example chosen, the transmitter power is 150 W and the antenna is 7 x 3.37 m. With the 7-m "vertical" aperture the vertical resolution in the rain is satisfactory. The 3.37-m horizontal aperture allows obtaining a horizontal resolution of 3.5 km. The parameters of the radar are shown in Table 2.

TABLE 2. FORWARD-SCAN REAL-APERTURE RAIN RADAR

Altitude	500	km
Frequency	15	GHz
Angle of Incidence	32°	
Scan Angles (From Flight Direction)	±45°	
Swath Width	441	km
Antenna	7 x 3.37	m
Elevation Beamwidth	0.16°	
Azimuth Beamwidth	0.34°	
Range Resolution	30	m
Horizontal Resolution	3.5	km
Peak Transmission Power	150	W
SNR for 0.5 mm/hr to 50 mm/hr Rain Top and Bottom	10	dB

The effect of different antenna heights for this case is shown in Fig. 9 as a function of the selected angle of incidence. The results are quite satisfactory for antennas of 7 m and larger up to 35° and, in fact, would be reasonable beyond that.

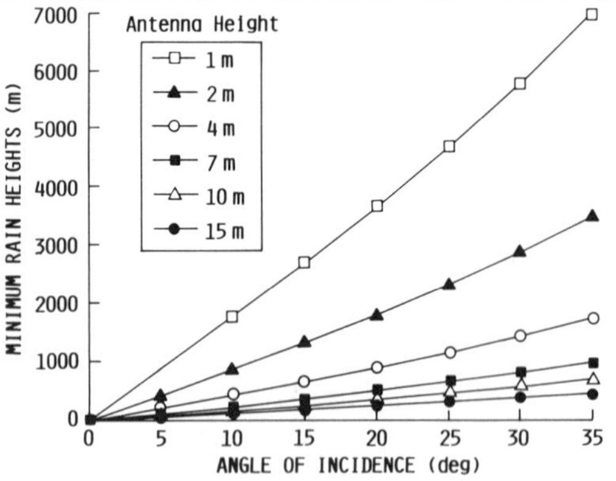

Figure 9. Minimum detectable rainfall heights (or resolutions) vs. angle of incidence.

Figure 10 shows the signal-to-noise ratio for this example, using 32° angle of incidence. Both top and bottom of the 5-km-high rain are shown. The signal-to-noise ratio exceeds 0 dB for each pulse as long as the rain rate exceeds 0.5 mm/hr at the top and for rain rates up to about 60 mm/hr at the bottom.

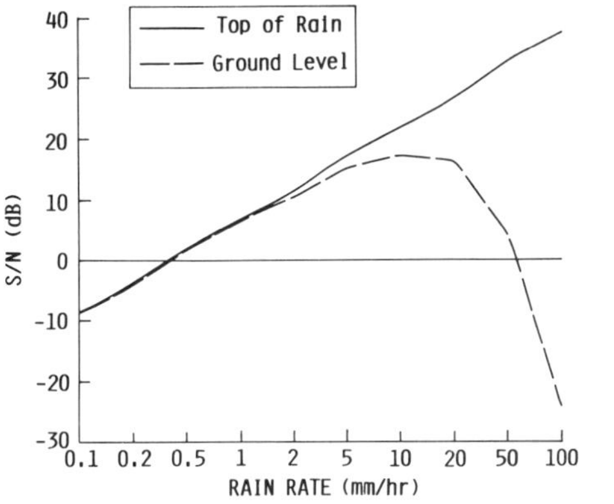

Figure 10. Signal-to-noise ratio for scanning RAR with 150-watt peak power, 3.37 x 7-m antenna, 500-km height, and 32° incidence.

5. CONICAL-SCANNING SAR EXAMPLE

A conical-scanning SAR can be made using either the entire antenna for transmitting and receiving or a narrow antenna for transmitting and synthetic subdivision for receiving. For the example selected, the receiving antenna is made somewhat larger, but the entire area is illuminated at one time by the transmitting antenna. The antenna selected is 8 m high and the receiving aperture is 3.5 m. The frequency again is 15 Ghz, and the scan is from 10° to the side of the flight track to 45° to the side. There is little advantage in going beyond 45° to the side because of the geometry of the conical scan. The peak transmitter power in this case needs only to be 70 W. The parameters of the radar are shown in Table 3.

TABLE 3. CONICAL-SCANNING SAR EXAMPLE

Altitude	500	km
Frequency	15	GHz
Angle of Incidence	32°	
Scan Angles (From Flight Direction)	10 to 45°	
Antenna (Receive)	8 x 3.5	m
(Transmit)	8 x 0.0065	m
Elevation Beamwidth	0.28°	
Azimuth Beamwidth (Receive)	1.16°	
(Transmit)	35°	
Range Resolution	30	m
Peak Transmission Power	70	W
SNR for 0.5 mm/hr to 50 mm/hr Rain Top and Bottom	10	dB

The signal-to-noise ratio is shown in Fig. 11 for 33° as a function of rain rate. The radar in this example has a signal-to-noise ratio exceeding 10 dB for rain rates exceeding 0.5 mm/hr for the top of the rain and up to about 50 mm/hr for the bottom. The precision of measurement is shown in Fig. 12, where the error is quite small over the range from 0.5 mm/hr to almost 100 mm/hr. For small rain-rates it is degraded because of weak signals from both top and bottom, and for large rainrates it is degraded because of the attenuation of the signal from the bottom.

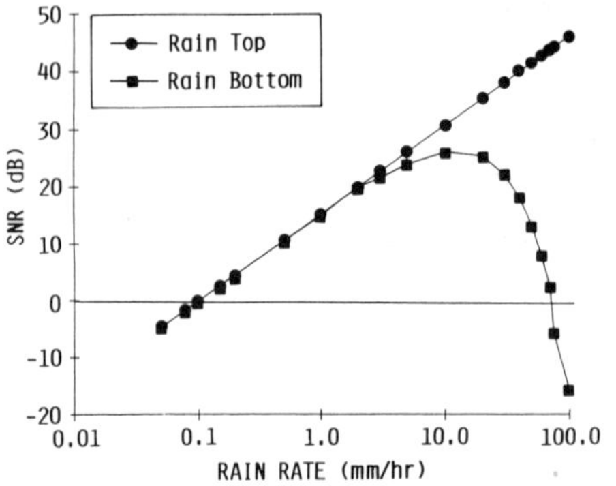

Figure 11. Signal to-noise ratio performance of rain echoes measured by conical scan SAR at incident angle of 33.2°.

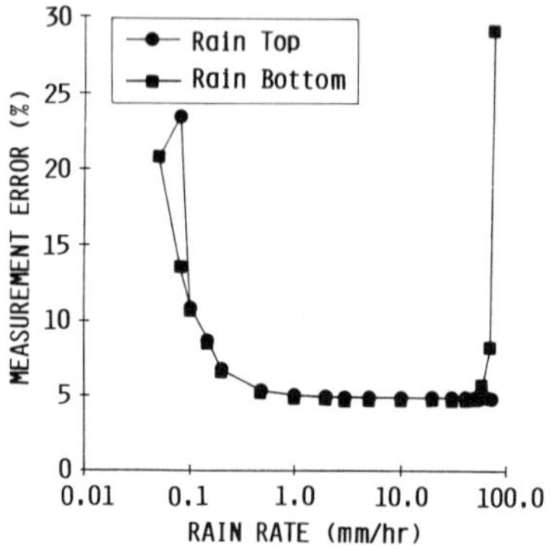

Figure 12. Power measurement precision of rain echoes measured by conical scan SAR (peak power of 100 watts, incident angle of 33.2°).

6. CONCLUSIONS

We have shown here that the use of a fine range resolution and a narrow vertical beamwidth allows profiling rain rate as a function of height. Scanning synthetic-aperture methods allow wide swathwidths with reasonable antenna sizes, as illustrated here. A cross-track SCANSCAR is relatively easy to build, but suffers because the cell sizes are a function of position in the scan as a result of the differing angles of incidence.

With a conical scanning radar, the cell sizes are all essentially the same. The conical scan may be obtained either with a real aperture scanning from side to side about the subsatellite track or with a conical-scanning synthetic-aperture radar that leaves a gap near the subsatellite track. Processing for the conical-scanning SAR is more complicated than for the side-to-side scanning SAR. The forward-scanning real-aperture radar is quite simple and allows a wider swath than the conical scanning SAR, but it requires a bigger antenna. The power levels and range sensitivities obtained for the examples are reasonable, with peak powers from 70-150 W for 15 GHz.

These same scanning concepts could be applied for other types of rain radars. They might be useful in some surface-reference techniques and in the double-bounce method proposed by Atlas (Atlas and Matejka, 1985).

Although it appears too late to use such a method for the TRMM mission, we believe it is quite important to consider these methods and variations of them for future rain-measuring missions where more time is available for study and design.

REFERENCES

Atlas, D., and T. J. Matejka, 1985: Airborne Doppler Radar Velocity Measurements of Precipitation Seen in Ocean Surface Reflection, J. Geophys. Res., 90, 5820-5828.

Im, K. E., F. K. Li, W. J. Wilson, and D. Rosing, 1987: Conceptual Design of a Spaceborne Radar for Global Rain Mapping. In M. C. Dobson (Ed.), 1987 IEEE IGARSS Digest, Vol. I, IEEE, New York City, New York, 299-304.

Moore, R. K. , J. P. Claassen, and Y. H. Lin, 1981: Scanning Spaceborne Synthetic Aperture Radar with Integrated Radiometer, IEEE Trans. Aerospace Elec. Sys., AES-17, 697-708.

RAIN RETRIEVAL ALGORITHMS FOR SPACEBORNE RADAR

David Atlas[1] *and Robert Meneghini*
NASA Goddard Space Flight Center
Greenbelt, Maryland 20771, USA

ABSTRACT

We reviewed a family of algorithms proposed for the estimation of rain rates by a spaceborne radar. Virtually all depend upon the measurement of rainfall attenuation, either through the use of a surface reference echo, the relative attenuation of the precipitation echoes at two wavelengths, or the single wavelength echo loss in vertically uniform stratiform rain. The mirror-image method allows the simultaneous estimate of rain rate and surface properties at one wavelength but is most demanding in total power. Preliminary aircraft experiments show that several methods are promising.

1. INTRODUCTION

In this paper we discuss the essential features of the algorithms proposed for retrieving rainfall estimates with a spaceborne radar or jointly with a microwave radiometer. Some of these algorithms have been tested in aircraft trials; preliminary results are treated by Meneghini and Nakamura (1987).

2. BACKGROUND

We cannot use measurements of the reflectivity factor, Z, of rain and its relationship to the rainfall rate, R, because such a method requires long non-attenuating wavelengths and excessively large antennas; or the beams are so large that it is difficult to separate the surface clutter from the precipitation; and the empirical relations between Z and R are dependent upon the drop size distribution (DSD) and vary greatly with the precipitation type. The alternative is to use the shorter attenuating wavelengths in the 1 and 3 cm bands where the attenuation is almost perfectly linear with rain rate and is essentially independent of the DSD (Atlas and Ulbrich, 1977).

Most of the algorithms which we shall describe are based upon the measurement of attenuation, A, and its relation to rain rate, R. When R becomes small, however, the attenuation also becomes small relative to the measurement errors; then we shall resort to the use of reflectivity as our only alternative. In general, attenuation by rain is given by

1 Independent consultant under NASA Contract

$$A = \int_0^H kR\,dH \qquad (1)$$

where A is the total path attenuation in dB, R is rain rate in mm/hr, k is the attenuation coefficient in db/km per mm/hr, and dH is the incremental path length. In most of what follows we assume that k is independent of R so that we may write $A \approx kRH$ where R is now the path averaged rain rate, and H is the total path through rain. In order to deduce R we, therefore, need to measure or estimate H. For most purposes this can be taken as the slant range to the height of the freezing level, which corresponds to the top of the bright band in stratiform rain.

In order to measure the attenuation we need to use some reference to determine what the signal level would have been in the absence of rain. For this purpose we can use either the surface echo in a rain-free area adjacent to the storm, the relative attenuation of the surface echo at two wavelengths over which the dependence of the surface cross-section is a known function of wavelength, the relative attenuation of the precipitation itself at two wavelengths, or, when over the ocean, we may use the brightness temperature seen by a microwave radiometer. The latter is a useful proxy for attenuation provided that the corresponding attenuation does not exceed about 10 dB. When this is the case, the radiometer may be used to estimate the total attenuation while the radar provides the rain path length, thus providing the path-averaged rain rate. The radar and radiometer are complementary in many respects.

The various methods are illustrated schematically in Fig. 1. At the top we show the surface echoes with and without the presence of rain. The difference is the desired attenuation. We also see that the range-normalized echo power in the rain itself decreases at some average slope. When the rain can be assumed to be constant with distance, then that slope is simply $dA/dH = kR$ (from Eq. 1). This is one of the simplest algorithms, but the assumption of constant rain with height is valid mostly in stratiform rain, and not always even then. Also shown at the top of Fig. 1 is the mirror image of the rain echo seen through a double reflection from the ocean surface at ranges beyond the surface echo. This too can be used as an estimator as will be shown.

At the bottom of Fig. 1 is shown the relative increase in spread of the range-normalized echo powers at two wavelengths, the higher frequency attenuating more strongly than the lower one with distance into the storm. The rate of increase of the difference is proportional to R. Note that we also have the differences in the surface echoes at the two wavelengths so that the dual wavelength system includes the methods shown at the top. Indeed, we require at least two wavelengths to attain sufficient dynamic range in R.

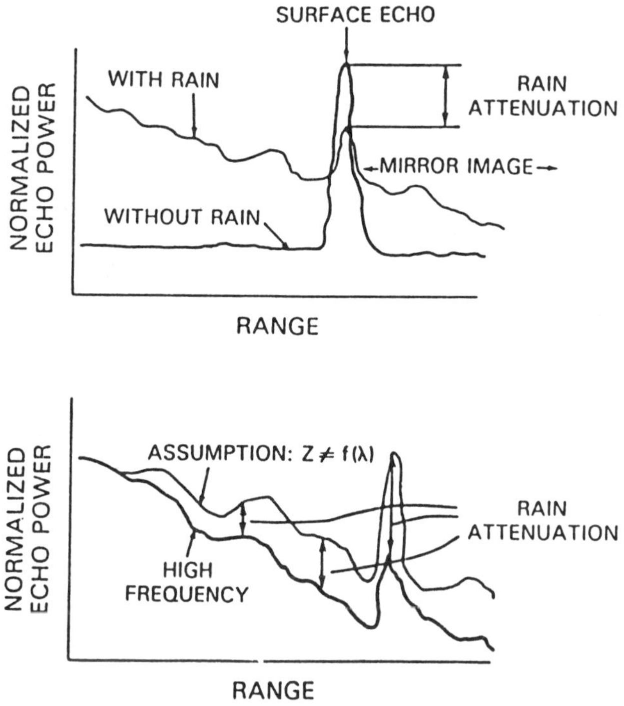

Figure 1. Top: Schematic of single wavelength algorithms; surface in and out of rain; mirror image; and negative slope of rain echo. Bottom: Dual-frequency algorithms for surface reference and relative attenuation in rain at two wavelengths.

We note that high-resolution microwave systems may provide useful measures of total rain volume simply by measuring the storm areas with only crude estimates of rain rate, assuming that geosynchronous satellite cloud observations can furnish the data on the evolution of the storm areas between successive orbits (Doneaud et al.; 1984).

3. ALGORITHMS

3.1 SURFACE REFERENCE - SINGLE WAVELENGTH

The first approach is the surface reference method. This and the dual-wavelength analogue have been used with some success in the analysis of airborne radar data obtained from the Radio Research Labs (RRL), Japan, and NASA/RRL rain measurement experiments (Meneghini and Nakamura, 1987). The basic principle of the method is simple. If measurements of the surface return power are made in the presence and absence of rain, the ratio of the powers will be proportional to the path attenuation as long as the surface scattering properties do not change significantly between the two measurements. There are two main disadvantages: 1) Only the total path attenuation is estimated. Since the attenuation may include contributions from precipitating and non-precipitating hydrometeors, some other method is needed to separate out the precipitating component of the attenuation. 2) Another drawback is that the rain itself may alter the ocean surface cross-section. As with most attenuation methods, the surface reference is limited at light rain rates where the uncertainties in the measurement are comparable to the quantity to be estimated. The method is also limited at high rain rates when the signal is attenuated so strongly that the surface return is no longer detectable. The method has been simulated by Meneghini et al. (1983).

3.2 SURFACE REFERENCE - DUAL WAVELENGTH

The dual wavelength algorithm to estimate the path attenuation can be formed by applying the single wavelength surface reference to each wavelength and then taking the ratio of the results. For example, when expressed in decibel form, then the difference in echo power involves differences in: 1) The radar constants, which may be regarded as a calibration constant itself, and 2) The difference in the normalized radar cross-section (NRCS). The wavelength difference in the NRCS is either zero over the oceans for nadir angles less than about 20-25 deg (i.e., where the reflections are quasi-specular) or reasonably well-known under other conditions. The residual is then simply the difference in attenuation at the two wavelengths. Analysis of the method shows that the accuracy of the estimate improves relative to the single wavelength counterpart if a moderate degree of correlation exists between the surface return powers at the two wavelengths.

3.3 THE MIRROR IMAGE METHOD

This approach owes its origin to the observations by Atlas and Matejka (1985) showing virtual images of precipitation "below" the surface of the sea due to a double reflection involving energy reflected from the sea surface to the precipitation and back to the radar via a second reflection from the sea. Meneghini and Atlas (1986) adapted it to a space configuration as shown in Fig. 2a and b with a nadir beam. Fig. 2c shows a vertical profile of the precipitation echoes seen directly down to the surface and those from the surface. Below the surface we show the mirror image of the precipitation echoes. The latter echoes are reduced by the double surface reflection and further by the additional two-way propagation through the attenuating medium.

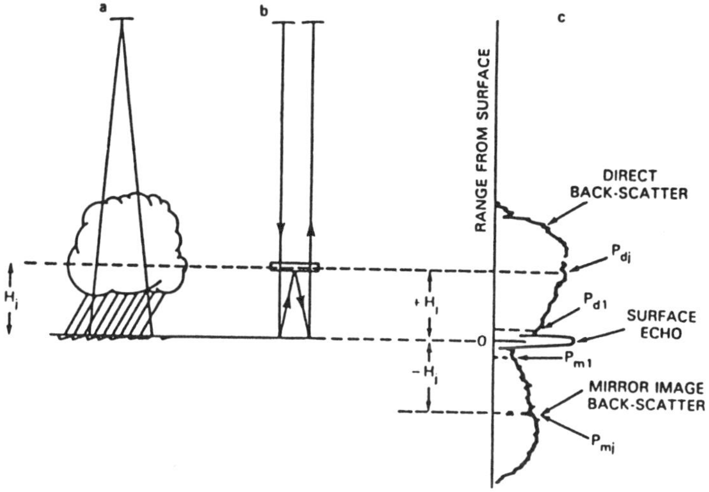

Figure 2. Schematic of the mirror image algorithm for nadir incidence. See text for details (After Meneghini and Atlas, 1986).

There is a general solution for any geometry, but the essence of the method may be understood qualitatively in this way. The difference in the directly scattered echo power from any height H_j and its mirror image at $-H_j$, involves essentially two components: 1) The loss in the two reflections from the surface (i.e., the Fresnel reflectivity coefficient) and any additional loss due to the diffusion by surface roughness which is related to the NRCS; and 2) The additional two-way loss through rain. The crucial point is that, within reasonable assumptions, the ratio of the mirror echo power to the direct echo power from the rain in the first gate, (P_{m_1}/P_{d_2}), accounts for the first of these factors, thus allowing the measurement of the rain attenuation and rate.

The method is applicable to all range gates so that one may obtain the vertical profile of A and R. Moreover, unlike the dual wavelength precipitation profiling algorithm discussed below, the method does not require that the precipitation reflectivity be independent of wavelength. Having calculated either the attenuation profile or its integral, one may then compute the rain-free echo power and deduce the surface NRCS and associated parameters such as the mean square surface slope of the ocean and the windspeed (Fedor and Brown, 1982).

The power of this technique lies in the fact that, in principle, we need not make any assumptions about the nature of the surface or invoke the assumptions involved in the previous methods. Moreover, it makes no difference if the rain itself is altering the surface NRCS. Indeed, this approach allows us to explore such rain effects. Of course, the method is not without its limitations. For one, it is probably restricted to use within a relatively small angle around nadir. It should work best over oceans, but should also operate over land with large enough NRCS. Secondly, because it involves an additional two-way path through rain, it will lose signal at smaller rain rates than will the other methods. Other potential problems are discussed by Meneghini and Atlas (1986). Meneghini and Nakamura (1987) discuss some experimental results.

3.4 CONVENTIONAL DUAL WAVELENGTH ALGORITHM

The essence of this algorithm, depicted at the bottom of Fig. 1, is that at any range, the ratio of the return powers at two wavelengths is proportional to the differential attenuation out to the range. If we write the equation for the echo power at an attenuating wavelength in decibels, it appears as

$$10 \mathrm{Log} P_r = 10 \mathrm{Log} C + 10 \mathrm{Log} Z - 2\int_0^H kR\,dH \qquad (2)$$

where C is the radar constant, Z the reflectivity factor, and the last term is the attenuation as previously defined. If we subtract the power at a highly attenuating wavelength from that which is less so, then we get

$$10 \mathrm{Log}[P_1/P_2] = 10 \mathrm{Log}[C_1/C_2] + 10 \mathrm{Log}[Z_1/Z_2] - 2\int_0^H (k_1-k_2)R\,dH \qquad (3)$$

We may account for the first term on the right by calibration in a non-attenuating region. If $Z_1 = Z_2$ the difference in echo power is simply due to the differential attenuation over the total path. We may then profile the attenuation by letting the end point of the path vary.

The primary weaknesses in this method are twofold. First, in order to obtain a measurably large attenuation, it is necessary that the wavelengths be spaced fairly wide apart (e.g., say 15 and 35 GHz). In this case, the rain reflectivities at both wavelengths are no longer equal (Wexler and Atlas, 1963) and the error in rain rate is of the order of this difference. On the other hand, if one uses wavelengths which are sufficiently close such that the Z's are equal, the attenuation coefficients, k, approach one another, and the differential attenuation becomes negligible. The same is true if the range increment is too small. Secondly, in order to measure the attenuation with some degree of confidence, the magnitude of the differential attenuation must exceed the "noise" in the ratio of the average echo powers. This requires that we average a large number of independent samples and leads either to large dwell times or reduced swath widths, or to very high pulse repetition frequencies and excessive radiated power. The method was developed further by Eccles and Mueller (1971).

3.5 SINGLE WAVELENGTH STRATIFORM AND BRIGHT BAND ALGORITHMS

The assumption that the rain rate is vertically uniform may sometimes be valid in stratiform precipitation. In such instances the rate of decrease of range-normalized echo power (in dB) is simply the range derivative of Eq. (1) or -2kR. This simple idea has been revived by Goldhirsh and Walsh (1982) for application to measurements from space with a modified Seasat A radar. Caution must be exercised because decreases in echo power may result from real gradients in reflectivity.

Just above the bright band (BB) in stratiform precipitation the reflectivity of snow closely approximates that of the rain just below on average. Thus, when the rain below is no longer detectable our last resort is to use the reflectivity of the snow as a rough proxy. Alternatively, one may use the BB itself because its reflectivity is typically 7 to 10 dB greater than that of the snow or rain on either side.

4. SPECIAL REQUIREMENTS

4.1 ADAPTIVE SCAN

In any realistic scenario for the spacecraft, the available power will be limited. Also, the confidence with which we can measure the echo powers from noisy targets, such as precipitation, depends largely upon the dwell time or the number of independent samples which can be averaged. At the same time we must scan the widest possible swath on each orbit to obtain a sufficiently large number of storm samples during the monthly averaging period. Thus, there is a conflict between the scan angle or swath width on one hand, and the number of independent pulse samples and overall power consumption on the other.

One solution to this conflict is adaptive scan, a concept first proposed by Atlas (1982). In its original form he proposed that a microwave radiometer scan in the forward direction to map the positions of precipitation cells, and that an electronically scanned radar be programmed to view only those cells seen by the radiometer. Assuming that on average only a fraction of the swath is filled, the radar can then concentrate all of its pulses and power on the precipitation, and not waste any energy on the echo-free regions. In the worst case scenario in which the entire swath may be filled, as in a hurricane, then the radar would be programmed to sub-sample the region in some logical fashion.

Elachi and Fuk Li (Personal communication, Jan. 1986, Jet Propulsion Laboratory) proposed that the radar itself could be designed to scan adaptively. The idea would be to use the radar about 10 percent of the time to scan the entire swath rapidly. During this scan the regions would be classified into several priority classes according to the first-order estimate of their echo intensity and area. The beam would then be programmed to return to dwell on each of the cells in priority order. As above, sub-sampling would be used whenever the swath was completely filled.

Clearly, adaptive scan requires that the beam be moved electronically rather than mechanically. This increases costs and complexity both in the on-board computer and in data processing. However, it appears to be the only way to achieve the ultimate objective of a usefully accurate rainfall measurement.

4.2 DISCRIMINATION OF PRECIPITATION FROM CLUTTER

If one wishes to measure rainfall close to the surface at relatively large nadir angles, one must be able to discriminate the precipitation echoes from those due to surface clutter. Im and Atlas (1987) have shown that this is possible at the shorter wavelengths and moderate to high rain rates such that the precipitation echoes exceed the clutter echoes, assuming that the NRCS of the surface is not too large. When the excess is sufficient, one can estimate the rain rate. Moreover, when the precipitation and the surface NRCS are uniform across the beam, the Im-Atlas algorithm permits the estimation of rain rate even at lower rates.

REFERENCES

Atlas, D., 1982: Adaptively pointing spaceborne radar for precipitation measurements. J. Appl. Meteor., 21, 429-431.

Atlas, D. and T. J. Matejka, 1985: Airborne Doppler radar velocity measurements of precipitation seen in ocean surface reflection. J. Geophys. Res. 90, 5820-5828.

Atlas, D. and C. W. Ulbrich, 1977: Path- and area-integrated rainfall measurement by microwave attenuation in the 1-3 cm band. J. Appl. Meteor., 16, 1322-1331.

Doneaud, A. A., S. Ionescu-Niscov, D. L. Priegnitz, and P. L. Smith, 1984: The area- time integral as an indicator for convective rain volumes. J. Climate Appl. Meteor., 23, 555-561.

Eccles, P. J. and E. A. Mueller, 1971: X-band attenuation and liquid water content estimation by a dual-wavelength radar. J. Appl. Meteor., 10, 1252-1259.

Fedor, L. S. and G. S. Brown, 1982: Wave height and windspeed measurements from the Seasat radar altimeter. J. Geophys. Res., 87, 3254-3260.

Goldhirsh, J. and E. J. Walsh, 1982: Measurement of Precipitation with a TOPEX-Class dual frequency satellite altimeter. Proceedings of URSI Commission F, Multiple Parameter Radar Measurements of Precipitation. Bournmouth, UK.

Im, K. E. and D. Atlas, 1987: The estimation of precipitation intensity in the presence of surface backscatter and the converse using a spaceborne radar. This volume.

Meneghini, R. and D. Atlas, 1986: Simultaneous ocean cross-section and rainfall measurements from space with a nadir-looking radar. J. Atmos. and Oceanic Tech. 3, 400-413.

Meneghini, R., J. Eckerman, and D. Atlas, 1983: Determination of rain rate from a spaceborne radar using measurements of total attenuation. IEEE Transactions on Geoscience and Remote Sensing, GE-21, 34-43.

Meneghini, R. and K. Nakamura, 1987: Some characteristics of the mirror image retorn in rain. This volume.

Wexler, R. and D. Atlas, 1963: Radar reflectivity and attenuation of rain. J. Appl. Meteor., 2, 276-280.

RAIN RETRIEVAL ALGORITHMS BY SPACEBORNE RADAR

Kenji Nakamura
Radio Research Laboratory
Ministry of Posts and Telecommunications
Koganei-shi, Tokyo 184, Japan

and

Robert Meneghini
NASA Goddard Space Flight Center
Greenbelt, Maryland 20771, USA

ABSTRACT

Spaceborne rain radar receives not only rain echoes but also surface echoes. While this fact causes a problem in discriminating rain echoes from surface echoes it also allows the use of rain retrieval algorithms using rain attenuation. The algorithms use surface echoes, rain echoes and mirror images of rain echoes. These new algorithms have features that reduce the error due to variations in raindrop size distribution and make absolute calibration unnecessary. The former feature provides good rainfall rate estimation while the latter will be important to spaceborne radar since the calibration of spaceborne radar will be difficult.

INTRODUCTION

Rainfall data over tropical regions is indispensable in clarifying the mechanisms of global-scale atmospheric variations. However, the rain data are very scarce since the oceans occupy a large part of the tropical regions. Hence, spaceborne rain sensors are thought to be the only means of providing tropical rain data. Spaceborne rain sensors such as visible/infrared radiometers or microwave radiometers are already operating and provide data on cloud activities. However, those sensors measure rain only indirectly. For example, the visible/infrared radiometer observes only clouds and not the rain itself. The microwave radiometer can estimate integrated rain in a column, but it has problems with accuracy and in observing rain over land areas, due to the high brightness temperature of the land. Spaceborne rain radar which directly detects rain and has a range resolution and vertical profiling capabilities is expected to provide invaluable data on cloud and precipitation activities.

ISBN 0-937194-14-4

Spaceborne radar observes the earth downward, receiving not only rain echoes but also surface echoes. The surface echoes are generally much stronger than the rain echoes. This causes a problem in discriminating the rain echoes from the surface echoes. On the other hand, this facilitates the use of new rain retrieval algorithms which use the rain attenuation of the radio waves along the path in the rain region.

The rain retrieval algorithms and the result of an airborne rain radar experiment are described below.

2. RAIN RETRIEVAL ALGORITHMS

2.1 CONVENTIONAL METHOD

The conventional rain retrieval method used for ground-based radar is to derive rainfall rates from the return power of the rain. The study of the relation between the radar reflectivity of rain and rainfall rate has a long history and empirical relations have already been presented. One of the difficulties of this method is that the relation between the radar reflectivity of rain and the rainfall rate is sensitive to the size distribution of raindrops. To get a better estimation of the rainfall rate, the rain type including thunderstorm, drizzle, or orographic rain should be taken into account.

2.2 ATTENUATION METHOD

High-frequency radio waves suffer rain attenuations. The rain attenuation is stronger for a strong rain and weaker for a weak rain. The relation of the rain attenuation and the rainfall rate is much less sensitive to the variations of the raindrop size distributions. When the transmitted waves from a radar are scattered by scatterers and the back-scattered waves are caught by the radar receiver, the rain attenuation along the round trip from the radar to the scatterers is measured by comparing the return power with that at no rain (Meneghini, 1983) The return power at no rain is obtained by real observations in no rain or is theoretically predicted. Figure 1 schematically shows this method when the surface echoes are used as scatterers. When it is not raining, the radar echo has a strong peak at the surface. When it is raining, the echo consists of both rain echoes and the surface echo. The strength of the surface echo during rain is weaker than that in no rain. Rain attenuation is measured from the amount of the return power reduction. One problem of this method is that the

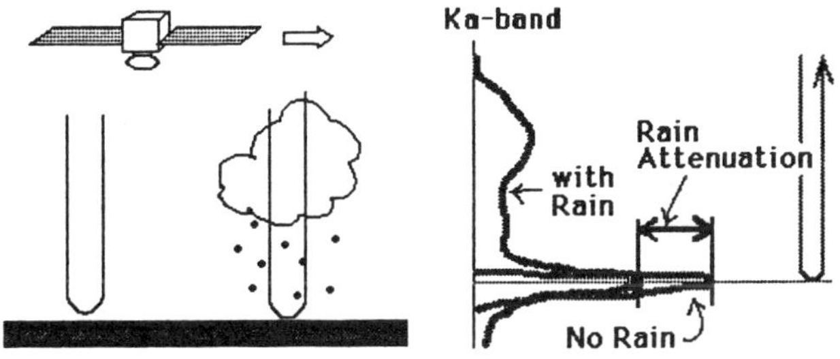

FIGURE 1. Rain Retrieval using surface echoes.

backscattering characteristics of the surface are not necessarily constant but depend on many physical conditions. The use of dual-wavelength radar may mitigate this problem. Assuming that one radio wave suffers negligible rain attenuation and the other suffers much rain attenuation, the difference of the surface return at the two frequencies depends on the rain attenuation. If the surface backscattering characteristics vary in the same way at the two frequencies, taking the ratio of the return power at both frequencies cancels the variation of the surface characteristics, and the result might be much less sensitive to the surface condition.

The rain echoes can also be used as the scatterers in the rain attenuation method (Eccles and Mueller, 1971). Think of two rain echoes at two radar frequencies, as shown in Figure 2, assuming again that one suffers small rain attenuation and the other suffers much. If the rain scattering is approximately the Rayleigh one, the backscattering cross-section can be obtained by the return power of the rain at the less attenuated frequency, since the backscattering cross-section is proportional to the fourth power of the frequency in Rayleigh scattering. The difference of the frequency-compensated return power at two frequencies can be interpreted as the rain attenuation. One extension of this method gives the range resolution (Nakamura, 1983). Taking the difference of the differences of the return powers at two range bins results in the rain attenuation between the two observed points, as shown in Figure 2.

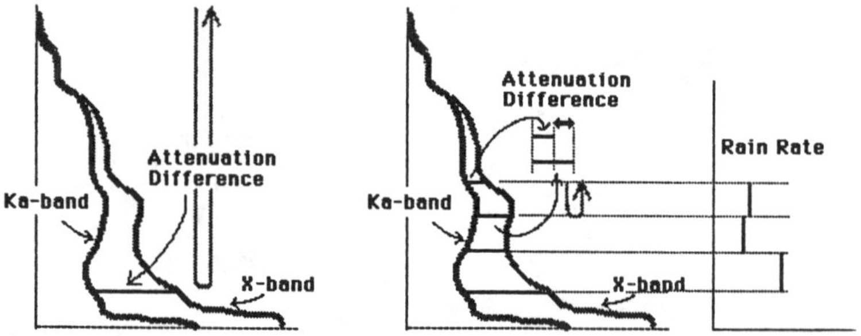

FIGURE 2. Rain retrieval using rain echoes.

Another method is to use the mirror image of rain. The ocean surface is a very good reflector of the radio waves and the rain over the ocean has its mirror image (Figure 3). Comparing the direct rain echoes and the mirror-reflected rain echoes, the rain attenuation along the double round trip between the surface and the observed point in rain can be obtained (Meneghini and Atlas, 1986).

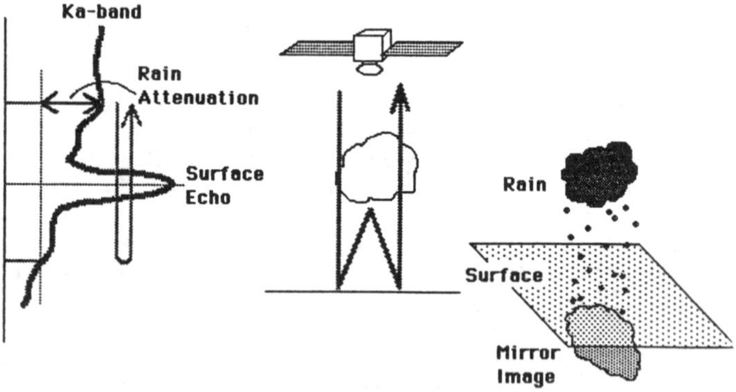

FIGURE 3. Rain retrieval using mirror image.

3. EXPERIMENTAL RESULTS

A rain observation experiment using an airborne dual-frequency rain radar was conducted in 1985 and 1986 as a cooperative program between NASA Goddard Space Flight Center and the Radio Research Laboratory, Japan (Meneghini et al., 1986). The radar applies X-band and Ka-band radio waves. Rain attenuation on the X-band radio wave is negligible compared to that on the Ka-band radio wave.

Figure 4 shows the comparison between the attenuation method using the surface echoes and the conventional method applied to the X-band radar echoes. The rain was a heavy one and the altitude of the airplane was around 1.5 km. The two rain rates show close agreement, suggesting that both are reliable.

The correlation coefficients between rainfall rates derived by conventional and attenuation methods were calculated for each frequency, but are not shown. The result indicates that the attenuation methods work for a strong rain but not for a weak rain. This is because rain attenuation is significant for strong rain but not for weak rain.

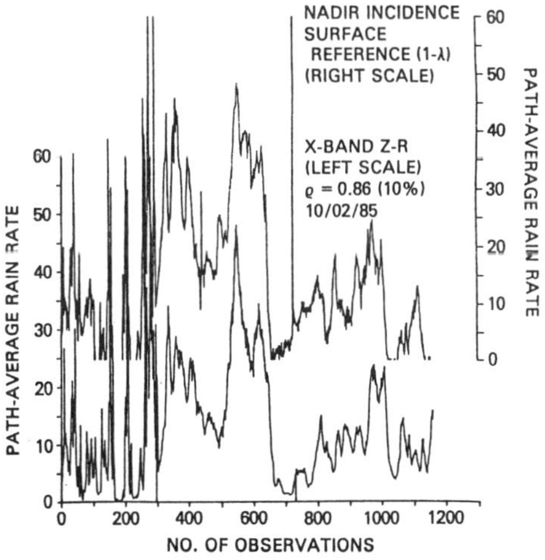

FIGURE 4. A comparison of estimated rain by conventional method and attenuation method.

4. SUMMARY

The rainfall rates can be retrieved using the relation between the rainfall rate and the rain attenuation. Rain attenuation can be measured by a radar using the surface or rains as reference scatterers. An airborne experiment showed that rain retrieval algorithms based on a rain attenuation work for a strong rain.

The attenuation method has an advantage in that it is insensitive to radar calibration errors and raindrop size distributions. On the other hand, the method provides little or no range resolution. The conventional method would be the primary method even for spaceborne rain radar. However, it should be emphasized that the attenuation method can provide rain estimations which are independent of those derived by the conventional method. This estimation can then be used to deduce other rain parameters or be used for radar calibrations.

ACKNOWLEDGMENTS

The authors are grateful to Dr. D. Atlas, University of Maryland, for many invaluable discussions. They wish to express their thanks to Mr. D. Clem and their colleagues at NASA Goddard Space Flight Center for successfully conducting the flight experiment. Thanks are also extended to Dr. K. Okamoto, Radio Research Laboratory, for supporting the experiment.

REFERENCES

Eccles, P. J. and E.A. Mueller, 1971: X-band attenuation and liquid water content estimation by a dual-wavelength radar. J. Appli. Meteor., 10, 1252-1259.

Meneghini, R., J. Eckeman and D. Atlas, 1983: Determination of rain rate from a spaceborne radar using measuring of total attenuation. IEEE Trans. Geoscience and Remote Sensing, GE-21, 34-43.

Meneghini, R. and D. Atlas, 1986: Simultaneous ocean cross section and rainfall measurements from space with a nadir-looking radar. J. Atmos. Oceanic Tech., 3, 400-413.

Meneghini, R. et al., 1986: Comparisons of methods using data from an air-borne dual wavelength radar. 23rd Conf. Radar Meteor., 100-103.

Nakamura, K. et al., 1983: Simultaneous rain observation by C-, X-, Ku- and Ka-band radars. 21st Conf. Radar Meteor., 213-220.

RAINDROP SIZE DISTRIBUTIONS IN WARM RAIN AT PENANG

Roland List, D. Hudak, and R. Nissen
University of Toronto
Toronto, M5S 1A7, Canada

and

N.P. Tung, S.K. Soo, and T.S. Kang
Malaysian Meteorological Service
46667 Petaling Jaya, Selangor, Malaysia

ABSTRACT

Raindrop spectra measurements were carried out on the island of Penang in October 1986 with a Joss-Waldvogel disdrometer and a 2-dimensional PMS Grey Scale Laser Spectrometer. Two different types of spectra were obtained, one reflecting steady state rain origins, the other fluctuating sources. No difference was found in the spectra patterns for warm and warm/cold (from clouds containing ice) rain. Main peaks were observed at diameters of about 1 and 2 mm for both steady state and "streaky" precipitation, with an indication of another peak around 200-300 μm - as predicted in general by laboratory based numerical models.

Previously proposed parameterizations by gamma distributions are describing observed raindrop number distributions at the ground better than Marshall-Palmer (MP) equations.

Large drops measured by the disdrometer cause dead times for the instrument which suppress the smaller drops at higher rainfall rates, thus affecting the usefulness of the instrument for microphysical measurements.

1. INTRODUCTION

A joint experiment on the drop size distribution in rain was carried out in October 1986 in Penang by the Malaysian Meteorological Service and the University of Toronto. The main purpose was to obtain a clearer picture about the most "simple" type of rain, warm rain, by the establishment of size spectra which then can be used for the modelling of the rain formation process, its prediction and modification. At the sampling site the freezing level is at about 5 km. Thus the chance to observe rain from clouds which do not contain ice is substantial. On the other hand, rain from clouds containing ice particles is also to be expected.

The actual data collection period of the experiment started on October 20 and ended on November 1, 1986. This corresponds to the latter part of the intermonsoonal period (which is the time of the highest rainfall for that area) and the transition to the winter monsoonal circulation. The sounding, aircraft observations and

ISBN 0-937194-14-4

radar data were used to categorize the precipitation events according to cloud type, stratiform or convective, rain from clouds not containing ice particles (warm rain), and that from clouds with ice (warm/cold rain). Nine events, primarily in the first week of observations, were related to convective situations, another nine, mostly in the second week, to stratiform clouds. Ten of the 18 events are categorized as warm rain, the other eight as warm/cold rain.

Since the thrust of this report is directed towards radar interpretation, no data will be shown as measured by the PMS equipment.

1.1 LISTING OF CONTENTS

Table 1, containing the list of headings and subheadings, is given as a quick reference guide of the paper's contents.

TABLE 1. LIST OF HEADINGS AND SUBHEADINGS

2. FACILITIES

The operations center for the experiment was the Bayan Lepas weather office of the Meteorological Service of Malaysia, located at Penang airport (about 5°N and 100°E, off the west coast of Peninsular Malaysia). Radiosonde ascents were carried out twice daily at 0800 and 2000 LST, and standard meteorological hourly observations, a recording raingauge, weather maps and forecasts were provided. Pictures from the GMS geosynchronous satellite were available for post analysis. An S-band radar, located on the mainland at the Butterworth Air Force Base, 23.5 km northeast of the Penang airport, produced PPI and RHI pictures, which could also be relayed to a console in the operations center in near real-time. A twin-engine Super Kingair was available for special studies. Planning and coordination of the experiment were shared among the scientists from the Malaysian Meteorological Service and the University of Toronto.

Special equipment was provided by the University of Toronto and consisted of a Joss-Waldvogel (1967) disdrometer and a 2-dimensional Grey Scale Laser Spectrometer by Particle Measuring Systems Inc. [PMS] (Joe and List, 1987), both for ground measurements of raindrop spectra in the respective diameter range of 0.3-4.5 mm and 15-960 µm. In both rain spectra systems the final processing stages were taken over by a specially equipped IBM PC AT. The hard disk had a memory of 20 MB, which was sufficient to store PMS data in heavy rainfalls up to 30 min. Thus the PMS probe was switched off on occasion. After rain events the information was transferred to 1.2 MB floppy disks for further processing. The main equipment problems were connected to intense solar radiation and the high humidities with related condensation. Moisture caused effects in both electronic sensing systems after two weeks of operation. Attention to voltage stabilization was also critical due to the large fluctuations in the power grid.

3. DATA DISPLAYS

3.1 SPECTRAL TABLES

All the drops received by the Joss-Waldvogel disdrometer were collected in terms of their analog impact information as measured by the instrument head. This raw data was conditioned and then assigned to bins of size 100 µm (for diameters < 2 mm) and 200 µm (for diameters > 2 mm) and grouped into one minute intervals. The number of all drops in each bin was printed out in the form of a table. Figure 1 represents such a display, together with isolines 0/1 and 9/10.

Figure 1 represents a streaky warm/cold rain (0.4 mm/h^{-1}) from multi-layered clouds with a radar top temperature of -15°C. It is typical for "packaged" rain, where the larger drops arrive first, followed by the smaller ones. New packages may start to arrive before the last drops of the previous one have arrived. Typical times between the onset of new packages are 8-10 min while the maximum diameter within a package decays with a rate of about 0.07 mm/min^{-1} for drops < 2 mm.

Figure 1 shows the effect of humidity on the electronics of the receiver head of the disdrometer. Drying out removed the noise which is so clearly indicated at diameters of 0.4 mm.

3.2 DOT DIAGRAMS

It was found that the display shown previously by Johnson et al. (1986), where each drop is represented by its diameter value when it arrives, gives a good conceptual picture about a rain event. In particular, such a representation is the most reliable one to indicate peaks because the eye is excellent at pattern recognition. Figure 2 represents a steady state warm rain (2.9 mm/h^{-1}, echo top at -4°C, with two maxima around 1.1 and 1.9 mm). On the basis of the time-dependent modelling studies by List et al.

DISDROMETER DATA
SPECTRUM PERIOD = 60 SEC.
OCTOBER 27, 1986
DIAMETERS IN mm, ELAPSED TIME IN MINUTES, BIN NUMBERS BELOW DIAMETERS
T = 0 AT DAY 27; TIME 12:18:59

		0.5					1.0					1.5					2.0		
1	2	3	4	5	6	7	8	9	10	11	12	13	14	15	16	17	18	19	TE mn
8	54	1	0	0	1	6	3	4	2	0	1	0	1	0	0	0	0	0	15
6	77	1	0	3	7	7	6	4	0	0	1	0	2	1	1	0	0	2	16
30	151	0	1	5	8	5	8	7	1	0	2	1	1	2	0	0	0	0	17
31	162	0	1	6	2	2	2	2	1	1	2	1	1	1	1	0	0	0	18
12	105	0	1	4	6	5	4	2	3	4	3	2	2	4	1	2	2	4	19
10	46	1	0	2	5	0	5	5	3	13	7	8	1	5	6	5	1	4	20
29	134	0	1	2	3	4	4	14	12	9	11	3	4	2	3	2	1	0	21
10	78	1	1	0	1	3	12	20	10	7	2	1	0	1	0	0	0	0	22
9	52	0	0	1	2	7	13	18	6	3	1	0	0	0	0	0	0	0	23
13	75	0	0	3	9	16	30	26	8	0	0	0	0	0	0	0	0	0	24
5	38	0	1	10	19	18	28	11	1	1	0	0	0	0	0	0	0	0	25
14	35	0	3	11	17	23	16	2	0	0	0	0	0	0	0	0	0	0	26
14	72	2	11	20	21	24	13	0	0	0	0	0	0	0	0	0	0	0	27
11	55	3	11	18	21	13	5	1	0	0	0	0	0	0	0	0	0	0	28
6	22	6	8	11	12	6	0	1	0	0	0	0	0	0	0	0	0	0	29
13	99	6	10	0	3	4	3	5	1	0	0	0	0	0	0	0	0	0	30
11	71	0	2	8	6	6	9	4	0	1	0	0	0	0	0	0	0	0	31
9	59	5	1	5	14	8	1	3	3	1	0	0	0	0	0	0	0	0	32
4	25	2	1	8	5	8	6	17	2	0	0	0	0	0	0	0	0	0	33
8	64	2	1	6	14	29	35	15	1	2	0	0	0	0	0	0	0	0	34
17	88	4	8	4	24	30	39	14	1	0	0	0	0	0	0	0	0	0	35
3	44	3	6	18	29	20	20	15	0	0	0	0	0	0	0	0	0	0	36
11	79	4	11	20	22	15	21	3	1	0	0	0	0	0	2	1	3	1	37
6	55	5	21	20	19	12	6	1	0	2	3	4	7	3	7	4	0	0	38
5	41	3	20	18	13	10	6	4	6	10	8	3	3	2	1	0	0	0	39
4	22	3	17	10	7	3	7	12	7	5	2	0	0	0	0	0	0	0	40
5	17	1	9	8	4	7	12	15	3	0	0	0	0	0	0	0	0	0	41
0	8	3	7	7	11	12	9	6	4	0	0	0	0	0	0	0	0	0	42
6	27	2	1	10	12	12	12	3	0	0	0	0	0	0	0	0	0	0	43
2	18	2	6	6	10	4	2	0	0	0	0	0	1	0	0	1	0	2	44
3	21	1	7	11	9	4	6	0	3	4	0	0	1	0	6	1	6	2	45
1	5	0	1	4	11	6	2	3	6	1	0	4	1	1	2	1	1	0	46
3	19	1	6	9	5	7	3	6	3	5	1	3	3	1	0	3	4	1	47

Figure 1. One-minute spectra of fluctuating warm/cold rain from multi-layered clouds; R = 0.4 mm/h-1, top -15°C.

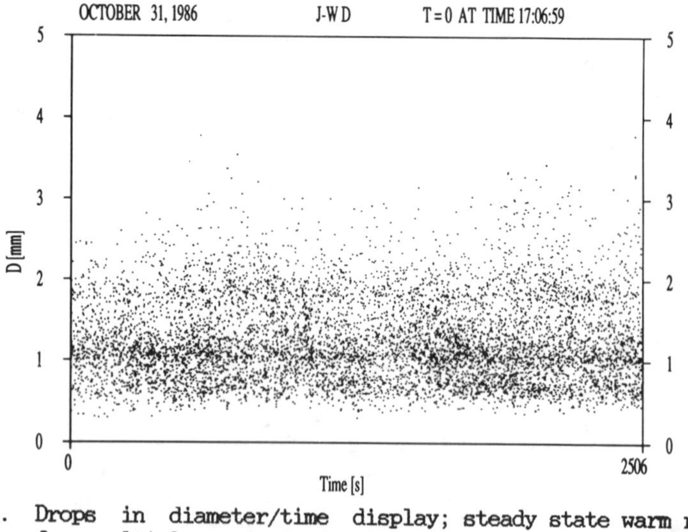

Figure 2. Drops in diameter/time display; steady state warm rain from multi-layered clouds; R = 2.9 mm/h-1, top -4°C.

(1987) longer evolution times for maxima (order of hours) might be anticipated than those available during the fall of the drops of this light rain. However, we are dealing here not with an original MP distribution, based on a drop spectrum as obtained by melting of ice particles. The drops in this warm rain evolved as an ensemble and, in all probability, were never far from equilibrium.

4. RESULTS

Data are represented for three cases: (1) heavy warm rain from a convective cloud (Figure 3a, b and c, continuous rain at 18.3 mm h^{-1}, radar echo top at -5°C, with no visual glaciation), (2) intermittent, streaky warm/cold rain from a multiple stratiform deck (Figure 4, as in Figure 1), and (3) a moderate, continuous rain from a warm, stable multi-layered cloud (Figure 5, also see Figure 2).

Figure 3a represents particle size distributions in terms of concentration per logarithmic size interval. It contains three different curves. The first is representing the data collected by the disdrometer. The total concentration is then used to fit the corresponding Marshall-Palmer (1948) curve and the one based on the equilibrium distribution (List, 1986; List et al., 1987). The constants for the MP distributions were assumed to be N_o=0.08 cm^{-4} and $\Lambda = 41\ R^{-0.21}$ [cm^{-1}], where R is the rainfall rate in mm/h^{-1}. The equilibrium distribution consists of three gamma distributions with the multiplicative factors 73.1, 28.8 and 20.2, respectively, and the corresponding power factors of 10, 13 and 19.

Figure 3b contains the resulting radar reflectivities for the same three sets of information, and Figures 3c, 4, 5 and 6 present number concentration in typical MP displays. Figure 6 definitely shows the drop-off in droplet concentration at diameters of < 1.5 mm. In this case continuous rain fell at a rate of 43.7 mm/h^{-1} from a convective cloud having radar tops at -30°C. The corresponding PMS data show that there is no lack in drops with sizes up to 1 mm. Indeed, their concentrations were the highest measured. This anomaly will be discussed below.

The first observation on the number concentrations, relating to the MP distributions, is obvious. The MPs have no structure, no peaks, whereas, in general, all measurements have one peak each in the neighborhood of 1 and 2 mm. The exception is the case of the very heavy rain (Figure 6) where the 1-mm peak was not recorded by the instrument because of saturation problems. In this case there is a slight hint of a peak or shoulder at a diameter of about 3 mm.

The disdrometer, while being an excellent instrument to measure rainfall in terms of radar reflectivities, is not designed for unbiased cloud physical measurements. It favors larger drops at the expense of smaller ones, particularly during heavy rain where the instrument dead-time may even prevent the counting of drops

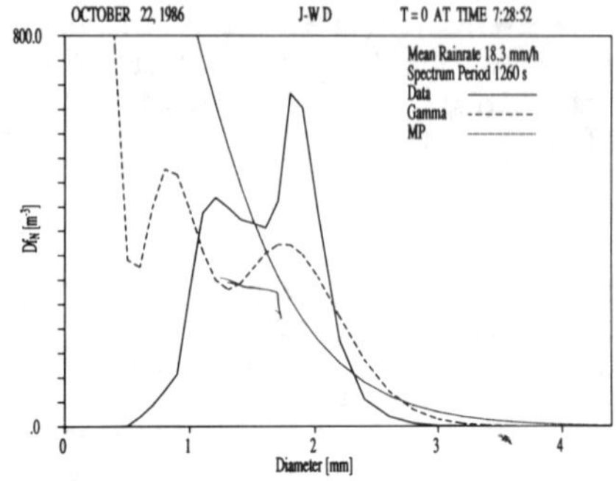

a) # conc./ln interval,

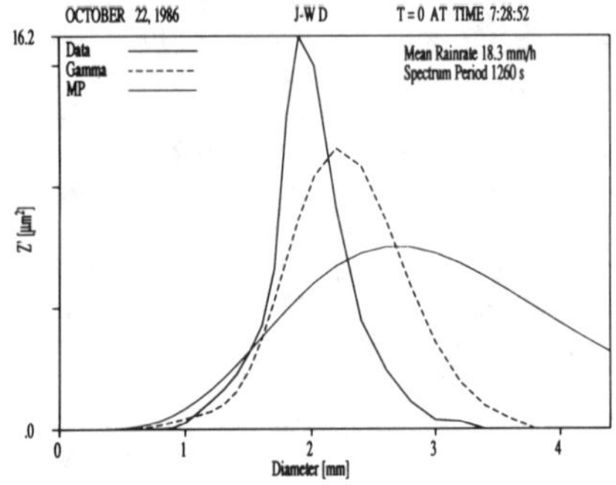

b) radar reflectivity,

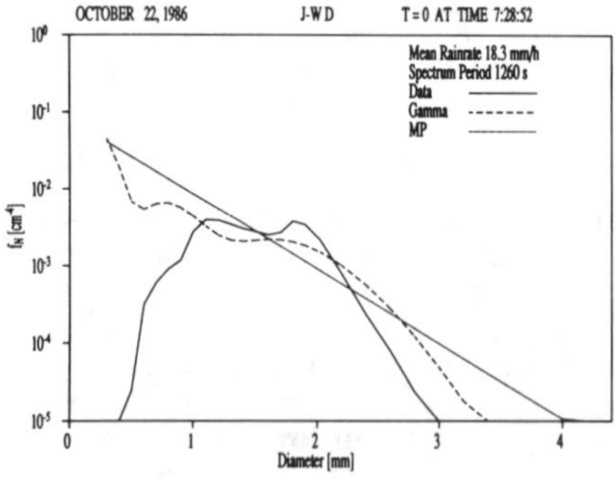

c) # concentration [cm-4].

Figure 3. Cont. heavy warm rain spectra from convective cloud; R = 18.3 mm/h-1, top -5oC;

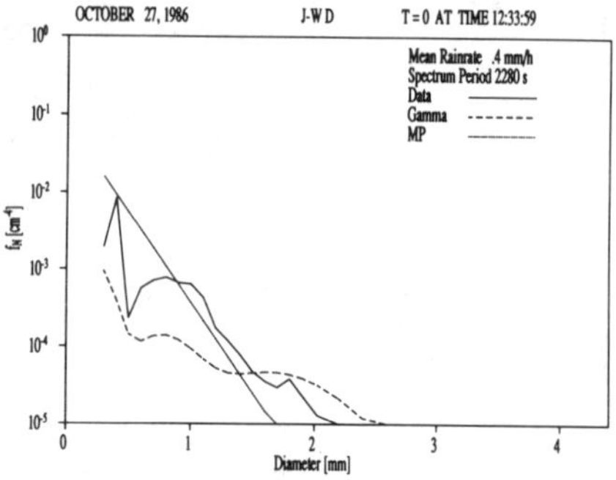

Figure 4. Average spectra, conditions see Figure 1.

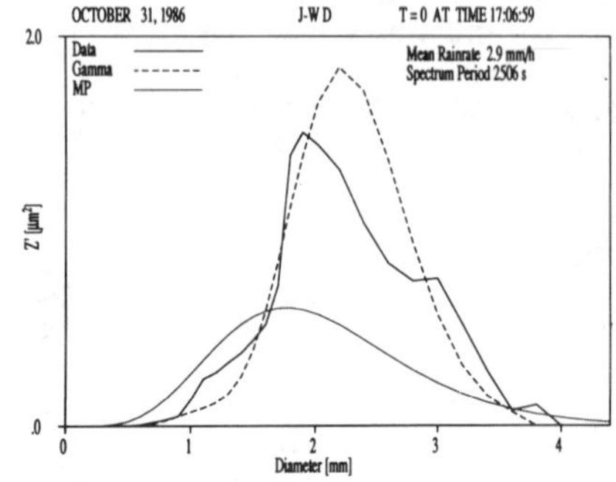

Figure 5. Average radar reflectivity (see Fig.2).

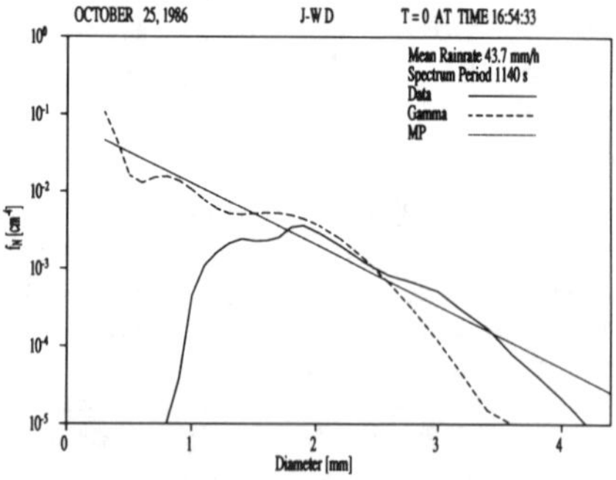

Figure 6. Spectra from continuous warm/cold rain from convective cloud; R = 43.7 mm/h-1, top -30oC.

smaller than about 1 mm (Figure 6). This effect also leads to shifts to larger diameters of peaks and to reductions in peak concentrations for the 1-mm maximum (Figure 3). In light rain, the disdrometer seems to correctly account for the small drops down to diameters of 0.3 mm. (Due to this rainfall-dependent bias disdrometer data should not be used to establish average or climatological dropsize distributions.)

The model produces distributions which have an identical shape (but not concentrations) and peaks at 264 µm (beyond the sensitivity of the Joss-Waldvogel disdrometer and the concern of this paper), 0.79 mm and 1.76 mm. The measurements in Malaysia indicate a slight shift of peaks to higher diameters (by about 200 µm for the two larger diameters). The model predicts a higher peak height for the 0.79- mm drops than for the 1.76 mm ones. The measurements show this trend only for the moderate and light rain, the disagreement for heavy rain could be explained by instrument characteristics.

5. CONCLUSIONS AND COMMENTS

The "Joint Warm Rain Experiment" has provided an important data set for raindrop size distributions at the ground with peaks in the size distributions at diameters of about 1 and 2 mm (and at 300 µm as measured with thc PMS probe - not shown here). This is very similar to the predictions by the equilibrium three-peak models by Valdez and Young (1985) and List et al. (1987). While there was never any great expectation to find an agreement between the measurement of natural drop size distributions with the simple model predictions based on the laboratory experiments by Low and List (1982), the data nevertheless show that there seems to be a better agreement between the equilibrium peak model and the real raindrop size distributions than between the latter and a Marshall-Palmer distribution. In particular, the multi-peak character is more or less as predicted but with the measured peaks at somewhat higher diameters. The relative heights of the peaks are not in agreement for heavy rain, probably due to the dead-time effects in the disdrometer. There is no difference at ground level between warm rain and rain from clouds containing ice particles, and between continuous rain from convective and layered clouds.

It is clear that the development has to follow three directions:

(1) The evolution of the raindrop size distributions aloft and their evolution during growth and fall to the ground needs to be measured by aircraft. Maybe the same equilibrium type distributions can be found as detected above cloud base in Florida by Willis (1984).

(2) The rain spectra measurements in the field need to be carried out with equipment which more realistically depicts the concentration of smaller drops in heavy rain.

(3) The modelling has to become more sophisticated.

A preliminary Z-R relationship for remote sensing of some types of tropical rain may be established with the available data set. However, more extensive measurements need to be undertaken to provide a better base.

ACKNOWLEDGMENTS

Special thanks go to the Minister for Science and Technology of Malaysia, Datuk Amar Stephen Yong, for his support of this experiment. The financial assistance by Professor G.E. Connell, President of the University of Toronto and by Dr. Jingjai Hanchanlash, Director of the Canadian International Development Research Center, Singapore Office, was instrumental in bringing the Toronto group and its equipment to Penang. Thanks go also to Dr. T. B. Low, President of KelResearch in Toronto for his assistance.

REFERENCES

Joe, P., and R. List, 1987: Testing and performance of two-dimensional optical array spectrometers with greyscale. J. Atmos. Ocean. Techn., 4, 139-150.

Johnson, D., K.V. Beard and D. Baumgardner, 1986: Airborne observations of raindrop distributions in Hawaiian clouds. Preprints Conference on Cloud Physics, American Meteorological Society, Snowmass CO, Sept. 22-26, pp. 48-51.

Joss, J., and A. Waldvogel, 1967: Ein Spektrograph fuer Niederschlagstropfen mit automatischer Auswertung. Pure Appl. Geophys., 68, 240-246.

List, R., 1986: Equilibrium raindrop size distributions in warm rain. Preprints 23rd Conference on Radar Meteorology and Conference on Cloud Physics, American Meteorological Society, Snowmass CO, Sept. 22-26, pp. 142-145.

List, R., N.R. Donaldson and R.E. Stewart, 1987: Temporal evolution of drop spectra to collisional equilibrium in steady and pulsating rain. J. Atmos. Sci., 44, 362-372

Low, T., and R. List, 1982: Collision, coalescence and breakup of raindrops. Parts I and II. J. Atmos. Sci., 39, 1591-1618.

Marshall, J.S., and W.M. Palmer, 1948: The distributions of raindrops with size. J. Meteor., 9, 327-332.

Valdez, M.P., and K.C. Young, 1985: Number fluxes in equilibrium populations: A Markov chain analysis. J. Atmos. Sci., 42, 1024-1036.

Willis, P.T., 1984: Functional fits of observed drop size distributions and parameterization of rain. J. Atmos. Sci., 41, 1648-1661.

MEASUREMENT OF SIZE DISTRIBUTION TRANSITION OF THE PRECIPITATION PARTICLES USING A TWO-WAVELENGTH RADAR METHOD

Tatsuo Yokoyama
Meteorological Research Institute
Tsukuba, Ibaraki 305, Japan

ABSTRACT

The size distribution transition of precipitation particles in and near a melting layer was measured by a two-wavelength radar method. The utility of the method was confirmed by calculations using a precipitation model. Based on the size distribution transition obtained from the measurement, we can conclude that the transition due to equivalent coalescence is predominant in the melting layer.

1. INTRODUCTION

Microphysical process such as size distribution transition due to coalescence or breakup of the precipitation particles is one of the important problems for radar meteorology and telecomunication engineering as well as cloud physics. The problems have been, so far, investigated by many workers using direct sampling methods and remote sensing with microwave radar. The latter is a convenient method to detect precipitation particles, which are distributed over a wide area. In the melting layer associated with the stratiform precipitation, however, it is difficult to detect the size distribution transition with a single-wavelength radar, because the backscatter of the radar wave changes under the influence of the melting phenomenon of the snowflakes as well as the size distribution transition.

Therefore, the proposed technique is a two-wavelength radar method to detect selectively the size distribution transition, even in the melting layer. In this paper, the method will be confirmed by the calculation using simple precipitation models and applied to the observation of the stratiform precipitation.

2. PRINCIPLE OF THE METHOD

The technique utilizes the following principle: changes in the radar reflectivity factor corresponding to the size distribution transition occur differently for each wavelength. In other words, the radar sensitivity to the size distribution transition depends on the wavelength. Considering a case of a spherical water droplet at 0 °C, for example, when the droplet grows from 0.3 cm to 0.4 cm in diameter for some reason, the backscattering cross-section becomes approximately 5 times, 10 times and one－third for wavelengths of 5.6 cm, 3.2 cm and 0.86 cm, respectively.

ISBN 0-937194-14-4

3. MODEL CALCULATION

We calculated how the radar reflectivity factors of the two wavelengths (0.86 cm and 3.2 cm) respond differently to the change of size distribution curves. In this model, the following microphysical process is assumed; when a spherical snowflake falls across 0-°C altitude, the snowflake begins to melt at the surface and the melted water covers the particles. Consequently, the melting snowflake can be described by two concentric spheres with water and snow.

Three types of size distribution transition are considered for the calculations, which are schematically shown in Fig. 1. We assume that, in all the cases, the initial size distribution curve at the upper altitude (before the change) obeys Marshall and Palmer (1948) distribution (hereafter MP type) shown by the solid curves in each figure. And we assume that the size distribution shifts to that denoted by either the dotted or broken curves at the lower altitude (after the change). The size distribution is assumed to change while the particles fall from the upper altitude to the lower one.

In Case-A (see Fig. 1a), the mass of each particle becomes P times and consequently the number density changes and initial diameter range shifts so as to hold a constant precipitation rate. If P is larger than unity, the transition corresponds equivalently to a coalescence process, since the larger particles increase and the smaller ones decrease. If P is smaller than unity, the transition represents an equivalent breakup process. P can be a measure of the activity of the equivalent transition. In Case-B (Fig. 1b), the number density of the particles is assumed to become M times only in a specified size range (between a and b in the figure). In Case-C (Fig. 1c), we assume that MP type is always kept and the shift is given by changing only the precipitation rate. Precipitation rate is not conserved in Cases-B and C.

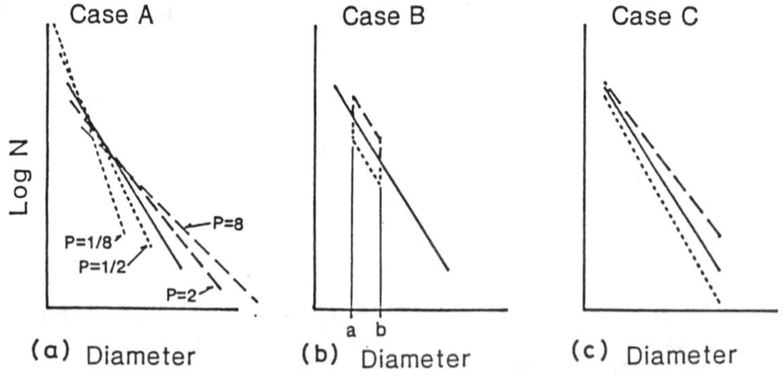

FIGURE 1. Schematic figure of the size distribution transition for three types of Cases-A, B and C. Ordinates denote number density of the particles in logarithmic scale. Abscissas denote melted diameter. Solid curve in each figure represents MP type used as an initial distribution before the change (at the upper altitude). Broken and dotted curves represent the shifted distribution after the change (at the lower altitude).

We calculated the radar reflectivity factors for the two wavelengths (0.86 cm and 3.2 cm) before and after the size distribution transition for every case, applying Mie's theory for two concentric spheres as was described in Yokoyama and Ishizaka (1985) in detail. The difference of the radar reflectivity factors between the upper and the lower altitudes was calculated. We can regard the difference as the change of the radar reflectivity factor corresponding to the transition between the two altitudes. The difference was calculated for each wavelength. In this paper, we select the two altitudes in the following way; center of the bright band where the radar reflectivity factor is maximum, and an altitude where the melting of the snowflakes is finally completed. Therefore, the difference corresponds to the change in the lower half region of the melting layer.

The differences between the two altitudes obtained by model calculation are shown in Fig. 2 as the correlation between the two wavelengths. Even if the size distribution does not change between the two altitudes, the differences never disappear owing to the effect of the melting (closed circles in each figure). When the distribution transition of Case-A occurs between the two altitudes (open and closed squares in Figs. 2a and 2b), the 3.2-cm wavelength is more sensitive to the transition than 0.86 cm. For example, in the case that the precipitation rate is 0.5 mm/hr and P=8 in Fig. 2b, the discrepancies of the difference between the unchanged case (closed circle) and the changed case (closed square) are 7 dBZ for the 3.2-cm

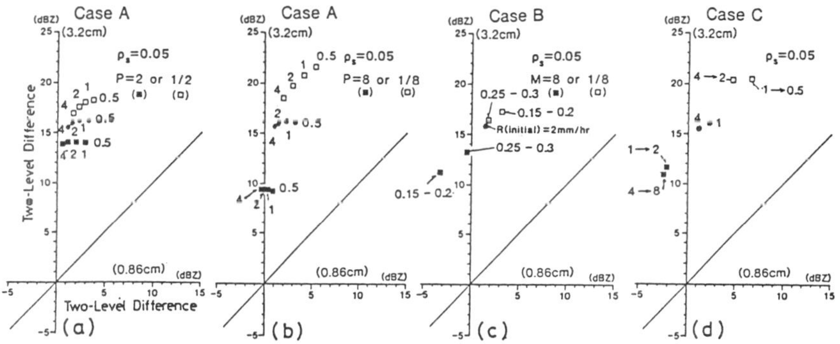

FIGURE 2. Calculated differences of the radar reflectivity factors between the two altitudes (center of the bright band and raindrop region). Abscissas and ordinates represent the two-altitude difference for 0.86- and 3.2-cm wavelengths, respectively, in dBZ. Closed circles denote the differences when the distribution transition does not occur between the two altitudes. Open and closed squares denote the differences when the size distribution changes from solid curves to the dotted and broken curves shown in Fig. 1, respectively, between the two altitudes. The slant lines denote the points on which the values of abscissa and ordinate are equal. In (a) and (b), parameters beside each point represent precipitation rate in mm/hr. In (c) and (d), the parameters denote the shifted diameter ranges in cm and the change of the precipitation rate in mm/hr, respectively.

wavelength and 2 dBZ for the 0.86 cm. In Figs. 2a and 2b, the two-wavelength discrepancy of the differences, which is represented by the distance from the point to the slant line, depends on P. P denotes the activity of the equivalent size distribution transition. On the other hand, in Cases-B and C, the two-wavelength discrepancies of the differences are approximately constant for both the unchanged and the changed cases. The result means an equal sensitivity for the two wavelengths in Cases-B and C. Eventually, we can conclude that locations of the points on Fig. 2 provide the most important information on the size distribution transition between the two altitudes.

4. OBSERVATION

We observed the stratiform precipitation by the two-wavelength radar technique. This observation was carried out at Nagoya (137.0 °E, 35.2 °N) in "Baiu" season. In this season, long-lasting stratiform precipitations covering broad areas often occur and the 0-°C level occurs between about 3- and 5-km altitudes in the Japan Islands. Two vertically pointing radars with 0.86- and 3.2-cm wavelengths were used to measure the radar reflectivity factors. Because of the attenuation of the microwave energy by the precipitation particles, the radar reflectivity factor of 0.86-cm wavelength was corrected as was described in Yokoyama (1985) using the data of 3.2-cm radar, precipitation rate and size distribution measured on the ground.

Figure 3 shows the 20-minute averaged two-level differences of the radar reflectivity factor obtained from the observations. During the period shown in Fig. 3, the precipitation satisfies the condition of a typical stratiform precipitation with stable and clear bright

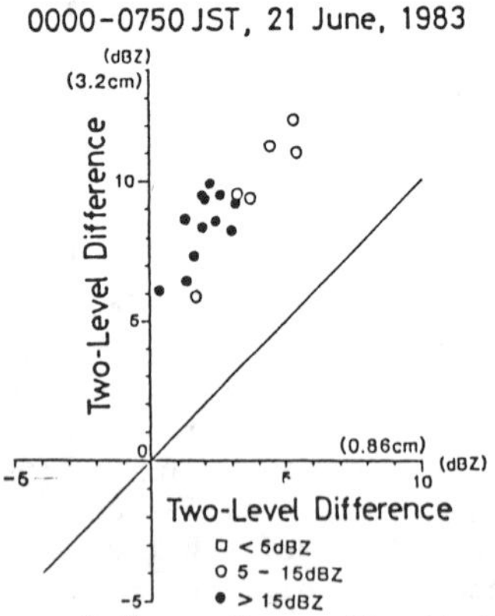

FIGURE 3. The differences of the radar reflectivity factor between the two altitudes (center of the bright band and an altitude 1 km below). Abscissa, ordinate and slant line have the same meaning as in Fig. 2. Each point was derived from the profile along the streak line by taking average during 20 minutes and is classified by the radar reflectivity factor of the 3.2-cm radar at 3-km altitude as three symbols.

band. From the results shown in Fig. 3, we can find the following facts; the differences of 3.2-cm wavelength (ordinate) are always larger than those of 0.86 cm (abscissa) and the two-wavelength discrepancies range from 3 to 9 dBZ. Comparing Fig. 3 with Fig. 2, the location of the points in Fig. 3 is similar only to the case of P=8 in Fig. 2b. Furthermore, the two-wavelength discrepancies are approximately constant and are independent of the radar reflectivity factor. Based on these results, we can conclude that the equivalent coalescence occurs predominantly in the lower half region of the melting layer for typical stratiform precipitation. And, we can find the independence of the activity of the equivalent distribution transition, which is defined as P in this paper, on precipitation intensity.

5. CONCLUSIONS

The availability of the two-wavelength radar method to detect the size distribution transition was confirmed from the results of the model calculations. Based on the observations using this method, it is concluded clearly that the transition due to equivalent coalescence is predominant in the lower-half region of the melting layer. The results agree with those obtained by direct sampling method (Yokoyama et al., 1985). Furthermore, the two-wavelength method is available in the raindrop region below the melting layer, too (Yokoyama and Ishizaka, 1985). In this region, the transition shown as Case-B is found by both the direct sampling and the two-wavelength radar methods. Because of the attenuation of 0.86-cm wavelength, the pair of 3.2- and 5.6-cm wavelengths is available in the upper-half region and the snowflake region as shown in Yokoyama and Tanaka (1984).

REFERENCES

Marshall, J. S., and W. M. Palmer, 1948: The ditribution of raindrops with size. J. Meteor., 5, 165-166.

Yokoyama, T., 1985: Two-wavelength (0.86 cm and 3.2 cm) radar observation of microphysical processes in stratiform precipitation. Part I. Attenuation of radar wave due to precipitation particles and correction method. J. Meteor. Soc. Japan, 63, 1097-1108.

———, and H. Tanaka, 1984: Microphysical processes of melting snowflakes detected by two-wavelength radar. Part I. Principle of measurement based on model calculation. J. Meteor. Soc. Japan, 62, 650-667.

———, ———, K. Akaeda, T. Ohtani, N. Yoshizawa, M. D. Yamanaka, A. Mita, Y. Ishizaka, and A. Ono, 1985: Observation on microphysical processes in the stratiform precipitations including melting layers at Mt. Fuji. J. Meteor. Soc. Japan, 63, 100-111.

———, and Y. Ishizaka, 1985: Two-wavelength (0.86 cm and 3.2 cm) radar observation of microphysical processes in stratiform precipitation. Part II. Study on size distribution change of precipitation particles. J. Meteor. Soc. Japan, 63, 1109-1125.

TOWARDS TRISPECTRAL SATELLITE RAINFALL MONITORING ALGORITHMS[1]

E.C. Barrett, G. D'Souza, Clare H. Power, and C. Kidd
Remote Sensing Unit
University of Bristol
Bristol, BS8 1SS, United Kingdom

ABSTRACT

There has been an active interest in the University of Bristol since 1970 in the use of satellite imagery for improved rainfall monitoring, particularly within the tropics. At first attention was concentrated on the development of manual, then interactive "cloud indexing" techniques based on visible (VIS) and/or infrared (IR) imagery. More recently, IR temperature thresholding and VIS/IR bispectral techniques have been developed and tested over Africa. Today, we are actively involved in attempts to exploit the rainfall information in passive microwave (PMW) data for further refinement of such techniques. Preliminary results are promising, and suggest that the best approach to satellite rainfall monitoring which can be conceived at present will combine VIS, IR and PMW data, and exploit both polarization and frequency information from the PMW. The TRMM satellite should provide invaluable extra experience in respect of the synergisms between imagery obtained from these three spectral regions.

1. INTRODUCTION

The University of Bristol took its first interest in satellite rainfall monitoring in 1970, with a nephanalysis-based study of monthly rainfall in the Tropical Far East. This led to the development of a manual cloud-indexing method for satellite-improved rainfall monitoring (the 'Bristol Method') for meteorological and hydrological use (Barrett & Martin, 1981).

The AgRISTARS Program in the USA sponsored subsequent development of an interactive version of the 'Bristol Method', namely BIAS (the

[1]Parts of the work summarised herein have been sponsored by contracts or grants from the European Space Agency, and EEC within its programme "War against Hunger in the World", NOAA/NESDIS under the provisions of a Co-operative Agreement between its Satellite Applications Laboratory and the Remote Sensing Unit, and the Public Authority for Water Resources in the Sultanate of Oman.

ISBN 0-937194-14-4

Bristol/NOAA InterActive Scheme). This is described in more detail below. (See also Barrett et al., 1986). It was applied to parts of the USA and South America as part of NOAA's prematurely-terminated AgRISTARS Large Scale Applications Test in 1985. It has been applied additionally in support of this and other programs to parts of Western Europe, the USSR, the Middle East (especially the Sultanate of Oman), and the western Sahel.

Work for the Food & Agriculture Organisation of the United Nations in support of food crop harvest forecasting and associated food security planning has led to the formulation of an objective, bispectral, geostationary technique called ADMIT (Agricultural Drought Monitoring Integrative Technique). (See Barrett & D'Souza, 1986). A simplified variant of this, for objective use with either polar-orbiter or geostationary satellite data has since been developed for use over selected parts of Africa. PERMIT (the Polar-orbiter Effective Rainfall Monitoring Integrative Technique) is based on IR imagery alone, using temperature thresholding to identify potential rain clouds (Barrett, Power et al., 1986).

2. BASES OF THE BRISTOL METHODS

2.1 GENERAL CONSIDERATIONS

Although the methods introduced above differ from each other in important ways, they are all based on a number of common premises. First, it is assumed that the best possible analyses of rainfall will be achieved if both satellite and conventional data are evaluated together whenever possible. Second, that the satellite imagery provided are most useful for revealing the spatial distribution of potential rain clouds. Third, that climatological records can be used to provide first approximations of rain amounts. Fourth, that these records are best deployed in detail both to help differentiate relatively light or heavy rain areas, and to differentiate locally between relatively wet and dry seasons of the year. And fifth, that synoptic meteorological reports, if available, can be used beneficially to help establish final or adjusted rainfall estimates. Brief summaries of the machine-based Bristol methods follow.

2.2 BIAS

Figure 1 illustrates the stages in BIAS, as they appear on the image processing system screen.

A functional equation for BIAS can be written thus:

$$R = f(C_t, C_a, C_d, C_c, S_w) \qquad (1)$$

where R is accumulated rainfall (commonly for 6-, 12- or 24-hr periods), C_t is rain cloud type according to a menu developed from theory and empiricism, C_a is fractional rain cloud type area(s), and C_d is rain cloud duration. Products of C_a, C_t and C_d are the "cloud

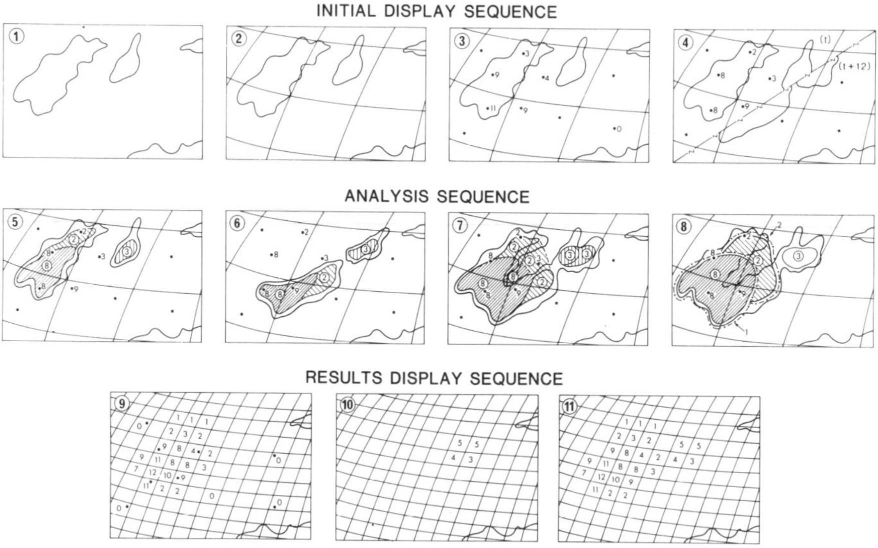

Figure 1: Stages in BIAS. Satellite imagery are mapped (1), gridded (2), overlain with SYNOP data (3), for both ends of the analyses period (4); cloud type fields are then analyzed for t_1 (5), t_2 (6) and the time between, t_i (7), before 'areas of influence' are mapped so SYNOP data can influence final estimates (8); results are then computed for parts of the area (9) and (10), and the whole (11). (From Barrett, et al., 1986).

indices" from which this type of technique derives its generic name. The cloud indices are translated into initial estimates of rainfall via a set of regressions developed for this purpose. Every pixel in the study area is accorded a climate category (Cc) for identification of the most appropriate cloud index/estimated rainfall regression in the absence of synoptic weather reports. However, if these are available, they may be used to evaluate S_w, synoptic weather "floats", which may suggest that other regressions should provide more accurate results.

Independently executed verification tests over the USA carried out for AgRISTARS indicated that, when real-time calibration station raingauge data are quite numerous (mean station spacing c.300 km) and fairly regularly spaced, BIAS gives improvements in categoric rainfall monitoring results of usually between 5 and 25% (Hubanks, 1985).

2.3 PERMIT

At its simplest this may be represented by:

$$R = f(C_T, C_d, M_c, S_w) \quad (2)$$

where C_T is cloud top temperature, areas below a selected value or values being deemed potential rain clouds. PERMIT is designed to provide estimates of rain for periods of 10 days and upwards. Thus, initial estimates can be obtained by multiplying C_d (here the number of satellite-observed rain cloud days for an individual pixel) by the 'morphoclimatological weight' (M_c). This is a climatological statistic designed to act as a combined climate/terrain weight, calculated for monthly or (preferably) decadal periods. Specifically, it is mean rainfall per rain day. Once initial rainfall estimates have been evaluated for all pixels in the study area, any suitable reports of observed rainfall can be regressed against corresponding initial estimates, and the resulting curve used as a transform function for all other locations. In these ways "final" or "adjusted" rainfall fields can be prepared. (See Figure 2).

2.4 ADMIT

In order to reduce the problem of thick cirrus which can masquerade as potential rain cloud in PERMIT, a bispectral approach may be preferable in some areas of the world. ADMIT is such a technique, using synchronized VIS/IR image pairs during the daytime to establish rain/no rain thresholds (see Figure 3a), and relating the analysis of subsequent nighttime IR-only images to them by a controlled dilation process to prevent rain clouds "blowing up" where extensive cirrus is found. (See Barrett & D'Souza, 1986 for fuller details). Thus the functional equation for ADMIT may be written as:

$$R + f(C_B, C_T, C_d, M_C\ S_w) \qquad (3)$$

where C_B is cloud brightness in the visible, and other terms are as above.

For Africa as a whole (see Figure 3b) verification studies indicate that the satellite-based rain/no rain categorisations of individual pixels between 80-85% are correct. For 14 verification stations in the state of Niger, it has been shown that final ADMIT estimates averaged 86.4 mm against 87.5 mm reported from the ground. Corresponding mean standard errors were 13.8 and 12.1 mm, respectively.

3. COMPARISONS OF PASSIVE MICROWAVE METHODS

Recognising that all VIS/IR rainfall monitoring methods are handicapped by the physical indirectness of the relationships between cloud top parameters and the rain which falls from their bases the Bristol Remote Sensing Unit has begun to evaluate existing algorithms for rainfall estimation over land from passive microwave data. Using image data from the Nimbus-7 SMMR (Scanning Multi-channel Microwave Radiometer) we have analysed a number of meteorolgical situations using 3 passive microwave techniques.

The first is a Frequency Algorithm, in which the vertical polarization temperature difference at 18 and 37 GHz frequencies is taken to be some function of the instantaneous rain rate. The second is a Polarization Algorithm which utilizes the correlation between emissivity at different frequencies and their two polarizations to

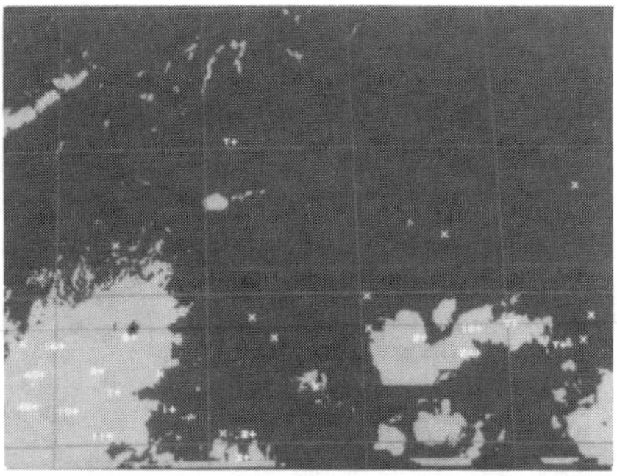

(a)

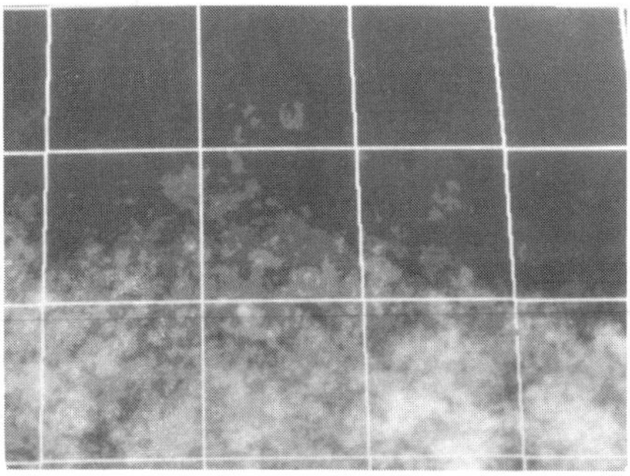

(b)

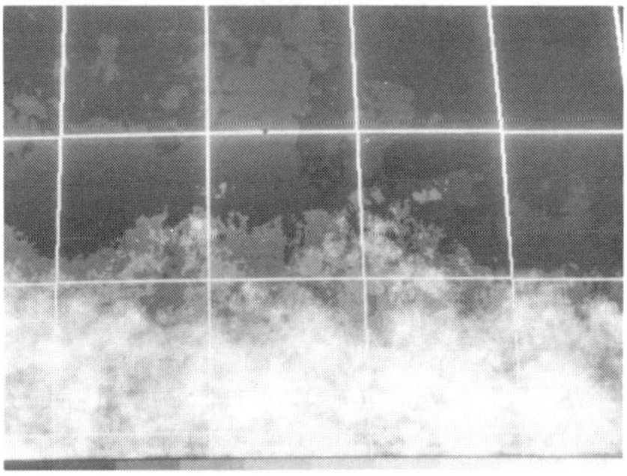

(c)

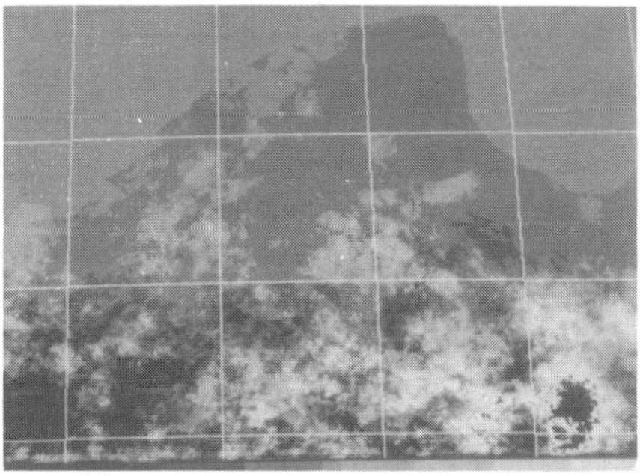

(d)

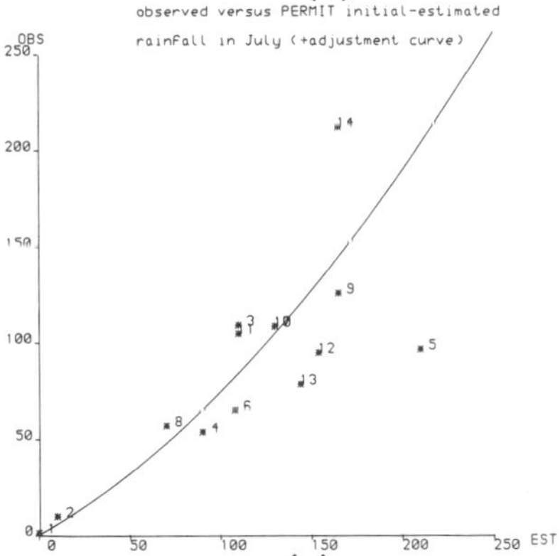

(e)

41-80	0	0	0	0	0	0
21-40	0	0	0	1	4	2
11-20	2	9	8	17	9	2
6-10	1	12	12	13	4	0
1- 5	7	12	8	6	0	0
> 1	6	3	0	0	0	0
Est. / Obs.	>1	1-5	6-10	11-20	21-40	41-80

Summary Statistics

Obs.: n = 137; mean = 31.68 mm; s.d. = 28.52 mm.
PERMIT Ests.: mean = 27.90 mm; s.d. = 18.32 mm.
Ests. vs. Obs.: corr. = 0.618; rms. = 22.66 mm.

(f)

Figure 2: PERMIT applied to the W. Sahel (between 10-25°N and 5°W-15°E): (a) Meteosat-derived raincloud area map, 1 July, 1985; (b) Meteosat-derived total numbers of days with rain, July, 1985; (c) final estimates of rainfall, July, 1985; (d) PERMIT-derived rainfall anomalies, July, 1985 (above normal, white; below normal, dark grey/black; normal, mid-grey); (e) regression of observed on estimated rainfall, W. Sahel, July 1985; (f) contingency table and correlation statistics for 10-day estimates, W. Sahel, July 1986. (From Barrett, Power et al. 1986).

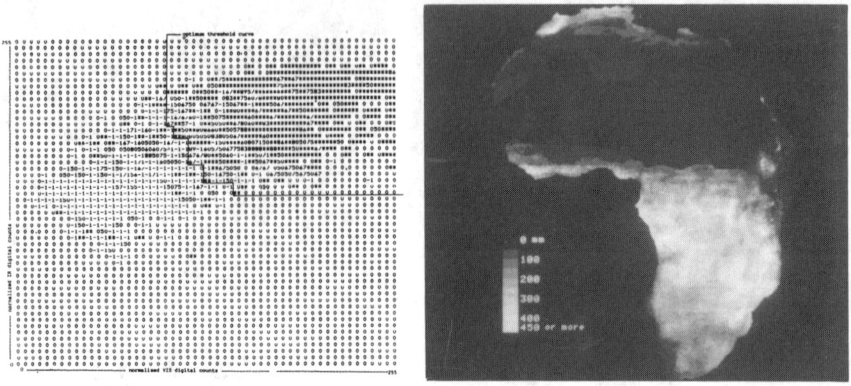

(a) (b)

Figure 3: ADMIT applied to the whole of Africa, Jan. 1985: (a) analysis of bispectral feature space (x-axis, VIS; y-axis, IR) to establish best rain/no rain boundary (rain area, top right); (b) map of final rainfall estimates for the month.

isolate rain areas, and the resultants to provide some measure of rain rate. (For both see Grody, 1984). The third is a Modified Frequency Algorithm which includes calibration constants derived from digitized radar-based regressions (after Spencer, pers. comm.) Figure 4 compares results obtained when applying two of these techniques to a selected event over the central USA. Clearly they produce somewhat different rain areas, and further tests have shown that they also yield rather different histograms of rain rates.

In attempting to assess the relative performances of the three techniques some positive and negative points have emerged in each case. It is clear that the Polarization Algorithm often gives the noisiest backgrounds with some cold land surfaces masquerading as rain. Unfortunately it is difficult to see how these might be removed both easily and objectively. Further, the β value which features prominently in the Polarization Algorithm, and has been generally reckoned to be treatable as a quasi-global constant, seems to be quite widely variable (Barrett, Kidd & Bailey, 1987). In the meantime, the Frequency Algorithms sometimes over-represent rain areas where rivers, lakes and marshes are found; fortunately in such cases simple polarization screening is a great help. Less fortunately, Frequency Algorithms have under-represented probable rain areas over some (e.g. mountain) areas, though physical insight for this may be available. Overall the Modified Frequency Algorithm seems to carry the best balance of advantages, and the wider range of frequencies available on the SSM/I should further improve overland PMW rainfall monitoring techniques.

4. OPERATIONAL USE OF PMW DATA

Although PMW more directly evidences rain areas than VIS or IR

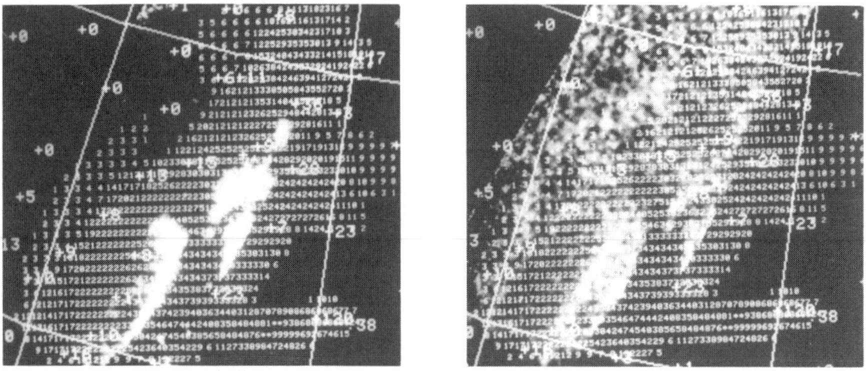

(a) (b)

Figure 4: BIAS rainfall estimation fields for south-central USA, 1800Z, 23 Nov. to 0600Z, 24 Nov. 1983, overlain with SMMR-derived rain areas: using (a) the simple Frequency Algorithm, and (b) the Polarization algorithm, described in the text.

imagery, PMW data are less fully researched, less easily procured, less frequent, and of lower spatial resolution than the VIS/IR. Consequently, it seems likely that VIS/IR imagery will continue to be used for rainfall monitoring for the foreseeable future, but also that PMW data should become increasingly useful to improve rain area delimitation, and to help calibrate rain rates. Our initial experiments in overlaying BIAS and ADMIT analyses with SMMR rainfall algorithm outputs already suggest that PMW is useful in four ways. First, in evidencing rain areas should VIS/IR imagery be missing; second, in revealing "under-fit" areas, where rain areas have not been so analyzed; third, in locating and assessing centres of relatively heavy rain within broadly overcast areas; and fourth, in recognizing thick, non-raining cirrus, and distinguishing this from rain clouds, e.g. where thick cirrus sheets are found. It is our belief that PMW has considerable potential for operational rainfall monitoring over land, though most immediately for days or longer periods. Thus, for the foreseeable future, PMW will be most usefully deployed within trispectral (VIS/IR/PMW) techniques, which must now be developed. Finally, climatological and meteorological weighting schemes will still be necessary until the PMW sensors are more refined, and associated physical theory is much more advanced.

REFERENCES

Barrett, E. C., and D. W. Martin, 1981: The Use of Satellite Data in Rainfall Monitoring, Academic Press, London, 340pp.

Barrett, E. C., M. J. Beaumont, A. Harrison, and T. S. Richards, 1986: BIAS I^2S Users Guide, Final Report to the U.S. Department of Commerce, Contract No. NA85AA-H-RAI10, Washington D.C., RSU, University of Bristol, 72pp.

Barrett, E. C., and G. D'Souza, 1986: Meteosat Dry Day, Drought and

Rainfall Products for Africa: An Introduction, Proceedings of 5th Meteosat Scientific Users Meeting, Rome, 28-30 May 1985.

Barrett, E. C., C. Kidd, and J. O. Bailey, 1987: The Use of SMMR Data in support of the Bristol/NOAA Interactive Scheme (BIAS) for Satellite/Improved Rainfall Monitoring, Annual Report to the U.S. Department of Commerce under Co-operative Agreement No. NA86AA-H-RA001, RSU, University of Bristol, 77pp.

Barrett, E. C., C. H. Power, M. J. Beaumont, and T. S. Richards, 1986: Cloud Cover and Rainfall in the Western Sahel, Final Report to the Natural Environmental Research Council, Contract No. F60/C1/19, U.K. 85pp.

Grody, N. C., 1984: Precipitation Monitoring Over Land from Satellites by Microwave Radiometry, Proceedings of IGARSS 1984 Symposium, Strasbourg, 27-30 August, 1984, 417-422.

Hubanks, P. A., 1985: Evaluation of Interactive Precipitation Estimation Techniques Using Polar-orbiter and GOES Satellite Data, Final Report to the U.S. Department of Commerce under Contract No. NA-83-SAC-00117, Research and Data Systems, Inc., Gaithersburg, Md., 110pp.

ON A NEW APPROACH FOR CONVECTIVE RAIN AREA DELINEATION WITH NOAA-7 SPLIT WINDOW DATA

Toshiro Inoue
Meteorological Research Institute
Tsukuba, Ibaraki 305, Japan

ABSTRACT

The split window data (11 μm and 12 μm) are used for convective rainfall area delineation. Optically thick clouds, classified by the split window data, whose cloud tops are over 700 mb level are regarded as the rain-producing clouds. Concurrent radar data was compared with the satellite rain-clouds.

1. INTRODUCTION

Many studies of convective rainfall estimation have been done using satellite data, since satellite data are expected to implement a rainfall estimation over areas where conventional rain gauge data are sparse.In the technique which uses single channel infrared data, a threshold value (usually -20°C of brightness temperature) is used as rain-cloud identification. However, some erroneous estimations of rainfall amount due to the long-lived anvil (non-precipitating cirrus) and to low-level clouds were reported (Stout et al.,1979; Augustine et al.,1981). This implies shortcomings of the rain area delineation using the single channel infrared threshold technique.

A multi-channel method, using both visible and infrared data, for rain area delineation has been developed. This method shows better results in rain area delineation than the single channel infrared method. However, the visible data cannot be used at night and further correction for the variation of the sun angle is not a simple task.

Inoue (1987 b) has shown a method using the split window data of AVHRR on board NOAA-7 for convective rainfall area delineation. He showed that the method is better than the conventional single channel infrared method by 12% for the probability of detection and 13% for the false alarm ratio. The improvement of the false alarm ratio by his method over the single channel infrared method is comparable to that shown by other authors who use the method combining visible and infrared data. In this paper, the active convective case is studied according to his method.

2. DATA

The satellite and radar data are mapped onto the 0.1° x 0.1° grid over the area 33° - 39°N and 135° - 141°E. All of the grid points within this area are covered by both satellite and radar data.

The calibrated channel 4 (11 μm) and channel 5 (12 μm) data of AVHRR on board NOAA-7 are used. The brightness temperature (hereafter referred as TBB) and the brightness temperature difference between the split window data (hereafter referred as BTD and defined by TBB in 11 μm minus TBB in 12 μm) are computed for the cloud type classification and the rain area delineation.

The composite Radar-AMeDAS (Automated Meteorological Data Acquisition System) precipitation map is used in this study. The Radar-AMeDAS map indicates the

ISBN 0-937194-14-4

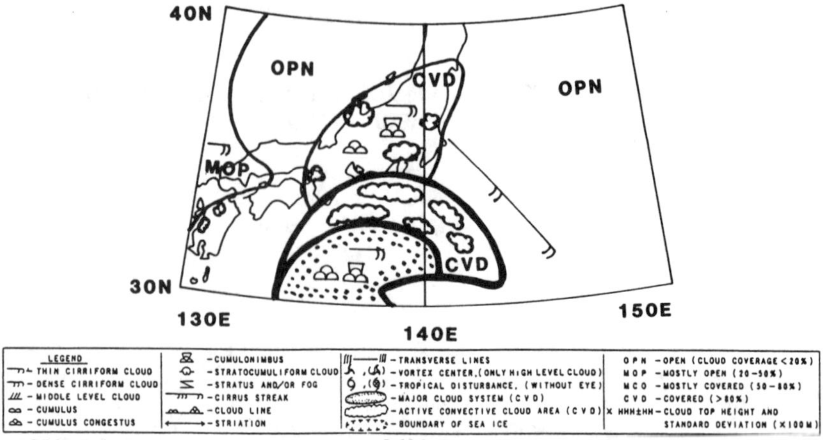

FIGURE 1. The nephanalysis chart at 0600 GMT July 28,1984.

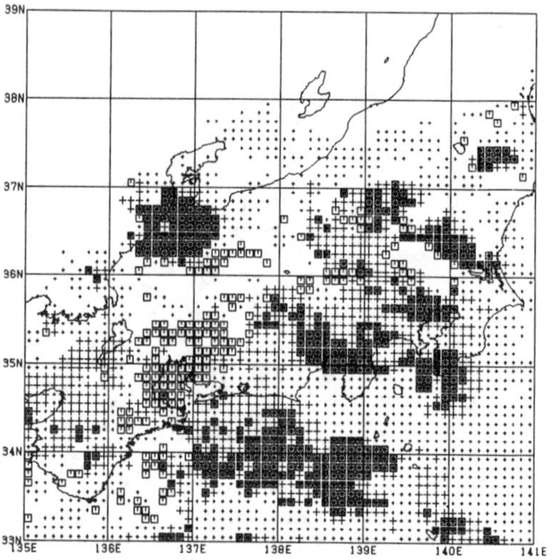

FIGURE 2. The cloud type classification map by the split window method around 0540 GMT. Cirriform cloud is indicated by small +. Cumulonimbus type cloud and dense cirriform cloud are indicated as ▩ , + ,respectively. Cumulus type cloud is shown as ▫ .

hourly precipitation estimated by integrating eight radar echo intensity maps within the hour and calibrating them by rain gauge observations of AMeDAS.

3. RESULTS AND DISCUSSION

The case on July 28, 1984 is studied in this paper. The nephanalysis chart analyzed by meteorologists using GMS images of 0600 GMT at Meteorological

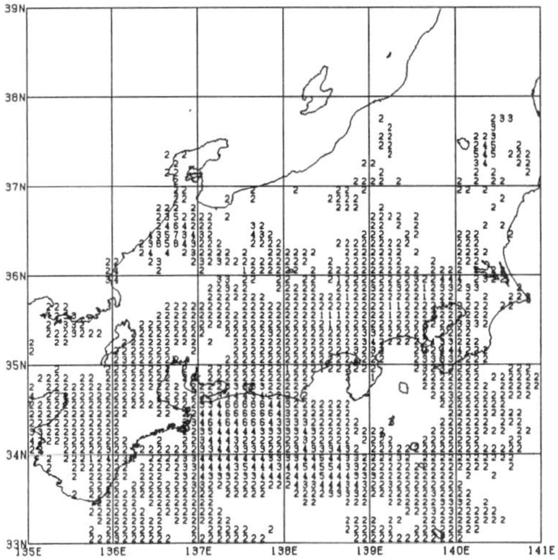

FIGURE 3. The composite Radar-AMeDAS hourly precipitation map at 0600 GMT July 28, 1984. The digit shows the level of the table below.

Level	1	2	3	4	5	6	7	8
Hourly Precipitation (mm)	0<	1<	5<	10<	20<	30 <	40<	60< 80

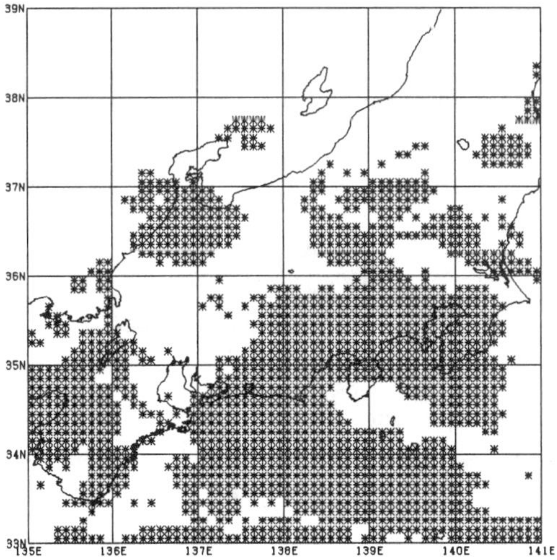

FIGURE 4. The map of cloud whose TBB (11 μm) is colder than -20°C.

TABLE 1. EVALUATION OF THE PERFORMANCE

The performance of split window method (SW) and single channel infrared method (IR). The probability of detection (POD), false alarm ratio (FAR) and error factor (ERF) are shown in percent. (After Inoue, 1987 b)

Date (1984)	Observation Time (GMT)	# of RADAR Rain Grid	POD (SW)	POD (IR)	FAR (SW)	FAR (IR)	ERF (SW)	ERF (IR)
Jul. 11	0552	33	45	52	60	86	2	4
Jul. 16	0631	807	87	67	14	38	13	22
Jul. 25	0620	878	82	93	23	27	20	16
Jul. 26	0608	436	91	78	23	23	9	10
Jul. 27	0556	249	84	33	23	39	5	7
Jul. 28	0540	1395	75	92	4	16	27	17
Aug. 2	0622	236	93	97	28	36	5	7
Sep. 5	0604	182	84	29	38	55	6	6
	Average		80	68	27	40	11	11

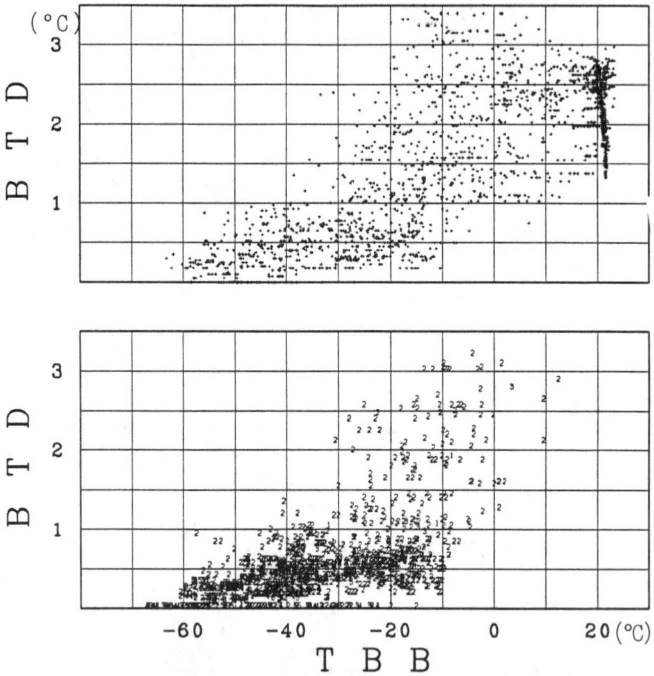

FIGURE 5. The scatter BTD-TBB diagram of satellite data corresponding to radar rain grid points (bottom) and no radar rain grid points (top). Digits(bottom) indicate the level in Fig.4.

Satellite Center in Tokyo is shown in Fig.1. The study area is located north of the typhoon 8407 and covered by active convective clouds with cirriform clouds .

The cloud type classification map based on Inoue (1987 a) is shown in Fig.2. Fig.3 indicates levels of the hourly precipitation in the Radar-AMeDAS map (hereafter referred to as radar rain) at 0600 GMT. Active convective cloud areas in Fig.1

correspond well with the cumulonimbus-type cloud areas in Fig.2. And these cumulonimbus-type cloud areas also correspond well with radar rain area in Fig.3.

However, both cloud areas A and B in Fig.2 are associated with radar rain in Fig.3. The cloud area A consists of dense cirriform clouds, while the cloud area B consists of cumulus type clouds. The map of cloud whose TBB(11 μm) is colder than -20°C is shown in Fig.4. The cloud area A corresponds to the clouds in Fig. 4 but the cloud area B does not correspond to the satellite rain cloud obtained by using the single channel infrared method.

Inoue(1987 b) defined satellite rain clouds as those optically thick clouds whose tops are over the 700mb level. Therefore, these radar rain areas are very critical for both the split window method and the single channel infrared method. Table 1 shows the statistics of the split window method and conventional infrared method. The cloud areas classified as dense cirriform clouds are ignored in the rain area delineation by the split window method. This results in the poor probability of detection for the case of multi-layered cloud system with cirrus clouds as is this case.

Fig.5 represents the scatter diagram of BTD and TBB (11 μm) in this case. The bottom figure represents the diagram for the radar rain grids, while the top shows the diagram for the no-rain grids. Numerical values in this figure exhibit the level of radar rain. The BTD and TBB are determined as the minimum and mean of clouds, respectively, within the nine adjacent grid points of the satellite map with the radar rain grid point as center. We did not compare the radar grid point and satellite grid point directly, chiefly because of the lower mapping accuracy of the satellite data. In this diagram, 75% of the radar rain grid points are associated with the existence of optically thick clouds ,i.e., cumulus or cumulonimbus-type clouds, within the nine adjacent grid points, while no radar rain grid points are covered by relatively warmer TBB and larger BTD satellite data than those of radar rain grid points. However, 13% of the no-radar rain grids are covered by cumulus or cumulonimbus-type clouds near the grid points. This diagram shows the feasibility of the split window method.

The method developed by Inoue (1987 b) is very simple and can be used for night observations. And the method shows relatively good scores for single layer cloud systems. The threshold values for satellite rain clouds in this method can be changed for different climate regimes.The coincident radar and satellite data can be processed in the TRMM program. These data will surely improve the rainfall estimation by the current satellite data including split window data.

REFERENCES

Augustine, J.A., C.G. Griffith, W.L. Woodley and J.G. Meitin, 1981: Insights into errors of SMS-inferred GATE convective rainfall. J. Appl. Meteor., 20, 509-520.

Inoue, T., 1987 a: A cloud type classification with NOAA-7 split window measurements. J. Geophys. Res., 92: 3991-4000.

Inoue, T., 1987 b: An instantaneous delineation of convective rainfall areas using split window data of NOAA-7 AVHRR. J. Meteor. Soc. Japan , 65,469-481.

Stout, J.E., D.W. Martin and D.N. Sikdar, 1979: Estimating GATE rainfall with geosynchronous satellite images. Mon. Wea. Rev., 107, 585-598.

A RELATIONSHIP BETWEEN UPPER CLOUDINESS AND PRECIPITATION IN THE WESTERN PACIFIC

Takashi Eguchi
University of Tokyo
Tokyo 113, Japan

ABSTRACT

In the western Pacific and Southeast Asia, precipitation data observed at surface raingauge stations are insufficient to clarify precipitation distribution. The data derived from meteorological satellites are expected to be useful in estimating precipitation in this region. The relationship between upper cloudiness and precipitation is examined in the western Pacific during FGGE period using the 10-day total precipitation observed at sixteen surface stations and the 10-day mean upper cloudiness derived from the geostationary meteorological satellite (GMS). The correlation coefficient between the precipitation P (unit of mm) and the upper cloudiness C (unit of 1/10) using all(544) data, is 0.66, and the coefficient and constant of the regression line between them by the least-squares best fit is 30.12 and 12.74, respectively. The precipitation ratio per cloudiness(P/C) is 36 when all data are used. However, the coefficient of the regression line and the precipitation ratio per upper cloudiness calculated at each station show great differences between stations. One of the reasons for the large variations among the stations is the difference of topographic conditions among the stations. Therefore, a regional topographic coefficient is necessary to estimate precipitation only using the upper cloudiness data of GMS.

1. INTRODUCTION

In the western Pacific and Southeast Asia, the sea area is widely distributed, and the land area is mountainous. The precipitation distribution is complex and the precipitation data observed at surface raingauge stations are insufficient for clarifying the seasonal and interannual changes of precipitation distribution. Therefore, the observations made by meteorological satellites are expected to be useful in filling the sparse data areas. The study of precipitation estimation focused on this region is examined by Barrett(1970) and Maruyama et al.(1985). Barrett estimates the monthly rainfall by the cloud indexing method using ESSA nephanalysis. Maruyama et al. estimate the monthly rainfall by the diagnostic method using the high cloud amount of the geostationary meteorological satellite(GMS) in the western Pacific. The monthly data are too long for clarifying the seasonal change. Therefore, the author analyzed the relationship between the 10-day total precipitation and the 10-day mean upper cloudiness in the western Pacific.

ISBN 0-937194-14-4

2. DATA

The cloudiness data used in this study are the upper cloudiness derived from the geostationary meteorological satellite(GMS) at the Meteorological Satellite Center in Japan. The threshold value of the upper cloudiness is the climatic 400mb temperature and the upper cloudiness are determined from the number ratio of cloud pixels with equivalent black-body temperature lower than the threshold value within a 1-degree latitude-longitude box using the GMS IR data. The precipitation data used in this study are derived from FGGE level II-c precipitation and snow data set.

The period of this study is from December 2nd, 1978 to December 1st, 1979 and almost corresponds to the FGGE period. This period is divided into 36 10-day periods, i.e. the 1st 10-day period is from 2nd to 11th December, 1978 and the last(36th) period is from November 22nd to December 1st, 1979. Only the 27th period includes 15 days from August 23rd to September 2nd. The mean upper cloudiness of this period consists of 15-day means, and the 10-day total precipitation is two-thirds of the 15-day total precipitation. Both the 6th period(from 21st to 30th in January) and the 22nd(from June 30th to July 9th) are missing due to the lack of satellite data. Therefore, 34 10-day periods are used in this study. The stations used in this study are 16 in the western Pacific(Fig. 1). The relationship between the precipitation and the upper cloudiness in the box where each station is included is examined.

3. RELATIONSHIP BETWEEN UPPER CLOUDINESS AND PRECIPITATION

The regression line of the least-squares best fit between the upper cloudiness C (in unit of 1/10) and precipitation P (mm) using all stations and periods(544 data) is as follows(Fig. 2):

$$P = 12.74 + 30.12\ C$$

with the correlation coefficient 0.66. The coefficient and the constant of the regression line calculated at each station using 34 period data are shown in Table 1. One of the most important criteria in estimating precipitation on the basis of satellite data is the value of upper cloudiness determined from the threshold between precipitation and no precipitation. However, it is difficult to

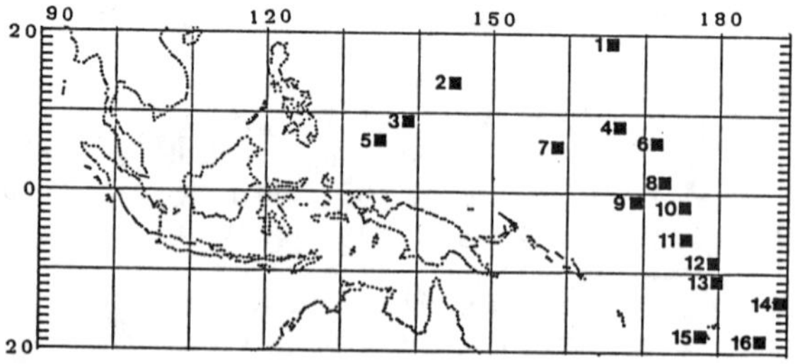

FIGURE 1. Location map of 16 stations used in this study. Station name is shown in Table 1.

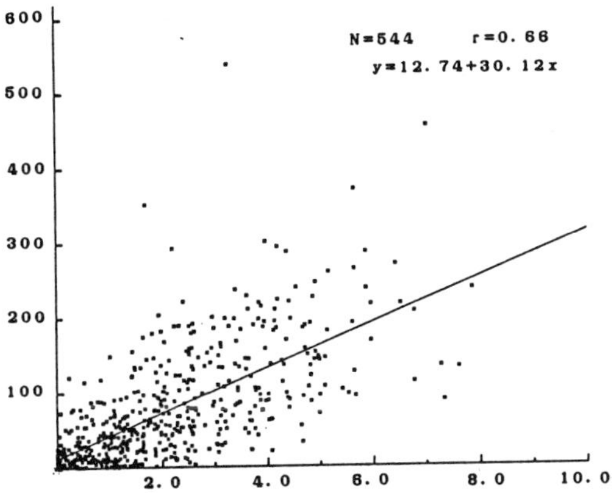

FIGURE 2. Scatter diagram and regression line between precipitation(Y-axis, unit of mm) and upper cloudiness(X-axis, unit of 1/10).

TABLE 1

Correlation coefficient, and coefficient and constant of regression line at 16 stations. Station number is the same as Fig.1

STATION NO.	STATION NAME	CLOUDINESS (unit: 1/10)	PRECIPITATION UNIT: mm	P/C RATIO	CORRELATION COEFFICIENT	REGRESSION LINE (Y=a+bX)	
						a	b
1	WAKE	0.62	25.58	41.3	0.82	5.05	33.23
2	GUAM	1.49	45.16	30.3	0.74	8.93	24.34
3	YAP	2.18	78.59	36.1	0.80	19.63	27.09
4	KWAJALEIN	1.98	73.92	37.3	0.62	-0.29	37.51
5	KOROR	2.26	109.40	48.4	0.55	34.07	33.40
6	MAJURO	2.67	83.21	31.2	0.58	20.44	23.49
7	PONAPE	2.40	131.43	54.8	0.72	30.68	41.98
8	TARAWA	1.95	61.04	31.3	0.62	10.28	26.09
9	OCEAN IS	2.31	78.10	33.8	0.65	12.63	28.38
10	BERO	1.63	34.00	20.9	0.73	-1.11	21.55
11	NANUMERA	2.66	87.90	33.0	0.82	5.51	30.96
12	FUNAFUTI	2.71	84.34	31.1	0.70	20.72	23.45
13	NURAKITA	2.50	95.21	38.1	0.70	11.08	33.64
14	APIA	2.01	81.48	40.5	0.52	40.62	20.36
15	NANDI	1.76	50.46	28.7	0.70	-16.72	38.23
16	VAVAU	1.66	71.06	42.8	0.44	24.48	28.01
TOTAL		2.05	74.43	36.3	0.66	12.74	30.12

determine the threshold value of upper cloudiness of GMS. The intercepts shown in Table 1 differ greatly between stations and most of them are above 0, although this constant is expected to be nearly 0 or below 0. This is the limit of precipitation estimation only using the upper cloudiness data of GMS.

The coefficient using all data is 30.12 and the precipitation ratio per upper cloudiness(P/C ratio) is 36.3 using all data. These values show the precipitation change per upper cloudiness derived from GMS. The relationship between the recorded precipitation and the estimated precipitation is shown in Fig. 3. Precipitation tends to be overestimated when the recorded precipitation is below 100mm and underestimated when the recorded precipitation is above 100mm. In particular, the estimated precipitation is frequently smaller than the

		ESTIMATED PRECIPITATION (mm)								
		0	0–25	25–50	50–75	75–100	100–150	150–200	>200	TOTAL
RECORDED PRECIPITATION (mm)	0		2	1						3
	0–25		57	60	25	13	3			158
	25–50		11	34	29	14	7	1		96
	50–75		4	19	23	17	14	2		79
	75–100		2	10	15	10	19	5	1	62
	100–150		1	3	15	14	22	12	3	70
	150–200				5	12	19	9		45
	>200				2	2	14	8	5	31
	TOTAL		77	127	114	82	98	37	9	544

FIGURE 3. Relationship between recorded and estimated precipitation.

recorded precipitation when the recorded precipitation is above 200mm. The coefficient and the P/C ratio calculated at each station show large differences between stations. In the Northern Hemisphere, the coefficient is large and the correlation coefficient is high along the area between 5 N and 10 N east of 170 E where the upper cloudiness is relatively large. On the contrary, the coefficient is low along the equator east of 170 E where the upper cloudiness is small. In the Southern Hemisphere, the relationship between the upper cloudiness and the precipitation is relatively good between the equator and 10.75 S. However, the correlation coefficient is relatively low at Apia and at Vavau. One cause for the large differences of the coefficient and the P/C ratio from those mean values is expected to be the topographic effect. For example, the largest value of the coefficient and the P/C ratio occurs at Ponape located in the windward area of the mountain whose altitude is 791m, although the correlation coefficient is high(0.70). Apia is also located in the island with the high mountain and the P/C ratio is recorded above 40. This station is also expected to be affected by topographic effects. However, the correlation coefficient is low and the coefficient is below 30. These values of Apia are different from those of Ponape. The precipitation amount at Apia is expected to be affected not only by the topographic effects but also by the South Pacific Convergence Zone(SPCZ).

A large overestimation of precipitation occurs from September to October in the Northern Hemisphere and during the summer monsoon(from December to February) in the Southern Hemisphere. On the contrary, underestimation occurs during the period when the developed tropical cyclones bring precipitation.

4. CONCLUDING REMARKS

The five geostationary meteorological satellites proving global coverage are able to take hourly data. Therefore, it is possible to estimate precipitation in tropical areas where the diurnal variation is dominant. Precipitation estimation based on the geostationary meteorological satellites is expected to be useful in clarifying precipitation distribution in the tropical areas. The simple estimation method and the satellite data for estimation which is compiled by the same criteria are necessary in order to clarify seasonal and interannual changes of precipitation distribution. Although the author has attempted to estimate precipitation in the ocean areas, this precipitation estimation method is expected to be also useful over land areas where it is difficult to observe precipitation.

Precipitation distribution is largely affected by topography, particularly in the tropical areas. However, it is difficult to clarify topographic effects using only GMS data. Therefore, it is necessary to clarify the topographic effects not only on the basis of the GMS data, but also the polar-orbiting meteorological satellite data. Both regional and topographic coefficients are necessary for precipitation estimation using the GMS data.

CLOUD CLASSIFICATION IN SATELLITE IMAGERY: A LATITUDINAL AND SEASONAL VARIATION OF CLOUD CHARACTERISTIC PARAMETERS

Tadashi Aso, Kazuhiro Oosawa, and Yoshiaki Takeuchi
Meteorological Satellite Center
Japan Meteorological Agency
Kiyose-shi, Tokyo, Japan

ABSTRACT

Latitudinal and seasonal variations of cloud characteristic parameters occurring mainly in the midlatitudes result in systematic failure in cloud classifications. These variations should be considered when extracting clouds relating to precipitation using cloud classification or a brightness temperature (TBB) threshold value.

1. INTRODUCTION

Cloud classification in satellite imagery is a powerful tool for extracting clouds related to severe weather such as cumulonimbus, or mapping each cloud type for a cloud climatology. It is necessary that cloud classification should work over a wide area and for all seasons.

Parikh (1977) showed that it was possible to classify clouds into four classes with reasonable accuracy from IR and VIS imagery data. Recently Garand(1986) performed a detailed cloud classification and tried to retrieve atmospheric conditions from data obtained over the Atlantic Ocean.

Although there are many cloud characteristic parameters, a cloud temperature and an albedo are always required to classify clouds. The former distinguishes the height of the cloud layer and the latter the cloud thickness. So it is very important to know the nature of these cloud features to understand the results of cloud classification.

In this note we present some results on the variation of a cloud top temperature. The reader should refer to Aso et al.(1988) concerning a cloud albedo.

2. CLOUD CLASSIFICATION

We classified clouds(nine classes) by the discriminant analysis in the GMS imagery mainly in the midlatitudes, ranging 5～50 N (see Aso et al., 1987 in detail). The "true" cloud class was determined for the training and an independent data set. We used the same computer programme by Kato and Ishikawa(1982) and details of that classification procedure are given there.

The nine cloud classes are CB, CID, CIM, CIT, AS, CU, SC, ST and CLR. Their explanations are in the caption in Fig. 2. Two important characteristic parameters are a cloud top temperature (a 16 % level in the IR accumulated histogram) and a cloud maximum albedo (a 84 % level in the VIS accumulated histogram).

TROPICAL RAINFALL MEASUREMENTS
John S. Theon and Nobuyoshi Fugono (Eds.)

ISBN 0-937194-14-4

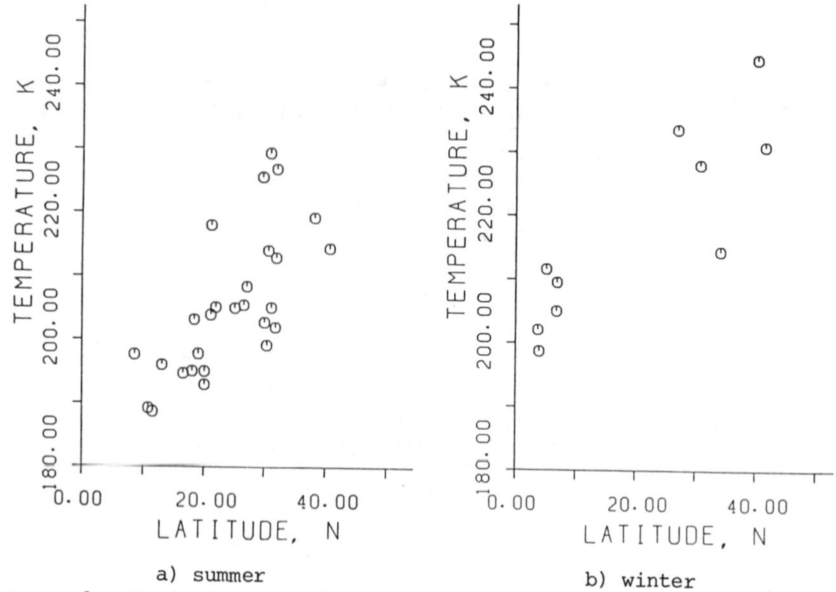

FIGURE 1. Latitudinal variation of the cloud top temperature for a class CB.

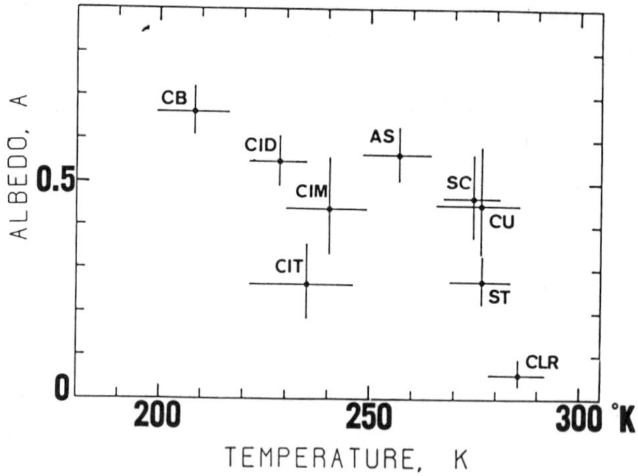

FIGURE 2. Distribution of 9 cloud classes.
Cloud class CB: Cumulonimbus, CID: Dense cirrus with low and middle level clouds, CIM: Cirrus with low level clouds, CIT: Cirrus only, AS: Middle level clouds, CU: Cumulus, SC: Stratocumulus, ST: Stratus and fog, CLR: No clouds. (After Aso et al., 1988)

3. LATITUDINAL AND SEASONAL VARIATIONS OF THE CLOUD TOP TEMPERATURE

Cloud top temperature versus latitude for a class CB is shown in Fig. 1 for summer and winter. The temperature difference between the 10 N and 40 N is about 30 K both in summer and winter. The cloud top temperature in summer is colder by about 20 K than in winter on the average. As we selected only mature CB clouds with an anvil cirrus, these differences roughly correspond to the change of height of the tropopause. In lower latitudes, atmospheric conditions depend largely on convective activity and the location of the ITCZ. So we would have a different variation of the cloud top temperature, when we analyse other images. It should be stressed that there are variations of cloud top temperatures even in lower latitudes.

In addition, we should notice that a tendency of the variation of the cloud top temperature is specific to each class. For example, fig. 4 indicates that for low-level cloud classes, cloud top temperature decreases with higher latitude, contrary to a class CB.

There are also variations of cloud maximum albedo (Aso et al., 1988), which are related to a change of the solar zenith angle.

4. EFFECTS OF THE TEMPERATURE VARIATION ON THE CLOUD CLASSIFICATION

The distribution of nine cloud classes is shown in Fig. 2. An examination of Fig. 2 shows that the accuracy of classification for a class CB and CLR would be higher than for other class, because their mean location in this diagram is sufficiently different from other classes. For example, the effects of temperature change for the CB class are less than that of same temperature change for low level classes.

Fig. 3 shows cloud classification for a class CB using only IR characteristic parameters (a cloud top temperature and one texture parameter) and indicates failure in classification (△). The failures occur at temperatures warmer than about 220 K which separate class CB and CT groups. Since a cloud top temperature contributes greatly to the classification, this procedure of using only IR features is about the same as using a fixed TBB threshold value.

We show the results of cloud classification in winter for low level classes CU, SC and ST in Fig. 4 using only IR parameters. To show the effects of the temperature variation clearly, we gather these three classes together and indicate failures (△, +) in Fig. 4. Wrong classifications to higher level clouds occur at temperatures colder than about 265 K and to no clouds at temperatures warmer than about 280 K. Low level clouds north of about 35 N are all classified into the middle or high level cloud classes.

These two cases indicate the importance of temperature variation to cloud classification.

5. CONCLUDING REMARKS

For the CB class there is a latitudinal and seasonal variation of cloud top temperature. The temperature difference between 10 N and 40 N is about 30 K both in summer and winter. The cloud top temperature in summer is colder by about 20 K than in winter on the average. For low level classes cloud top temperature decreases towards higher

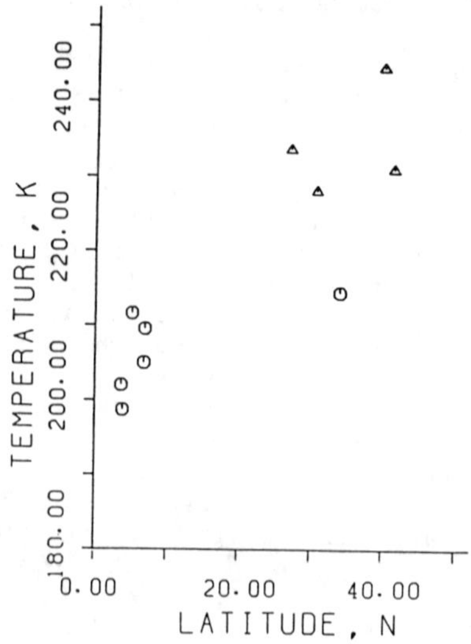

FIGURE 3. Cloud classification for a class CB in winter.
⦶: rightly classified to a class CB.
△: classified to classes CID and CIM.

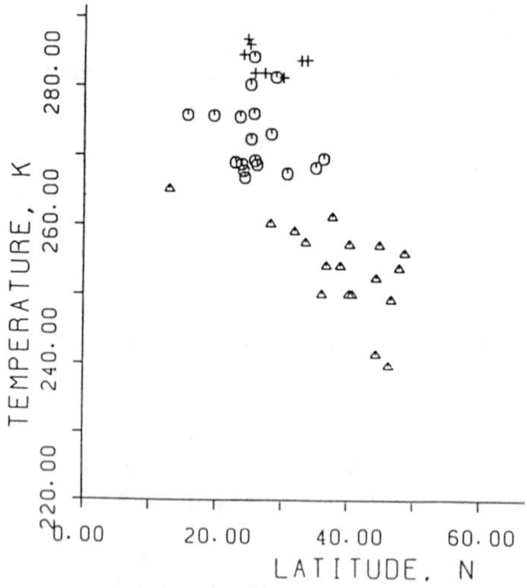

FIGURE 4. Cloud classification for classes CU, SC and ST in winter.
⦶: classified to these classes. △: classified to middle or higher level cloud classes. +: classified to CLR.
(After Aso et al., 1988)

latitudes on the contrary.

These variations strongly affect the accuracy of cloud classification. For the CB class, the failures occur at temperatures warmer than about 220 K. For low-level classes clouds north of about 35 N are all classified into the middle or high-level classes in winter.

It is desirable to take account of these variations, when we extract clouds in a class CB or other class relating to precipitation using cloud classification or TBB threshold value.

ACKNOWLEDGMENTS

The authors would like to thank Mr. S. Osano for helpful discussions and encouragement. The authors would also like to thank Mr. K. Kato and K. Ishikawa for helpful comments on using cloud classification programs.

REFERENCES

Aso, T., K. Oosawa, and Y. Takeuchi, 1988: Latitudinal and seasonal variation of the cloud characteristic parameters and their effects on the cloud classification, Meteorological Satellite Center Technical Note, 16, 9-20.

Garand, L., 1986: Automated recognition of oceanic cloud patterns and its application to remote sensing of meteorological parameters, Ph. D. Thesis, University of Wisconsin-Madison.

Kato, K., and K. Ishikawa, 1982: Objective cloud top-height estimation system for GMS imageries, Meteorological Satellite Center Technical Note, 5, 1-24.

Parikh, J. A., 1977: A comparative study of cloud classification techniqunes, Remote Sensing of Environment, 6, 67-81.

AN APPROACH TO STUDIES OF INTERANNUAL VARIATIONS OF RAINFALL USING MULTIFREQUENCY DUAL POLARIZATION MICROWAVE RADIANCES

B.B. Hinton and W.S. Olson
Space Science and Engineering Center
University of Wisconsin-Madison
Madison, Wisconsin 53706, USA

ABSTRACT

We provide the rationale for rain estimation from microwave radiances observed by Nimbus-7 over oceans. We conclude that two dual polarization frequencies are sufficient to account for several classes of cloud structure.

1. INTRODUCTION

We are working on an interannual comparison of rainfall in the Indian Ocean region. Resolution of the study is 2° Lat, 2° Lon, and one month. The data are the multi-year multi-channel observations from SMMR, the passive microwave radiometer on the Nimbus-7 satellite.[1] Algorithm development, the subject of this paper, has been guided by studies of Weinman and Guetter (1977), Wilheit (1979), Wu and Weinman (1984), and Olson (1987). If we can construct an approximate, but realistic, "complete set" of cloud models having only two or three parameters (one of which is rain rate, R), we hypothesize these model parameters can be found from the SMMR observations (see Table 1). To make this plausible we must show observations have independent projections onto the model parameters. Our models are realistic, but not exact, so the final step is to calibrate, or re-tune, using real-world truth data, primarily from radar observations.

TABLE 1. CHARACTERISTICS OF THE NIMBUS-7 SMMR

f(Ghz)	Polarization		Footprint(km)
6.63	hor.	vert.	121 by 79
10.69	hor.	vert.	74 by 49
18	hor.	vert.	44 by 29
21	hor.	vert.	38 by 25
37	hor.	vert.	21 by 14

[1]We are using the "TCT" tapes provided by the National Space Science Data Center through the courtesy of Per Gloerson.

ISBN 0-937194-14-4

2. SUMMARY OF MODELLING RESULTS

Maxwell's equations govern the radiation. Using the dielectric and conductivity properties of water, one can derive Fresnel's reflection formulas. The emission coefficient for water follows from Kirchoff's law. Emissivity, e, and reflectivity, r, vary with the polarization and direction relative to the surface. Since electromagnetic properties of water vary with temperature and salinity so do e and r. Due to resonance-relaxation effects of water e and r also depend upon frequency, f. For the 50° incidence angle of SMMR $e \sim 0.4$. The temperature of the tropical oceans is ~ 300 K, so the surface emission brightness temperature will be ~ 120 K. Additionally, e increases with wind roughening and surface foam (Wilheit, 1979).

There are non-negligible absorption-emission effects of gaseous H_2O and O_2, but these are approximated using approximate profiles of temperature, T, pressure, p and moisture, q. Extinction rates, k_{ext} (km^{-1}), for clear air at 100 % humidity are: 0.03 at 18 GHz, 0.09 at 21 GHz, 0.05 at 37 GHz. From symmetry considerations gaseous extinction and emission are independent of polarization.

The effects of liquid drops are greater, variable and directly associated with rainfall through the drop size distribution (e.g. Marshall-Palmer, 1948, or Merceret, 1974. Small diameter (d < 100 μm) non-precipitating cloud drops are Rayleigh particles at SMMR frequencies. From the k_{ext} coefficients given by Battan (1973) scattering is negligible relative to absorption-emission. Extinction by absorption, k_a, ranges from < .004 km^{-1} at 6.6 GHz to < 0.15 at 37 GHz. By Kirchoff's law 37 GHz sensors "see" clouds as a slight increase of brightness temperature, T_B, lower channels do not.

Rain drops (d > 100 μm) are Mie particles at 37, or even 18 GHz. Olson (1987) calculated extinction coefficients for absorption and scattering, k_s, as a function of R using a realistic drop size distribution family. The ratio k_s/k_{ext} and k_{ext} are shown in Fig. 1. For uniform liquid precipitation reaching to various heights, the upwelling brightness temperature, T_B, seen by SMMR would resemble Fig. 2. As the rainfall rate (and thus optical thickness, τ) increases, low intensity polarized surface emission is replaced by higher intensity unpolarized cloud emission. This explains the initial slope of the curves. At higher R, τ increases further, so only radiation nearer the top of the cloud is seen causing a decrease in the slope. Increasing R further increases scattering causing T_B to decrease by scattering out of the upward stream of radiation. This energy is only replaced slightly by backscattering of cold (< 3 K) downwelling "cosmic" background into the upward direction. This effects a change in the sign of the slope, resulting in T_B values corresponding to two R values. The slope transitions mentioned above occur for lower R values as f increases, and to a lesser degree as the thickness of the layer increases.

Most rain systems will have vertical development above freezing level, so frozen cloud particles (d < 100 μm) and frozen precipitation particles (d > 100 μm) may be present. Due to the dielectric-conductivity properties of ice, the k_{ext} are less than 0.001 km^{-1} for frozen cloud particles (Olson, 1987). For the larger particles k_{ext}(ice) « k_{ext}(liquid), assuming the same size distribution and water content. For $f \gtrsim 18$ GHz, $k_s \approx k_{ext}$. According to Olson (1987), a one to two km layer of ice affects T_B noticeably above 18 GHz. (See Fig. 3.)

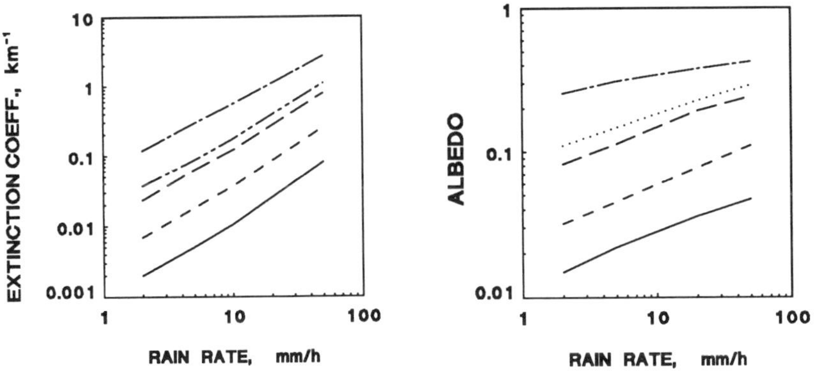

Figure 1. Extinction and albedo variation with rainfall rate. From bottom to top, in both panels, the curves represent 6.6, 10.7, 18, 21, and 37 GHz. Curves are from Olson (1987).

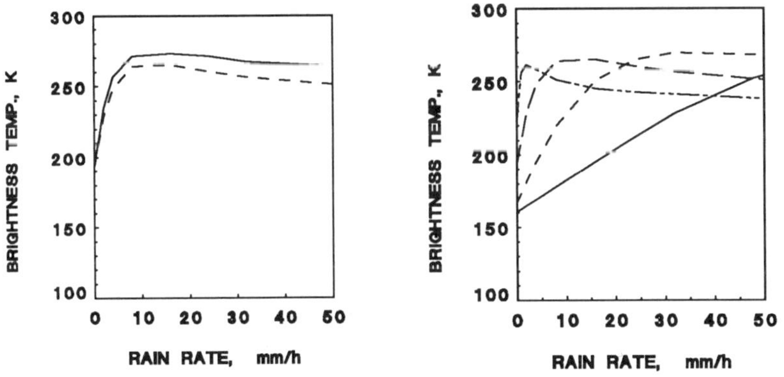

Figure 2. Examples of brightness temperature as a function of rain rate for liquid-only clouds (Olson, 1987). All curves are for 50° incidence angle. The left panel is for 18 GHz. The solid curve and the short dashed curves are for v-pol. radiation from 3.8 and 6.8-km thick layers of precipitation respectively. The right panel shows variation for one thickness (6.8 km) and one polarization (v), but for all the SMMR channels except 21 GHz. Solid denotes 6.6, short dashes 10.7, long dashes 18 and dash-dots 37 GHz.

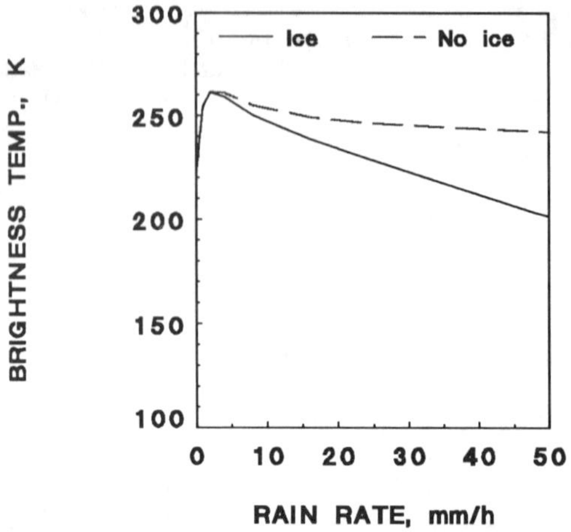

Figure 3. Influence of ice on the variation of brightness temperature with rainfall rate. The example shown is for v-polarization and 37 GHz. The liquid precipitation layer is 5.8 km thick, the frozen layer is one km thick (Olson, 1987).

3. APPROACH TO A PRACTICAL ALGORITHM

Three cases are considered in Table 2. The table is constructed so that higher altitudes are represented nearer the top. Asterisks represent ice, while "oo" implies liquid drops.

TABLE 2. CLOUD STRUCTURE CASES TO BE CONSIDERED BY THE ALGORITHM

Case Number	1a	1b	1c	2a	2b	3
Frozen cloud particles			**		**	**
Frozen precipitation						**
Freezing level	.....	.....	.....	.....	.....	.....
Liquid cloud particles		oo		oo	oo	oo
Liquid precipitation				oo	oo	oo
Rain rate	---no	rain---		lt.		>lt.

The simplest algorithm distinguishing the three cases would use two frequencies and two polarizations. The reader can follow our arguments by examining the modelling results above. For convenience, we refer to the mean of the two polarizations as *the* brightness temperature, or just brightness, and to the difference as polarization.

Case 1a: Both frequencies show brightnesses and polarizations characteristic of the unobscured ocean. Over a 2° square in a month rain- and cloud-free data are almost certain. Therefore, levels can be established statistically from the *distributions* of the T_B in the 2° square for the month, on the assumption that sea surface temperature and wind speed stay near their average values over area and time.

Case 1b: This is similar to Case 1a, except slight warming and depolarization of the higher frequency would be tolerated.

Case 1c: Again, the observations would be similar to 1a.

Case 2a: Noticeable warming and depolarization would be seen on the high *f* channel, but very little to a slight change should be detected on the lower channel. In this case *R* is best estimated from the degree of warming of the high *f* brightness.

Case 2b: Indistinguishable from Case 2a, so the estimation of *R* should follow from warming of the high *f* brightness.

Case 3: This case is distinguished by definite warming and depolarization of the lower *f*. The latter provides the best estimate of rate, unless it is as warm as 250 K, or so. In this event the high *f* channels are at rates well above their maximum brightness, so *R* can be estimated from the depression below that value. The maximum brightness would be obtained statistically from the month's data for the latitude-longitude cell, but constrained to be near a model value around ~ 260 K if insufficient data with warm low *f* brightnesses are found that month.

Which channels should we use? One set might be: 37 GHz h & v and 18 GHz h & v, while another set would be 18 h & v plus 10.7 h & v. The 37 GHz channels appear to be too sensitive to handle Case 2 quantitatively over a useful range of *R*, the maximum brightness sets-in at too low*R*. However, for $R > 8$ mm/h^{-1}, they provide information on areal rain coverage within the 2° by 2° boxes. On the other hand, 18 GHz would be provide rain intensity information up to 15 mm h^{-1} for case 2. Companion 10.7-GHz data are useful in sensing by warming up to about 25 mm/h^{-1}, handling most of case 3. Above 25 mm/h^{-1}, 18 Ghz is past its maximum brightness and could be used in the temperature depression mode. At the same time, 10.7 GHZ seems sensitive enough to detect widespread Case 2 up to roughly 8 mm/h^{-1}.

The drawback of the 10.7-18-GHz combination is the depletion of signal from rain by non-raining areas within the same large field of view of the 10.7-GHz channel. This means working at lower pixel average *R*, and poorer signal-to-noise ratios. On the other hand, much more of the range of *R* would be retrieved in linear portions of the $T_B(R)$ curves. In the present study, the poor spatial resolution of 10.7 is not itself a serious problem.

Naturally, more elaborate schemes incorporating more channels are possible. Perhaps because of intrinsic redundancy of information, noise, and varying resolution, or noncoincidence of footprints, additional empirical studies by the authors suggest that, among all the channels, there are no more than about three degrees of freedom. In our scheme we have to retrieve at least two parameters: (1) Case number (1,2,3, used to select the appropriate cloud structure) and (2) *R* averaged over 2° by 2° boxes. More channels are unlikely to allow one to retrieve any more parameters. To make our scheme robust, we have used the artifice of statistical analysis to avoid the necessity of simultaneously retrieving sea surface temperature, wind speed, humidity, and other parameters having lesser impact on upwelling T_B. To retrieve more parameters would result in an unstable algorithm as their number would equal or exceed the degrees of freedom of the data.

REFERENCES

Battan, L.J., 1973: Radar Observations of the Atmosphere. The University of Chicago Press, Chicago.

Marshall, J.S., and W.M. Palmer. 1948: The distribution of raindrops with size. J. Meteor., 5, 165-166.

Merceret, F.J., 1974: On the size distribution of raindrops in Hurricane Ginger. Mon. Wea. Rev., 102, 714-716.

Olson, W.S. 1987: Estimation of Rainfall Rates in Tropical Cyclones by Passive Microwave Radiometry. PhD. Thesis, University of Wisconsin-Madison, Madison, Wisconsin 53706, USA.

Weinman, J.A., and P.J. Guetter, 1977: determination of rainfall distributions from microwave radiation measured by the Nimbus-6 ESMR. J. Appl. Meteor., 16, 437-442.

Wilheit , T.T., 1979: A model for microwave emissivity of the ocean's surface as a function of wind speed. IEEE Trans. Geoscience Electron., GE-17, 244-249.

Wu, R., and J.A. Weinman, 1984: Microwav e radiance s from precipitating clouds containing aspherical ice, combined phase, and liquid hydrometeors. J. Geophys. Res., 89, 7170-7178.

TRMM DATA SYSTEM STUDY

Daesoo Han
NASA Goddard Space Flight Center
Greenbelt, Maryland 20771, USA

ABSTRACT

A data system that will receive, process and distribute Tropical Rainfall Measuring Mission (TRMM) science data is studied. An end-to-end data flow scenario is developed. Estimation on the data throughput time is discussed for two cases. The results show that level 0 through level 2 of certain parameters may be available to scientist within 26 hours after the satellite observation. Science data products identified by the Science Steering Group are used to estimate storage space requirements. Currently available technologies to accommodate the requirements are studied. It is feasible to store all Levels of one year's TRMM data on-line with current technology. Data distribution to users is also studied. Using T3 communication lines, it is possible to download one days global microwave data in 30 minutes for remote site users in USA. Finally, issues to be resolved are listed.

1. DATA PRODUCTS

Five levels (level 0 to level 4) of data will be produced from TRMM measurements. Definitions of these Levels closely follow the convention used by Earth Observing Satellite Data Panel Report edited by R. Chase (1986), but add unique features of TRMM. Levels 1 and 2 will have two types of data and are called levels 1A, 1B, 2A, and 2B. Definitions of these levels are:

Level 0: Reconstructed (redundancy removed and time-ordered) unprocessed instrument data at full resolution.

Level 1A: Reconstructed unprocessed instrument data at full resolution, time-referenced, and annotated with ancillary information, including radiometric calibration coefficients and georeferencing parameters computed and appended but not applied to the level 0 data. It is possible to reconstruct level 0 data.

Level 1B: Level 1A data that has been processed to sensor units (i.e., radar backscatter cross section, brightness temperatures, etc.). May not be possible to reconstruct lower level data.

Level 2A: Derived environmental variables at the same resolution and location as the level 1 data.

Level 2B: Applied averaging process (spatial/temporal) on the level 2A data. May combine level 2A data from different sensors (e.g., rainrates from radar and microwave radiometers).

Level 3: Variables mapped on uniform space-time grid scales.

Level 4: Model output or results from analyses of lower-level data.

TRMM also includes ground truth measurement experiments. These data will be included in the standard TRMM data products. Other data sets which will be needed to process and/or validate TRMM data, are also included in the list. The list of level 2 physical parameters identified so far by the TRMM Science Steering Group (Personal communication, 1987) is given in Table 1.

TABLE 1. LEVEL 2 PHYSICAL PARAMETERS

AVHRR:	(1) Cloud Cover (Daytime)
	(2) Cloud Cover (Nighttime)
	(3) Cloud Height
	(4) Optical Thickness
	(5) Albedo
	(6) Rain Estimate
ESMR/SSM/I:	(7) Rainrate (19 GHz)
	(8) Rainrate (37 GHz)
	(9) Rainrate (85.5 GHz)
	(10) Composite Microwave Radiometer Rainrate
RADAR:	(11) Two Frequency Backscatter Rainrate (Vertical Structure)
	(12) Path Averaged Rainrate
	(13) Differential Attenuation Rainrate
	(14) Liquid Water Content
	(15) Composite Radar Rainrate
	(16) Storm Height
	(17) Well-defined Bright Band (Y/N)
TRMM:	(18) Composite Rainrate (All Instruments)
	(19) Vertical Profile
GROUND TRUTH AND OTHER CORRELATIVE DATA:	(20) Radar Rainrate
	(21) Rain Gauge - Rainrate
	(22) Composite Rainrate (Each Site)
	(23) Experimental - Instruments - Rainrate
	(24) Other Satellite Rain Estimates
	(25) Special Meteorological Data

2. END-TO-END DATA FLOW

An end-to-end data flow is developed and shown in Figure 1. TRMM collects data from the earth surface and atmosphere, receives ephemeris data from Global Positioning System (GPS), and sends those data to the Tracking and Data Relay Satellite System (TDRSS) twice per orbit.

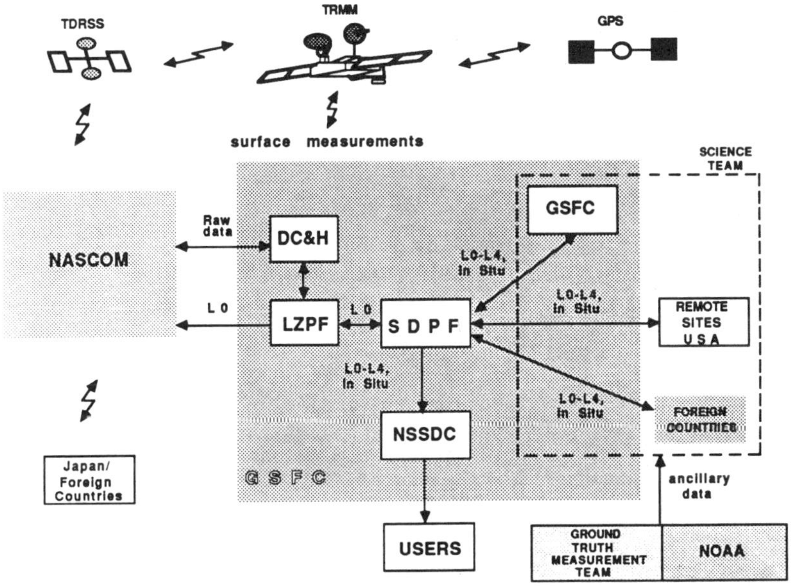

FIGURE 1. TRMM END-TO-END DATA FLOW

The data will be relayed to the White Sands Ground Terminal (WSGT) in the United States. From there the data are transmitted to the Data Capture and Handling (DC&H) Facility at the NASA Goddard Space Flight Center (GSFC) in Greenbelt, Maryland, USA. TRMM data will be processed at the same location to level 0. Level 0 Radar data will be transmitted to Japan via NASA Communication Networks (NASCOM). Also level 0 data will be transmitted to the TRMM Science Data Processing Facility (SDPF), located at GSFC, through a local area network. The SDPF will further process the data and generate level 1, 2, and 3 products. Level 4 data may be produced by TRMM Science Team members or by SDPF when the Team members provide models and analysis tools. Preprocessed data from ground truth experiments, which are part of the mission, will be processed at SDPF. Other auxiliary data needed for the data processing and/or data validation will be stored at SDPF. Data processed by SDPF will be available to the TRMM Science Team members in the local area as well as remote sites including outside of the USA via telecommunication lines or through communication satellites. Once the TRMM Science Team approves the release of data, level 1 through level 4 data will be transferred electronically to the National Space Science Data Center (NSSDC) at GSFC for archival and distribution to general users.

3. DATA PROCESSING

In Figure 2, the TRMM science data processing is shown. In this section, the throughput time to generate through level 2 data will be estimated based on available information and some assumptions. The assumptions are:

a. It will take about 12 hours to process one day's data to level 0.

b. It will take about 8 hours to process one day's data from level 0 to level 2A which do not require data other than TRMM.

These assumption are not difficult to meet with today's technology. Two cases are considered here. The first case is that TRMM has GPS receiving capability and the other case is that TRMM has not the capability.

For the first case, there is no time delay to get ephemeris information because it is part of the data stream from the satellite. There will be two data transmissions from TRMM to TDRSS per orbit; the oldest data are measured 45 minutes before the transmission time. The time required to process the raw data through level 2 is less than 40 minutes based on the above assumptions. In addition, there is less than 30 minutes of transmission time from the TRMM satellite to SDPF. Based on this analysis, these data will be available to users within two hours. **An underlying assumption of this analysis is that data quality is good all the time. There are, however, bit errors occurring during transmission. In particular, the bit error rate of TDRSS is 1/100,00, which is very high. Therefore, requests for data retransmissions from DC & H to WSGT will be required. Time estimation for this extra request is difficult to make at this time. Past experience shows that it takes somewhere between 1 day to 1 week. So the throughput time really depends on the requirements for the completeness of data set by the TRMM Science Team.**

For the second case the ephemeris data are available through ground tracking. In this case, it takes about 24 hours to receive ephemeris data from all the ground tracking stations. So it will take one day longer than the first case, that is, 26 hours after satellite observations.

Another aspect of GPS is that it gives more accurate earth location than the ground tracking data. The position accuracy from GPS is better than 100 m, while data from grounding tracking gives about 1-km earth location error. In addition to this, there are uncertainties in the spacecraft attitude and pointing accuracies.

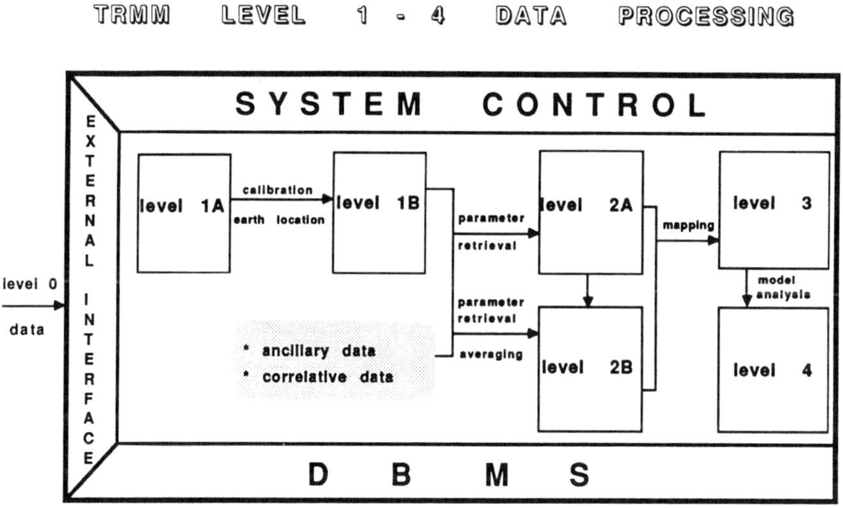

FIGURE 2. TRMM DATA PROCESSING ELEMENTS

Table 2 shows a summary of location error in terms of percent of the pixel size when using GPS information.

TABLE 2. EARTH LOCATION ERROR

Pixel Location	Error (% Pixel)
Nadir	67
Mid-Scan	73
End-Scan	83

The error estimates are made based on 1-km resolution at nadir point and attitude knowledge uncertainty of 0.1 degrees. The most sensitive parameter for the pixel location computation is the attitude knowledge. Without GPS, there is an additional 1-km error in addition to the above error.

4. DATA STORAGE AND DISTRIBUTION

The current plan of TRMM includes 3 instruments; AVHRR, Radar, and TRMM Microwave Radiometer (TMR). Data rates for these instruments are listed in Table 3.

Based on the above data rates of Table 3, and level 2 parameters listed in Table 1, data volume per day for level 0 through level 2A are computed and listed in Table 4.

TABLE 3. DATA RATES FROM INSTRUMENTS

Instruments	AVHRR	RADAR	TMR
Data Rates	120 kbps	170 kbps	50 kbps

TABLE 4. DATA VOLUME PER DAY (Gigabytes)

Data Level	AVHRR	RADAR	TMR
Level 0	1.3	1.9	0.5
Level 1A	1.7	2.5	0.7
Level 1B	2.9	5.1	1.0
Level 2A	7.8	5.1	0.5
Total	13.7	14.6	2.7

It is assumed that each sensor data will be 12 bits wide and converted to 16 bits in level 1 processing. Also it is assumed that earth location of each instantaneous field of view (IFOV) is computed at level 1B. The above data volume may be reduced significantly, especially for AVHRR and Radar, if some fraction of IFOV's earth locations are computed. This needs further analysis to meet scientific requirements.

The data volumes of level 2B and level 3 will be much smaller than lower level data volumes and was not estimated in here. Level 0 data will not be stored, except a small fraction of the total data to check calibration and other instrument-related events. Excluding these data, the total data per day is about 31 Gigabytes. TRMM will be operating every day for 3 years. The total volume for the 3 year time period is approximately 34,000 Gigabytes.

For a comparison, Table 5 shows data volumes per day when only 2% of IFOV locations are computed.

TABLE 5. DATA VOLUME PER DAY (Gigabytes)

Data Level	AVHRR	RADAR	TMR
Level 0	1.3	1.9	0.5
Level 1A	1.7	2.5	0.7
Level 1B	1.7	2.5	0.7
Level 2A	2.6	2.5	0.4
Total	7.3	9.4	2.3

Total data volume to be stored for one day is 19 Gigabytes and about 20,000 Gigabytes for 3 years.

Data storage media currently available can hold up to 3,000 Gigabytes. According to the recent study by H. Ramapriyan et al. (1984) on a data compression technique applied on Landsat data, a compression ratio of 3 is not difficult to achieve without any loss of information. Assuming 3-to-1 data reduction by data compression, one year of TRMM data can be stored on-line with current technologies.

The TRMM Science Team members will be located not only at GSFC but also at other research sites. Large amounts of data will be distributed with high-density storage media, such as optical disks or videocassette tapes. Users may need to transfer large amounts of data over communication lines. An important factor in designing a data system is to estimate how much data can be transferred to other research site users within a reasonable time. For GSFC users, there will be a local area network by 1992 that will have data transfer rates of 100 megabytes per second (Mbps). Using this network, a user can download 1 day's global microwave data in less than 3 minutes. For other research users, T1 lines which have a data transfer rate of 1.544 Mbps, are available. T3 lines which have a 40 Mbps data transfer rate will also be available. Using T1 communication lines, it will take about 3 hours for a user to download 1 day's global microwave radiometer data. On the other hand, with T3 lines, it will take less than 10 minutes to download the same amount of data.

5. ISSUES

There are several issues to be resolved in designing an optimum data system. The issues are listed in Table 6.

TABLE 6. ISSUES CONCERNING DATA SYSTEM DESIGN

Issues	Impact Area
1. Throughput Time	- Processor - Software System
2. Amount of Data Gap	- Processor - Software System
3. Earth Location Accuracy	- Processor - Software System - Data Storage - Network
4. Remote Sites	- Processor - Software system - Data Storage - Data Network

REFERENCES

Chase, R. (Ed.), 1986,Report of the EOS Data Panel, NASA TM 87777

Ramapriyan, H. K., J. C. Tilton, and E. J. Seiler, 1985: Impact of Data Compression on Spectral/Spatial Classification of Remotely Sensed Data. In A. Deepak, H. E. Fleming, and M. T. Chahine (Eds.), Advances in Remote Sensing Retrieval Methods, A. Deepak Publishing, Hampton, VA, 685-705

A RADIATIVE TRANSFER MODEL OF MICROWAVE RADIANCES FROM HORIZONTALLY FINITE CLOUDS CONTAINING ICE AND LIQUID HYDROMETEOR LAYERS

J.A. Weinman[1] *and C.D. Kummerow*[1,2]
[1]University of Wisconsin
Madison, Wisconsin 53706, USA

[2]University of Minnesota
Minneapolis, Minnesota 55455, USA

ABSTRACT

Microwave radiances that would be measured from satellite-borne radiometers have been computed as a function of rainfall rates from horizontally finite precipitating clouds that contain ice and liquid hydrometeors. Ice at the top of the precipitating clouds depresses the brightness temperatures. The brightness temperatures depend on rainfall rates because the ice hydrometeor concentrations are assumed to be related to the rainfall rates at the cloud base. It is also found that the brightness temperatures of footprints that are partially covered by precipitation cells are a non-linear function of the rainfall rate averaged over the footprint. Thus the average brightness temperatures depend on the peak rainfall rates and the size of the precipitating cloud within the radiometer footprints.

1. INTRODUCTION

Satellite borne microwave radiometers can be used to retrieve underlying rainfall rates in reasonable agreement with radar measurements; see for instance Spencer et al. (1983) and Rodgers and Siddalingaiah (1983). Although some rainfall retrieval techniques are empirical, most depend on radiative transfer models which theoretically predict brightness temperature for a given rainfall rate; see Wilheit et al. (1977), Weinman and Davies (1978), Wu and Weinman (1984). None of these models however, consider the problem fully. Weinman and Davies used a vertically homogeneous cloud to show that the representation of a moderately dense precipitating cloud by a plane parallel model is questionable. More recently, Wu and Weinman showed that plane parallel clouds, with frozen hydrometeors in the top of the cloud depress the brightness temperatures. This agrees with observations of brightness temperatures in severe thunderstorms reported by Spencer et al.

An analytical model for horizontally finite clouds was developed which considers an ice as well as liquid water layer within the cloud. The hydrometeors were assumed to be oblate spheroids with deformations suggestd by Oguchi (1983).

TROPICAL RAINFALL MEASUREMENTS
John S. Theon and Nobuyoshi Fugono (Eds.)

ISBN 0-937194-14-4

A Marshall-Palmer drop size distribution is assumed for both the rain and the ice layer. The extinction coefficient, k, the albedo for single scattering, a, and the asymmetry factor, g, cited in Table 1 were used. The emissivity of water was obtained from Fresnel's equation. The emissivity depends on the polarization of the measured radiances and upon the measuring angle.

2. THEORY

We will consider the transfer of radiances, I, through a two-layer, horizontally finite cloud. Hydrometeors are assumed to be homogeneously distributed in each layer. The temperature in the cloud is allowed to vary smoothly from the surface to the cloud-top height of either 4.77 km for the 'liquid only' clouds or 6.77 km if an ice layer is added. The lapse rate is assumed constant.

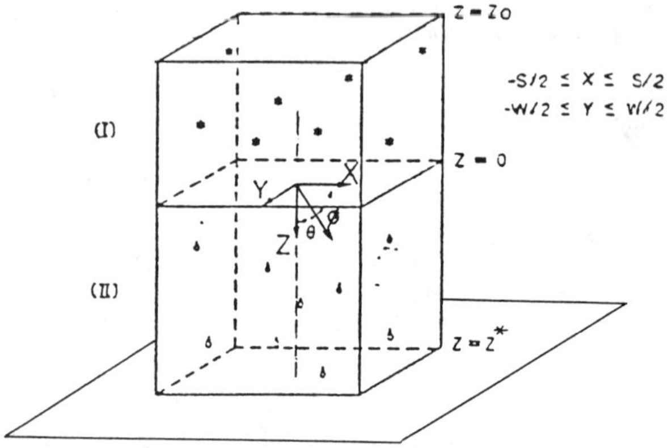

FIGURE 1. Schematic view of the cloud in the radiometer footprint.

The radiances are calculated using Eddington's second approximation to the equation of radiative transfer for each of the two layers. The total diffuse radiance is obtained by requiring continuity of the radiances across the interface of the two layers. The basic equation of radiative transfer, generalized to three dimensions, is given by

$$\sin\theta\cos\phi\frac{\partial I}{\partial x} + \sin\theta\cos\phi\,\frac{\partial I}{\partial y} + \cos\theta\frac{\partial I}{\partial z} = -k(I-J) \qquad (1)$$

Details of the geometry are shown in Fig. 1. Here, z = 0 was chosen to be at the interface between the two layers for computational simplicity. Note also that the lack of superscript (I) or (II) on the radiances in Eq. (1) is meant to indicate that this equation is equally valid in the two regions in question. This formalism will be maintained throughout the report.

The source function, J, is given by

$$J[x,y,z,\theta,\phi] = (1-a)(B_0+B_1 z) + \frac{a}{4\pi}\int_0^{2\pi}\int_{-1}^{1} P(\theta,\phi;\theta',\phi')I(\theta',\phi')d(\cos\theta')d\phi' \quad (2)$$

where $P(\Theta,\phi;\Theta',\phi')$ is the phase function for scattering from direction Θ',ϕ' into Θ,ϕ. The Planck blackbody source function is represented by the Rayleigh-Jeans approximation at microwave frequencies. The temperature profile is imbedded in the constants B_0 and B_1. This term represents the contribution to the radiances due to thermal emission inside the cloud.

The radiances may be represented by a series of Legendre polynomials and associated Legendre functions

$$I(x,y,z,\theta,\phi) = I_0(x,y,z) + I_x(x,y,z)\sin\theta\cos\phi + I_y(x,y,z)\sin\theta\sin\phi + I_z(x,y,z)\cos\theta + \ldots \quad (3)$$

If the scattering angle from Θ',ϕ' to Θ,ϕ is represented by $\cos\Theta$ and the phase function $P(\cos\Theta)$ is similarly expanded as

$$P(\cos\Theta) = \sum_{l=0}^{n} \tilde{\omega}_l P_l(\cos\Theta) = 1 + \tilde{\omega}_1 P_1(\cos\Theta) + \ldots \quad (4)$$

then, to the approximation shown, the source function can be written as

$$J[x,y,z,\theta,\phi)] = (1-a)(B_0+B_1 z) + a[I_0+g(I_x\sin\theta\cos\phi + I_y\sin\theta\sin\phi + I_z\cos\theta)] \quad (5)$$

where g can be shown to equal $1/3\tilde{\omega}_1$. Equation (5), when substituted back into Eq. (1) can be shown to lead to the Laplacian for I_0

$$\nabla^2 I_0 = \lambda^2(I_0-B_0-B_1 z) \quad (6)$$

where $\lambda^2 = 3k^2(1-a)(1-ag)$.

I_x, I_y and I_z can be found by

$$I_x = -\frac{1}{k(1-ag)}\frac{\partial I_0}{\partial x} \tag{7}$$

$$I_y = -\frac{1}{k(1-ag)}\frac{\partial I_0}{\partial y} \tag{8}$$

$$I_z = -\frac{1}{k(1-ag)}\frac{\partial I_0}{\partial z} \tag{9}$$

The boundary conditions required to solve this equation can be obtained from consideration of the irradiances on the faces of the cloud. The downwelling irradiance at the top of the cloud is due to the backround radiation. If the layer is assumed to be isothermal and non-scattering, one obtains

$$F_{\downarrow}(z_0)=(I_0^I + \frac{1}{h}\frac{\partial I_0^I}{\partial z})_{z_0} = I(0) \tag{10}$$

where $h = (3/2)k(1-ag)$, $I(0)$ is the background radiation.

The irradiances directed inward on the cloud sides consist of a surface, a non-precipitating layer contribution and the emission from the gaseous atmosphere, T_{mol}. These can be expressed as

$$F_{x\pm}(\pm 1/2) = (I_0 \pm \frac{1}{h}\frac{\partial I_0}{\partial x})_{\pm 1/2} = \frac{\varepsilon T^*}{2} + \frac{1}{2}\{B[T_0](1-\beta)+\beta I(0)\} + T_{mol} \tag{11}$$

$$F_{y\pm}(\pm w/2) = (I_0 \pm \frac{1}{h}\frac{\partial I_0}{\partial x})_{\pm w/2} = \frac{\varepsilon T^*}{2} + \frac{1}{2}\{B[T_0](1-\beta)+\beta I(0)\} + T_{mol} \tag{12}$$

where β is given by

$$\beta = \frac{4}{\pi}\int_0^1 e^{-\tau/\mu}(1-\mu^2)^{\frac{1}{2}}d\mu \tag{13}$$

$\tau = k\cdot\rho\cdot s$, k is the mass extinction coefficient, ρ is the density of the ice in the layer in (gm/kg air), s is the vertical projection of the path in the layer and ε is the surface emissivity while T^* is the surface temperature.

At the cloud base, the upwelling irradiance is the sum of the irradiance emerging from the surface plus the reflected irradiance from both the cloud and the gaseous atmosphere.

$$F_{\uparrow}^{I}(z^*) = 2\pi \int_0^1 \varepsilon_p(\mu) T^* \mu d\mu + 2\pi \int_0^1 I_{\downarrow}^{II}(\mu)(1-\varepsilon_p(\mu))\mu d\mu + (1 - \varepsilon_p(\mu))F_{\downarrow mol} \quad (14)$$

which can be written in terms of I_0 as

$$\left(I_0^{II} - \frac{1}{h}\frac{\partial I_0^{II}}{\partial z}\right)_{z*} = \varepsilon_{eff}T^* + (1-\varepsilon_{eff})\left(I_0^{II} + \frac{1}{h}\frac{\partial I_0^{II}}{\partial z} + F_{\downarrow mol}\right)_{z*} \quad (15)$$

The boundary condition at the interface can be found by requiring continuity in the up- and downwelling irradiances

$$F_{\uparrow}^{I}(0) = F_{\uparrow}^{I}(0) \quad (16)$$

$$F_{\downarrow}^{I}(0) = F_{\downarrow}^{I}(0)$$

The solution to Eq. (6) subject to the above mentioned boundary conditions can be implemented using Fourier transformations, see Sneddon (1951). We thus define

$$\psi(\zeta_i,\zeta_j,z) = \int_{-w/2}^{w/2}\int_{-1/2}^{1/2} \left[I_0(x,y,z) - \frac{\varepsilon T^*}{2} - (1-\beta)\frac{B[T_0]}{2} - \beta\frac{I(0)}{2} - T_{mol}\right] \cdot \cos(\zeta_i x)\cos(\zeta_y)dxdy \quad (17)$$

If the eigenvalues are defined by

$$\zeta_i = h \cot\left(\frac{\zeta_i 1}{2}\right) \quad (18)$$

$$\zeta_j = h \cot\left(\frac{\zeta_j w}{2}\right)$$

then the transform of Eq. (6) reduces to

$$\frac{\partial^2\psi}{\partial z^2} - \Lambda^2\psi = \chi_0 + \chi_1 z \quad (19)$$

where

$$\Lambda^2 = \lambda^2 + \zeta_i^2 + \zeta_j^2 \quad (20)$$

$$\chi_0 = \frac{4\lambda^2}{\zeta_i\zeta_j}\left(\frac{\varepsilon T^*}{2} + (1-\beta)\frac{B[T_0]}{2} + \frac{I(0)}{2} - B_0 + T_{mol}\right)\sin\left(\frac{\zeta_i l}{2}\right)\sin\left(\frac{\zeta_j w}{2}\right)\beta \quad (21)$$

$$\chi_1 = -\frac{4\lambda^2}{\zeta_i\zeta_j} B_1 \sin\left(\frac{\zeta_i l}{2}\right)\sin\left(\frac{\zeta_j w}{2}\right) \quad (22)$$

and ψ is of the form

$$C_{+i,j}e^{\Lambda_{i,j}z} + C_{-i,j}e^{-\Lambda_{i,j}z} - \frac{X_0}{\Lambda^2_{i,j}} - \frac{X_1 z}{\Lambda^2_{i,j}}$$

I_0 can be obtained from the inverse transformation

$$I_0 = \sum_i \sum_j \frac{4\zeta_i\zeta_j\cos(\zeta_i x)\cos(\zeta_j y)}{[\zeta_i l+\sin(\zeta_i l)][\zeta_j w+\sin(\zeta_j w)]} \quad (23)$$

$$\cdot\ \psi(\zeta_i,\zeta_j,z) + \frac{\varepsilon T^*}{2} + (1-\beta)\frac{B[T_0]}{2} + \beta\frac{I(0)}{2} + T_{mol}$$

The details of the mathematical manipulations are described more comprehensively in Kummerow and Weinman (1988).

Once $C^I_{+i,j}$, $C^I_{-i,j}$, $C^{II}_{+p,q}$, $C^{II}_{-p,q}$ are known, I_0, I_x, I_y and I_z can be found. I_0 can then be obtained by employing the inverse transformation as defined in Eq. (23).

Fourier decomposition implies that the function will be weighted with $\cos(\zeta_i l)$ and $\cos(\zeta_j w)$ in the inverse transform summation. Thus, an optically thick cloud, which one would expect to have a very homogeneous radiation field, will converge less quickly than an optically thin cloud which should have a peak at the center, where the cosines peak as well. Thus, for a small cloud with low rain-fall rates, one eigenvalue is often enough to assure convergence, whereas the optically denser clouds may need as many as ten eigenvalues in i and j to converge.

We confine our attention to the $\phi = 0$ case since the azimuthal orientation of the cloud is arbitrary. The earth is assumed Lambertian so that the upwelling radiances from a point x_1, y_1, z_1 on the surface of the cloud are given by

$$I(x_1,y_1,z_1) = \varepsilon T^*(e^{-k_2S_2^*(0)}\, e^{-k_1S_1^*(0)}) + \tag{24}$$

$$[\int_{x_1^{II}(\theta)}^{x_u^{II}(\theta)} J^{II}(x,y_e,z(z,\theta),\theta,0)\cdot e^{-k_2S_2(\theta)} \cdot \frac{k_2dx}{\sin\theta}]\cdot e^{-k_1S_1^*(\theta)} +$$

$$[\int_{x_1^{I}(\theta)}^{x_u^{I}(\theta)} J^{I}(x,y_e,z(z,\theta),\theta,0)\cdot e^{-k_1S_2(\theta)} \cdot \frac{k_1dx}{\sin\theta}] +$$

$$(1-\varepsilon_{eff})[I_0^{II}(x_S,y_1,z^*) - \frac{1}{h^{II}}\frac{\partial I_0(x_S,y_1,z^*)}{\partial z}]\; e^{-k_2S_2^*(\theta)}\cdot e^{-k_1S_1^*(\theta)}$$

where $x_u(\Theta)$, $x_1(\Theta)$, $S(\Theta)$, $S^*(\Theta)$ and $z(x,\Theta)$ are parameters that depend strictly on the geometry of the cloud. The different possible geometries for a two-layered cloud is (2n + 1) and each cloud has (2n + 1) geometrically distinct regions. See Kummerow and Weinman (1988) for details of this analysis.

3. COMPUTATIONAL RESULTS AND DISCUSSION

A model that employed the parameters which include gaseous absorption/ emission outside the cloud and a non-percipitating liquid water cloud layer above the entire area of the footprint area was implemented. Table 1 gives the parameters used for these calculations at 37 Ghz.

TABLE 1. Parameters used in all calculations

$k_{liq,v}$	$= 0.0841R^{0.911}$
$k_{liq,h}$	$= 0.0904R^{0.916}$
a_{liq}	$= 0.3070R^{0.108}$
g_{liq}	$= -0.0443+0.00066R+0.026 \log_{10}R$
$k_{ice,v}$	$= 0.00450R^{1.260}$
$k_{ice,h}$	$= 0.00762R^{1.180}$
a	= 1.000
g	$= 0.170+0.00029R+0.159 \log_{10}R$
T^*	= 300.15
$T_{cloud\ base}$	= 298.15
B_0	= 273.15
B_1 (lapse rate)	= 5.42°/km
$k \cdot z_{mol}$	= 0.07
Liquid water content of non-precip layer	= 0.5 gm/m³

Figures 2-5 represent the results of these calculations. Figures 2 and 3 present the results of vertically and horizontally polarized radiances emerging over land. Figures 4 and 5 present the same quantities over water. The solid curves represent liquid water clouds (ht. = 4.77 km, $T_{cloudtop}$ = 273°K), for the given filling fraction, while the dashed curves represent the respective clouds with a 2-km ice layer added to the top so that ht. = 6.77 km.

As can be seen in Figs. 2 and 3, the ice layer at the top 2 km depresses the brightness temperatures of the cloud considerably, especially for the heavier rainfall rates, where scattering becomes the dominant extinction mechanism. The maximum of the brightness temperatures around 4 mm/hr in Figs. 4 and 5 are due to the competition between extinction and emission. Although the light rain clouds do not scatter or absorb much of the surface radiation, they also do not emit very much. In a recent paper, Spencer (1986), had reasonable success in estimating rainfall rates through empirical radiance/ polarization relations. Figures 2 - 5 support the assumptions that rainfall rates are correlated to the polarization. Radiances with different polarizations remain quite distinct for heavy rainfall rates over water as well as land. This residual effect which is due to the polarization-dependent scattering and extinction properties of aspherical ice hydrometeors. With only water in the cloud, the same effect persists, but it is markedly less pronounced. This occurs because the scattering and extinction

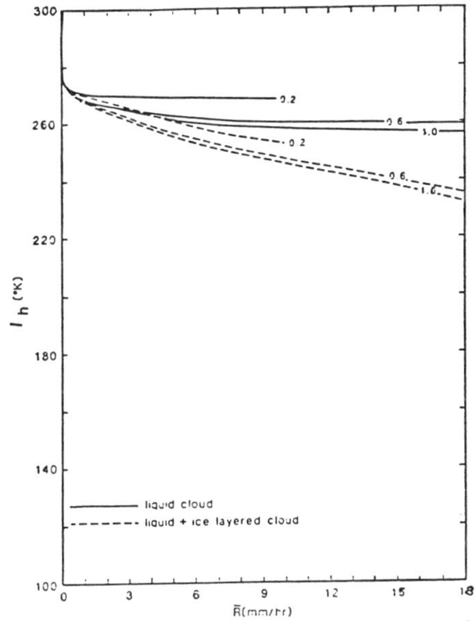

FIGURE 2. Horizontally polarized brightness temperatures emerging from a precipitating cloud over land. The brightness temperature and rainfall rate are averaged over the 22 x 22 km area of the radiometer footprint. The numbers cited are the fraction of the footprint area that is covered by precipitation.

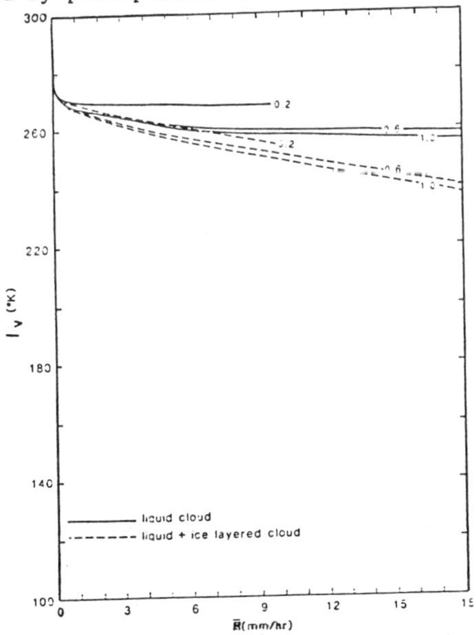

FIGURE 3. Vertically polarized brightness temperatures emerging from a precipitating cloud over land.

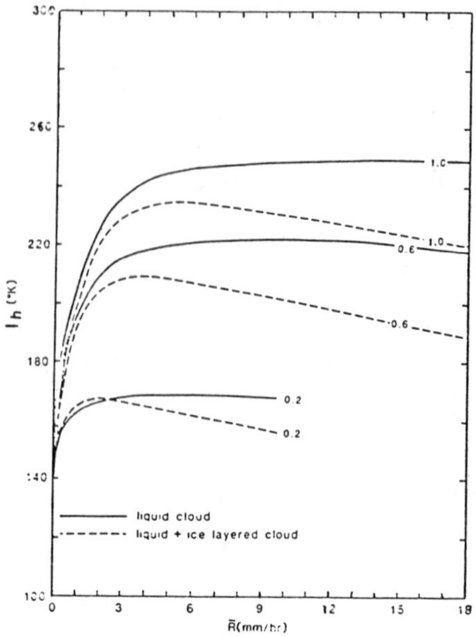

FIGURE 4. Horizontally polarized brightness temperatures emerging from a precipitating cloud over water.

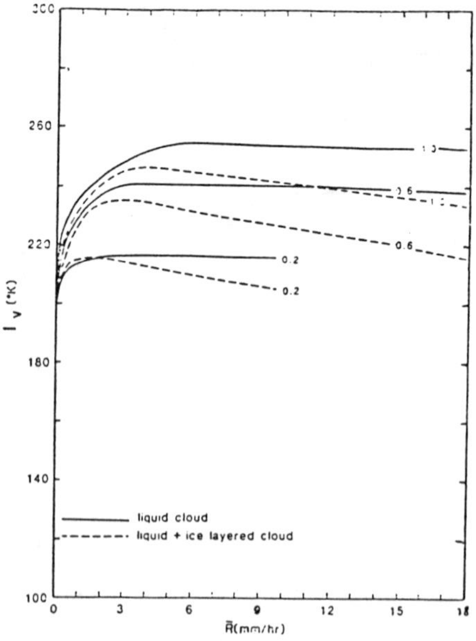

FIGURE 5. Vertically polarized brightness temperatures emerging from a precipitating cloud over water.

parameters for liquid hydrometeors vary much less than the same quantities for frozen ones. Furthermore, although the relation between brightness temperatures and the average rainfall in the footprint is highly non-linear, a reasonably linear relationship does exist between the average brightness temperature and the difference in the brightness temperatures of the two polarizations. The average rainfall rate can be fitted empirically to:

$$R \cong 62.5 - 16 \ln (I_v - .48\, I_h - 91.6) \qquad (25)$$

The results compare well with those in a recent theoretical paper by Adler and Mack (1986). Although the overall magnitude of the brightness temperatures differs sightly from those in that report, it must be kept in mind that the overall magnitude of the brightness temperatures depends on a variety of parameters such as the footprint area, the molecular absorption coefficients or the surface temperature and especially upon the relationship between ice above the freezing layer and liquid precipitation below the freezing layer.

4. CONCLUSIONS

In summary, the analytic model developed in this paper provides microwave brightness temperature estimates fast enough to run cloud models with numerous parameters that may be allowed to vary. This study has shown that a number of factors: non-precipitating ice layers, ice at the top of the cloud and polarization all play a significant role in determining the upwelling brightness temperatures. Unless these features are incorporated into a retrieval scheme, an inherent error in the estimation is apparent. The deviation from a vertically homogeneous finite cloud model is also found to be enough to produce a consistent bias. This study has shown that the use of plane parallel solutions is questionable. Brightness temperatures from finite clouds give lower brightness temperatures than those obtained by using a finite slice from a plane parallel model. The value of using polarization measurements is shown. Unfortunately, the shortcomings of operating without the necessary spatial resolution were also evident. Analysis of multispectral data, however, can reduce this difficulty.

ACKNOWLEDGMENT

Funds for this research were provided from NASA grant NAG5-580 and U.S. Navy grant NA00014-86-K2001.

REFERENCES

Adler, R.F. and R. Mack, 1986: Field of view effects in estimating precipitation from satellite microwave radiances. Second Conf. on Satellite Meteorology/Remote Sensing and Applications.

Kummerow, C.D. and J.A. Weinman, 1988: Microwave radiances from horizontally finite-vertically structured precipitating clouds, (accepted for publication in J. Geophys. Res.).

Oguchi, T., 1983: Electromagnetic wave propagation and scattering in rain and other hydrometeors. Proc. IEEE, 71, 1029-1078.

Rodgers, E. and Siddalingaiah, 1983: The utilization of Nimbus 7 SSMR measurements to delineate rainfall over land. J. Clim. Appl. Meteor., 22, 1753-1763.

Sneddon, I.N., 1951: Fourier Transforms, McGraw-Hill, New York, 552 pp.

Spencer, R.W., W.S. Olsen, W. Rongzhang, D.W. Martin, J.A. Weinman, and D.A. Santek, 1983: Heavy thunderstorms observed over land by the Nimbus 7 scanning Multichannel Microwave Radiometer. J. Clim. Appl. Meteor., 22, 1041-1046.

Spencer, R.W., 1986: A satellite passive 37 Ghz scattering based method for measuring oceanic rain rates. J. Clim. Appl. Meteor., 25 754-766.

Weinman, J.A. and R. Davies, 1978: Thermal microwave radiances from horizontally finite clouds of hydrometeors. J. Geophys. Res., 83, 3099-3107.

Wilheit, T.T., A.T.C. Chang, M.S. Rao, E.B. Rodgers, and J.A. Theon, 1977: A satellite technique for quantitatively mapping rainfall rates over the ocean. J. Appl. Meteor., 16, 551-560.

Wu, R. and J.A. Weinman, 1984: Microwave radiances from precipitating clouds containing a spherical ice, combined phase and liquid hydrometeors. J. Geophys. Res., 89, 7170-7178.

SURVEY OF SAMPLING PROBLEMS FOR TRMM

*Gerald R. North**
Texas A&M University
College Station, Texas 77843, USA

ABSTRACT

Prerequisite to TRMM mission planning is an assessment of the errors in monthly rain rate estimation assuming the instruments make perfect measurements. Since a low earth orbiting satellite at any inclination, with respect to the equator, visits a given averaging box on the earth's surface at most a few times a day for snapshots of instantaneous rain rates, there will be large temporal gaps in coverage. To gauge the importance of these gaps in the construction of space-time averages of rain rate, many studies have been conducted using both real data taken from the GATE experiment and using stochastic models of rain rates based upon the data. The studies suggest that for a 300-Km orbit with inclination 30 degrees, averaging boxes 600 by 600 Km, and for temporal smoothing of a month, the rain rate errors over oceans lie between 8 and 15% with a best guess at about 10%. The latitude dependence of the errors is weak. Data from other tropical areas are needed to assess the representativeness of these conclusions.

1. INTRODUCTION

The determination of space-time smoothed rain rates from low-altitude orbiting satellites is limited by the sampling rate, which is about once or twice per day. Hence, even perfect instruments can be expected to err randomly due to these temporal gaps in the measurements. In fact, the averaging box, nominally taken to be 600 by 600 Km, will usually only be partially visited by the swath of a scanning microwave radiometer. These so-called sampling errors have been studied extensively in the last few years and were recently reviewed by North (1987). We shall summarize in this paper the studies of special interest to the TRMM planning community.

1.1 SAMPLING DESIGNS

Figure 1 shows a schematic diagram of a rain rate function in a space-time volume (rain rate greater than

---

* Supported in part by a grant from NASA Headquarters.

ISBN 0-937194-14-4

zero stippled). A particular sampling design may be represented as in Figure 2a or 2b. In 2a we have the typical rain gauge, discrete in space and continuous in time. In Figure 2b we have the design of a low earth orbiting spacecraft which returns to the averaging box at finite intervals through the space-time slab. The sampling errors to be expected from each of these designs are dependent upon the space-time correlation characteristics of the rain fields to be encountered.

Two hypothetical extreme cases are interesting to examine. First, if the correlations lagged in space and time are large over space-time distances, comparable to the dimensions of the space-time volume, then we can perform the integral over the rain-rate function in the space-time volume easily since it will be slowly varying over the space-time range of the integral. In this case the limitations are that of a numerical approximation (design becomes the finite difference grid) to an integral with a smoothly varying integrand. A second interesting extreme is when the integral time and length scales are small compared to the dimensions of the space-time volume, *and* the statistics of the field are homogeneous (space) and stationary (time). In this latter case, one looks at the sampling problem associated with estimating the value of some population parameter (e.g., a moment such as average instantaneous rain rate) by collecting a large ensemble of individual estimates and taking their arithmetic average. Of course, real rain is rarely either of these hypothetical extremes. In fact, it is debatable whether a horizontal length scale exists in rain fields (Cf., Lovejoy and Schertzer, 1985).

1.2 CLASSES OF SAMPLING STUDIES

The sampling studies for TRMM fall into several categories. Some are preliminary feasibility studies, where actual rainfall data are available but the area is not as large as a TRMM box, and the duration of the data set in weeks is not long enough for true comparisons to the TRMM case. These studies are also marred by the fact that the data are from one specific location in the tropical oceans along the ITCZ. This is possibly not representative of other ITCZ areas. None the less, the data are real and were collected under nearly ideal conditions.

The data set available and most appropriate for the present use comes from the GATE (GARP Atlantic Tropical Experiment; GARP = Global Atmospheric Research Programme). These data were taken in the summer of 1974 from an array of ships centered at 8.5N and 23.5W (Hudlow and Patterson, 1979). The experiment was conducted in three phases of 18, 15 and 18 days and denoted Phase 1, 2 and 3. During Phases 1 and 2 instantaneous radar-inferred rain rates were collected every 15 minutes and averaged into 4 by 4 km bins. The data are in a circle inscribed within a 400 by 400 km square. A square inscribed within the circle is 70 by 70

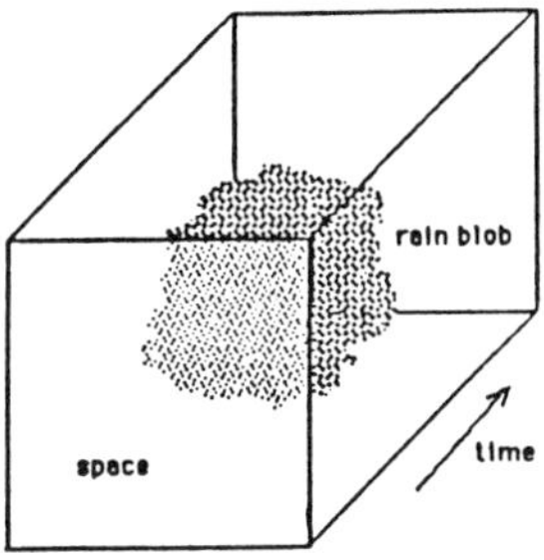

SPACE–TIME SAMPLING

Figure 1. *Schematic diagram of a volume in space-time over which an average of the rain rate is to be taken. The stippled region depicts a volume where the rain rate is nonvanishing.*

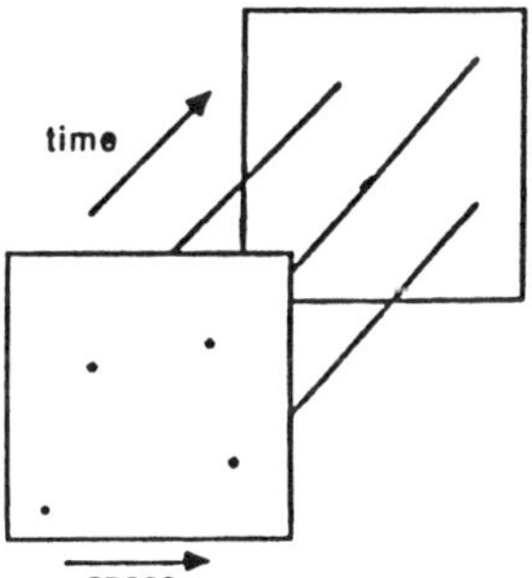

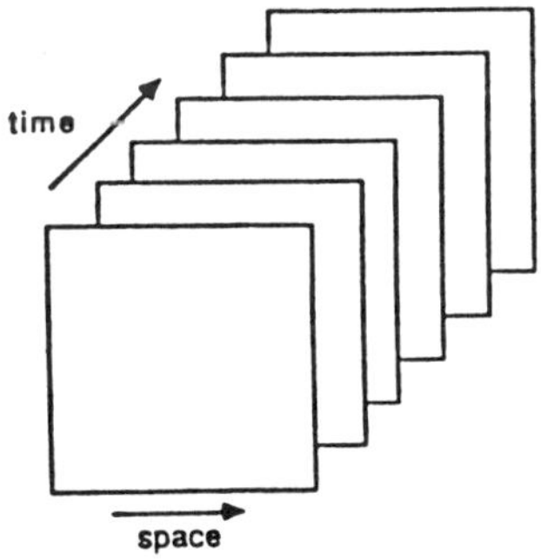

Figure 2. *a) This diagram indicates the sampling design corresponding to rain gages; i.e., discrete points in space and continuum in time. b) The sampling design for a low earth orbiting satellite; i.e., snapshots taken at discrete intervals in time.*

bins or 280 by 280 km. This latter square array is usually taken as the 70 by 70 multivariate time series to be considered in our statistical baseline studies.

Several elementary properties of the GATE data are worth discussing before we enter the sampling studies specific to the mission analysis.

(1) The probability density function for rain rates (4 by 4 km bins) is rather well-described by a mixed lognormal distribution (Kedem, Chiu and North, 1987). By mixed distribution, I mean a finite probability that it is not raining in the pixel, $1 - p$, and conditional upon rain a lognormal distribution with mean a and standard deviation σ. In looking through the scenes one can estimate the average rain rate as the average value of pa. Roughly speaking for GATE, the probability of rain is about 0.12 (this figure obviously depends upon the threshold taken for rain/no rain) and the mean rain rate when raining is about 5 mm/hr; hence, the average rain rates are about 0.4 mm/hr for the whole of Phase 1. For bins the order of 12 by 12 km, the size of a small TRMM FOV, the figures will be different. In this case the probability of rain is of order 0.20 and its area-average value when raining is about 2.0 mm/hr.

Experience with the GATE data (Chiu, private communication) suggests that from scene to scene 78% of the variance in log rain rate is explained by variations in log p, leaving only 22% to be explained by variations in the parameters internal to the continuum pdf. This suggests that estimating the area of rain is the most important variate to measure while details of the pdf, whether it is log normal or gamma are secondary.

(2) The spatial statistics of GATE rainfall rates are interesting. Bell (1987) suggests that the spatial autocorrelation function is described by a power law $s^{-4/3}$. This implies that the concept of a conventional horizontal integral length scale is questionable. This notion is further supported by the variance of area averages (Laughlin, 1981) as a function of averaging area size. If a horizontal length scale existed, one would expect this function to be an inverse first power law. Instead, the variance of area average rain rate decreases much more slowly; in fact, very nearly at a minus 1/3 power. This finding is consistent with the -4/3 spatial correlation power law (North, 1987). Figure 3 (Shin and North, 1987) indicates the functional dependence of variance of the area averages versus area.

(3) The time dependence of area average rain rate is surprisingly friendly compared to the spatial properties. Figure 4 (Laughlin, 1981) shows the autocorrelation function for a series of averaging areas. Note that these are reasonably close to exponential. Time scales are shown in Figure 5 as a function of averaging area. The largest areas have an

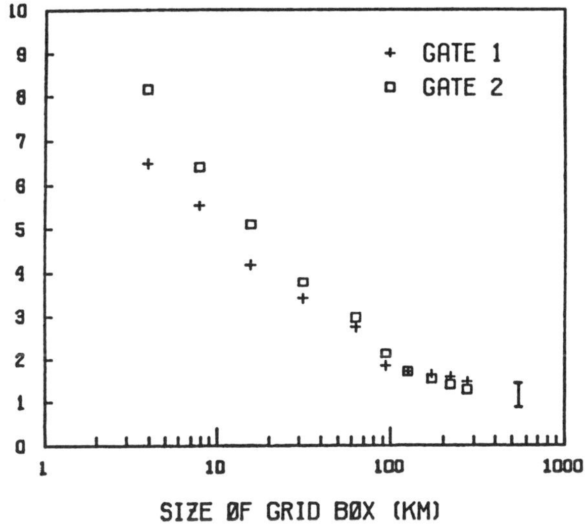

Figure 3. *The coefficient of variation (standard deviation divided by mean) of area average rain rate for various sizes of the averaging area for data taken from both phases of GATE. Note the uncertainty in extrapolation to a 550 km box.*

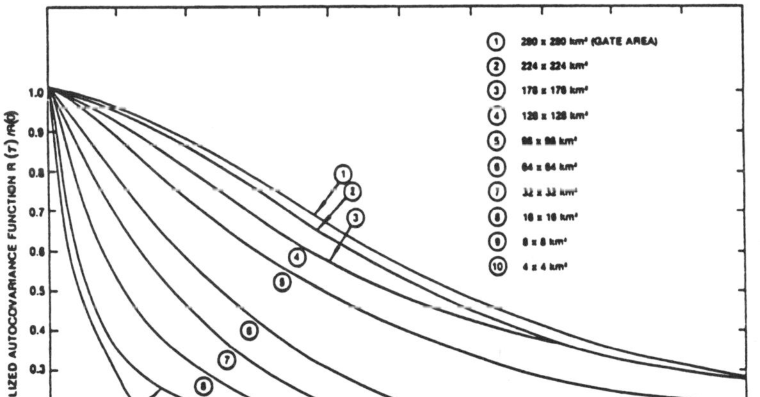

Figure 4. *Autocorrelation functions versus lagged time for area average rain rate (Laughlin, 1981) for various areas taken from GATE 1. Similar plots for GATE 2 are nearly identical.Note that for the GATE wide area the e-folding time is close to 6 hours.*

autocorrelation time of 6 hours for the GATE wide area and for a 600 by 600 km box we might even expect autocorrelation times of order 10 hours.

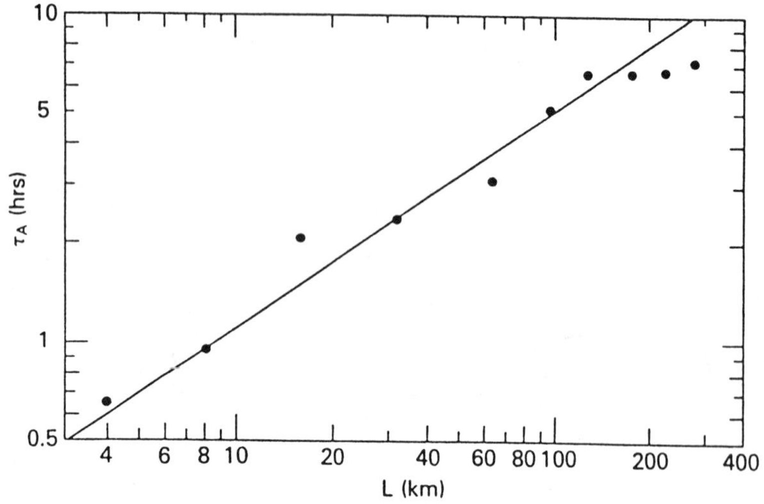

Figure 5. *A plot of the e-folding time for the autocorrelation function for various averaging areas; Cf., Figure 4.*

2. SAMPLING STUDIES RESTRICTED TO THE GATE AREA

Several studies have been conducted using the GATE area as the averaging area. This is an important class of studies since we have data covering that region and no extrapolation to larger areas is necessary. The studies are conducted as if the satellite in question returns to the GATE area after a fixed interval and reads the rain rate over the *entire* area; this is referred to as a flush visit. More realistic simulations would involve *partial* visits to the averaging area. The consequences of dropping the above restrictions will be considered in section 3.

2.1 MARKOV MODELING

Laughlin (1981) was the first to report a study of sampling errors for precipitation estimates from low earth orbiting satellites based upon a stochastic model. He examined the time series of area average rain rates and noted that the lagged autocorrelation functions were reasonably close to exponential as shown in Figure 4. Not shown are very similar curves for GATE 2. Laughlin constructed a first order Markov process using the autocorrelation time corresponding to a particular averaging area and proceeded to find an analytical expression for the sampling errors to be expected for space-time averages of a given time span and for different visiting (snapshot) intervals. The results of his study are shown in Figure 6 (results for GATE 2 are almost identical). Note that for the present example, 30-day averages and approximately 12 hour repeated snapshot visits, the sampling errors are about 8%. Laughlin's study differs from the TRMM case in two respects: the averaging square for TRMM is envisioned to be 600 km on a side and the visits will not be flush (100% coverage on every visit), but rather the visits will be partial. A partial visit to a box will occur at least twice per day but the total area visited will be equivalent to about 30 flush visits in a month. What is the appropriate scaling of Laughlin's results to the ones of interest to TRMM planning?

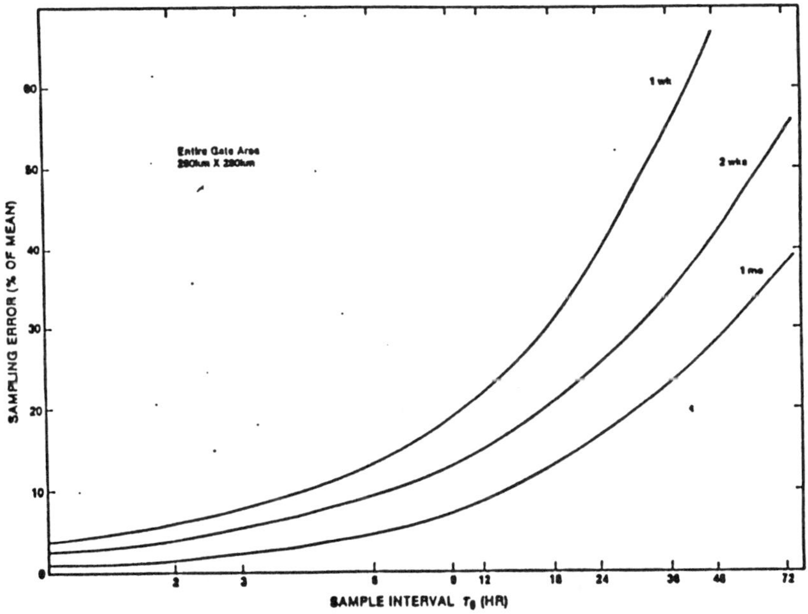

Figure 6. *Laughlin's (1981) results for estimates of time averages based upon temporally discrete flush visits to the full GATE area (280-km box).Generalization of the results to TRMM require extrapolation of the autocorrelation time (Figure 5) and coefficient of variation (Figure 3) to the larger area. In addition account must be taken of the partial visits of the tropical inclination satellite.*

2.2 PARTITIONED RATES AND ORBIT ENSEMBLES

McConnell and North (1987) used the GATE data in another type of sampling study. First, an imaginary satellite orbit was started and flown over the GATE area at, say, 8 AM, it returns at 7:30 PM and 11.5 hours later and so on throughout the GATE 1 phase. During each flush visit the individual GATE pixels are collected into 5 categories: no rain; 0 to 5 mm/hr; 5 to 10 mm/hr; 10 to 20 mm/hr; and 20 mm/hr and above. The average rain contributed from each of these categories is then estimated from the single imaginary orbit (realization). Realization number 2 is constructed the same as above except that the orbit is started at 9 AM on the first day of GATE. In this way an ensemble of satellite orbits can be constructed and the results can be compared. If the results differ wildly from one ensemble member to another, we can infer that significant storms are missed during the 11.5-hour gaps. Quite the contrary is found as shown in Figure 7 where the individual estimates of rain rate from the medium (5 to 10 mm/hr) category are shown for ten members of the ensemble of orbits shown. Similar depictions for the other categories (each of the raining categories contributes about equally to the total rain) for both GATE 1 and 2 have about the same appearance (McConnell and

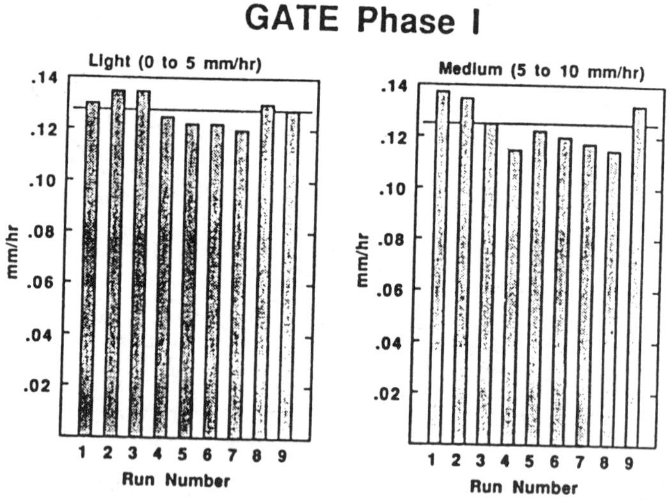

Figure 7. *Results for light (left) and moderate rain rate (right) contributions to the 18-day totals. The different vertical bars represent different members of an imaginary ensemble of orbiting satellites. That different ensemble members give about the same results indicates that sampling errors are small (less than 10%).*

North, 1987). In other words the spread of the estimates among the members of the satellite ensemble is about 10% for this category of rain rates. We do not find wildly different estimates for the different members. In fact, the eye can easily detect that successive members of the ensemble are correlated. Clearly this is because of the large lagged autocorrelation in the area-wide time series of rain rates (Figure 4). This study is so simple and assumption-free that it is very appealing.

2.3 MIXED DISTRIBUTION METHOD

Kedem, Chiu and North (1987) have considered another approach to the sampling question. The method involves use of a sparse sampling in space and time on a regular grid through a GATE phase. At the end of the phase, the pixel average rates are collected into a histogram of rain rates. These are then fitted to the mixed log normal distribution by a chi-square minimization procedure (Cf., also Kedem and Chiu, 1987). After determining the appropriate pdf for the data throughout the GATE phase the average rain rate is calculated by finding the expectation value of the pdf. One such design corresponds to the sampling design of a satellite orbit with flush visits (exactly like McConnell and North above). An ensemble of flush-visiting orbits gives an indication of the errors that can be expected due to the long sampling gaps. Using the mixed distribution method we find a standard error of between 8 and 10% for both GATE 1 and 2 (two to three week periods). As mentioned above, the main contributing variate is the area in a scene that is raining. The lognormal parameters hardly change from GATE 1 to GATE 2, while there are fairly large differences in the raining area parameter. The main advantage of the mixed distribution parameter estimation method is the fact that less than average rain rates almost completely determine the pdf parameters. A good knowledge of the histogram below 10 mm/hr is sufficient to pin down the parameters in the continuum part of the pdf. This latter could be important since some microwave channels saturate in the neighborhood of 10 mm/hr and it is important to be able to make maximum use of the data that will be available in these channels.

3. EXTRAPOLATIONS TO TRMM-SIZED AREAS

The above studies are restricted to a smaller area (280 by 280 Km) and flush visits by the orbital swath. Still, they give us some idea about the nature of our problem. We find, for example, that the dominant factor allowing the estimation of monthly averaged rain rate from low orbit satellites is the long autocorrelation time of area-averaged rain rates and its increase with area. Another important ingredient is that the variance of area-averaged rain rate decreases with area. As we consider areas four times as large as for the GATE area we must ask how these two quantities extrapolate to the larger area and how the sampling errors will be affected.

3.1 A MARKOV MODEL APPROACH

Shin and North (1987) report the results of a Markov model study similar to Laughlin's, but using an extrapolation to the larger area (autocorrelation time = 12 hours) and including (crudely) the effects of partial visits to the averaging box. They used low inclination orbit calculations to find the fraction f of the box intersected by the swath on each visit over a typical 30-day period. Each visit to the time series was weighted by the fraction of area intersected, f. No attempt was made in this study to optimally weight the visits using prior information about correlations across a box (frequent visits to only a small part of the box may be worth more than occasional large area visits because of distance correlations). Shin and North found that the combined effects of larger averaging box and partial visits increased the sampling errors to about 10 to 12%. This result was very sensitive to the choice of value taken for the autocorrelation time (they took 12 hours; see Figure 5) and the variance of the area average rain for 600-Km boxes (they took a value of 1.25 for the ratio of standard deviation to the mean; see Figure 3). These midrange estimates are reasonably conservative for the present planning purposes. For all extrapolation figures they found that the latitude dependence of the errors was negligible.

3.2 SIMULATIONS WITH RANDOM FIELD MODELS

Bell (1987) has devised a random field model that can be tuned to have many of the statistical characteristics of GATE. Once the model is calibrated to GATE statistics, it can be extended to larger areas, modified with artificial diurnal cycles, etc. The model is constructed by the following recipe: first a gaussian random field is constructed in Fourier spectral form over a grid (this can be thought of intuitively as a kind of turbulent vertical wind field). If the "wind" exceeds a certain threshold, we agree that it is raining. Such a procedure will lead to islands of rainy areas. The random variable representing the exceedence over threshold can now be converted to a lognormal distribution of rain rate. The horizontal spectrum of the original gaussian field can be adjusted until the rain field statistics look like the graph in Figure 3. Finally the time scales can be entered so that area average time scales look like those in 5. The fields are constructed and evolved in Fourier space with conversion by way of FFT's. The program is easily vectorized and runs on a Cyber 205.

For flush visits, Bell finds that Laughlin's Markov model is an excellent imitator of his own model for sampling error studies like those discussed above. Bell also finds using his model that the sampling errors for the mixed lognormal distribution method are almost identical to the simple arithmetic estimates of rain rate. Of course, for an instrument with limited range the mixed distribution method would be better. Bell and colleagues (private communication) have found that for the realistic partial orbital visits and the larger TRMM boxes the sampling error results are consistent with the Shin and North (1987) studies.

4. CONCLUSIONS

The results of several preliminary studies indicate that sampling errors of about 10% are to be expected for TRMM-sized boxes and for the low altitude, low inclination orbit. However, this estimate depends upon the extrapolation of two parameters from their values for GATE-sized areas up to TRMM-sized areas (factor of 4). The uncertainty in this extrapolation leads to a range (8% to 15%) in the estimated sampling errors. Further studies on sampling errors are needed in the diurnal cycle estimation and removal. Similarly, we need to conduct studies with rain rate data from other parts of the world to clarify representativeness questions.

REFERENCES

Bell, T. L., 1987: A space-time stochastic model of rainfall for satellite remote-sensing studies. J. Geophys. Res. Atmos.,92(D8), 9631-9644.

Hudlow, M.D. and V. L. Patterson, 1979: *GATE Radar Rainfall Atlas*. NOAA Special Report, available from U. S. Government Printing Office, Washington, DC 20402, 158pp.

Kedem, B. and L. Chiu, 1987: On the lognormality of rain rate. Proc. Nat. Acad. Sci., 84, 901-905.

Kedem, B., L. Chiu and G. North, 1987: Estimation of mean rain rate: Application to satellite observations. J. Geophys. Res., submitted.

Laughlin, C. R., 1981: On the effect of temporal sampling on the observation of mean rainfall. Precipitation Measurements from Space, Workshop Report, ed. by D. Atlas and O. Thiele. Available from Goddard Space Flight Center, Greenbelt, MD 20771.

Lovejoy and Schertzer, 1985: Generalized scale invariance in the atmosphere and fractal models of rain. Water Resour. Res. 21, 1233-1250.

McConnell, A. and G. North, 1987: Sampling errors in satellite estimates of tropical Rain. J. Geophys. Res., 92(D8), 9567-9571.

North, G. R., 1987: Sampling studies for satellite estimation of rain. Tenth Conf on Probability and Statistics in Atmospheric Science. Amer. Met. Soc., Edmonton, Alberta, Canada, 129-135.

Shin, K.-S. and G. R. North, 1987: Sampling error study for rainfall estimates by satellite using a stochastic model. Submitted to J.Clim. and Appl. Meteor.

PERFORMANCE OF A VISIBLE-INFRARED TECHNIQUE IN ESTIMATING MONSOON RAINFALL

David W. Martin and Brian A. Auvine
University of Wisconsin
Madison, Wisconsin 53706, USA

ABSTRACT

The technique known as grid history is tested on onset and active monsoon daily rainfall, measured in gauges along the west coast of India. Grid history is a semi-automated, intensive regression technique which employs sequences of visible and infrared geostationary satellite images. The present test covers 18 days of the June 1979 onset and active southwest monsoon and up to 11 gauge stations. Gauge rain averages 29 mm/d^{-1} for the 85 station-days of the sample; satellite rain, 17 mm/d^{-1}. Surprisingly, the under-estimate is largest at small rain rates. Gauge and satellite rains are correlated as 0.6; the standard error of estimate is 32 mm/d^{-1}. Considering residual separations between gauges and grid points, these results appear to demonstrate notable skill in estimating rain rates. A poor understanding of the physics of monsoon rain may be the biggest obstacle to further increasing that skill.

1. INTRODUCTION

"Grid history" is the name Martin and Howland (1986; henceforth, MH) gave to a technique for estimating daily rainfall in the tropics. By this technique MH aimed to achieve an accuracy of factor 2 for one day rainfall over a 100 km box. They did report meeting this goal, but only in the case of rain over the Arabian Sea during the onset of the 1979 monsoon.

In the context of 1979 monsoon rainfall over and near the Arabian Sea the present paper has two purposes: first, to reassess the MH calibration and, second, to describe a test of the grid history technique. Ultimately, the paper aims to answer the question, "Can the grid-history technique provide information on rates of daily rain falling on the Indian Ocean?"

2. GRID-HISTORY TECHNIQUE

Grid history employs geostationary satellite visible and infrared image sequences. It presumes, first, that rain is a rare event; further, that when it occurs we can distinguish three rates, or classes of rain; and, finally, that image pixels are highly correlated with their neighbors. Grid history makes estimates, or predictions, of rain rate on the scale of an image pixel for a grid of pixels.

TROPICAL RAINFALL MEASUREMENTS
John S. Theon and Nobuyoshi Fugono (Eds.)

ISBN 0-937194-14-4

2.1 ALGORITHM

The algorithm which makes the prediction is hierarchical. In order of importance the properties of the pixel, or of the image in the neighborhood of the pixel, which determine the prediction (i.e., rain class which is assigned) are brightness, trend, structure and texture. Brightness and trend are measured and evaluated by a computer. In that sense the technique is semi-automated. The prediction of daily rainfall $\hat{R}$ is simply the sum of the weighted frequencies of the three rain classes. Thus, if over a day f_1, f_2 and f_3 are, respectively, the frequencies of the light, moderate and heavy rain classes and r_1, r_2 and r_3 are their respective weightings, we have $\hat{R} = r_1f_1 + r_2f_2 + r_3f_3$. Ordinarily the interval between images is 1 h; thus $\Sigma f_i \leqq 24$.

2.2 COEFFICIENTS

The subjective values of the rain classes are quantified in a process we call calibration. Calibration has gone through two stages. The first was reported by MH. By means of least-squares multiple linear regression MH calculated weightings, or coefficients, for six areas of the tropics. There was considerable variability from area to area and, in the case of three areas, even an inversion in the values of coefficients. On the other hand, the coefficients calculated for the Arabian Sea (1.5, 8.8, and 19.9 corresponding to, respectively, light, moderate and heavy rain classes) were well behaved.

The second stage of calibration involves a re-determination of the Arabian Sea coefficients. This was provoked partly by an attempt to estimate rainfall in a Bay of Bengal cyclone, and partly by a need to extend the onset mapping into the active period.

MH based the Arabian Sea calibration on gauge measurements of rain from research ships as well as island stations. For their nine day period of analysis, 11-19 June 1979, MH accepted a total of 47 gauge-satellite rain rate pairs. Recently, we screened the entire MH calibration data set. Applying more stringent criteria for ship movement reduced the total number of pairs to 45. For these 45 pairs, (a) rain class daily frequencies were interpolated to actual rather than nominal ship locations, (b) rain rates were normalized to the actual number of hourly satellite images available each day, (c) for island stations the satellite day was shifted to correspond to the actual (∿0330 UTC) time of measurement of daily rainfall, (d) for one "outlier" day rain classes were freshly assigned, and (e) an inversion of gauge rainfalls was corrected.

To this screened calibration data set we added four days from late June, 1979. For each day, four Maldive island stations reported. This yielded an enlarged calibration data set of 61 pairs.

Finally, it occurred to us to impose conditions on the coefficients. The first ($r_1 < r_2 < r_3$) is a condition of order; the second ($r_1/r_2 < r_2/r_3$) is a condition of weight.

Screening the MH Arabian Sea calibration data set led to an inversion in the order of two of the coefficients from what we expected. Furthermore, overall, the screened coefficients had a lower weighting than the MH coefficients. Nevertheless, in terms of explained variance and standard error the screened coefficients performed about as well as did the MH coefficients. Enlarging the sample (case of n = 61) essentially resulted in a light/heavy algorithm. Performance remained about the same.

Conditions of order and weight then were imposed on the enlarged sample. We required as well that the bias be of the order of ten percent or less. These conditions narrowed the range of coefficients to a relatively small number. By trying various combinations we were led to a new set of coefficients, 1:5:15. The uncertainty in these values is thought to be less than 20 percent.

For the enlarged data set the performance of these coefficients is compared with that of the MH coefficients in Table 1. With other measures being similar, in the mean the imposed condition coefficients more accurately predicted changes in observed rain rates and in the total amount of rain. But were the new coefficients really superior to the old in predicting rainfall over the Arabian Sea? Could they meet the MH standard? To answer these questions we needed a set of independent gauge rainfalls.

TABLE 1. PERFORMANCE OF "MH86" and "IMPOSED CONDITION" SETS OF RAIN CLASS COEFFICIENTS IN PREDICTING RAINFALL

COEFFICIENTS	DATA SET	PERFORMANCE				
		a_0 mm h^{-1}	a_1	ρ^2	SEE mm h^{-1}	bias
MH86 (1.5/8.8/19.9	MH86 (n=47;corrected)	4.7	0.81	0.43	13.0	0.01
	enlarged sample (n=61)	7.5	0.36	0.42	13.7	0.51
imposed condition (1/5/15)	enlarged sample (n=61)	7.7	0.55	0.41	13.8	0.05

The model is $R=a_0+a_1\hat{R}$. In a perfect prediction a_0=SEE=Bias=0; $a_1=\rho^2=1$. SEE is standard error of estimate; n is sample size.

3. COASTAL EXPERIMENT

The best source of independent data was gauges on the Indian sub-continent. We assumed that some of India's coastal gauges reflected a marine regime of rainfall. Comparing daily rainfalls at two pairs of stations we concluded that a typical correlation distance for these gauges was 30 km. Translated into specific terms, this meant that in order to qualify for the coastal experiment a gauge had to meet the following conditions: (1) Site not more than 50 km from the coast, (2) Exposure toward the sea, (3) Altitude less than 100 m, and (4) Separation from the nearest grid point of 30 km or less. Up to eleven stations qualified on any given day. Over the 18 days of the calibration period the selection process yielded 85 station-days of data.

It is well known (e.g., Virji et al., 1982) that after the onset the west coast of India usually is shrouded in clouds. Well-known, too, (e.g., Das, 1968; Raman and Krishnan, 1960), are the heavy rains that fall along this coast. For these reasons we expected the coastal experiment to provide a conservative measure of grid history performance over the Arabian Sea and (in spite of the estimator's "tilt" toward heavier rain classes at grid points along the coast) an underestimate of coastal rainfall.

Observed and predicted daily rainfalls are plotted in Fig. 1. The scatter is large. Gauge rain tended to exceed satellite rain. However, ranges are comparable, the points form an elongated cluster and the slope of the major axis of the cluster is positive.

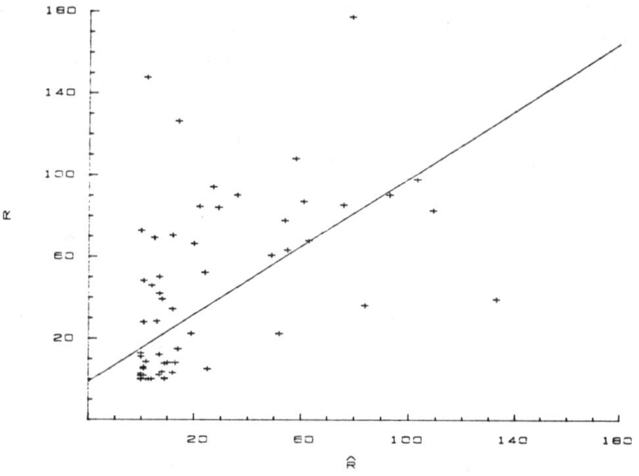

Figure 1. Scatterplot of gauge rainfall (R) and satellite rainfall ($\hat{R}$). Units are mm/d^{-1}. There are 25 points at (0,0). Also shown is the linear regression line of R on $\hat{R}$.

Statistically (Table 2), whereas the gauge mean was around 30 mm/d^{-1}, the satellite mean was 17. The bias this implies, 0.7, certainly was larger than expected. Gauge rainfalls ranged up to 177 mm; for this sample the standard deviation (29) was very large. Regressing R against $\hat{R}$ (Fig. 1, Table 3) yielded a positive intercept. This surprising result implies the grid history bias is largest for lighter rainfalls. Otherwise the slope of the regression line lies near unity, the correlation is moderately high and the standard error of estimate is smaller than the standard deviation of gauge rainfall.

TABLE 2. STATISTICS OF PREDICTED AND OBSERVED DAILY RAINFALLS FOR THE COASTAL EXPERIMENT

STATISTIC	SATELLITE PREDICTION	GAUGE OBSERVATION
mean	17.2	29.5
standard deviation	29.2	40.2
range	133	177

Sample size is 85; units are mm/d^{-1}.

TABLE 3. REGRESSION ANALYSIS OF PREDICTED AND OBSERVED DAILY RAINFALLS

STATISTIC	VALUE	
Intercept	15	mm d^{-1}
Slope	0.83	
Correlation coefficient	0.60	
Standard error of estimate	32	mm d^{-1}

4. CONCLUSIONS

We draw three principal conclusions from this assessment of the grid history technique. First, the 1986 coefficients of Martin and Howland probably overestimate rainfall over the Arabian Sea. Second, the 1986 claim for performance probable is inflated. Third, and in spite of one and two, the grid history technique can provide information on rates of daily rain falling on the Indian Ocean. We suspect that our poor understanding of the physics of monsoon rain is the biggest obstacle to further increasing that skill.

ACKNOWLEDGMENTS

Gary Krueger helped to reduce data. Jan Waite-Schuster typed the manuscript. This work was supported by the National Science Foundation under Grant ATM-8419735 and by the National Aeronautics and Space Administration under Grant NACW-920.

REFERENCES

Das, P.K., 1968: The Monsoon, National Book Trust, New Delhi, 168 pp.

Martin, D.W., and M.R. Howland, 1986: Grid history: A geostationary satellite technique for estimating daily rainfall in the tropics, J. Clim. Appl. Meteor., 25, 184-195.

Raman, P.K., and A. Krishnan, 1960: Runs of dry and wet spells during southwest monsoon and onset of monsoon along the west coast of India, Indian J. Meteor. Geophys., 11, 105-116.

Virji, H., W.L. Smith, A.J. Schreiner and L.D. Herman, 1982: Earth-Atmosphere Radiation Balance from Geostationary Satellite Data for the Summer Monsoon Onset Region. Space Science and Engineering Center, University of Wisconsin-Madison, Madison, Wisconsin 53706, 252 pp.

MODELING TROPICAL RAINFALL FROM RADAR AND RAINGAGE DATA — PARAMETER ESTIMATION ISSUES

Witold F. Krajewski
University of Iowa
Iowa City, Iowa 52242, USA

and

James A. Smith
Interstate Commission on the Potomac River Basin
Rockville, Maryland 20850, USA

ABSTRACT

A statistical space-time model of rainfall has been developed. The model is based on rain gage and radar data. The radar data required by the model are 0 - 1 mosaics: 0 - indicates no rain, 1 - indicates occurrence of rain. Such form of radar data can be easily obtained from radar echo images. The radar data are used to model the spatial patterns of rainfall and the raingage data are used to model the magnitude of rainfall. The potential applications of the model to study the so-called ground-truth techniques, useful in verification of rainfall estimation methods, are discussed. The focus of the paper is on the parameter estimation problem. An estimation method of model parameters is presented and investigated via a simulation experiment. The experiment provides means to study the properties of the parameter estimators under a wide-range of sampling and measurement error conditions. Also, other issues concerning sampling strategies of space-based observations and the model are addressed.

1. INTRODUCTION

Difficulties in monitoring rainfall on a global scale have been widely acknowledged. The technology that seems to be most promising for the future is spaceborne remote sensing. Satellite-based techniques of visible and infrared imagery have been developed and partially tested (Atlas and Thiele, 1981; Barrett and Martin, 1981). However, the performance of these methods is not adequate yet for real-time hydrologic and weather forecasting operations, especially in meso-scale convective-type situations prevalent in the tropics. One of the major problems with remote-sensing techniques, both spaceborne and ground-based, is their validation and testing. Validation methodology, which utilizes in situ data from rain gages, is not well developed due to many theoretical and practical problems. One approach to improve this situation is through the use of space-time

TROPICAL RAINFALL MEASUREMENTS
John S. Theon and Nobuyoshi Fugono (Eds.)

ISBN 0-937194-14-4

rainfall models. Numerical simulation studies based on such models can be especially useful in large-scale experimental design, in studying sampling strategies, and comparison of rainfall estimation methods.

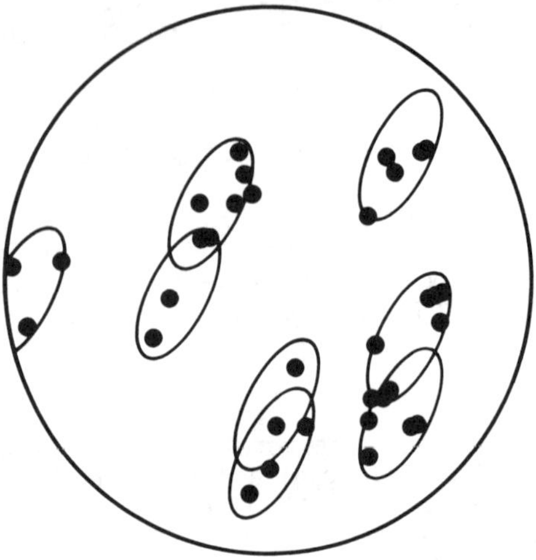

Figure 1. Schematic representation of the model structure.

In the past decade or so, there has been much progress in statistical modeling of rainfall fields. In this paper, we present one such model which is based on radar and rain gage data. The model is described in detail by Smith and Krajewski, 1987. Here we examine properties of parameter estimators, both theoretically and numerically.

2. THE MODEL

This model attempts to mimic time-integrated rainfall fields. Temporal evolution of the model is governed by a Markov chain. The spatial structure of the rainfall field is illustrated in Figure 1. Locations of "rain cells" follow a spatial Poisson process with rate γ. The circular rain cells with radius r are constrained to lie within elliptical "rain bands", the centers of which constitute a spatial Poisson process with rate λ. Geometry of rain bands is specified by three parameters, radius of the major axis a_1, radius of the minor axis $a_1 a_2$ (with $0<a_2<1$) and orientation of the major axis from north-south, θ. Rainfall depths for individual rain cells are assumed to be independent and identically exponentially distributed with parameter β. Smith and Krajewski (1987) have shown that a method of moments procedure could be used to estimate parameters of the model from radar and rain gage data. The radar data is used only in the 0-1 mosaic form, which means that the spatial structure of the modeled rainfall is based on radar data. It also means that the issues of radar-rainfall estimation are avoided and the only critical assumption concerns the ability of radar to delineate rain patterns.

It can also be shown (Krajewski and Smith, unpublished manuscript) that the parameter estimators have some very attractive properties such as asymptotic normality and consistency in mean square error. In addition the estimation procedure has an attractive feature that a simple statistic based on sample mean variance and probability of no rain can be precomputed to determine whether feasible parameter estimates exist.

Despite all these positive features of the model, many practical issues remain to be solved. Smith and Krajewski (1987) applied the model to GATE rainfall data and learned that light intensity background rainfall can mask structural features of the modeled field. Also, small sample properties of the parameter estimators and the required rain gage network density need to be studied.

3. NUMERICAL EXPERIMENT

In this paper, we present an experimental framework for a study of these issues, along with some preliminary results. To determine the minimum sample size required to apply the model for various rainfall regimes (or combinations of λ, γ, β, r, a_1, a_2, and θ), a Monte-Carlo study is proposed. In practical situations, the estimators may not exist in two basic cases:

1) If the model structure does not adequately describe the underlying process, or

2) If the sample size is too small.

Unfortunately, there is no easy way to distinguish these two cases. In our experiment, we generate the sample fields using our model, therefore, we know that only the second case may apply. It is hoped that the results of this study can offer some help in practical situations by examining appropriate rainfall patterns. Some minor simplifications were made on the model in order to facilitate efficient execution of the numerical experiment. It is assumed that the bands are circular. This assumption eliminates the need to estimate two parameters (a_2 and θ) and significantly speeds up the estimation procedure.

In the experiment, a limited number of realizations were generated from the model. These realizations are then sampled by a simulated network of rain gages and a radar. The parameter estimation procedure follows and the same process is repeated a number of times for various sets of control variables: (1) Model parameters (λ, γ, β, r, a_1), and (2) Sampling network parameters (the number of rain gages N_g and rain gage measurement error E_R).

Answers to the following questions are sought:

1. What is the minimum sample size required for parameter estimation under a wide range of model parameter values?

2. What is the effect of rain gage network density on the estimation process?

3. What is the effect of rain gage data measurement error on the estimation process?

The sampling domain is the same as for GATE, i.e., 400 by 400 km grided with resolution of 4 km by 4 km. The following sample sizes were used: 25, 50, 100, 250, 500, 1,000. One sample of 1,000 realizations provides, therefore, at least 40 samples of size 25 but only 1 of size 1,000. Also, rain gage network sizes of 25, 100, and 400 were investigated. The strategy adopted for the experiment was to start with such combination of the parameters which would be easy to estimate and then to examine how far one can depart from that combination under various sampling conditions. It seems that such easy-to-estimate combination should correspond to a "clean" structure of the model; spatial intensity of bands and spatial intensity of cells should allow for well-defined (separated) bands filled with rain. Also, the feasibility statistic should be in the middle of the allowable range. One combination which satisfies these conditions is $\lambda = 0.0025$, $\gamma = 1.0$, $\beta = 1$ mm/hr, $a = 20$ km, $r = 4.5$ km. The corresponding feasibility statistic is 1.5 (the allowable range is $\langle 1,2 \rangle$). Although such "clean" combination may not be very realistic its investigation proved to be very valuable for software testing and model learning. Preliminary results show small variance of the parameters even for the small sample size. They also revealed the existance of biases even for large sample size. Other interesting observation is relatively insignificant influence of raingage network density in the investigated range. Detailed results of the experiment will soon be ready for publication.

4. FINAL REMARKS

Successful solution of the parameter estimation problem of a space-time rainfall model is believed to be the determinant of model usefulness. A practical model must be based on available data in order to be useful. This requirement may pose a problem for the presented model. The model data requirements, although very "mild" may still be too much for the tropical zone where the availability of both raingage and radar data is rare. This is especially true over the oceans. As a result of this situation it is proposed that the sampling design of tropical rainfall measuring missions be investigated via a number of various models with common space and time scales. If the above-presented model is chosen to be one in such group of models it should be remembered that it is adequate (in terms of estimated parameter values but not the structure) only for areal extent corresponding to the size of the radar umbrella used in parameter estimation.

There are two major potential applications for the model. One is in the design study of satellite mission and the second in the analysis of performance of the satellite-based rainfall estimation techniques. In the first application, both the time and space components of the model are very important. The time component of the model, described in Smith and Krajewski (1987) as a Markov chain, can be easily substituted by a more complex mathematical structure. In addition, in such application the remote sensor measurement error model (in terms of the sensor's ability to delineate the spatial structure of rain) must be developed and used in a simulation experiment framework. The second application does not require the time component to be directly used provided, that the model parameters have been estimated for the appropriate time scale. In this application, however, the measurement error model of all the sensors involved must be developed. Such models should be able to quantify the observational error.

ACKNOWLEDGMENT

Most of the research leading to this paper was done while the first author was with the Hydrologic Research Laboratory of the National Weather Service. Its support is gratefully acknowledged.

REFERENCES

Atlas, D. and O.W. Thiele (Eds.), 1981: Precipitation Measurements from Space, Workshop Report, GSFC, Greenbelt, Maryland.

Barrett, E.C. and D.W. Martin, 1981: The Use of Satellite Data in Rainfall Monitoring, Academic Press.

Smith, J.A. and W.F. Krajewski, 1987: Statistical Modeling of Space-Time Rainfall Using Radar and Raingage Observations, Water Resources Research, 23(10), 1893-1900.

ESTIMATING AREAL RAINFALL FROM RAIN AREA

Long S. Chiu
Applied Research Corporation
Landover, Maryland 02139, USA

ABSTRACT

Because of the intermittent nature of rain fields, rainrate observations can be modeled by a mixed distribution which is completely described by the fraction of rainy observations (p) and other parameters, such as the mean and variance of the rainy portion of the distribution. Once the parameters are estimated, the mean and variance of monthly rainfall are obtained. The mean and variance of the mixed distribution are related to the parameter p, which can be interpreted as a measure of intermittency in the rain fields. With the proper space and time sampling, p can also be interpreted as an Area Time Integral which is intimately related to rain volume associated with convective storms. The parameters of the mixed distribution are dependent on the scales of observation. Analyses of GATE (GARP [Global Atmospheric Research Program]Atlantic Tropical Experiment) data show that at a resolution of 4 km, the rainy portion of the distribution can be modeled by the lognormal distribution and about 90% of the rain falls between 1 and 50 mm/hr. As the resolution is decreased, p increases. At a resolution of 40 km about 90% of the rain falls between 0.25 and 15 mm/hr.

The relation between areal rainrate and fractional rain area in the B-scale area of GATE is investigated. Linear regression analyses show that over 90% of the areal rainrate variance can be explained by the fractional rain area. The best regressor is the fractional area defined by rainrates greater than 5 mm/hr at a resolution of 4 km. At a resolution of 40 km, fractional rain area still explains over 90% of the variance if a threshold of about 3 mm/hr is used. For a threshold of 0 mm/hr, the variance explained is only about 30%.

1. INTRODUCTION

Because of the intermittent nature of rain fields, rainrate observations can be modeled by a mixed distribution. The mixed

This work is supported by the National Aeronautics and Space Administration through contract NAS5-30083.

ISBN 0-937194-14-4

distribution consists of a discrete component of (1-p) at zero rainrate and a continuous part whose integral is p for positive rainrates. Here, p is the probability of rainy observations and is analogous to intermittency in turbulence. Analysis of GATE (Global Atmospheric Research Program- Atlantic Tropical Experiment) data suggested that the lognormal distribution may be appropriate for the rainy observations (Houze and Betts, 1981; Kedem and Chiu, 1987a).

The space-time average rainrate, <[R]>, where < > and [ ] denote time and area averaging, respectively, is the mean of the distribution which is p multiplied by a, where a is the average rainrate of all rainy observations. If the a's estimated from different storms are fairly constant, then the rainfall amounts associated with these storms are proportional to p. If we consider sampling in space and time following a rain field with the rainy element embedded in it, p should be multiplied by the area and time of observation and hence can be identified as an area-time index. This forms the basis of the so-called "Area Time Integral" (ATI) technique which has found wide application in estimating rainfall amounts (Lopez, 1983; Doneaud et al., 1981, 1984). This approach requires continuous observation and the detection of rainy areas. Meteorological radars provide the best means of detecting rainy areas although satellite measurements in the visible and infrared channels have also been utilized (Lovejoy and Austin, 1979; Arkin, 1979; Doneaud et al., 1987).

The detection of rainy areas involves choosing a threshold which discriminates between rainy and dry areas. Since there is random noise in the field of observation, a low threshold may invariably include noise in the estimation. If the threshold is too high, the signal may be weak and the statistics may be unstable. This problem is further complicated by the varying resolution of the different sensors which involves different area averaging. This paper examines the sensitivity of resolution and thresholds in the detection of rainy area from satellite measurements for estimating rainfall amounts.

2. GATE RAINRATE STATISTICS

The 15-minute radar inferred rainrates in the B-scale area, for 4-km pixels from the first two phases of GATE, are used in this study. (See Patterson et al., 1979 for a detailed description of the data.) We have restricted our analysis on a square of 280 km by 280 km centered around the B-scale area of GATE. Table 1 shows the parameters of the mixed distribution estimated for GATE phases I and II. Note that the average rainrates for the rainy pixels (about 4 mm/hr) are almost identical for both phases. The difference in the average rainrate for the two phases is due only to the parameter p.

We can define the cumulative distribution function (C) and the cumulative contribution to the total rainfall (W) by rainrates lower than a rainrate R as:

$$C(R) = \sum_{Ri=0}^{Ri=R} p(Ri) \Big/ \sum_{Ri=0}^{Ri=Rmax} p(Ri)$$

$$W(R) = \sum_{Ri=0}^{Ri=R} Ri\ p(Ri) \Big/ \sum_{Ri=0}^{Ri=Rmax} Ri\ p(Ri)$$

where Rmax is the maximum rainrate observed and p(Ri) is the percent of rainy pixels with rainrate Ri. The normalization factors ensure that

$$C(0) = 0 = W(0) \quad \text{and} \quad C(Rmax) = 1 = W(Rmax)$$

Both functions C and W have been calculated for phases I and II and are presented in Table 2. They are almost identical at the 4-km resolution. There are about 36% of the rainy pixels with rainrates less than 1 mm/hr. They contribute to less than 4% of the total rainfall. Within the range of 1-50 mm/hr is contained over 90% of the total rainfall but only 50% of the rainy pixels.

TABLE 1. ESTIMATED PARAMETERS OF THE MIXED DISTRIBUTION

parameter	p(%)	a(mm/hr)	space-time average(mm/hr)
GATE I	12.4	3.98	0.49
GATE II	9.7	3.99	0.39

TABLE 2. CUMULATIVE DISTRIBUTION AND CONTRIBUTION TO TOTAL RAINFALL

RESOLUTION	4 KM				40 KM			
GATE	PHASE I		PHASE II		PHASE I		PHASE II	
RAINRATE	C	W	C	W	C	W	C	W
0.25 (mm/hr)	0.13	0.004	0.13	0.004	0.57	0.04	0.58	0.05
0.5	0.23	0.01	0.24	0.01	0.67	0.08	0.70	0.10
1	0.36	0.04	0.38	0.04	0.77	0.15	0.80	0.19
2	0.54	0.11	0.55	0.11	0.86	0.28	0.89	0.34
3	0.66	0.18	0.67	0.18	0.91	0.39	0.93	0.45
4	0.74	0.25	0.74	0.25	0.93	0.49	0.95	0.56
5	0.78	0.31	0.80	0.31	0.95	0.57	0.96	0.63
10	0.91	0.53	0.91	0.51	0.987	0.82	0.99	0.86
15	0.95	0.67	0.95	0.64	0.997	0.94	0.998	0.96
20	0.97	0.75	0.97	0.72	0.999	0.98	0.999	0.98
30	0.987	0.85	0.985	0.82	0.999	0.998	0.999	0.999
50	0.997	0.94	0.996	0.92	1.0	1.0	1.0	1.0
100	0.999	0.99	0.999	0.99	1.0	1.0	1.0	1.0

Parameters of the mixed distribution are functions of the spatial and time scales of observation and the probability of finding rain in an area (a pixel) increases with the size of the area (Kedem and Chiu 1987b). Table 3 shows the parameter p estimated for different pixel sizes for both phases of GATE. The rainrate for a 40 – km pixel, for example, is obtained by averaging the rainrates in 10x10 4-km pixels in a 40 km square. The functions C and W estimated for the 40 – km pixels are different from the 4-km pixels as can be seen from Table 2.

TABLE 3. ESTIMATED p FOR DIFFERENT PIXEL SIZES

pixel size(km)	4	8	20	40	80	140	280
GATE I	0.124	0.185	0.359	0.511	0.728	0.889	0.974
GATE II	0.097	0.175	0.331	0.485	0.724	0.869	0.984

The relation <[R]> = pa holds in general regardless of resolution. The a's corresponding to the different p's for different resolutions can be computed. Since p increases as the resolution is degraded, a decreases with decreased resolution. Hence, in applying the ATI technique or similar techniques to satellite rainfall estimation, which involve rain areas in the estimation procedure, the calibration constants are not only functions of the type of storm, but are functions of the resolution of the observations as well.

3. AREAL RAINFALL-RAIN AREA RELATION

If the averaging area is large compared to the lagrangian integral scale, it can be anticipated that the areal rainfall is significantly related to the rain area. Linear regression analyses between the instantaneous areal average rainrate [R] for each snap shot in GATE and the fractional rain area within the 280-km square box are performed using resolutions of 4 and 40 km respectively. The form of the regression is

$$[R] = b\ Pc + c + e$$

where Pc is the fractional area within the 280 km square with rainrates above Rc and b and c are the slope and intercept of the regression and e is the error the square of which we try to minimize. To obtain an optimal threshold, a sensitivity study is performed by varying the threshold Rc. Table 4 summarizes our results.

At a resolution of 4 km, the linear model explains over 97% of the variance at a threshold of 5 mm/hr for both phases of GATE. Fig. 1 shows the scattergram of [R] and Pc with Rc = 5 mm/hr for GATE I and II. At the lower thresholds of 0 and 1 mm/hr, the relation seems to be parabolic rather than linear whereas at the higher thresholds the points tend to fall off more rapidly than a linear curve (Chiu, 1988). At a resolution of 40 km, the maximum variance explained is about 93% (91%) at a threshold of 3 mm/hr for GATE I (II).

TABLE 4. REGRESSION COEFFICIENTS, MEAN SQUARE ERROR AND VARIANCES EXPLAINED

4 km PIXELS	GATE I				GATE II			
VARIANCE$(mm/hr)^2$	0.50				0.24			
THRESHOLD	b	c	VAR	MSE	b	c	VAR	MSE
0(mm/hr)	5.06	-0.13	77.6	0.113	4.94	-0.09	70.6	0.071
1	7.21	-0.01	88.6	0.057	7.52	-0.07	85.3	0.035
5	15.85	0.05	97.9	0.011	18.09	0.01	97.0	0.007
10	33.27	0.12	94.7	0.027	36.24	0.07	96.0	0.009
20	93.14	0.16	85.3	0.074	82.42	0.14	87.4	0.030
40 km PIXELS								
0 mm/hr	1.53	-0.27	30.7	0.350	0.98	-0.08	22.6	0.186
1	4.63	-0.04	84.5	0.078	4.28	-0.01	85.3	0.035
2	6.63	0.03	92.2	0.040	6.16	0.06	91.3	0.021
3	8.54	0.09	93.8	0.031	8.09	0.10	91.4	0.021
4	10.43	0.14	92.2	0.039	10.01	0.15	87.1	0.031
5	12.53	0.18	89.5	0.053	12.09	0.18	83.6	0.040

b and c are slope and intercept of the linear regression, VAR is the variance explained by the linear model and MSE the mean square error.

4. DISCUSSION

Why does the linear regression work best for a threshold of 5 mm/hr at the 4-km resolution? The ATI technique relies on a tight relation between convective rain area and convective precipitation. About 30-40% of the GATE rainfall is attributed to stratiform precipitation (Cheng and Houze, 1979) and they fall in the range of about 2-4 mm/hr (Leary and Houze, 1979). From Table 2, it can be seen that about 30% of the total rain is contributed by rainrates lower than 5 mm/hr. The areas with rainrates above 5 mm/hr can be considered mostly convective rain areas. Our results show that the it is the convective rain area that explains a large fraction of the areal rainfall variance in the GATE region.

Satellite observations do not measure rainrates directly, but contain covariant information such as radiances from different portions of the spectrum. Spencer (1986) developed an oceanic rain estimation technique using polarization at 37 GHz. Adler and Mack (1986) showed that the polarization is related to the fractional rain area at this frequency. This suggests that the rain area technique is independent of resolution. On the other hand, if we only need to know if a certain pixel has rainrate above a certain threshold, a model which predicts the probability of such an occurrence may be appropriate. A logistic regression model has been developed (Chiu and Kedem, 1986) which is capable of predicting the probability of rain above a certain threshold. By combining physical and statistical

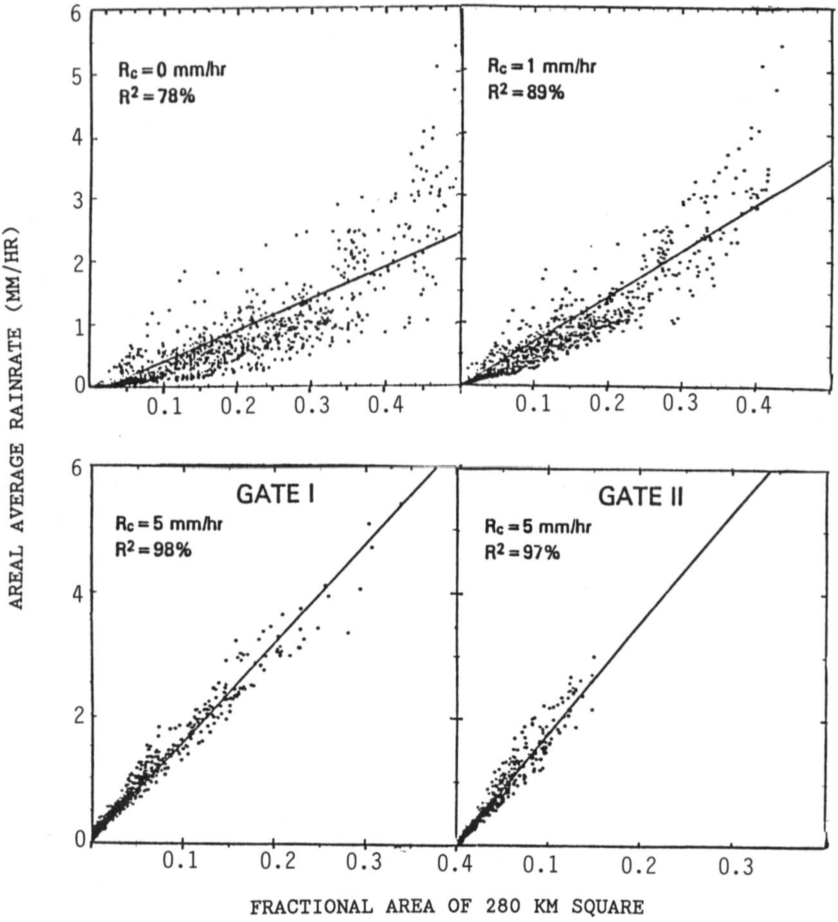

Figure 1. Scattergram of [R] versus Pc for thresholds of 0 mm/hr and 1 mm/hr for GATE I (upper) and for a threshold of 5 mm/hr for GATE I and II (lower panel).

models, it is hoped that important covariates in the rain estimation problem can be identified and our estimates of areal time average rainrates from space observations improved.

REFERENCES

Adler, R. F., and R. A. Mack, 1986: Field-of-view effects in estimating precipitation from satellite microwave radiances, 2nd Conf. Satellite Meteor. and Remote Sensing Appl., Williamsburg, VA, Amer. Meteoro. Soc., Boston MA.

Arkin, P. A., 1979: The relationship between fractional coverage of high cloud and rainfall accumulation during GATE over the B-Scale array area, Mon. Wea. Rev., 107, 1382-1387.

Cheng, C-P, and R. A. Houze, Jr., 1979: The distribution of convective and mesoscale precipitation in GATE radar echo patterns, Mon. Weath. Rev., 107, 1370-1381.

Chiu, L. S., 1988: Rain estimation from space observations: areal rainfall rain area relation, Preprint, 3rd conf. satellite Meteor. and Oceanogr., Anaheim, CA, Amer. Meteoro. Soc., Boston, MA 02139.

Chiu, L. S., and B. Kedem, 1986: Satellite rainfall retrieval by logistic regression, Preprint, 2nd Conf. Satellite Meteor./Remote Sensing and Applications, Williamsburg, VA, Amer. Meteoro. Soc., Boston, MA.

Doneaud, A. A., P. L. Smith, A. S. Dennis, and S. Senguita, 1981: A simple method for estimating convective rain volume, Water Resour. Res, 17, 1676-1682.

Doneaud, A. A., S. Ionescu-Niscov, D. L. Priegnitz, and P. L. Smith, 1984: The area-time integral as an indicator for convective rain volume, J. Clim. Appl. Meteor., 23, 555-561.

Doneaud, A. A., J. R. Miller, Jr., L. R. Johnson, T. H. Vonder Haar, and P. Laybe, 1987: The area-time integral technique to estimate convective rain volumes over areas applied to satellite data- a preliminary investigation, J. Clim. Appl. Meteor., 26, 156-169.

Houze, R. A., and Betts, A. K., 1981: Convection in GATE, Rev. Geophys. Space Phy., 19, 541-576.

Kedem, B., and L. S. Chiu, 1987a: On the lognormality of rainrate, Proc. Natl. Acad. Sci., USA, 84, 901-905.

Kedem, B., and L. S. Chiu, 1987b: Are rainrate processes self-similar? W.ater Resour. Res., 23, no. 10, 1816-1818.

Leary, C. A., and Houze, R. A., 1979: The structure and evolution of convection in a tropical cloud cluster, J. Atmo. Sci., 36, 437-457.

Lopez, R. E., J. Thomas, D. O. Blanchard and R. L. Holle, 1983: Estimation of rainfall over an extended region using only measurements of the area covered by radar echoes. Preprint, 21st Conf. Radar Meteor., Edmonton, Alberta, American Meteor. Soc., 681-686.

Lovejoy, S., and G. L. Austin, 1979: The delineation of rain areas from visible and IR satellite data for GATE and mid-latitudes, Atmo.-Ocean, 17, 77-92.

Patterson, V. L., Hudlow, M. D., Pytlowany, P. J., Richards, F. P., and Hoff, J. D., 1979: GATE Radar Rainfall Processing System, NOAA-TM-EDIS 26, NOAA, Washington, D. C.

Spencer, R. W., 1986: A Satellite Passive 37 GHz Scattering Based Method for Measuring Oceanic Rain Rates, J. Clim. Appl. Meteor, 25, 754-766.

TROPICAL RAINFALL MEASURING MISSION (TRMM)

Thomas Keating
NASA Goddard Space Flight Center
Greenbelt, Maryland 20771, USA

ABSTRACT

The National Aeronautics and Space Administration (NASA), the Radio Research Laboratories (RRL) of Japan, and the National Space Development Agency (NASDA) of Japan are studying the Technological and Programmatic features of a cooperative program to design, develop, and launch the Tropical Rainfall Measuring Mission (TRMM) spacecraft.

The TRMM program seeks a 3-year data set of monthly averaged rainfalls in grid boxes of 360,000 km^2 with an accuracy of better than 20 percent for the portion of the Earth between the latitudes of 30 degrees North and 30 degrees South.

The TRMM satellite will be a free-flyer spacecraft in a 300-km circular orbit with an inclination of approximately 30 degrees. The planned mission life of 3 years will operationally overlap the Tropical Oceans and Global Atmosphere (TOGA) program. It is further planned that the TRMM instrument package will, after the initial 3 years, be installed on the Space Station to allow a decade of observation.

This paper will present the status of the definition of: (1) The joint United States/Japan venture; (2) The design of the spacecraft; (3) The performance of the instruments; and (4) The interface requirements imposed upon the spacecraft by the instrument complement.

All discussions in this report that describe the United States/Japan cooperative agreements are "working level" understandings and are not officially sanctioned positions. It is understood by both the United States and Japan that the present studies are nonbinding and can be stopped at any time without prejudice.

1. INTRODUCTION

The measuring of the rainfall of the tropics will make possible the understanding of the heat engine that drives the global weather. TRMM's instrumentation complement includes radar, microwave radiometers, and visible and infrared sensors that will make possible, for the first time, both the in situ quantification of rainfall and the means of interpreting previously recorded visible and infrared data. A launch using the Japanese H-II vehicle and the possibility of extending the mission life, utilizing the Space Station, provide a unique opportunity for a joint United States and Japan cooperative venture.

1.1 MISSION PARAMETERS

The TRMM has the following parameters:

• Mission Life	3 years	• Orbit	Circular; 28.5 inclination
• Orbital Altitude	300 km	• Launch	September 1993 (earliest)

Reprinted, with permission, from *Proceedings of the 20th Annual Electronics and Aerospace Systems Conference*, Washington, DC, October 14-16, 1987.

1.2 SPACECRAFT DESCRIPTION

The TRMM spacecraft (Fig. 1) will be a free-flyer launched on an Expendable Launch Vehicle (ELV). The spacecraft bus will be an adaptation of existing spacecraft designs, available equipment, and modules to minimize the mission costs.

The instrument module will contain the Precipitation Radar, modified Advanced Very High Resolution Radiometer (AVHRR), and a microwave radiometer that may be the TRMM Microwave Radiometer, which is a new instrument, or a combination of instruments (the Electronically Scanning Microwave Radiometer (ESMR) and the Special Sensor Microwave Imager (SSM/I).

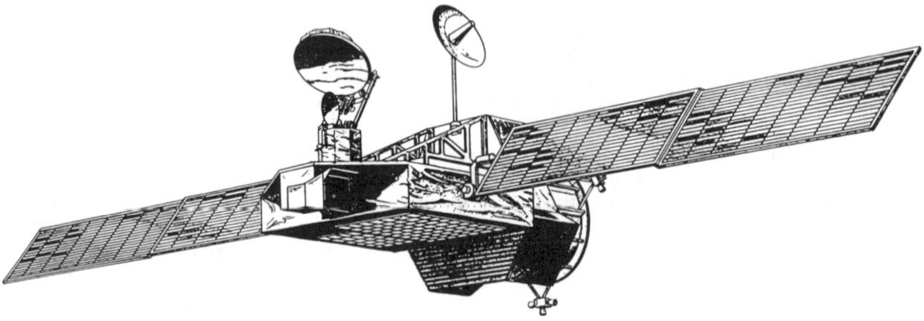

Figure 1. Tropical Rainfall Measuring Mission (TRMM)

2. INSTRUMENT COMPLEMENT

The complement of Earth-observing instruments required to meet the scientific objectives is shown in Fig. 2, which also shows the coverage path for each instrument. A brief description of each of the three instruments is discussed in the following paragraphs.

2.1 PRECIPITATION RADAR

The radar is a dual frequency (14 Ghz and 24 Ghz) linearly polarized electronically-scanning beam instrument. The tentative measurement requirements are:

• Rain Rate (minimum)	0.5 mm/hr
• Rain Rate (maximum)	50 mm/hr
• Scan Angle	±20 degrees
• Pulse Repetition Frequency (PRF)	Variable
• Vertical Resolution	250 m (at nadir)
• Sidelobe Level	25 dB down

2.2 ADVANCED VERY HIGH RESOLUTION RADIOMETER

The AVHRR is a flight proven instrument that will be modified for the unique requirements of the TRMM mission. The performance features of the AVHRR are listed in Table 1. The instrument will contain a daytime and a nighttime visible channel (0.5 to 0.7 μm) and one thermal channel (10.5 μm), a thermal channel (6.7 μm) for water vapor detection, and possibly thermal channels of 1.6 and 11.5 μm. The provision of a stereoscopic infrared (IR) channel is being evaluated.

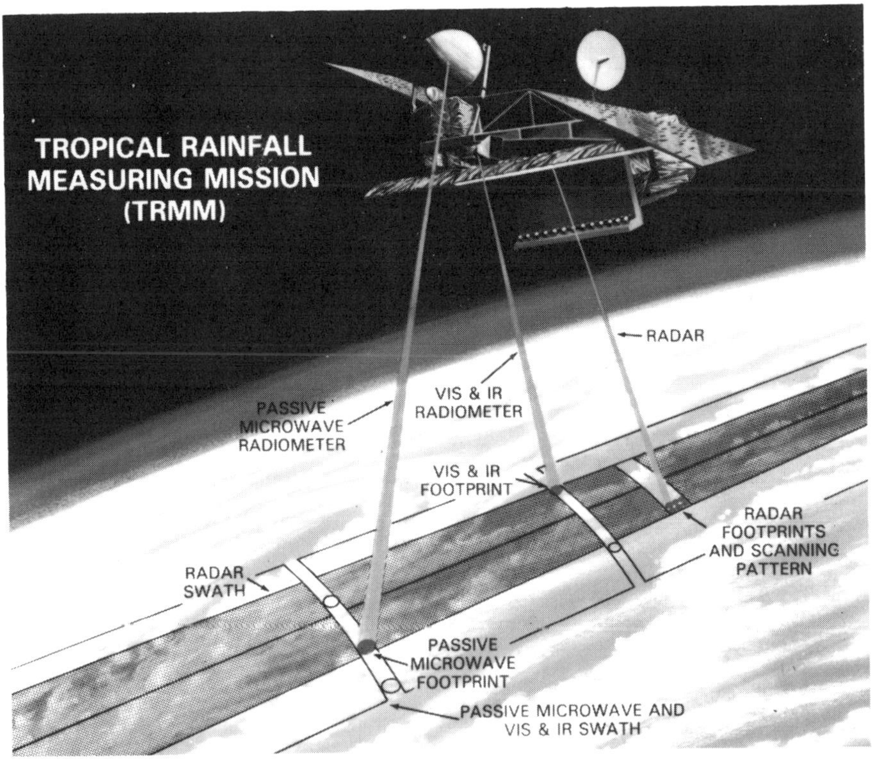

Figure 2. Spacecraft and Instrument Scanning

TABLE 1
PERFORMANCE FEATURES OF THE AVHRR

Parameter	AVHRR (Modified)
IFOV (Degree)	±0.105
(mrd)	±1.82
Footprint, Maximum (km)	3.7 × 3.7
Footprint, Minimum (km)	1.09 × 1.09
FOV (Degree)	±57.2

2.3 TRMM MICROWAVE RADIOMETER

The TRMM Microwave Radiometer (TMR) is a new instrument. It will have six channels: 6.6, 10.7, 18.5, 21, 37, and 92 GHz. The system will have dual polarization and will continuously scan with a 4-km instantaneous swath in a circular ground track (Fig. 2).

3. MISSION LIFE

3.1 ATMOSPHERIC DATA

The low orbiting altitude of the TRMM and the 3-year mission require that serious attention be given to the aerodynamic drag force on the spacecraft. A preliminary study has been made of the atmospheric density predictions and the drag force (Figs. 3 and 4).

3.2 FUEL CONSUMPTION

The fuel consumption is estimated by computing the booster fuel requirements to essentially maintain a constant altitude. By using this method and Eqs. 1 and 2, the computed fuel consumption for 3 years per one square meter of cross sectional area is 54.7 kg/m^2.

$$\text{Force of Drag} = F_D = 1/2\ C_D\ APV^2 \qquad (1)$$

$$\text{FUEL} = \frac{t \cdot F_D}{ISP \cdot G} \qquad (2)$$

C_D = Drag Coefficient = 2.2
A = Cross Section Area (m^2)
P = Density (kg/m^2)
V^2 = (S/C Vel)^2
G = 9.8 m/s^2
t = Burn Time (s)
ISP = Specific Impulse (s) = 220

3.3 EXTENDED LIFE

While the minimum requirement for the mission life has been set at 3 years, the meteorological and climate scientific community desire that the observations be continued for a complete solar cycle (10 to 12 years). Methods are being studied for the purpose of extending the life of the mission. Present consideration is that after 3 years of flight, the TRMM will be moved to the Space Station and the instrument module will be installed on an available positioning mount.

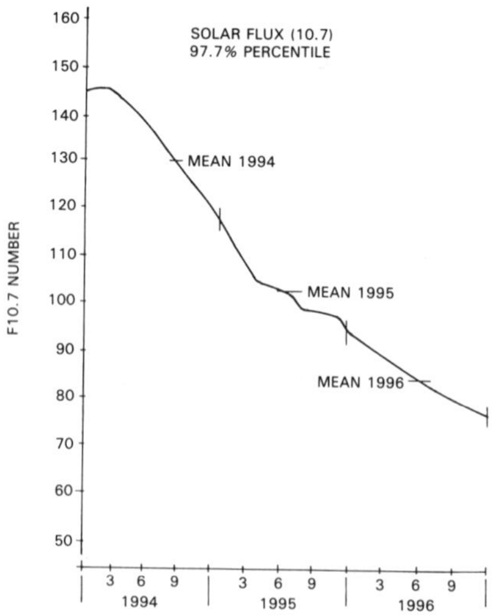

Figure 3. Solar Flux versus Year

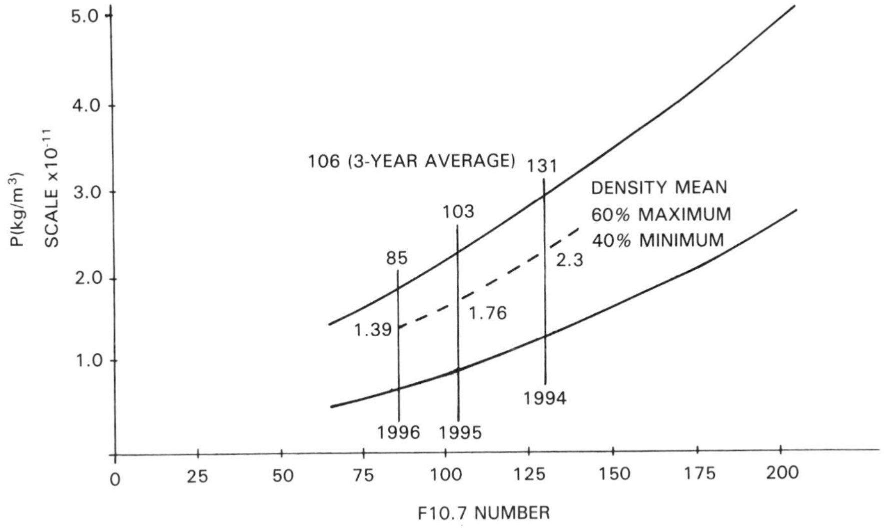

Figure 4. Solar Flux versus Atmospheric Density

4. SPACECRAFT INTERFACES

The present configurations (Fig. 5) are a combination of the Cosmic Background Explorer (COBE) propulsion subsystem, Multimission Modular Spacecraft (MMS) modules, a generic core structure, and an instrument package. Information for the spacecraft-to-vehicle interfaces is also shown in Fig. 5.

4.1 PAYLOAD CONFIGURATION

The packaging of the TRMM in the H-II shroud for launch with a Geostationary/Earth Orbit (GEO) satellite is shown in Fig. 6.

5. SPACECRAFT SUBSYSTEMS

The spacecraft subsystems: power, propulsion, and Command and Data Handling (C&DH) as well as the instruments (AVHRR and TMR) and the two radar antenna designs are shown in Fig. 6.

5.1 POWER

The solar array size and ±30 degree angular movement will provide in excess of 800 W for distribution by the Modular Power System (MPS). The 28 ±7 V system will utilize standard NiCd batteries for operation during the eclipse period. Since the arrays (Fig. 7) present a large drag area, their position will be adjusted as a function of the Sun angle and to minimize drag.

The arrays are driven over the ±30 degree range. The resultant total (S/C and arrays) drag area is 13.6 m^2 requiring 745 kg of fuel. The orbital average power requirements for the TRMM spacecraft are estimated to be 800 W. The MPS can easily handle this requirement.

5.2 PROPULSION

A maximum of 860 kg of hydrazine will be contained in the COBE fuel tank system.

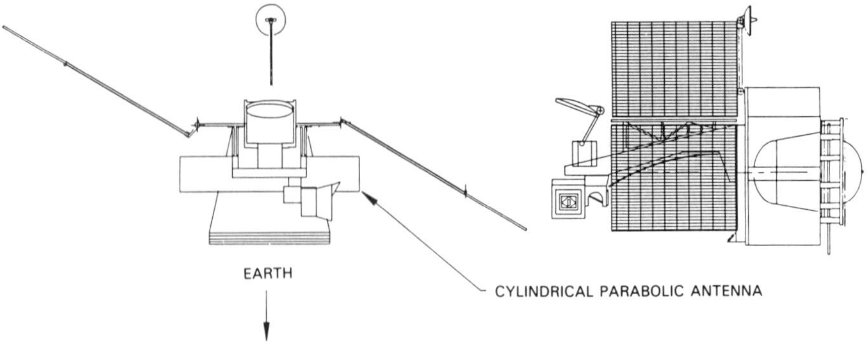

TRMM/H-II CONCEPT A; ORBIT CONFIGURATION

TRMM Structural Data and Interface Applicability

Parameter	Units	Value	Limit	Source
Mass (Dry)	kg	2,500	N/A	GSFC
Mass (Wet)	kg	3,360	3,500	GSFC
CG/Interface	m	UNK	UNK	GSFC
CG/Centerline	mm	UNK	50	H-II
Launch Moment	M-kg	UNK	6,000	H-II
Resonant Frequency (Latitude)	Hz	UNK	10	H-II
Resonant Frequency (Axial)	Hz	UNK	30	H-II
Shroud Diameter (TRMM)	m	3.7 maximum	3.7	H-II
Shroud Length (TRMM)	m	5.3	7.7	H-II
Attach Bolt Circle	m	2.0 diameter	UNK	H-II

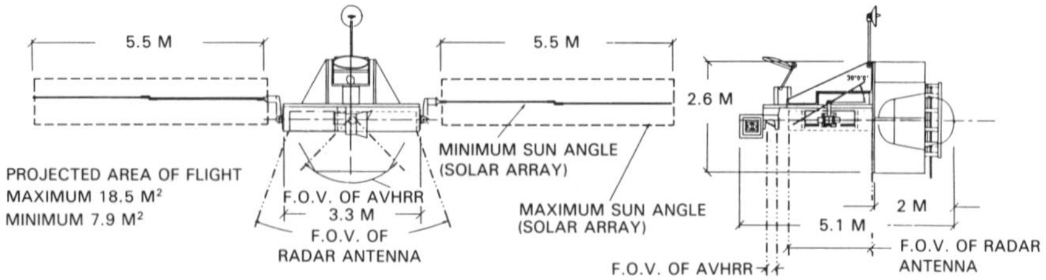

TRMM/H-II CONCEPT B; ORBIT CONFIGURATION

Figure 5. TRMM/MMS/COBE Configuration

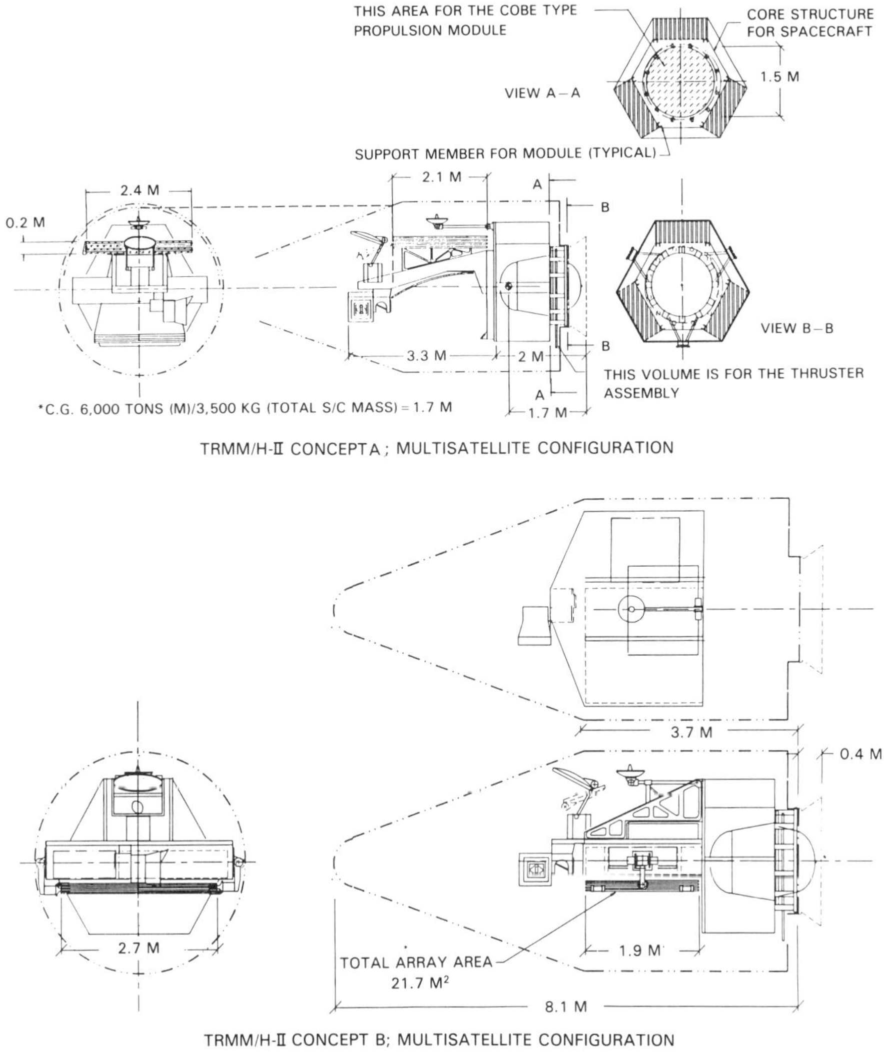

Figure 6. TRMM/H-II Launch Configuration

5.3 C&DH

The primary science data will be transmitted via the TDRS on the K-band Single Access (KSA) link. The C&DH subsystem will contain: two TDRSS transponders, four tape recorders (500-Mb capacity each), two fixed low-gain S-band antennas, a K-band and S-band 0.5-meter diameter antenna on a hemispherical coverage positioning mount, and the K-band and S-band transmitters. The science data rate is 370 kbps with a playback rate of 5 Mbps with a duration of 3.5 minutes every one-half orbit. S-band "ranging," which is required for both instrument orbit position determination and TDRSS operations, will be performed during a playback period.

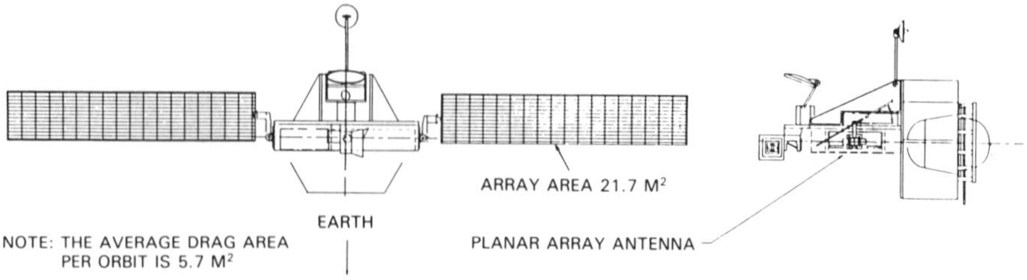

Figure 7. TRMM Orbit Configuration

6. DATA RELAY TRACKING SATELLITE TERMINAL (DRTST)

The Japanese-provided equipment will be designed to operate with their ETS-VI and EDRTS satellites. The operational forward link from the ETS and EDRTS to TRMM is 23 GHz and the return link is 32 GHz.

7. CONCLUSION

The preliminary TRMM design described in this paper provides a basic understanding of the mission. A more rigorous design will evolve over the next 6 months as will the agreements between the United States and Japan. If a NASA FY 90 new-start approval is obtained, measuring the rainfall of the tropics could begin in late 1993.

ACKNOWLEDGMENTS

The information presented in this paper has been obtained from numerous sources. The major contributors were:

Radar—Dr. Nobuyoshi Fugono, Mr. Toshiaki Kozu, and Dr. Ken'ichi Okamoto, Japan RRL.

DRTST—Mr. Makoto Kajii, Japan NASDA.

H-II—Dr. Masashi Mochizuki, Dr. Yoji Shibato, and Mr. Hitoshi Takatsuka, Japan NASDA.

Science—Dr. Joanne Simpson, NASA/GSFC.

Microwave—Dr. Thomas Wilheit, NASA/GSFC.

AVHRR—Dr. William Barnes, NASA/GSFC.

TMR—Dr. J. Shiue, NASA/GSFC.

Spacecraft—Mr. J. Cooley (Orbit Life), Mr. G. Frazier (Mechanical), Mr. J. Jermakian (Power), Mr. W. Hibbard, and Mr. G. Roach, NASA/GSFC.

BIBLIOGRAPHY

Keating, T., 1987a: Tropical Rainfall Measuring Mission (TRMM), Proceedings of the IEEE Conference, Washington, D.C. (For a copy of this paper contact Thomas Keating, Advanced Missions Analysis Office, Code 402, NASA/Goddard Space Flight Center, Greenbelt, Maryland 20771, U.S.A.)

ERROR ANALYSIS FOR THE TROPICAL RAINFALL MEASURING MISSION (TRMM)

Thomas T. Wilheit
Laboratory for Oceans
NASA Goddard Space Flight Center
Greenbelt, Maryland 20771, USA

ABSTRACT

In order to estimate the quality of the measurements which would result from a space mission such as the Tropical Rainfall Measuring Mission (TRMM), an error model is proposed. This model considers three sources of error: sampling, random error and correlated error (biases). Estimates of these errors are generated for a variety of different instrument configurations. Several general conclusions can be drawn from the results of this model. First, the random measurement error makes an insignificant contribution to the total mission error; the sampling and correlated errors dominate. Second, it is not particularly important to provide high quality measurements for rain rates below about 2 mm/hr. Third, and most important, several realistic potential instrument complements can, at least, approximate the required measurement accuracy.

INTRODUCTION

In proposing a mission such as TRMM, it is essential first to arrive at objectives for the mission in as quantitative a manner as possible and then to estimate how well any proposed combination of payload,orbit and other mission parameters will meet these objectives. It is generally quite difficult to estimate the mission requirements. Indeed, the results of the mission may be needed to provide the basis for such an estimate. It would serve no useful purpose simply to say that we cannot do a proper job of making these estimates and do nothing. So, difficult or not, we do as well as we can with what information is available to derive these requirements. It is entirely similar with the estimates of the mission performance. There are never enough data available before the mission to derive rigorous estimates of the errors in the measurements, but we do what we can with limited data and a great deal of judgement. Hopefully, the effort will yield a reasonable estimate of the mission performance and will certainly yield insight into the relative importance of various sources of error.

In this analysis we will concentrate on the primary mission goal, the determination of the rainfall on climatological time and space scales. In particular, we

A. DEEPAK Publishing
ISBN 0-937194-14-4

will derive estimates of the uncertainty of the monthly average rainfall in 5-degree squares. The results can be scaled to other space/time scales.

In trying to estimate the accuracy with which the TRMM can derive monthly rainfall estimates, we will only consider two classes of instruments, a two frequency radar and several radiometer options. Any benefit which could be derived from the use of the Visible/Infrared radiometer, in conjunction with the microwave instruments, will not be included because these concepts are still too immature even for this minimally quantitative analysis.

Estimates of the uncertainties in these monthly rainfall estimate averages will be generated for several different combinations of instruments. Attenuation-based algorithms for radiometers can only be used over an ocean background. Thus we must consider land and ocean backgrounds separately. Calculations will also be performed for the assumption of a perfect suite of instruments over either background. This assumption allows us to compare our results with the sampling studies, which assume perfect rain measurements, and to adjust a scale length to account for the sampling contribution to the total error. In all instrumental cases, a two frequency radar which meets the minimum requirements stated by the Science Steering Group is included. Along with the radar, several radiometer possibilities are examined. The simplest assumption is that only the Electrically Scanning Microwave Radiometer (ESMR) is included. This instrument is a space qualified spare of the ESMR which flew on the Nimbus-5 spacecraft (1972 launch). The instrument would observe radiation at 19.35 Ghz, and would image a swath of 600km from the proposed 300-km orbit of TRMM with a resolution at nadir better than 10km. A second instrument which can be included with the ESMR and the radar is the Special Sensor Microwave Imager (SSM/I), a copy of which was launched on the DMSP spacecraft in 1987. This instrument would image a swath nearly as wide as the ESMR. It would permit the use of scattering based algorithms at high rain rates over either land or ocean backgrounds, as well as extending the attenuation-based algorithms to somewhat lower rain rates over oceans. Its spatial resolution is decidedly poorer than that of the ESMR at their common frequency (19.35 GHz). The most optimistic case is that a special purpose microwave radiometer would be developed for this mission, the so-called TRMM microwave radiometer (TMR). This instrument would have channels at 6.6, 10.7, 18, 21, 37, and 90 GHz. The spatial resolution would be 10km at 18 and 21 GHz, 5km at 37 and 92 GHz 17km at 10.7 GHz and 27km at 6.6 GHz. The principal advantage this instrument would offer is the extension of the use of the attenuation algorithms over the oceans to the highest rain rates considered.

PROBABILITY DISTRIBUTION FOR RAINFALL INTENSITY

The logical beginning point for estimates of rain measurements is a determination of how hard it rains and how often. The log-normal distribution is often used for the probability distribution function (PDF) for rain (Kedam, Chou and North 1987). This form is:

$$P(r) = \frac{P_r}{\sigma(2\pi)^{1/2}} \exp\left[-\frac{(\ln(r/r_0))^2}{2\sigma^2}\right] d(\ln(r)) \tag{1}$$

Note that the normalization of the PDF contains, in addition to the usual factors, P_r, the probability of any non-zero rain rate. The average rain rate can be expressed:

$$\langle r\rangle = P_r\, r_0 \exp(\sigma^2/2) \tag{2}$$

As expressed here, $\langle r\rangle$ is the average rain rate, including the times when it is not raining, but if P_r is deleted it becomes the average rain rate when it is raining.

For the design of a measurement system, it is important to know what part of the total rain is contributed by which rain rates. For this purpose, we calculate r_{10}, r_{50} and r_{90}, the rain rates below which 10, 50 and 90 percent of the cumulative rain occur, respectively. These quantities can also be expressed in terms of the log-normal parameters from (1).

$$\begin{aligned} r_{50} &= r_0 \exp(\sigma^2) \\ r_{10} &= r_{50} \exp(-1.28\sigma) \\ r_{90} &= r_{50} \exp(1.28\sigma) \end{aligned} \tag{3}$$

There is a severe shortage of rainfall data which are appropriate for analyzing TRMM. Probably the best available data set was gathered by the precipitation radars on board the ships in the 1974 GATE experiment. These data are averaged into 4-km cells which makes them quite similar to the intended TRMM measurements. In Table 1 the data from phase I of GATE are expressed in terms of a log-normal distribution (Kedam, Chou and North 1987). Note that the percentage of time that it rains is rather high (7%) and that r_{50} is only 8.8 mm/hr. For comparison purposes, data are also shown which are derived from a 5-year record of 5-minute averages of a rain gauge at Wallops Island Va. (Goldhirsh 1983) and from 13 months of similarly averaged data from a rain gauge on an oil platform in the Gulf of Mexico near Lake Charles La. Both of these differ from the GATE data in that the probability of precipitation is much smaller and associated rain rates are somewhat larger. These differences are consistent with the area average versus point measurement distinction, the climate differences and the extreme drought in the Southeastern United States that coincided with the La. observation period (Aug 85 to Sept 86). For our analyses, we will use the GATE distribution; its comparison with the Va. and La. data sets suggests that it is not unreasonable. Judgement, on the other hand, would suggest that,if this data set is unrepresentative of what is to be measured by TRMM, it errs in the sense of a high probability of precipitation and low rain rates.

TABLE 1
LOG-NORMAL PARAMETERS FOR RAIN

DATA	P_r	r_0 mm/hr	σ	r_{10} mm/hr	r_{50} mm/hr	r_{90} mm/hr	<r> mm/day
GATE	0.07	3.1	1.02	2.4	8.8	32.4	8.8
VA.	0.021	2.2	1.21	2.0	9.5	44.8	2.3
LA.	0.012	4.1	1.19	3.7	16.9	77.5	2.4

STRUCTURE OF THE ERROR MODEL

Even with only two classes of instruments included, two domains must be considered. The swath of the radar must be narrow, compared to that of the radiometers, because of the geometry of the range measurements. In a central region, along the sub-satellite track, we have both radar and radiometer data, whereas, over a considerable swath outside of this, we have only radiometer data. The two regions will be labeled r/r and r/o respectively. For the purpose of this analysis we will assume a 200km swath for the radar and a 600km swath for the microwave radiometer. In the specific design, of the instruments, the engineers will try to extend both of these swath widths.

Within each region we will characterize the errors as three different types: sampling, random and correlated. The sampling errors are discussed in a companion paper by North. Simply put, the sampling error results from the approach of measuring only a small fraction of the rain and estimating the total by extrapolation. The random errors represent the noise in the measurement due to instrumental effects, such as thermal emission and Rayleigh statistics, as well as local departures from the assumptions on which the algorithms are based. Within the error model, these random errors are assumed to be unbiased. However, in the limit of a great number of observations, the errors will still not average to zero. There will still be a residual of error due to systematic error in the algorithm assumptions and instrumental calibration. To the extent that this error is known, it will be corrected by adjustments to the algorithms but this correction will never be perfect. The correlated error is the best available estimate of the magnitude of the unknown residual. Minimizing and estimating the correlated error will be the great challenge of the TRMM ground truth program.

It will be seen later that the correlated errors are the dominant source of the instrumental contribution to the total mission error budget. Fortunately even these correlated errors are random in a sense. Within the r/r portion of the swath, there will be a very intricate mosaic of algorithms used, each having its range of rain rates and incidence angles over which it is valid. The situation within the r/o portion of the swath is similar. It is reasonable to expect that the error residual of one algorithm is unrelated to that for any other algorithm. Moreover, in different portions of the valid range of any given algorithm, different problems can dominate and thus provide an additional degree of independence in the residual errors. For the purposes of this analysis, we will approximate this form of randomness by assuming that the correlated errors are independent from one octave (factor of two) of rain rate to the next and the correlated errors in the r/r swath are independent of those in the r/o swath.

Separate estimates of the contribution of each octave of rain rate, to the monthly total rainfall for a given 5° x 5° cell, R_j, can be generated using observations from the r/o and from the r/r portions of the swath. The two can then be averaged in an optimally weighted manner and then summed over the octaves of rain rate to arrive at a minimum uncertainty in the net estimate of the monthly total rainfall.

$$R_j = \frac{M}{N_{obs}} \sum_{obs} r \cong \frac{M}{N_{obs}} \sum_{octaves} r_i N_i \qquad (4)$$

where M is the length of a month, N_{obs} is the total number of observations in a month (including those for which there is no rain), r is the rain intensity associated with each observation, r_i is the mean value of each octave of rain rate, and N_i is the number of observations of rain rate within each octave. For our purposes, the octaves of rain rate will be 0.5 to 1 mm/hr, 1 to 2 mm/hr, ... 64 to 128 mm/hr for a total of 8 octaves.

The variance of R_j, $\sigma^2(R_j)$, can be expressed as:

$$\sigma^2(R_j) = \frac{M}{N_{obs}} \sum_{octaves} \sigma^2(r_i N_i) \qquad (5)$$

In turn, $\sigma^2(r_i N_i)$ can be expressed as a sum of three contributions:

$$\sigma^2(r_i N_i) = (r_i N_i)^2 \left\{ \frac{1}{\nu_i} + \frac{\rho_i^2}{N_i} + b_i^2 \right\} \qquad (6)$$

Here, the first term represents the sampling error and ν_i is the number of independent samples in the i'th octave. The second term is the random measurement error and ρ_i represents the fractional random error for measurements of rain rate in the i'th octave. The third term is the contribution to the variance of the underlying bias in the measurements and b_i is the fractional bias in the measurements in the i'th octave.

For each portion of the swath the total rainfall is the sum of the contributions from each octave of rain rate and likewise the variance of the total rainfall is the sum of the variances of the contributions from each octave. The best estimate of the total rainfall in the cell for the month is the average of the estimates from the two portions of the swath weighted inversely to their variance. The variance of this total rainfall is given by:

$$\frac{1}{\sigma^2_{tot}} = \frac{1}{\sigma^2_{r/o}} + \frac{1}{\sigma^2_{r/r}} \tag{7}$$

In order to estimate ν_i, the number of independent samples and N_i the number of observations in a given rain rate octave, we first note that the satellite will have a ground speed of approximately 7 km/s. The net width of the r/o portion of the swath is 400 km and the r/r portion is 200 km. Thus the satellite will image an area of 7.35 x 10^9 km^2/month in the r/o swath and an area of 3.68 x 10^9 km^2/month in the r/r swath. The total number of observations by a given instrument is given by:

$$N = A/(L_r)^2 \tag{8}$$

and the number of independent samples by:

$$\nu = A/(L_c)^2 \tag{9}$$

where A is the relevant imaged area, L_r is the instrumental resolution (expressed as a length) and L_c is the correlation length of the precipitation. We will use 10km as a typical radiometer resolution and 5km for the radar. A value of 69.9 km is chosen for L_c to match the present results to those of sampling studies for the assumption of a perfect instrument.

The variation of sampling with latitude is ignored; the samples are distributed uniformly over the 864 cells between ±30°. The samples are distributed among rain rate octaves according to the probability distribution function (1).

TABLE 2
ERROR CONSTANTS FOR TWO FREQUENCY RADAR + ESMR +SSM/I
OCEAN BACKGROUND

rain rate mm/hr	errors r/o random	errors r/o correlated	errors r/r random	errors r/r correlated
0.5 - 1	0.5	0.5	0.5	0.3
1 - 2	0.5	0.3	0.5	0.3
2 - 4	0.5	0.3	0.5	0.2
4 - 8	0.5	0.2	0.5	0.2
8 - 16	0.5	0.3	0.5	0.2
16 - 32	0.5	0.5	0.5	0.2
32 - 64	0.6	0.6	0.5	0.3
64 - 128	0.7	0.7	0.5	0.3

In table 2, the error constants are given for the combination of the two frequency radar, the Electrically Scanning Microwave Radiometer(ESMR) and the Special Sensor Microwave/Imager (SSM/I). For this table it is assumed that the sensors are observing rain over an ocean background. Similar tables have been generated for a variety of instrument and background combinations. For the assumption of a perfect instrument, the table contains all zeros for the error terms.

There is admittedly much judgement used in generating such a table, but some comparisons with the results from Nimbus-5 and 7 are possible. Based on the results of Wilheit *et al* (1977) and Austin and Geotis (1978), the r/o error estimates for the 2-4 and the 4-8 mm/hr octaves appear reasonable for the Nimbus 5 resolution. Similarly the three highest octaves (r/o) are consistent with an extrapolation of the results of Spencer *et al* (1983) using the Nimbus-7 data. The improved resolution of the TRMM payload would certainly improve the attenuation-based rainfall estimates but has not been used here to lower the error estimates. Thus, it can be argued that the estimates of these error constants are responsibly conservative.

RESULTS

Table 3 lists the results of applying the error model to the various instrument combinations. The 10% error listed for the perfect instrument is typical of the results from the sampling studies; if a significantly different number or a final value results from the sampling studies, then the correlation length, L_C, can readily be readjusted to match and the error model rerun. Note that for all cases with real instrument assumptions, the error estimates are much larger over land than over ocean. This results from the requirement of an ocean background for the attenuation-based algorithms for the radiometers. Although the case of radar only with no radiometers was not calculated, the ESMR over land case would represent a rather good approximation over both land and ocean backgrounds.

TABLE 3
TRMM ERROR ESTIMATES FOR MONTHLY 5° x 5° RAINFALL TOTALS
TWO FREQUENCY RADAR INCLUDED IN ALL CASES

BACKGROUND	PERFECT INST'S	ESMR ONLY	ESMR + SSM/I	TMR
LAND	10%	22%	19%	19%
	0.9	1.9	1.7	1.7 mm/day
OCEAN	10%	17%	13%	12%
	0.9	1.5	1.2	1.1 mm/day

In addition to the error totals, it is possible to partition the contributions to the variance from each portion of the swath, each octave of rain rate and from each of the three error types (sampling, random and correlated). Table 4 lists these contributions for the ESMR-SSM/I case.

TABLE 4
PERCENTAGE OF VARIANCE CONTRIBUTED BY EACH ERROR SOURCE
ESMR/SSM/I OVER OCEAN

RAIN RATE	RADIOMETER/ONLY SAM	COR	RAN	RADAR/RADIOMETER SAM	COR	RAN
½-1 mm/hr	<0.1	<0.1	<0.1	<0.1	<0.1	<0.1
1-2	0.2	0.6	<0.1	0.2	0.4	<0.1
2-4	0.5	2.2	<0.1	1.8	1.8	<0.1
4-8	3.0	4.8	<0.1	3.2	2.5	<0.1
8-16	4.5	8.4	<0.1	9.2	3.8	<0.1
16-32	5.0	8.3	<0.1	13.7	1.8	<0.1
32-64	6.6	3.2	0.1	6.5	0.4	<0.1
64-128	4.1	0.3	0.1	2.4	<0.1	<0.1

The most striking conclusion, which can be drawn from this table, is the lack of importance of the random error in the individual measurements. It is also clear that the very lowest rain rates (<2mm/hr) do not contribute in any meaningful way. If the total error is to be reduced, the measurements of rain rates above about 8mm/hr must be improved. This is the rationale for the TMR. By including channels at 10.7 and 6.6 GHz, the upper end of the dynamic range for attenuation-based algorithms is extended from 16 mm/hr up to 128 mm/hr. Further, if the GATE rain rates are lower than those actually encountered by TRMM, the measurement performance will not suffer with the TMR.

CONCLUSIONS

The Science Steering Group has approximated the mission measurement requirements at 1 mm/day uncertainty in the monthly mean rain rates for 5° x 5° cells. To the extent that either the requirement or the error analysis is exact, this requirement can be met over ocean areas by the TRMM with either the ESMR-SSM/I or the TMR options for the microwave radiometer. Over land areas, the performance of TRMM falls significantly short of the requirements; it will be necessary to supplement the satellite-borne measurements with land-based measurements in order to arrive at a complete picture of tropical precipitation.

The lack of importance of the lowest rain rates is also suggestive. The higher frequency of the two frequency radar is most important in the retrieval of low and moderate rain rates and is irrelevant to the retrieval of the very highest rain rates. Accordingly, the TMR and ESMR-SSM/I cases were re-examined with the assumption of only a single frequency (14 GHz) radar. The resultant degradation was too small to show at the two significant figure level given in Table 3.

REFERENCES

Austin, P.M., and S.G. Geotis,1978; Evaluation of the quality of precipitation data from a satellite-borne radiometer. Final report under NASA grant NSG 5024.

Goldhirsh, J.,1983; Yearly variations of rain rate statistics at Wallops Island and their impact on modelled slant path attenuation distributions. *IEEE Trans. Antennas Propagat.* AP31 918-921

Kedam,B., L.Chou, and G.North,1987; Estimates of mean rain rates: applications to satellite meteorology, Submitted to: *J. Geophys. Res.*

Spencer, R.W., D.W. Martin, B.B. Hinton and J.A. Weinman,1983; Satellite microwave radiances correlated with radar rain rates over land, *Nature* vol 304 141-143.

Wilheit, T.T., A.T.C. Chang, M.S.V. Rao, E.B. Rodgers and J.S. Theon,1977; A satellite technique for quantitatively mapping rainfall rates over the oceans, *J. Appl. Meteor.* vol 16 551-560.

ON THE DESIGN ISSUES FOR A SPACEBORNE RAIN MAPPING RADAR

F.K. Li, K.E. Im, W.J. Wilson, and C. Elachi
Jet Propulsion Laboratory
California Institute of Technology
Pasadena, California 91109, USA

ABSTRACT

The conceptual design and performance tradeoff issues for a spaceborne rain mapping radar for the Tropical Rain Measuring Mission are discussed. Potential special operation modes, such as adaptive scanning and nadir Doppler measurement, are also discussed.

1. INTRODUCTION

Rainfall has significant impact on the global energetic and hydrologic circulation, and climate structure. However, it must be observed frequently for a long period of time in order to gather sufficent data for any practical applications, such as large scale atmospheric modelling and long-term weather forecasting. One of the best ways to collect such a global rain data set is by remote sensing from the orbiting satellites. With the technological advance in recent years, the possibility of spaceborne precipitation measurements has been greatly enhanced. For instance, both the visible and infrared sensors and the microwave radiometers have demonstrated their feasibility to make precipitation measurements from a space platform (Arkin, 1979; Rao et al., 1976)

Another promising technique for rainfall remote sensing is the use of a microwave radar. Numerous ground-based rain radar measurements have been reported (e.g. Marshall et al., 1947; Joss et al., 1970) and it is natural to extend this technique to a spaceborne configuration. In addition to its capabilities in measuring both path-averaged rain rate and precipitation near ground level, they can potentially be used to explore the details of the rain characteristics and their interaction with the atmosphere. Based on these potentials, a rain mapping radar is planned as part of the Tropical Rainfall Measuring Mission (TRMM) (Thiele, ed., 1987). The prime objective of this mission is to obtain a 3-year record of monthly averages of rainfall over resolution cells of approximately 10^5 km^2 in size for various scientific applications.

This paper describes and discusses the design concept and tradeoff issues for such a spaceborne rain radar. The TRMM's science requirements on sampling strategy, spatial and height resolutions, observable altitude, rain rate measurement range and the corresponding accuracies are described. The radar design that specifically addresses these issues is provided.

2. RADAR PERFORMANCE REQUIREMENTS

The following radar performance requirements were derived based on the recommendations from the TRMM Science Workshop (Thiele, ed., 1987):

(1) Frequency : A dual frequency radar with matched antenna beams;

(2) Swath Coverage : A minimum radar coverage over a cross-track swath of 200 km;

(3) Range of Measurable Rain Rates : The radar must be capable of measuring rain rates between 0.5 mm/hr and 50 mm/hr at all appropriate altitudes. While greater rain rates may occur they are not expected to contribute significantly to the monthly average rainfall.

(4) Horizontal Resolution : The horizontal dimensions of the typical stratiform rain cells range from 5 to 10 km. Convective storms may have even smaller horizontal dimensions. Consequently, the minimum required horizontal resolution of the radar measurements is ~4 km.

TROPICAL RAINFALL MEASUREMENTS
John S. Theon and Nobuyoshi Fugono (Eds.)

ISBN 0-937194-14-4

(5) Range Resolution : Range resolution of 250 m or less is required so that the rainfall profile can be examined, and the thickness of the melting level can be resolved.

(6) Observable Range and Dynamic Range : The radar must have enough observable range and dynamic range so that the echoes backscattered from the melting level, rain cell, and earth surface at all appropriate incidence angles can be measured.

3. Conceptual Design

The basic design concept is graphically illustrated in Fig. 1. In the present design the radar antenna will electronically scan its beam across the sub-satellite track to achieve the desired swath. The radar will also provide contiguous along-track coverage. The ground resolution, which is determined by the antenna footprint size, will be achieved by using antennas of adequate sizes. The range resolution will be achieved by a pulse compression technique.

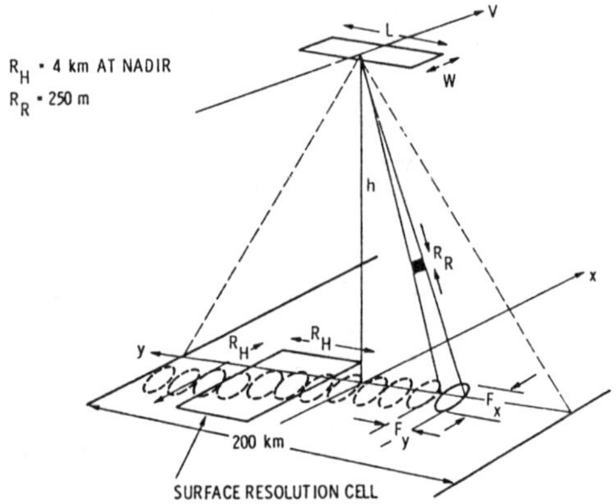

Figure 1. Design concept for the TRMM rain mapping radar.

(1) Frequency Selection : In order to attain a desired equivalent ground resolution of approximately 4 km from the TRMM's orbit altitude of 300 km, an antenna dimension of approximately 75 wavelengths is required. Antenna cost and size constraints rule out wavelengths much longer than about 4 cm (i.e., frequencies below 8 GHz). At frequencies above 40 GHz rain absorptions are prohibitive. Within these two limits the currently allocated bands for active earth remote sensing are around 10, 14, 17, 24, and 35 GHz. 14-GHz appears to be an excellent choice because the rain absorption is relatively small; this will allow precipitation measurements even in severe thunderstorm situations. For the selection of the second frequency, several considerations are relevant : (a) sensitivity at low rain rates, (b) signal attenuation at high rain rates, (c) algorithm sensitivity when applying any absorption technique. These criteria can be examined at each of the above frequencies by comparing the anticipated rain echo signal strengths at one frequency relative to that at 14 GHz. Using the well-known rain radar equation (Probert-Jones, 1962), it can be shown that the ratio of the averaged rain echo signal-to-noise ratios (SNR) of two frequencies f_1 and f_2, for the matched antenna beams case, is given by

$$\frac{\mathrm{SNR}(f_1)}{\mathrm{SNR}(f_2)} \simeq \frac{P_{T_1}\, T_2\, f_2^2\, \eta_1\, \kappa_1}{P_{T_2}\, T_1\, f_1^2\, \eta_2\, \kappa_2} \tag{1}$$

where P_{T_i} is the transmit peak power, T_i is the system noise temperature, η_i is the rain reflectivity, and κ_i is the round-trip signal loss due to precipitation and the ice melting layer absorptions. In deriving Eq. (1) we have made several simplifying assumptions, e.g., that the system bandwidth, system loss, etc., at the two frequencies were the same. Furthermore, in order to understand how the rain parameters affect this ratio, the transmit power and system noise temperature were assumed to be the same. This ratio has been evaluated for various frequencies versus 14 GHz and for a one-way rain pathlength of 5 km. The results are shown in Fig 2. Notice that the signal strengths of the rain echoes at 10 and 17 GHz are consistently less than that at 14 GHz for almost all rain rates of interest. At 35 GHz the SNR ratio is higher at low rain rates, which implies that 35 GHz would be a good frequency to use in light rain region. However, the severe signal loss at moderate to intense rain ($\geq$ 5 mm/hr) will significantly limit its potential usefulness. 24-GHz appears to provide a reasonable compromise between the low (10/17 GHz) and the high (35 GHz) frequencies.

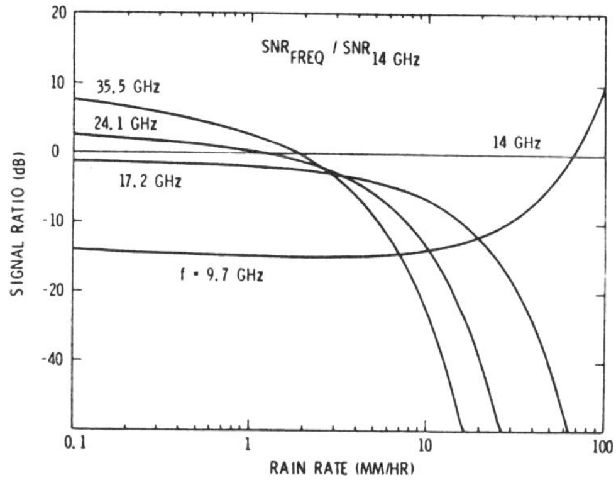

Figure 2. The SNR ratios of the rain echoes at different frequencies.

(2) Swath : In order to achieve large swath coverage, the antenna beams at both frequencies will be scanned in the cross-track direction. For the planned orbital altitude and the required 200-km swath, the maximum antenna scan angles of $\pm 20^\circ$ from nadir should be sufficient. However, it appears to be advantageous to increase the radar swath to 600 km in order to obtain measurements to supplement the radiometric measurements,such as by identifying the suspended cloud water from the precipitable rain particles. Unfortunately, any increase in the radar swath width would result in the reductions in both the radar dwell time and the effective vertical resolution of the rain echoes near the swath edge. As a compromise, we envision that the radar will be operated in two scan modes - a primary scan mode and a wide swath scan mode. The primary scan mode will operate for, say, 90% of the time in which the radar antennas will scan $\sim \pm 20^\circ$ from nadir. The measurements obtained in this mode constitute the primary data set for the rain rate retrieval. The wide-swath scan mode will operate for the remaining 10% of the time in which the radar antennas will scan $\pm 45^\circ$ to 600-km swath, and the measurements acquired in this mode will be used in conjunction with other sensor data for rain rate retrieval.

(3) Equivalent Ground Resolution : The radar footprint sizes in both the along-track and cross-track dimensions are inversely proportional to the antenna dimensions. It is of course desirable to have as large an antenna as possible in order to achieve fine resolution and high signal-to-noise ratios. However, in the spaceborne configuration, the antenna size is limited by the weight and size constraints set forth by the spacecraft. In this baseline design, we have assumed an effective antenna dimension of 2 m x 2 m for the 14-GHz antenna, and 1.2 m x 1.2 m for the 24-GHz antenna. For this set of antenna dimensions and at 300-km spacecraft altitude, the radar will illuminate a 3.7 km x 3.7 km footprint at nadir; a 4.3 km x 4.5 km footprint at scan angle $\theta = 20^\circ$, and a 6.1 km x 11.5 km footprint at $\theta = 45^\circ$.

When the radar beam is scanned off nadir, the vertical resolution is degraded because of the smearing in vertical extent of the pulse volume. In the present design this degradation amounts to ~1300 m at $\theta = 20^\circ$ and over 5600 m at $\theta = 45^\circ$. We consider this to be a serious degradation in system performance and this issue will be examined in the future.

(4) Range resolution : Although the required range resolution for the rain echo measure-ments is 250 m, it is, however, desirable to achieve an even finer radar's intrinsic range resolution in order to obtain independent echo samples for averaging, and at the same time maintainin sufficient echo power. This can be achieved by using a pulse compression technique. For ou baseline design, we chose a compressed pulse width of 0.28 μsec, and the corresponding intrinsi range resolution was ~42 m.

(5) Antenna Side-lobe Clutter Interference : When the antennas are scanned off nadir, un-desired surface and rain clutter signals from the antenna side-lobes will contaminate the desired rain echoes. The contamination from the surface clutters of the nadir-looking side-lobe is mor significant because the surface backscattering cross-sections at normal incidence, particularl for calm oceans, are usually much higher than that of the rain scattering volume. To obtain rain-signal-to-surface-clutter ratio above unity, the peak antenna side-lobe (β) must be

$$\beta \leq \left(\frac{\eta\, c\, \tau\, \kappa}{2\sigma^{o}}\right)^{1/2} \qquad (2$$

where τ is the compressed pulse width of the radar and σ^{o} is the surface backscattering co-efficient at nadir. For example, at the rain rate of 1 mm/hr and by adopting σ^{o} values c 15 dB at 14 GHz and 17 dB at 24 GHz, the required peak side-lobes are -32 and -28 dB respectively. These requirements are judged to be at the state-of-art limit of the phased arra antenna technology.

(6) Pulse Compression Side-lobe Clutter Interference : This interference arises when th Linear Frequency-Modulated (LFM) pulse compression side-lobes of a particular range bi that contains the surface echo contaminate the other range bins that contain the rain echoes This problem is most severe at nadir incidence and at light rain because of the large rada backscattering cross-section of the surface relative to that of the precipitation. In this case, th compression side-lobe interference ratio is given by

$$\frac{\tilde{P}_s}{P_r} \simeq \frac{2\sigma^{o} L_{\mathrm{SL}}}{\eta\, c\, \tau} \qquad (3$$

where $\tilde{P}_s$ is the surface backscattered power and L_{SL} is the side-lobe level at the rain volum of interest. At the rain rate of 1 mm/hr and for $L_{\mathrm{SL}} = 1$, this ratio amounts to 63 dB an 55 dB for 14 and 24 GHz, respectively. Consequently, the pulse compression side-lobes mus be suppressed to at least 60 dB below the main-lobe. In the present radar design the puls compression ratio is 480. For such ratio the conventional side-lobe reduction techniques are fa from satisfactory. As an illustration, a simulated LFM compressed signal, after weighting i the frequency domain by a 'cosine-square' function, is shown in Fig. 3.

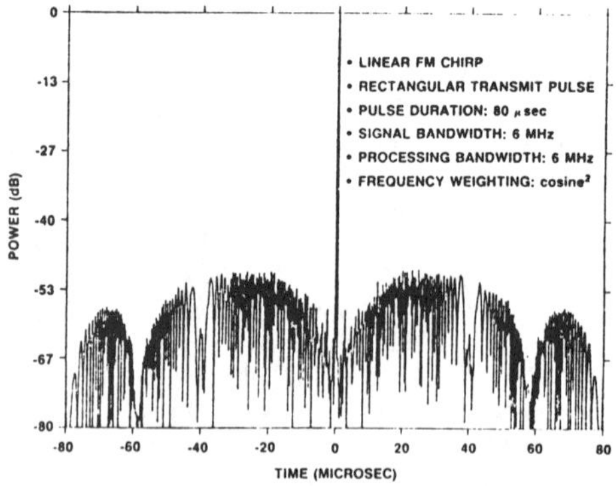

Figure 3. Linear FM chirp pulse for compression ratio=480 with cosine-square weighting in the frequency domain.

So far, we have studied two potential techniques to reduce the side-lobe levels. In the first technique the radar pulses are modulated in both amplitude and frequency prior to transmission. At the receiver end, the spectra of the match-filtered outputs are weighted by a 'cosine-square' function. An example of the compressed pulse signal using this technique is shown in Fig. 4. In the second technique the pulses are phase pre-distorted (Cook and Bernfeld, 1967) prior to transmission. And the detectable signals are processed in two different manners : (a) the full spectra of the match-filtered outputs are used without weighted; (b) the inner half of the match-filtered output spectra are weighted by a 'cosine-square' function. The processed signals from these two schemes are plotted in Fig. 5. We can see that by combining these results, most of the compression side-lobes can be reduced to the -60 dB level. We note that with either technique, a significant loss of signal-to-noise ratio will result. At present, our studies are based on computer simulation. We suggest detailed breadboard demonstration be conducted to fully evaluate the tradeoffs of these two approaches.

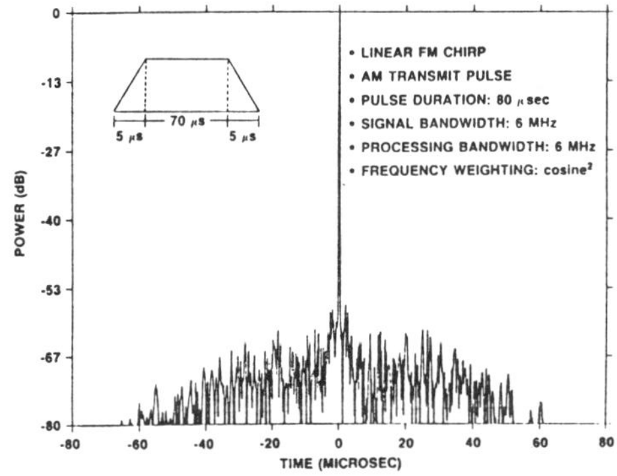

Figure 4. Linear FM chirp pulse for compression ratio=480 with cosine-square weighting in the frequency domain. The transmit pulse has been modulated in amplitude.

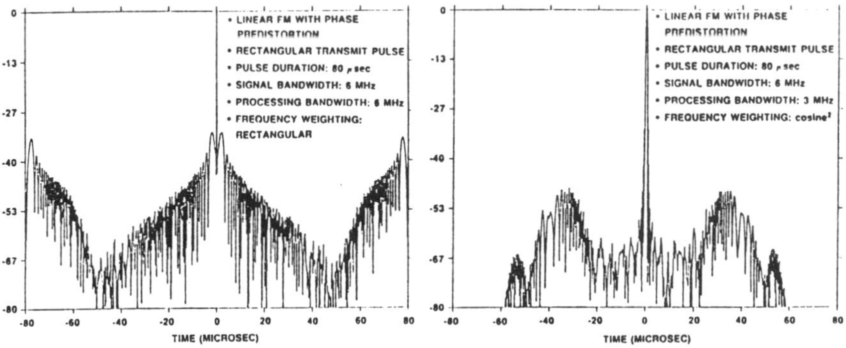

Figure 5. Linear FM and phase pre-distorted chirp pulse for compression ratio=480. (a) No weighting. (b) Cosine-square weighting over the inner half of signal spectrum.

(7) Antenna Main-lobe Clutter Interference : In addition to the antenna side-lobe clutter interference at non-nadir scanning, the unwanted surface clutters inside the main-lobe will also

interfere with the rain signals when the transmit pulse volumes encompass both the rain cell and the surface simultaneously. Im and Atlas (1987) have recently proposed a technique to separate the rain echoes from the clutter by assuming the geometric symmetry of the surface echoes at non-normal incidence. Other possible techniques include polarization and doppler filterings.

(8) Signal-To-Noise Ratio : Based on our design concept, a set of baseline system parameters are generated for the TRMM rain radar as listed in Table 1. Using these parameters, we have calculated the signal-to-noise ratio for the rain echoes (on a per pulse basis) over a 5-km rain cell. As an illustration the numerical results are plotted in Fig. 6 for the case of nadir incidence.

TABLE 1. SYSTEM PARAMETERS FOR THE TRMM RAIN MAPPING RADAR.

FREQUENCY (GHz)	13.8	24.1	
POLARIZATION	LINEAR (H/V)	LINEAR (H/V)	
ANTENNA			
EFFECTIVE APERTURE (m)	2 x 2	1.17 x 1.17	(2 x 2)
3-dB BEAMWIDTH (deg)	0.76	0.76	
GAIN (dB)	47	46.5	(53.4)
SCAN ANGLES (deg)			
PRIMARY MODE	±20° FROM NADIR	±20° FOR NADIR	
WIDE SWATH MODE	±45° FROM NADIR	±45° FROM NADOR	
TRANSMIT PEAK POWER (W)	125	100	
PRF (KHz)	3.5-3.8	3.5-3.8	
PULSE WIDTH (μsec)			
EXPANDED	80	80	
COMPRESSED	0.275	0.275	
BANDWIDTH (MHz)	6	6	
TRANSMIT LOSS (dB)	-1	-1.5	
RECEIVE LOSS (dB)	-1.6	-2.5	
PROCESSOR LOSS (dB)	-3	-3	
RECEIVER NOISE FIGURE (dB)	2.5	3.5	
SYSTEM DYNAMIC RANGE (dB)	50	40	

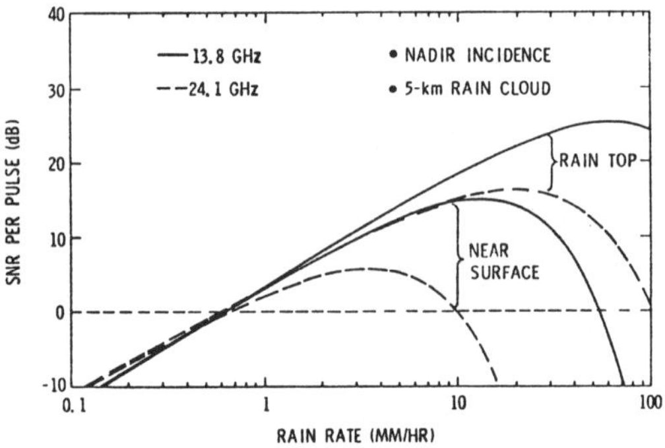

Figure 6. Expected SNR performance of the TRMM rain mapping radar. A 5-km rain cloud height and nadir incidence were assumed.

(9) Adaptive Scanning : Since the probability of raining at any given area in the Tropics is relatively small, a potential problem with the scanning scheme mentioned above is that most the time the radar will be dwelling on areas with no rain. This difficulty can be overcome by implementing an adaptive scan mode. There are a variety of methods to achieve the adaptive scan mode. The particular scheme that we are currently studying works as follows : For ~ 10% of the available scan times the radar antennas scan over a 600-km swath to search for the locations of the rain cells; and the remaining 90% of the scan times will be distributed over the identified rain cells such that a combined 200-km swath is achieved. Notice that in this scan

scheme the cross-track radar coverage will no longer be contiguous. Detailed design issues such as rain cell detection algorithm, sampling bias, etc., will be investigated in the future.

(10) Doppler Velocity Measurement : Due to the motion of the precipitation relative to the ground clutter, it is potentially possible to extract from the Doppler spectrum the mean motion of the precipitation. Unfortunately, due to the finite sizes of the antenna beams, there exists a spread of relative motion between the precipitation and the spacecraft. This Doppler spread is well-known in down-looking coherent radars, such as a synthetic aperture radar (see Li et al., 1985). Using the antenna configuration described above, we estimated that the 3-dB Doppler spread width is ~7.4 KHz. On the other hand, at 14 GHz the Doppler induced by a mean radial motion (rainfall velocity) of 10 m/s is ~930 Hz. In order to extract this shift from the Doppler spread, a special nadir Doppler measurement mode is needed. In this mode, the antenna dwells in the nadir direction for a duration that can generate a sufficient number of independent measurements in the Doppler spectrum to determine the mean shift (Li et al., 1985). Furthermore, the pulse repetition frequency must be increased to accomodate the inherent Doppler spread and the additional Doppler spread due to random precipitation motions caused by up/down draft variations. Details of this mode will be reported elsewhere.

4. SUMMARY

The conceptual design and tradoff issues for a spaceborne microwave rain mapping radar are presented in this paper. We have identified a number of design issues and technical challenges that must be fully evaluated. The data provide by this radar, in conjunction with those from the other instruments of the TRMM payload, will undoubtedly provide significant improvements in our knowledge of the precipitation characteristics in the tropics.

ACKNOWLEDGMENTS

The research described in this paper was performed by the Jet Propulsion Laboratory, California Institute of Technology, under contract with the National Aeronautics and Space Administration.

REFERENCES

Arkin, P.A., 1979; The relationship between fractional coverage of high cloud and rainfall accumulations during GATE over the B-scale array, Mon. Wea. Rev., 107, p.1382.

Cook, C.E. and Bernfeld, M., 1967; Radar Signals - an introduction to theory and application, Academic Press.

Im, K.E. and Atlas, D., 1987; The estimation of precipitation intensity in the presence of surface backscatter and the converse using a spaceborne radar, Proceedings of International Symposium on Tropical Precipitation Measurements, Tokyo, Japan, 1987.

Joss, J., Schram, K., Thams, J.C., and Waldvogel, A., 1970; On the quantitative determination of precipitation by radar, Wissenschaftliche Mitteilung Nr. 63. Zurich: Eidgenössische kommission zum studium der Hagelbildung und der Hagelabwehr.

Li, F.K., Held, D.N., Curlander, J.C., and Wu., C., 1985; Doppler parameter estimation for spaceborne Synthetic-aperture radars, IEEE Trans. Geos. and Remote Sensing, GE-23, p.47.

Marshall, J.S., Langille, R.C., and Palmer, W.M.K., 1947; Measurements of rainfall by radar, J. Meteor., 4, p.186.

Probert-Jones, J.R., 1962; The radar equation in meteorology, Quart. J. Royal Meteor. Soc., 88, p.485.

Rao, M.S.V., Abott, W.V. III, and Theon, J.S., 1976; Satellite derived global oceanic rainfall atlas (1973 and 1974), NASA SP-410.

Thiele, O.W., Ed., 1987; On requirements for a satellite mission to measure tropical rainfall, NASA Ref. Pub. 1183.

LAUNCH VEHICLE FOR THE TRMM SATELLITE

H. Takatsuka and Y. Shibato
National Space Development Agency of Japan
Tokyo 105, Japan

ABSTRACT

This paper describes the present status of the H-II launch vehicle of the National Space Development Agency of Japan (NASDA), and the results of the preliminary study on launch capability for the Tropical Rainfall Measuring Mission (TRMM) satellite of NASA's Goddard Space Flight Center (GSFC).

1. INTRODUCTION

Based on the experience gained through the H-I project, the H-II is being developed as a main launch vehicle of NASDA in the 1990's. Vehicle design activities and development tests are proceeding smoothly toward a first test flight in early 1992. The configuration was selected so as to be a high-performance, cost-effective and reliable vehicle to be offered for worldwide usage. The payload capability is 4.0 tons to geostationary transfer orbit (GTO). The two-stage configuration and second-stage restart capability make it a versatile launcher capable of fulfilling a wide range of missions including multi-satellite launches. These characteristics of the H-II essentially provide enough capability of launching the TRMM effectively. In July 1987, the first TRMM/H-II technical interface meeting was held with the GSFC, the integrator of the TRMM, and the interface adjustment and feasibility study have started under the following conditions;

- February 1994 launch schedule
- Satellite mass of 3 to 4 tons
- Dual launch with a GEO satellite

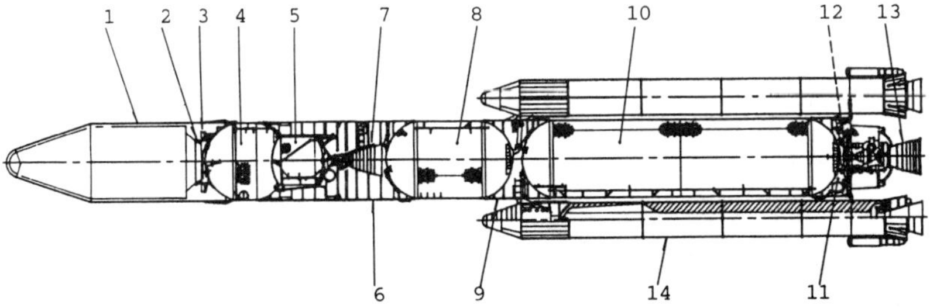

1. PAYLOAD FAIRING
2. PAYLOAD ATTACH FITTING
3. GUIDANCE SECTION
4. SECOND-STAGE LH_2 TANK
5. SECOND-STAGE LOX TANK
6. INTER-STAGE
7. SECOND-STAGE ENGINE (LE–5A)
8. FIRST–STAGE LOX TANK
9. CENTER BODY SECTION
10. FIRST-STAGE LH_2 TANK
11. FIRST-STAGE ENGINE SECTION
12. AUXILIARY ENGINE
13. FIRST-STAGE MAIN ENGINE (LE–7)
14. SOLID ROCKET BOOSTER (SRB)

FIGURE 1. H-II Launch Vehicle

ISBN 0–937194–14–4

TABLE 1. MAIN CHARACTERISTICS OF THE H-II

Overall length	:	49 m
Diameter	:	4 m ϕ
Total weight	:	260 t
Standard fairing * Diameter	:	4.1 m ϕ (Payload envelope 3.7 m ϕ x 10 mL)
Standard fairing * Length	:	12 m
Guidance system	:	Strap-down inertial guidance system

	* 1st stage	* SRBs(2 Units)	* 2nd stage
Propellant	LOX/LH_2	Composite	LOX/LH_2
Propellant weight	85 t	118 t	14 t
Thrust	93 t(Sea level)	320 t(Sea level)	12 t(Vac.)
Total burn time	315.8 s	95 s	527 s
Specific impulse	449 s(Vac.)	271 s(Vac.)	451.9 s(Vac.)
Total weight	97 t	140.5 t	16.8 t

2. H-II SYSTEM AND DEVELOPMENT STATUS

2.1 SYSTEM DESCRIPTION

The H-II launch vehicle consists of cryogenic first and second stages of 4 meters in diameter and a pair of strapped-on solid rocket boosters (SRBs). The outline and main characteristics are shown in Fig. 1 and Table 1.

The first stage is a newly developed stage and powered by a staged-combustion cycle engine, LE-7, in which specific impulse of 449 seconds is achieved by adopting high combustion pressure of 147 kg/cm^2a. The SRB is equipped with a movable nozzle for thrust vector control. The second stage is a scaled-up version of the current H-I second stage, and the LE-5 engine of the H-I is also replaced by an improved engine, the LE-5A. The LE-5A utilizes a newly-devised hydrogen bleed cycle, where the turbopumps are driven by gaseous hydrogen which cools the nozzle. A strap-down inertial guidance system with ring laser gyros is used for the guidance and control. A lateral acceleration measurement unit is also used for load relief attitude control during first-stage flight of the H-II. The payload fairing is an aluminum honeycomb-core/skin structure. The separation mechanism consists of springs, notch bolts, and mild detonating fuses.

2.2 DEVELOPMENT STATUS

Since the H-II program started in 1985, vehicle design activities and development tests are proceeding smoothly toward the first test flight of early 1992. A vehicle baseline was established at preliminary design review (PDR) held in May 1987. At present, in the Phase D study the engineering model (EM) trial and prototype model (PM) trial manufacturing are in progress.

The LE-7 engine is one of the most critical components of the H-II. PM component tests are near to completion overcoming various difficulties such as turbopump rotor whirls and turbine blade cracks. Now under way are engine system firing tests to verify the compatibility of components and to establish a reliable start/stop

sequence. Guidance and control system tests were carried out in early 1987 using breadboard models of the inertial measurement unit and guidance computer to verify hardware/software interface.
Separation tests of a EM payload fairing were performed successfully in April 1987. To verify the acoustic environment for the satellites, an acoustic vibration test of the PM fairing and acoustic noise tests using 1/15 scale model engines were carried out in May through August 1987. The results have now been carefully examined.
Full size domes of the first stage tank were assembled and pressure-tested in September 1987. A EM SRB motor will soon be loaded with propellant to be subjected to the first ground firing test in March 1988. The construction of the test and launch facilities is progressing according to schedule.

2.3 PERFORMANCE

Using a standard payload fairing, single payload capabilities of the H-II are 4.0 tons into GTO with inclination of 28.5°, 10 tons or more into low earth orbit, 4.6 tons into sun-synchronous orbit of 800 km in altitude. However, to meet a variety of mission needs such as a large volume satellite or multi-satellite launches, NASDA will prepare various types of payload fairings. It is possible to increase up to 5 m in diameter and 16 m long, but it involves some payload loss. The two stage configuration of the H-II and second-stage restart capability make it a versatile vehicle capable of fulfilling wide range of missions, such as launching LEO and GEO missions simultaneously.

The flight environment conditions, shown in Table 2, are almost on the same level as the Ariane or the Space Shuttle. As most of them are estimated or calculated values with enough safety margins, these conditions are expected to be favorably revised due to results of test flights.

Over the past 13 years, NASDA has had a success rate of 100 % for 17 satellite launches. The H-II is also expected to keep

TABLE 2. ENVIRONMENTAL CONDITIONS

Sinusoidal vibration	* Longitudinal axis : 1.0 Go-p (5-100 Hz) * Lateral axis : 0.8 Go-p (5- 45 Hz) : 0.5 Go-p (45-100 Hz)
Acoustic	* Overall sound pressure level : 141 dB
Quasi-steady acceleration	* Longitudinal axis : at lift-off 3.2 G (1.7+1.5) : at MECO 5.0 G (4.0+1.0)
Shock	* S/C separation pyro shock : 2000 Go-p
Thermal environment	* Air conditioning temperature (prelaunch) : 15-25° C (adjustable) * Radiation from fairing : less than 500 W/m^2
Cleanness	* Class 100000

high reliability of 0.96.

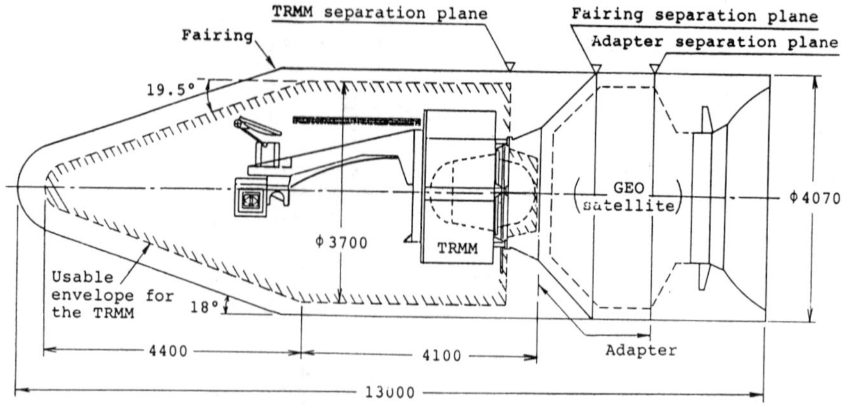

FIGURE 2. H-II dual launch system

3. LAUNCH CAPABILITY FOR THE TRMM

As a typical mission of the H-II is GEO satellite, we set a mission model in which the TRMM will be launched with a 800-kg class GEO satellite (1.5 tons on GTO) as a partner. In this case, the TRMM will be carried as an upper satellite in the payload fairing for dual launch as shown in Fig. 2. Tentative payload envelope and satellite separation plane are also depicted in the figure.

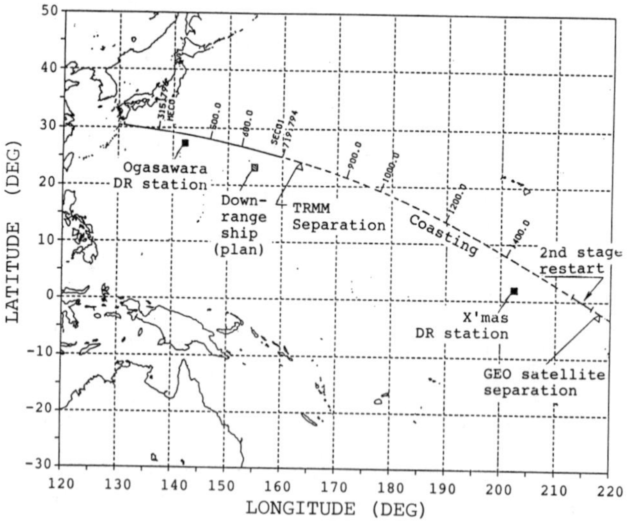

FIGURE 3. Flight trajectory (TRMM + GEO mission)

Currently TRMM's mission orbit is a 350-km altitude circular orbit and 28.5° in inclination. This orbit is directly achieved with initial launch azimuth of about 100° and subsequent dog-leg maneuver to adjust the orbit inclination. Another way is to adopt an interim low-altitude orbit for TRMM to allow for increased satellite mass. In this case, the TRMM will have to utilize its own propulsion module to boost up to the mission orbit from the initial orbit. On the other hand, a GEO satellite can be placed into GTO applying the second-stage restart capability of the H-II following the TRMM separation. The above flight sequence of events is shown in Fig. 3 along with the flight trajectory. Resultant payload capability of the H-II is shown in Fig. 4, and the TRMM weight on the mission orbit can be chosen from 3.7 to 4.0 tons depending on the flight modes.

Typical 3 σ orbit accuracies in circular orbit are <10 km altitude and $<0.05°$ in-orbit inclination.

4. CONCLUSIONS

Development of the H-II is proceeding smoothly toward the first test flight in early 1992. As a result of the preliminary study, the TRMM satellite can be launched simultaneously with another GEO satellite.

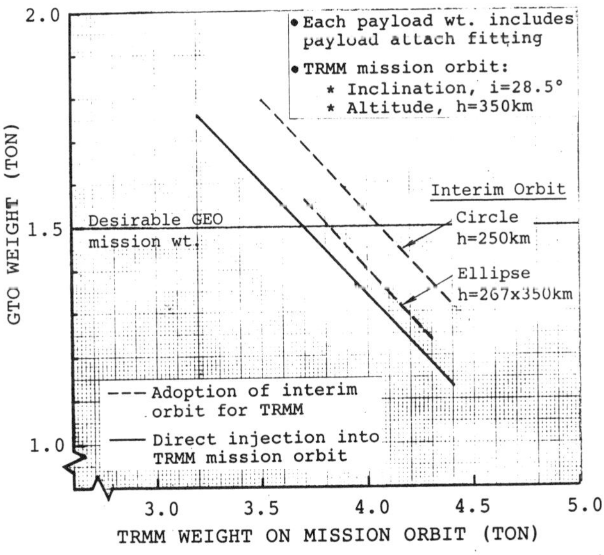

FIGURE 4. H-II dual launch capability (TRMM + GEO mission) Adoption of an interim low orbit for the TRMM can increase the TRMM mission weight.

DATA RELAY & TRACKING EQUIPMENT ON TRMM

N. Kawajiri, T. Mito, A. Awasawa, Y. Tsujino, and T. Doura
Tsukuba Space Center
National Space Development Agency of Japan
Ibaraki, Japan

ABSTRACT

This paper describes the telecommunication payload on TRMM for Japanese Data Relay and Tracking Satellite(DRTS) such as Engineering Test Satellite VI(ETS-VI) and Experimental Data Relay and Tracking Satellite(EDRTS). The purpose of this experiment is described, and significant design features of the electronic equipment as well as the performance requirements are presented.

1. INTRODUCTION

For the deployment of DRTS system, NASDA plans three phases: technology development phase, pre-operational phase, and operational phase. In the technology development phase, NASDA intends to develop key technologies of DRTS onboard equipment and system technology of DRTSS using S/Ka-bands intersatellite communication equipment on ETS-VI in 1992 and user spacecraft such as Advanced Earth Observing Satellite(ADEOS). In the pre-operational phase, NASDA will have to verify operation of DRTSS and plans to launch EDRTS in 1994 to provide experimental services to mainly ADEOS which is the earth observation satellite, and Japanese Experiment Module(JEM) which is a part of Space Station.

TRMM spacecraft is planned to be launched in summer 1993, so the life period of TRMM is overlapped with technology development phase and pre-operational phase.

NASDA plans to construct a receiving station of ADEOS imaging data for ETS-VI and EDRTS. So the imaging data such as rain radar and/or AVHRR and/or TMR can be transmitted to the Earth Observation Center(EOC) via ETS-VI and EDRTS.

The launching schedule of both Japanese DRTS and user spacecraft is shown in Table 1.

2. PURPOSES

Purposes of intersatellite communications equipment on TRMM are considered as follows:

(1) One candidate of user spacecraft for data relay and tracking experiment via ETS-VI
(2) Experimental operation via EDRTS using ground stations such as EOC and DRTS ground stations
(3) Interoperability experiment for space network cooperation

ISBN 0-937194-14-4

Furthermore, EDRTS has more than two large antennas and its attitude control is very difficult due to turbulence caused by the antenna movement, especially by simultaneous movement of two antennas. NASDA can do this experiment using TRMM and ADEOS.

TABLE 1 LAUNCHING SCHEDULE OF DRTSS

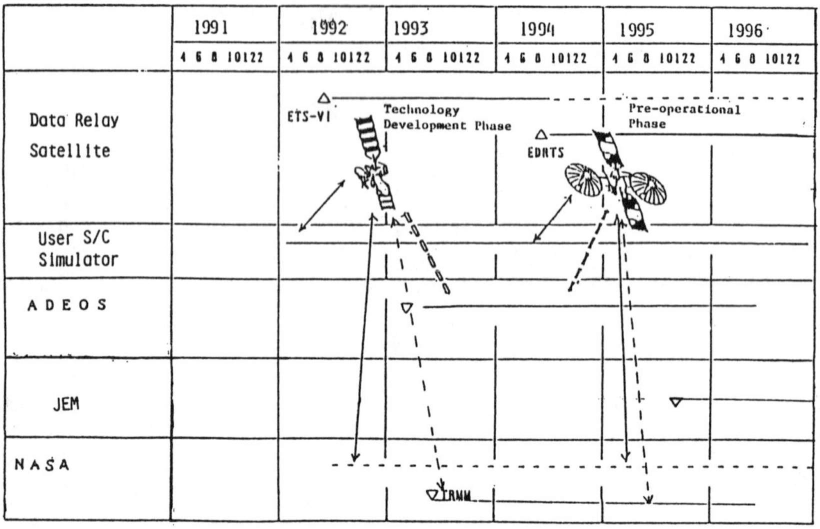

3. GENERAL DESCRIPTION OF EQUIPMENT

Data relay and tracking equipment on TRMM consists of an RF compartment, a gimbal drive assembly and a signal processing unit. There are two classes of equipment, depending on whether NASA communication equipment for TDRS on TRMM is used commonly or not.

The independent type of equipment, that is, equipment for the Japanese DRTS used independently from that for TDRS, is shown in fig. 2. Common use equipment, that is, an antenna, gimbal drive assembly and signal processing unit used commonly is shown in fig. 2. In each case, an S-band return beacon is necessary to point EDRTS toward TRMM.

Data rates of mission equipment are assumed to be those shown in Table 2. ETS-VI has limited capability of data transmission, but EDRTS has almost comparable capability with TDRS. So, NASDA plans to design the communication system such that all mission data can be transmitted only via EDRTS. Ka-band link analysis between ETS-VI and TRMM is shown in Table 3, and that between EDRT and TRMM is shown Table 4.

RF compartment contains the Ka-band antenna, diplexer, up-converter, power amplification components, and down-converter. Gimbal drive assembly contains Antenna Pointing Mechanism(APM) and Antenna Pointing Electronics(APE). Signal processing unit contains modulator and demodulator.

TABLE 2 CANDIDATE OF MISSION DATA RATE

Equipment	Data Rate(1)	Data Rate (2)	Data Rate(3)
AVHRR	120 kbps	120 kbps	------
Rain Radar	170 kbps	170 kbps	170 kbps
TMR etc.	80 kbps	80 kbps	------
Total	370 kbps	5 Mbps	170 kbps

4. CONCLUSION

This paper has described the design features and performance requirements of data relay and tracking equipment on TRMM associated with Japanese DRTS system. However the design of EDRTSS and ADEOS is under phase A study, so these plans are not fixed yet.

Further studies are needed before we can provide a detailed specification of data relay and tracking equipment needed for TRMM. Construction of some of the breadboard units is planned in the near future.

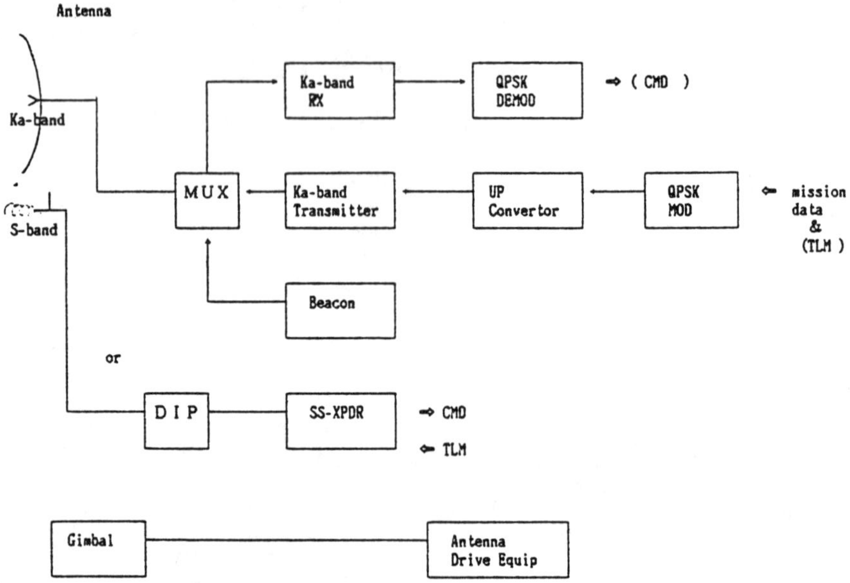

FIGURE 1 BLOCK DIAGRAM OF DATA RELAY & TRACKING EQUIPMENT ON TRMM (INDEPENDENT TYPE)

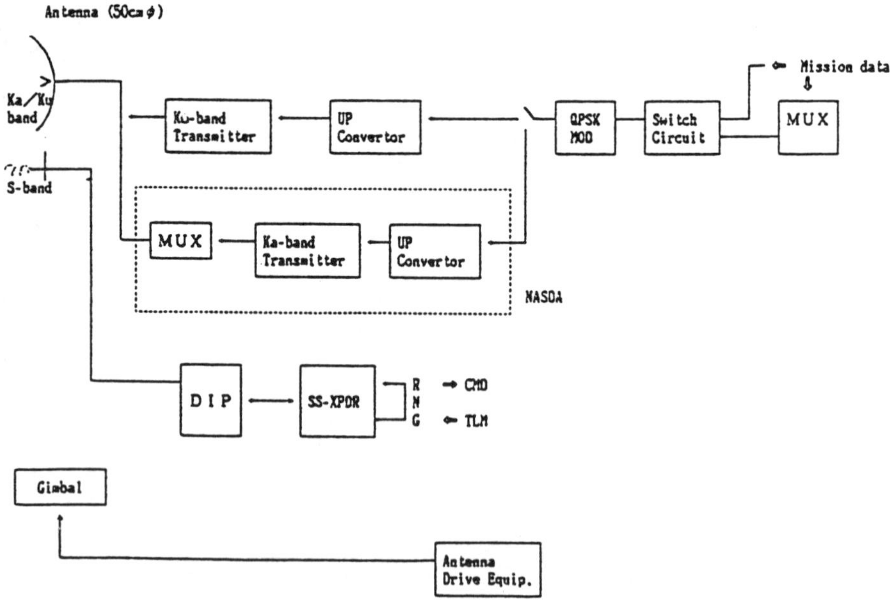

FIGURE 2 BLOCK DIAGRAM OF DATA RELAY & TRACKING EQUIPMENT ON TRMM (COMMON USE TYPE)

TABLE 3 LINK ANALYSIS BETWEEN TRMM AND ETS-VI

	RETURN LINK (32.555 GHz)	FEEDER LINK (19.900 GHz)
Tx Prameters		
TX POWER	10.0 dBW(10.0W)	6.0 dBW(4.0W)
TX CIRCUIT LOSS	-2.0 dB	-3.5 dB
TX ANTENNA GAIN	41.4 dBi(0.5m 47%)	51.2 dBi(3.5m 25%)
E.I.R.P.	49.4 dBW	53.8 dBW
POWER SHARING LOSS		-6.7 dB (30.0MHz)
Path Parameters		
POINTING LOSS	-1.0 dB	-1.0 dB
POLARIZATION LOSS	-0.5 dB	-0.5 dB
FREE SAPCE LOSS	-215.6 dB (44000 Km)	-209.8 dB (37286 Km)
ATMOS.RAIN ATTENUATION	0.0 dB (0.0%)	-3.5 dB (1%.47°EL)
Rx Prameters		
RX ANTENNA GAIN	41.3 dBi(0.8m 18%)	49.4 dBi(2.0m 50%)
RECEIVED POWER	-126.4 dBW	-118.4 dBW
ANTENNA NOISE TEMP.	300.0 K	70.0 K
RX CIRCUIT LOSS	-3.4 dB	-0.5 dB
RX NOISE FIGURE	5.0 dB	2.3 dB
SYSTEM NOISE TEMP.	33.0 dBK	25.2 dBK
RX NOISE SP. DENSITY	-195.6 dBW/Hz	-203.4 dBW/Hz
C/No	69.1 dBHz	85.0 dBHz
TRANSPONDER LOSS	-0.1 dB	
UQPSK LOSS	0.0 dB	
DEMODURATOR LOSS	-1.5 dB	
PN LOSS	0.0 dB	
MARGIN	3.0 dB	
REQUIRED Eb/No	9.9 dB	
	54.6 dBHz (288.6 Kbps)	

TABLE 4 LINK ANALYSIS BETWEEN TRMM AND EDRTS

	RETURN LINK (32.555 GHz)	FEEDER LINK (19.900 GHz)
Tx Prameters		
TX POWER	2.0 dBW(1.6W)	9.0 dBW(8.0W)
TX CIRCUIT LOSS	-2.0 dB	-3.5 dB
TX ANTENNA GAIN	41.4 dBi(0.5m 47%)	40.4 dBi(2.0m 40%)
E.I.R.P.	41.4 dBW	53.9 dBW
POWER SHARING LOSS		-0.8 dB (30.0MHz)
Path Parameters		
POINTING LOSS	-1.0 dB	-1.0 dB
POLARIZATION LOSS	-0.5 dB	-0.5 dB
FREE SAPCE LOSS	-215.6 dB (44000 Km)	-209.8 dB (37286 Km)
ATMOS.RAIN ATTENUATION	0.0 dB (0.0%)	-10.9 dB (1%.20°EL)
Rx Prameters		
RX ANTENNA GAIN	61.6 dBi(5.0m 50%)	68.5 dBi(18.0m 50%)
RECEIVED POWER	-114.0 dBW	-100.7 dBW
ANTENNA NOISE TEMP.	300.0 K	70.0 K
RX CIRCUIT LOSS	-3.4 dB	-0.5 dB
RX NOISE FIGURE	5.0 dB	2.3 dB
SYSTEM NOISE TEMP.	33.0 dBK	25.2 dBK
RX NOISE SP. DENSITY	-195.6 dBW/Hz	-203.4 dBW/Hz
C/No	81.5 dBHz	102.7 dBHz
TRANSPONDER LOSS	-0.0 dB	
UQPSK LOSS	0.0 dB	
DEMODURATOR LOSS	-1.5 dB	
PN LOSS	0.0 dB	
MARGIN	3.0 dB	
REQUIRED Eb/No	9.9 dB	
	67.1 dBHz (5.1 Mbps)	

OPERATIONAL PRECIPITATION OBSERVATION SYSTEM IN JAPAN METEOROLOGICAL AGENCY

K. Takase, Y. Takemura, K. Aonashi, N. Kitabatake, Y. Makihara, and Y. Nyomura
Japan Meteorological Agency
Tokyo 100, Japan

ABSTRACT

JMA has been promoting National Weather Watch (NWW) project to modernize JMA's weather information processing system for improved meteorological services by integrated utilization of various observational data.

Objective precipitation analysis of radar and surface raingauge data has been executed on an operational basis in JMA since Jun. 1983.

Accuracy of this analysis and radar observation is presented in this paper.

1. OBSERVATION SYSTEM OF JMA

The JMA observational system consists of conventional synoptic scale observatories for surface weather and aerological data, radar, satellite, marine observation ship and marine data buoys etc. Only high density observational systems are presented in this paper.

1.1 AMeDAS

Meso-scale surface weather observation system called AMeDAS (Automated Meteorological Data Acquisition System) was developed for real-time around the clock monitoring of weather throughout Japan. This system consists of an automated operation center and automatic observation stations which are deployed throughout Japan. Observed data elements and their average spacing are shown in Tab.1. A time interval of normal data acquisition is one hour and users use observational data within 8 minutes after observation every hour.

Table 1 OBSERVED ELEMENTS AND AVERAGE SPACING OF AMeDAS

element	specifications	station no	spacing
rainfall	hourly precipitation	1316	17km
wind	10 min.-average wind	841	21km
sun shine	hourly sun shine duration	841	21km
temperature	temperature	841	21km
snow cover	depth of snow cover	222	50km

1.2 RADAR SYSTEM

The JMA radar observation network consists of 20 conventional

ISBN 0-937194-14-4

radar observatories whose locations are shown in Fig.1. There are 7 more airport weather radars.

Under the NWW project conventional radar observation systems are to be equipped with observation control units which can control radar operation, digitize observational data and transmit these data to the JMA head office or related users.

Ground clutter superimposed on precipitation echoes is suppressed by Moving Target Indicator (MTI) filter for quantitative estimation of rainfall amount from radar observations. 10 radars have already been automated as of March 1987 as shown in Fig. 1.

For further improvement of precipitation measurement, especially over the sea area, data obtained by ship-borne meteorological radar will be transmitted through GMS telecommunication system to JMA data processing center after March 1988.

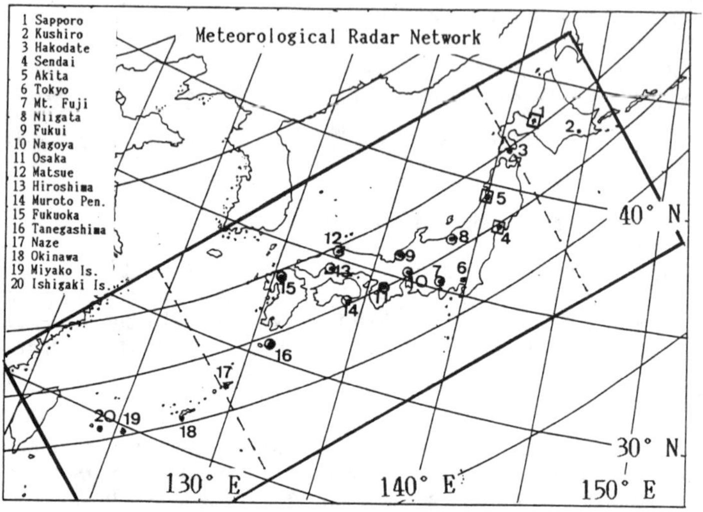

Figure 1 METEOROLOGICAL RADAR NETWORK IN JAPAN
○ automated radars as of Mar. 1987
□ radars to be automated by Mar. 1988
Diagonal frame is the analysis area of radar data. (as of 1987, this area is limited between broken lines)

Digitized radar provides 3 types of observational data : echo intensity, echo top height, and hourly accumulated radar precipitation amount. The detection range is 500 km× 500 km, and 5.0×5.0 km resolution data are transmitted to JMA data processing center.

Table 2. PROCESSED DATA AND THEIR GRADES IN AUTOMATED RADAR

GRADE	1	2	3	4	5	6	7	8	9	10	11	12	13	14	15	.....	40	41	42	....	62	63
intensity(mm/h)	< 1	< 2	< 4	< 8	< 12	<16	<24	<32	<40	<48	<56	<64	<80	<96	96≤							
top height (km)	< 2	< 4	< 6	< 8	< 10	<12	<14	14 ≤														
precip. (mm)	<0.5	<1.0	<1.5	<2.0	<2.5	< 3	< 4	< 5	< 6	< 7	< 8	< 9	< 10<	12<	14<	... <	64<	68<	72<	...	<152	152≤

These data are graded as in Tab.2. This table has been revised in Mar. 1988 to transmit more detailed information in order to allow better data analysis.

Observed data of individual radars are transmitted to JMA data processing center hourly during rainfall (precipitation) events and 3-hourly in fine weather.

1.3 GEOSTATIONARY METEOROLOGICAL SATELLITE (GMS)

Since the computer system in Meteorological Satellite Center (MSC) was upgraded in March, 1987, MSC can regularly disseminate imagery data hourly through satellite telecommunication system. Imagery in facsimile is represented in polar-stereographic projection with effects of landmark fluctuation eliminated. Stretched VISSR (Visible and Infrared Spin Scan Radiometer) data will be disseminated after next year. Cloud motion wind vectors are produced 6-hourly and satellite cloud information charts are also disseminated 3-hourly.

In addition to the real-time delivery of this imagery, MSC provides climatological information such as 5-day and monthly mean cloud amount, climatological equivalent Black–body Temperature, cloud retrieval and VISSR grid data, clear sky radiance data, etc.

It also provides selected information for special international projects such as International Satellite Cloud Climatology Project, Global Precipitation Climatology Project, etc.

Efforts have been made by MSC to provide precipitation intensity index from satellite infrared (IR) and visible (VIS) information for real-time actual analysis and very short range forecast of precipitation. (Detailed presentation will be given by other authors later in this symposium.).

2. PRECIPITATION ANALYSIS

2.1 OBJECTIVE PRECIPITATION ANALYSIS SYSTEM

Detailed precipitation information, covering almost all of Japan (see the frame in Fig.1) with 5 × 5 km resolution 1-hour precipitation amount, is processed on an operational basis by objective analysis.

Data from automated radars and raingauges of AMeDAS are collected at the data processing center and are objectively analyzed on a real–time basis by a successive correction method. (as for analysis scheme, Takemura et al.,1984)

One of the main purposes of this information is to detect severe localized rainfall events for disaster prevention activity. These products have been disseminated since June 1983 in the form of high-speed facsimile (CDF:coded digital facsimile) shown in Fig. 2.

For advanced usage, precipitation data at every 5 × 5 km resolution are available for data processing by computer.

2.2 VERIFICATION OF PRECIPITATION ANALYSIS

The performance of the analysis was investigated using data

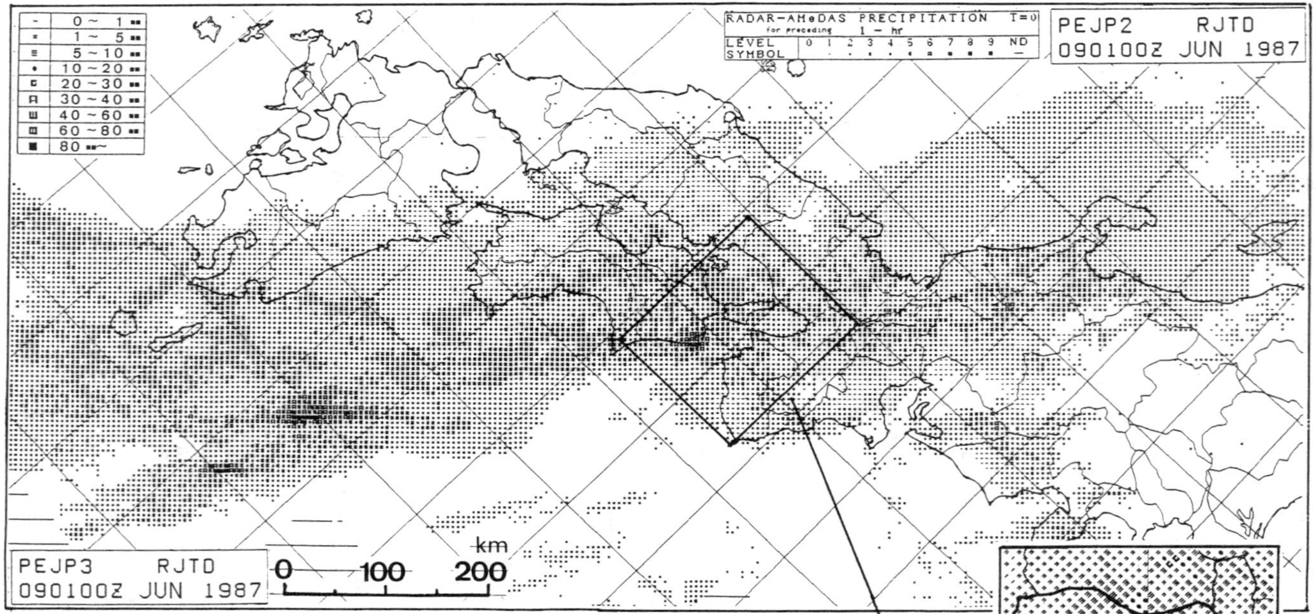

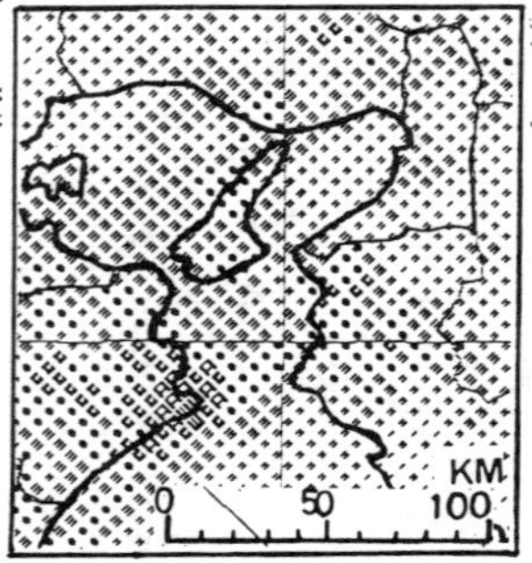

Fig. 2 EXAMPLE OF RADAR AMeDAS COMPOSITE PRECIPITATION CHART
At 01 GMT,JUN.09,1987, disseminated by facsimile.

Analyzed precipitation value for preceding hour is depicted by 10 graded expression symbols in the upper left. Each symbol represents 5 × 5 km area. The chart can reveal localized heavy rainfall as well as the wide range precipitation distribution.

Data from 10 automated radars are included in this chart.

obtained by raingauge network with an average spacing of 5 km, which is installed by Tokyo Metropolitan government and is independent of JMA observation network (This network is called TM-net hereafter).

The data used in the verification are coincident raingauge and radar data. A total of 830 data points were obtained for 12 cases of localized convective rainfall.

In the sentences below, relative error E_r means that,

$$E_r = | R_{true} - R_{anal} | / R_{true}$$

Here observation by TM net is assumed to be R_{true}.

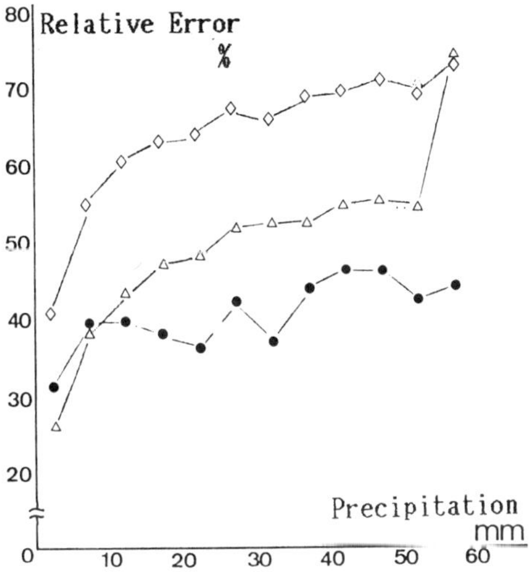

Fig. 3 Relative Error of Each System in detecting precipitation.
Observed (or analyzed) amount of precipitation is compared with the rainfall observation data of TM-net.

◇ radar observation
△ raingauge observation as provided by AMeDAS
● Radar-AMeDAS Analysis

As shown in Fig. 3 for localized rainfall events, relative error of JMA's raingauge (AMeDAS) and radar-estimated precipitation amount were 74% and 72% respectively in case of rainfall events of 60mm/hr or more. This result seems to reflect the error of absolute value in radar observation and spatial sparseness in AMeDAS observation.

As a result of bivariate objective analysis of radar and AMeDAS data, relative error(Er) in detecting precipitation is considerably reduced to 43%, about half as much as that of the individual observation. The difference between these two data(R_{true} and R_{anal}) is confined within 10mm/hr in 73% of the cases.

Furthermore, as shown in Fig. 4, localized heavy rainfall areas which could not be detected either by JMA's raingauge network or by radar network could be identified by this analysis.

According to other investigations for the case of wide-spread rainfall, the difference or relative error is much smaller than obtained in this case.

2.3 ACCURACY OF DETECTING PRECIPITATION BY AUTOMATED RADAR

In order to keep performance of objective analysis of good quality, accuracy of radar precipitation data as a field variable is an important factor. We have been investigating the relation

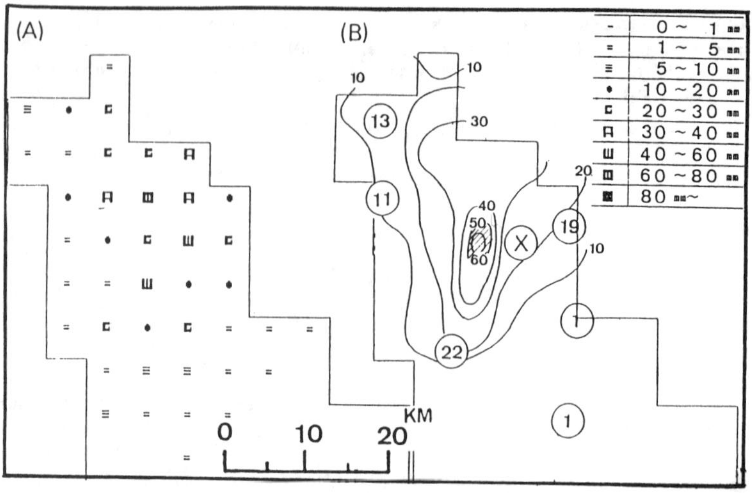

Fig. 4 PERFORMANCE OF RADAR-AMeDAS COMPOSITE PRECIPITATION CHART (Nyomura, 1985)
(A) Radar-AMeDAS chart
(B) Manual analysis of precipitation with data provided by Tokyo Metropolitan Government raingauge network(TM-net). Contour is drawn on the data by TM-net, and data with a circle is provided by AMeDAS.

between radar precipitation data and rain gauge data. The data set of 1-hour radar precipitation (see 1.2) and raingauge data by AMeDAS (see 1.1), collected in real-time basis, are used to verify the accuracy of precipitation detection of automated radar.

As has been studied by many authors, considerable deviations from standard Z-R relation are found in individual rainfall events caused by size difference of the precipitation particles.

It is also found in long-term average, accuracy of detecting precipitation has seasonal difference and areal variation (after Takemura et al., 1984), which is presumably caused by the interrelation in the radar beam height and depth of precipitation particles.

It is noteworthy that radar systems, even though they have the same specifications, have considerable difference in observational accuracy. Figure 5 shows the correspondence between radar-estimated and AMeDAS precipitation amounts.

3. CONCLUSION

Improved observed and processed data has become available in Japan for real—time detection and climatological research in precipitation. Observational data by remote-sensing method such as radar or satellite can better be utilized with appropriate ground truth data.

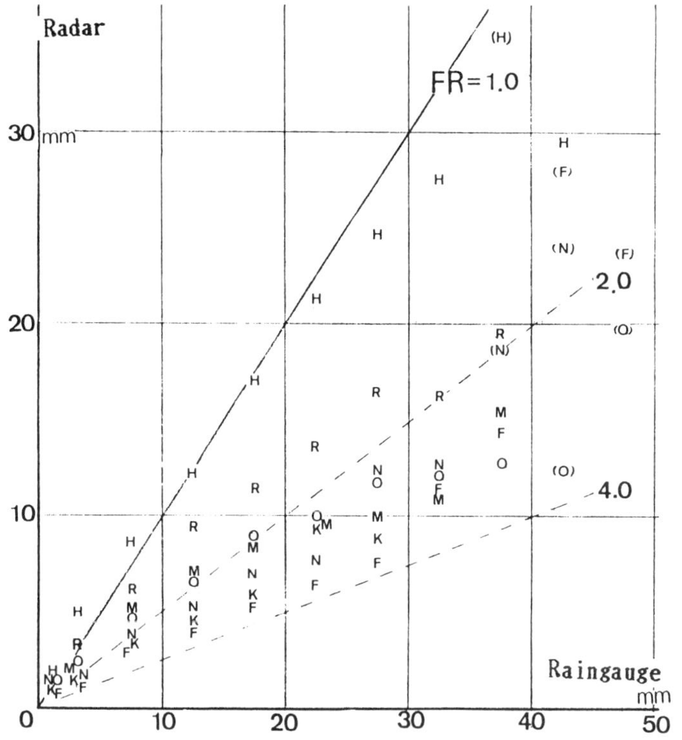

Fig. 5 CORRESPONDENCE OF PRECIPITATION BETWEEN RADAR AND SURFACE RAINGAUGE(AMeDAS).
Alphabetical symbols represent each radar observatory.
FR means radar calibration factors, that is,
FR = (rainfall amount detected by raingauge) / (rainfall amount detected by radar)

Precipitation observation by radar is verified by surface raingauge data as provided by AMeDAS. There are considerable differences of measurement accuracy from radar to radar. As can be seen, Radar(H) has a good correspondence with surface raingauge, although radar(F) shows a poor observational performance.

Parentheses attached to letters mean that the number of events in that correspondence is less than 5.

REFFERENCES

Takemura, Y., et. al. 1984 : Operational Precipitation Observation System Using Digitized Radar and Raingaues, Proc. of Now-casting II Symposium. Norrköping. Sweden. ESA SP-208, 411-416.

Nyomura, Y., 1985 : Verification ofJMA's Observation System Using Digitized Radar and Raingauges(AMeDAS) by Local Dense Rain Gauge Network of Tokyo Metropolice, Journal of Meteorological Research 37, 1-12 (in Japanese).

VALIDATING SPACE OBSERVATIONS OF RAINFALL

Otto W. Thiele
NASA Goddard Space Flight Center
Greenbelt, Maryland 20771, USA

ABSTRACT

It was recognized at the outset that observing precipitation from space would be a formidable objective and that validation would have to be a fundamental element of a tropical rainfall measuring mission. The technology for detecting precipitation through remote sensing is reasonably well in hand. However, the real challenge is in interpreting the spaceborne measurements because of the kind of sampling that results and the extreme variability of rainfall in space and time. Furthermore, to be meaningful in terms of latent heat release, the vertical structure of the precipitation is almost as critical as the horizontal distribution. Thus, for a space-based rainfall measurement project to be truly successful, well-instrumented and calibrated validation sites at locations representing significant rainfall regimes must be established within the area of satellite coverage. Such data can also make an important contribution to climate diagnostics as well as enhancing the space data set. To meet this end objective, however, a substantial pre-mission program is necessary because there remain serious limitations regarding the measurement of rainfall by conventional means, particularly over extended areas. The pre-mission program objectives must also include understanding of physical processes associated with rainfall, its regional characteristics and variability, sampling and statistical studies, climatologies of key validation sites, supporting measurements for instrumentation and algorithm development, and equally as important, establishing a transfer standard of measurement technology and procedures for the mission validation sites.

1. INTRODUCTION

The problem of validating space observations is not new. Deriving geophysical parameters through remote sensing, even when based on physical relationships, depends on many assumptions with as many levels of uncertainty. Thus, many attempts have been made and are continuing to be made to verify a wide range of parameters developed from earth-observing satellites, both research and operational. In some instances this is rather straight forward with easily obtained in-situ observations. On the other hand, the task is often difficult because adequate observational techniques are non-existent or arrangements for making them were not planned for and developed in a functional way before the requirement existed. These concerns were recognized and addressed at a major workshop on the Measurement of Precipitation from Space in 1981 (Atlas and Thiele, 1981). There is no question that a space mission to measure rainfall will depend crucially upon a strong and well-developed surface-based rainfall observational program, if credible rainfall data products are to be expected.

Rainfall, unlike other key atmospheric parameters is extremely variable in space and time. Not only is rainfall unevenly distributed in occurrence, but the rate is so highly variable even over a single convective rain system. In the tropics, of course, convective rain is primarily what we are dealing with. For these reasons, rain has always been an elusive parameter to measure. There still remain substantial uncertainties, with the best available rain gauges even for point measurements, and, when attempting to quantify areal rainfall, the uncertainties begin to magnify significantly. Thus, a fundamental task

A. DEEPAK Publishing
ISBN 0-937194-14-4

of the "ground truth" strategy is precipitation measurement research. Better ways must be found to obtain reliable area-averaged rainfall to properly verify the quality and usefulness of space measurements.

2. STRATEGY

There are two main components of the validation (often referred to as "ground truth") program envisioned for TRMM, i.e. pre-mission and mission (Thiele, 1987). The ultimate objective is to provide reliable comparative data during the mission for area and time-averaged rainfall representative of tropical or near-tropical regions. Of course, instantaneous comparisons will be attempted as well. To achieve the end objective, however, a strong pre-mission program is critical and must be organized and implemented without delay. At this point, the emphasis is on the pre-mission phase. Not only must we be prepared to acquire reliable validation data when the mission begins, but we must also learn how to do it beforehand. Support to mission development activities is a requirement as well. Elements of the validation program are outlined in Figure 1.

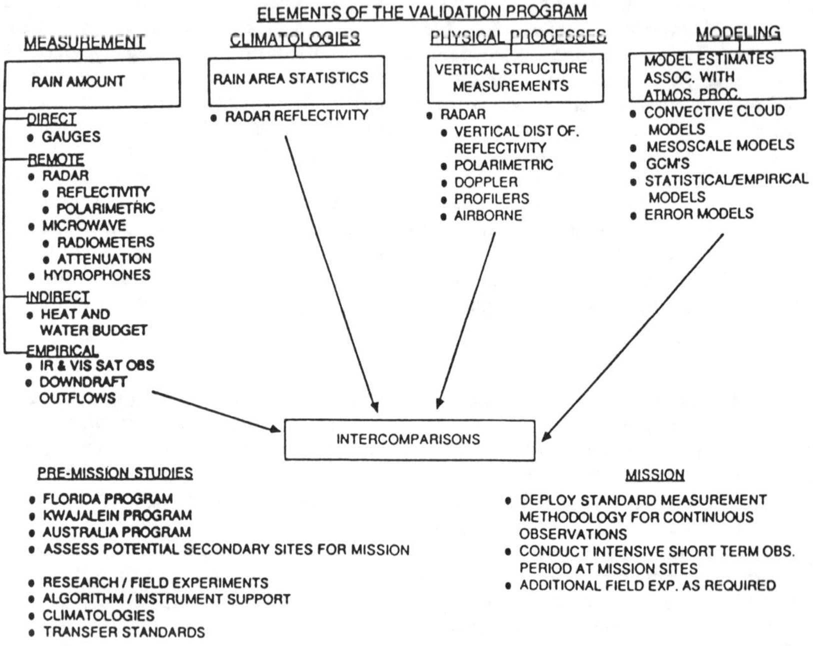

Figure 1. Hierarchy of activities necessary to validate TRMM observations

2.1 PRE-MISSION PHASE

The pre-mission phase will include a variety of activities; (1) development of primary rain measurement standards at a central first order test facility, (2) development of transfer standards, and (3) the establishment of second order test sites in representative rainfall regimes. More explicitly, the pre-mission phase must focus on rain measurement research, state-of-the-art measurements for rain processes and rain variability studies and development of measurement and procedural techniques for calibrating the mission ground truth sites. Other important functions of the test sites are to support flight instrumentation and algorithm development and to collect long-term climatological rainfall data bases for their respective regimes.

It is the concensus of the TRMM Science Steering Group (SSG) that "no adequate methods exist for the measurement of rainfall at the surface and in the atmospheric column. Concurrently, estimates of absolute standards and estimates of error of the measurement of rainfall are deficient or absent. These deficiencies and the objectives of TRMM can best be served by the establishment of a world-class test facility for the calibration of both new ground-based, aircraft and TRMM instruments and the algorithms used with them to give rainfall estimates."

The SSG also believes that "no single method for the measurement of rainfall will provide an adequate estimation of rainfall on the range of time and space scales that are required and that a hybrid system must evolve. The development of a hybrid system will begin with the best recognized techniques, i.e. rain gauges for point measurements and radar reflectivity for areal coverage." However, a substantial improvement in the reflectivity-rain rate (Z-R) relationship is critical if we expect to derive reliable quantitative rainfall amount from radar reflectivity measurements. It is becoming increasingly evident that a dedicated coherent radar with polarization diversity is required at the primary test facility for this improvement. In addition, rain measurement research such a radar is of fundamental importance for all components of precipitation research, especially the micro-physics in rain systems, and for transfer technologies as well.

In designing a world-class test facility, all identifiable measurement techniques, including potential ones, must be taken into consideration. The sampling strategy must receive equal attention so that the deployment of the measurement system will be optimum to produce reliable areal statistics. Of course, optimum configurations can be achieved only through progressive tuning. This is accomplished by continuing statistical analysis beginning with initial deployment locations. The sampling studies and related statistical analysis are an important part of developing the standard transfer technology. However, one must recognize that the geographical and rain system differences at other sites will dictate modifications in the deployment schemes. Nevertheless, concepts learned at the primary test site will be relevant.

2.1.1 Primary Test Site

A number of considerations were taken into account when selecting the Kennedy Space Center (KSC)/Cape Canaveral, Florida area as the primary test site. Among the most important was taking advantage of the existing and currently expanding observational network developed for weather forecasting support for launch and shuttle landing operations. Two other essential features are that this central Florida region has probably the highest incidence of convective rainfall in the world. The coastal location provides for rain measurements over the ocean with the same radar that can be properly calibrated over land.

The key functions of the primary test site can be categorized in the following way:

- o Rain measurement technique development

- o Precipitation research e.g. physical processes/variability/distribution characteristics, etc.
- o Sampling and statistical studies
- o Data assimilation methodologies
- o TRMM scale climatology development
- o Support to instrumentation and algorithm development
- o Development of a hybrid system as a transfer standard
- o Major "ground truth" site during the mission (maximum overflights occur at the latitude boundaries of the orbit)

Obviously, TRMM cannot accomplish all of these objectives alone. A key goal is to establish the essential elements of a baseline test site that will encourage other agencies and institutions to participate in a world-class test facility in pursuit of their own research or operational objectives while sharing some of the cost. Mutual research benefits are bound to result as well.

Although an attempt will be made to begin developing an area climatology using available radar and rain gauge data, the improvement of measurement techniques must be started early as well to lay the foundation for the other functions to be accomplished at the test site. Indeed, the other functions depend heavily on progress in measuring rainfall more accurately.

TABLE 1. EXPERIMENTAL RAINFALL METHODS

DEVICE	ADVANTAGES
1) Optical Rain Gauge	Fast response, wide dynamic range
2) Zenith pointing VHF-UHF Doppler Rain Gauge	Explicit vertical velocity, drop size distribution, rainfall rate, level of diabatic heating
3) Scanning Doppler radar	Mesoscale vertical motion, profiles of condensate production and fallout
4) Multiple frequency/polarization radar	Observations over water, semi-independent estimates of rainfall
5) Microwave attenuation with alternating horizontal and vertical polarization	Calibration of radar statistics over land and observations over water
6) Satellite beacon attenuation	Sampling over remote (ocean) areas
7) Airborne Doppler radar	Sampling over oceans/remote areas
8) Underwater hydrophones	Sampling over oceans, integration of rainfall over a surface area
9) Profilers	Vertical structure in convective cloud systems

There are a number of experimental techniques currently under study (Table 1). Some of these, notably Doppler and polarimetric radar and profilers, are expected to be important measurement tools for investigating the vertical structure of rainfall. In the validation of TRMM observations, the vertical structure is essentially as important as area-averaged rainfall. To model the effect of released latent heat on global circulation, the height at which it occurs must be known. Thus, for vertical structure measurements, including phase, and to develop reliable reflectivity relationships, the need for a dedicated multi-parameter radar for the primary test site cannot be overemphasized. Plans are already underway for a profiler to be installed at KSC to support weather forecasting for shuttle operations. The profiler also promises to be an important tool in vertical structure studies.

For improving surface measurement techniques, a small scale experimental rain measurement research facility, Fig. 2, is being developed to determine the best methods and equipment to deploy at the primary test site. A polarized multi-frequency microwave attenuation link will be the key component of the facility. Attenuation of horizontally propagating microwave radiation is strongly related to the intensity of intervening precipitation and less sensitive than radar Z-R to rain drop size distribution (Crane, 1985; Ulbrich and Atlas, 1985). If this technique can be successfully developed it will provide an accurate way of determining mean rain rates over large areas.

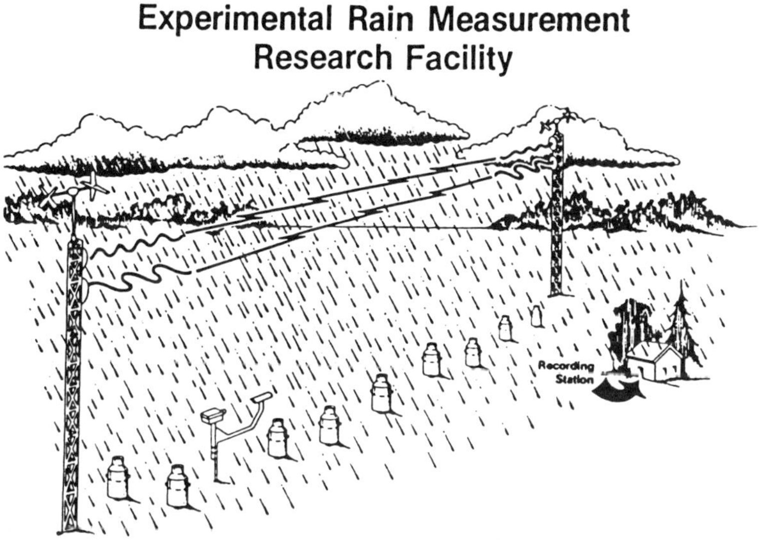

Figure 2. Artist's concept of attenuation link indicating vertical and horizontal polarization and supporting rain gauges. Planned frequencies are 8, 25, 38, 80 and 240 GHz. RRL Japan plans to furnish the 80 and 240 GHz systems.

Under the microwave transmission path will be a number of state-of-the-art rain gauges for calibration. Also, at least two new "gauge" techniques will be investigated, i.e. an optical rain gauge and an upward looking Doppler rain gauge. The Radio Research Laboratory (RRL) of Japan is participating in this research.

Two other methods of obtaining corroborative rain data is through inference from mass, heat and moisture budgets derived from surface wind measurements or rawinsonde data. In one technique, convective outflow at the surface is related to rainfall (Ulanski and Garstang, 1978a, 1978b, 1978c). The downward vertical velocities are integrated over time and space to estimate downward mass transport which in turn can be directly related to rainfall. Over much larger areas, surface fluxes of precipitation are inferred from heat and moisture budgets calculated from upper air soundings describing a large volume of the atmosphere (See, for example, Brummer, 1978). This budget method is one of the few techniques that can provide an estimate of rainfall amount on the largest scales of interest to TRMM.

2.1.2 Secondary Test Sites

An important parallel task to the primary test site activity during the pre-mission phase is to identify and begin to develop secondary ground truth test sites. These secondary test sites, of course, will become primary "ground truth" sites during the mission. Three tropical areas representing important rainfall regimes have been identified as secondary sites to provide (1) sampling strategies as needed in different regimes; (2) baseline climatologies; and(3) to support specialized field experiments. The three are:

- Kwajalein, M.I. (Ocean-ITCZ)
- Darwin, Australia (Monsoon)
- Amazon Basin - Brazil (Continental ITCZ)

There is evidence that rainfall characteristics are significantly different among convective systems in different regions of the tropics. The amount of rain produced in the core and in the anvils of these systems shows large variation (Houze, 1981; Houze and Hobbs, 1982). Thus, sampling different rain regimes is crucial if we are to understand the requirements correctly for designing the optimum observational system for each region.

Geographical differences will not only influence the observational configuration, but may affect sampling and data assimilation procedures as well. While it is unrealistic to expect a uniform rain measurement and sampling system at different sites, we must strive for a strategy that will provide a uniform rain data set for each regime. Failure will lead to considerable uncertainty and, perhaps, serious error in the interpretation of the TRMM measurements.

The sites chosen are not proposed as being the most representative of the significant rain regimes. For instance, there may very well be more representative monsoon locations in Asia. Island effects in the case of Kwajalein may reduce its representativeness as a Pacific ocean regime. However, as in the case of the primary test site, we must take advantage of existing facilities, accessibility and the mutual interests of other parties. Otherwise, the costs of even approaching the scale of activities proposed would be prohibitive.

2.3 MISSION PHASE

If the pre-mission objectives are satisfactorily accomplished, there will be at least 3 or 4 key ground truth sites already in place and functional to begin providing validation data when the mission begins. However, it is intended to pursue cooperative efforts to obtain "ground truth" observations from other representative areas as well. The TRMM program must depend heavily on contributions from other countries and their institutions. Providing "ground truth" support is a very direct way of participating in TRMM, i.e., in a cooperative exchange for TRMM data products. The project could not support all the desired "ground truth" effort alone.

The components of the validation program for the mission phase are summarized as follows:

- Primary (first order) ground truth sites (e.g. Central Florida; Northern Australia, Western Pacific (Kwajalein))

- o Second order ground truth sites (e.g. Central South America, India, Burma, Africa, etc.)
- o Coordinated near instantaneous as well as integrated measurements for validation between latitudes 30°N and 30°S
- o Use of airborne Doppler radar for instantaneous validation measurements over oceans
- o Infrared estimates of rainfall in the tropics and to high latitudes from geosynchronous satellites.

Validation of the TRMM observations is expected to be accomplished in two ways: (1) Average the TRMM rainfall data over time periods of a month or more for large areas (500 x 500 km) and compare with similarly averaged ground truth measurements; (2) Compare both the rainfall and height distribution data from TRMM with the instantaneous values observed at one or more "ground truth" stations.

To be successful, these strategies are dependent on high-quality "ground truth" data. The averaged TRMM rainfall data could conceivably be validated with only rain gauge data, but instantaneous calibration will require four-dimensional precipitation data from at least one location in the tropics. The second strategy must have priority early in the mission because verification would be more rapid than with the climatological approach.

3. FIELD EXPERIMENTS

Pre-mission field experiments are required to serve a multiplicity of research and mission development objectives, i.e., (1) aircraft flights for radar and radiometer instrument/system development, (2) algorithm development and testing, (3) basic research on physical processes related to rainfall initiation and distribution characteristics, (4) verification of cloud models, and, not least, (5) the development of primary and transfer standards leading to establishment of large area "ground truth" observational systems for the mission.

In addition to limited and specialized field studies tailored to TRMM needs, the program will take advantage of and participate in other major field experiments such as the Florida Area Mesoscale Experiment (FAME)/Second Thunderstorm and Microburst Project (STAMP) tentatively planned for the summer of 1990 and a western Pacific experiment presently being discussed for the early 1990's under the sponsorship of the Tropical Oceans and Global Atmosphere (TOGA) program.

During the mission, carefully designed field experiments will be necessary to assess, understand, and possibly enhance the data sets that are to be developed.

4. CURRENT STATUS

A number of thrusts are already underway to pursue the validation program. However, only the major activities are highlighted here.

The Cape Canaveral weather radar (WSR-74C) was recently digitized for range operational requirements. Taking advantage of this, we have begun installing networks of recording rain gauges in the Kennedy Space Center and Cape Canaveral area to begin a climatological record and to calibrate the radar. Initially six gauges were installed in the near field and 6 at the outer effective range of the radar for its calibration. The radar can then be in a position to collect larger area rainfall climatological statistics and to support rain research, algorithm development, etc. The ground work has already been laid for the TRMM validation program to participate in a 1990 major field experiment also involving FAME and STAMP.

The recording rain gauges are of the tipping bucket type and were made available to us by the Environmental Research Laboratory (NOAA). The 100 gauges supplied were those left over from the 1979-1980 Florida Areal Cumulus Experiment (FACE). Eventually, most of these will be deployed in nested grids of varying dimensions. A few of the gauges will be used with the Experimental Rain Research Facility (Figure 2) and initially, 10 or 12 will be deployed at Kwajalein Atoll.

Kwajalein is in the Marshall Islands approximately 2500 miles southwest of Honolulu, Hawaii. The missile test range there operates a WSR-74S (S-Band) weather radar. The radar is being upgraded at this time to provide digital recording of the reflectivity data. The recording rain gauges will be distributed among nearby islands of the Kwajalein Atoll within the effective range of the radar. Data collection is expected to begin in the next few months.

The TRMM study program has contributed significantly to the modification of a TOGA Doppler radar for use onboard a P.R.C. ship in the Gulf of Carpentaria during the Equatorial Mesoscale Experiment (EMEX). The purpose was to obtain another tropical rainfall data set for further progress in the TRMM sampling and statistical analysis studies (McConnell and North, 1987). Thus far, such studies have relied entirely on the GARP Atlantic Tropical Experiment (GATE) data which are unlikely to be representative of Pacific tropical data. Analysis of these data have just begun.

More recently, a cooperative agreement has been reached with the Australian Bureau of Meteorology to operate the same radar at Darwin for two rainy seasons to conduct research and to provide monsoonal climatologies (Keenan, et al., 1987). This schedule will permit return of the radar in time for its use during the FAME/STAMP/TRMM Ground Truth experiments in 1990. The plan is to operate the radar some distance off-shore as part of a network of other Doppler radars on land. Often, convective systems form over the ocean off the Cape Canaveral area and move on-shore. And, of course, there is similar interest in tracing the cycle of those systems forming over Central Florida and moving eastward out over the ocean.

Finally, the TRMM validation program is cooperating in a validation program being developed to support the World Climate Program's International Global Precipitation Climatology Project and expects to gain considerably from it. This project seeks to develop an interim 10-year data set in the form of a precipitation index inferred from operational satellite (visible infrared/microwave) data. A major contribution that NOAA is making for that validation program is in ocean rain measurements. Their program includes optical rain gauges on an ocean meteorological data buoy and the use of hydrophones for acoustic detection of rainfall. Plans are to test these systems off the Coast of Cape Canaveral in June 1988. If the tests are successful, it is anticipated that this technique will be expanded for use elsewhere. This can be a very important asset to the primary test facility if the systems can be maintained operationally there, especially during the intensive field experiments planned for the summer of 1990, described earlier.

ACKNOWLEDGMENTS

The author wishes to acknowledge the keen interest, participation and contributions of the following individuals in the planning and development of the TRMM validation program: R. Houze, U. of Wash.; M. Garstang, U. of Virginia; A. Jameson, Applied Research Corporation; R. Carbone, NCAR, M. Hudlow, NOAA; J. Wilkerson, NOAA; C. Ulbrich, Clemson U.; R. Crane, Dartmouth College; S. Geotis, MIT; G. North, Texas A & M; G. Austin, McGill U.; D. Atlas, U. of Maryland; J. Vogel, Illinois St. Water Survey, and R. Rasmussen, Bu. of Land Reclamation.

REFERENCES

Atlas, D. and O. W. Thiele, eds., 1981: Precipitation measurements from space. Workshop Report, D. Atlas and O. W. Thiele, Editors, NASA/Goddard Space Flight Center, Greenbelt, MD.

Brummer, B., 1978: Mass and energy budgets of a 1 kilometer high atmospheric box over the GATE C-scale triangle during undisturbed and disturbed weather conditions. J. Atmos. Sci., 35, 997-1011.

Crane, R. K., 1985: Comparative evaluation of several rain attenuation prediction models. Radio Science, 20, 843-863.

Houze, R. A., Jr., 1981: Structure of atmospheric precipitation systems: a global survey. Radio Science, 16, 671-689.

Houze, R. A., Jr. and P. V. Hobbs, 1982: Organization and structure of precipitating cloud systems. Advances in Geophysics, 24, 225-315.

Keenan, T., G. Holland, and M. Manton, ,1987: TRMM ground truth in a monsoon environment: Darwin, Australia. Proceedings of the International Symposium on Tropical Precipitation Measurements, Tokyo, Japan.

McConnell, A. and G. R. North, 1987: Sampling errors in satellite estimates of tropical rain. J. Geophys. Res., 92, 9567-9570.

Thiele, O. W., Ed., 1987: On requirements for a satellite mission to measure tropical rainfall. NASA Ref. Pub. 1183, 49 pp.

Ulanski, S. L. and M. Garstang, 1978a: The role of surface divergence and vorticity in the life cycle of convective rainfall, Part I: Observations and analysis. J. Atmos. Sci., 35, 1047-1062.

Ulanski, S. L. and M. Garstang, 1978b: The role of surface divergence and vorticity in the life cycle of convective rainfall, Part II: Descriptive Model. J. Atmos. Sci., 35, 1063-1069.

Ulanski, S. L. and M. Garstang, 1978c: Some aspects of Florida convective rainfall. Water Resour. Res., 14, 1133-1139.

Ulbrich, C. and D. Atlas, 1985: Extinction of visible and infrared radiation in rain: comparison of theory and experiment. J. Atmos. and Oceanic Tech., 2, 331-339.

ESTIMATION OF TROPICAL RAINFALL

Michael Garstang, Claire Cosgrove, Robert Swap and Steven Greco
University of Virginia
Charlottesville, Virginia 22903, USA

ABSTRACT

Convective systems produce the bulk of tropical rainfall but are variable in space and time both in terms of the occurrence of the rain system and the nature of the rain in the system itself. A single example of a continental equatorial squall line over the Amazon Basin is used to illustrate the variability in the rainfall rates and amounts for such systems. Convective core rainfall falling at 0.75 mm/min produces 50 to 70% of the rain; the remaining 30 to 50% occurs in multilayered and single layered cloud regions which vary in both space and time for the same system.

1. INTRODUCTION

Tropical rainfall between the latitudes of 30°N and S constitutes a significant part of the global rainfall (Fig. 1). The greater part of the tropical rainfall occurs in organized meso- to synoptic-scale systems but with large spatial and temporal variability. For example, a pervasive result found by many authors (Riehl, 1954), shows that 50% or more of the rainfall falls in about 10% of the time. Convective rain systems which produce an average of $\geq$ 15 mm of rain per station, typically produce more than 60% of the total annual rainfall. But rain systems with high rain rates, producing the large amounts of rainfall, occur only intermittently. This means that we must measure the high rain rates accurately and we must be able to detect and delineate organized convective meso- to synoptic-scale systems.

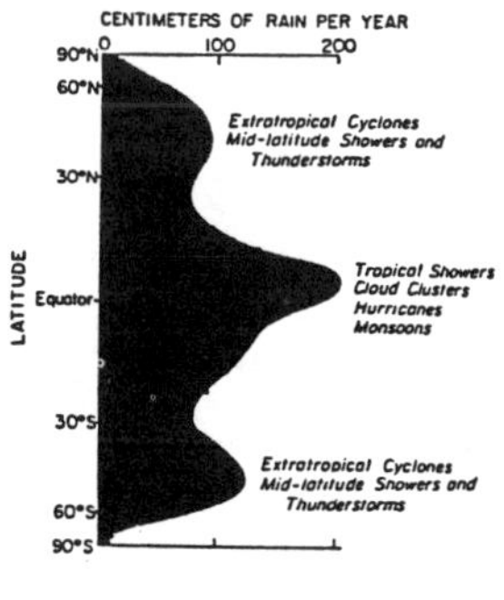

FIGURE 1. Globally averaged annual precipitation and associated rain-producing weather systems. (After Houze, 1981.)

TROPICAL RAINFALL MEASUREMENTS
John S. Theon and Nobuyoshi Fugono (Eds.)

ISBN 0-937194-14-4

In this paper, we take the position that organized meso- to synoptic-scale systems, as seen over the Amazon Basin of Brazil, produce a large fraction of the rainfall of that water-rich region. We further wish to establish whether the rainfall characteristics of these equatorial continental systems are similar or dissimilar to their tropical oceanic counterparts.

2. MEASUREMENTS

The NASA Amazon Boundary Layer Experiment (ABLE-2B) (Harriss et al., 1988) made surface meteorological, radar and satellite measurements over the Amazon Basin during the wet season months of April and May 1987. Four Portable Automated Mesonet (PAMs) stations were deployed with sensors on towers approximately 5 m above the canopy of the rain forest. Tipping bucket rain gages at this level measured accumulated rainfall with a resolution of 0.25 mm every minute. The location and spacing of the PAM towers is shown in Fig. 2. Temperature, humidity, pressure, and horizontal wind velocity were also measured at each tower site. Accumulated rainfall (an 8-inch gage) for each rain event was also measured at the surface in a clearing within 3 km of each of the three outer PAM towers.

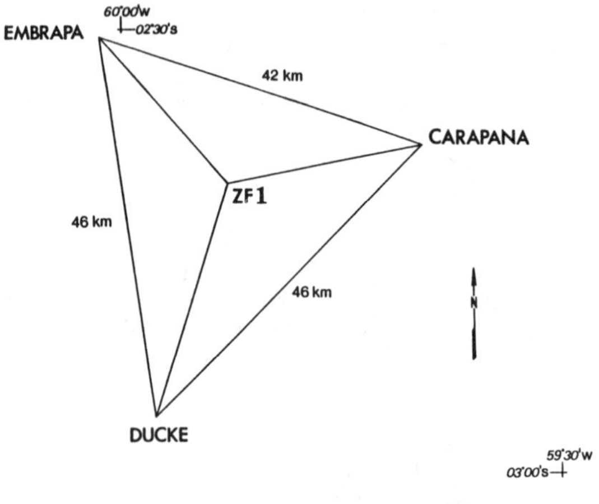

FIGURE 2. Location and spacing of the 4 PAM towers.

A 3 cm radar with digital recording capability was located approximately 15 km SSW of the most southerly PAM tower. GOES (Geostationary Environmental Satellite) satellite imagery (visible and infrared) was collected routinely at Manaus and at the NASA Langley Research Center in Hampton, VA.

Squall lines, at times up to 3000 km in length, were a relatively frequent phenomena occurring on at least 10 of the 45 days of ABLE-2B. The squall line which passed over the measurement network on the 26th of April 1987 was chosen for this study. The 26th of April system produced a total of 33.02 mm of rain at the Carapana station. Twenty-one rain systems passed across the measurement network in the 45 days of ABLE. The 26th of April ranked 7th and represented 4.51% of the total rain measured over the 45 days.

3. FINDINGS

Figure 3 shows the time distribution of rain amount at each of the 4 recording gage stations. Rain accumulation is plotted every three minutes. The total amount of rain recorded at each station is shown in Table 1. Table 1 also shows the non-recording rain gage amounts and the differences between the recording and non-recording gages.

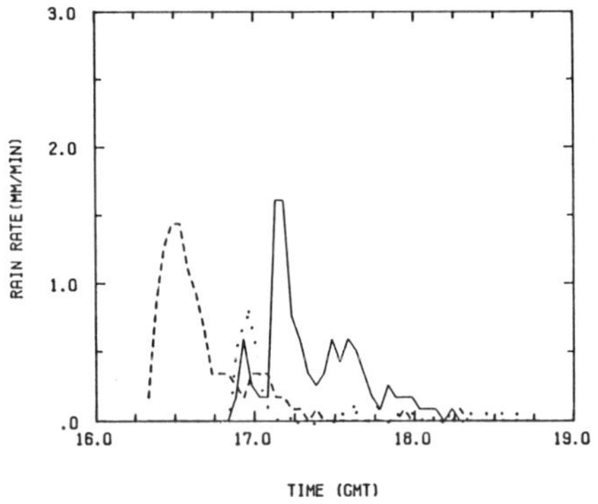

FIGURE 3. Distribution of 3 min totals of rainfall at each of the three perimeter stations: Carapana (dashed), ZF-1 (dotted) and Embrapa (solid).

TABLE 1. TOTAL AMOUNT OF RAIN RECORDED AT EACH STATION

Station	Recording Gage	Non-Recording Gage	% Difference
Carapana	33.02	52.32	37%
Embrapa	32.01	35.30	9%
ZF-1	15.49	-	-
Ducke	0.51	-	-

Figure 4 shows a time series of equivalent potential temperature at the top of the forest canopy. Table 2 compares the time of the θ_e drop with the occurrence of rain at each site. We note that significant θ_e changes are not related to the amount or intensity of measured rainfall but that θ_e drops are always indicative of convective rain process.

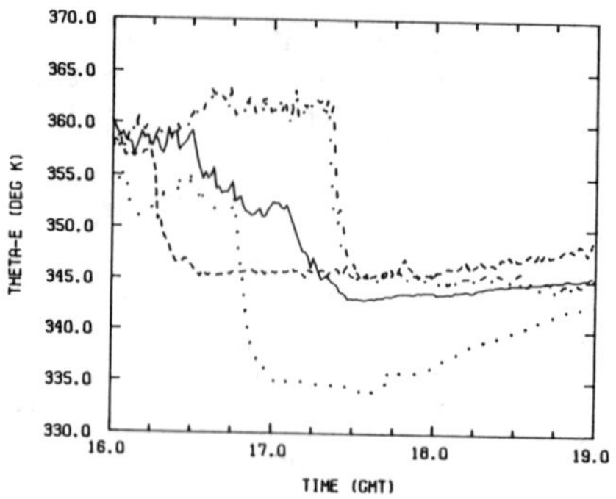

FIGURE 4. Equivalent potential temperature at 5 m above the forest canopy at Carapana (dashed), ZF-1 (dotted), Embrapa (solid) and Ducke (dash-dot).

TABLE 2. COMPARISON OF THE TIME OF θ_e DROP WITH THE OCCURRENCE OF RAIN AT EACH SITE

Station	Time of θ_e Drop	Start of Rain
Carapana	1614 (1.7°K in 1 min)	1616
ZF-1	1647 (2°K in 1 min)	1649
Embrapa	1630 (1.4°K in 1 min)	1651
Ducke	1722 (5°K in 1 min)	1722 (.25 mm)

Figure 5 shows the variability of rainfall intensities at Carapana and Embrapa. At Carapana (Fig. 5a), the greatest variability occurs near the onset of the rain event which coincided with the higher rainfall intensities. The subsequent relative deviations then tend to be similar for both high and low intensities alternating from positive to negative values but within the limits of one standard deviation of the variability. Rain amounts decrease sharply after the vertical dashed line on the abscissa. We note a similar distribution for Embrapa (Fig. 5b) except that we see three distinct regimes separated by variations in intensity and amount. We interpret this change as a change from convective rain to stratiform rain in the sense described by Houze and his collaborators (Houze, 1977; Leary and Houze, 1979; Gamache and Houze, 1982, 1983, 1985; Wei and Houze, 1987). We note in agreement with Smull and Houze (1987) that there is probably a transition zone between the convective and stratiform region of the storm. We do, however, believe that this transition zone is a product of rain from multiple stratiform cloud layers rather than residual convective clouds. We designate this rainfall as a multilayer rainfall and regard it as distinct from the single layer or anvil rainfall which signals the end of the system.

Based upon Fig. 5 and visual inspection of the change in rainfall rates evident in Fig. 3 we select 0.75 mm/min as the boundary between

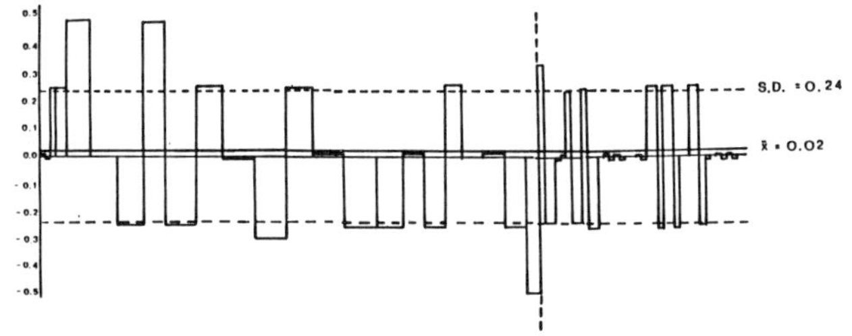

FIGURE 5a. Rainfall variability in intensity and amount is shown for Carapana. Variations in 1 min intensities are shown on the ordinate while the width of the bar shows the amount (or rate x time). The horizontal dashed line represents one standard deviation in intensity.

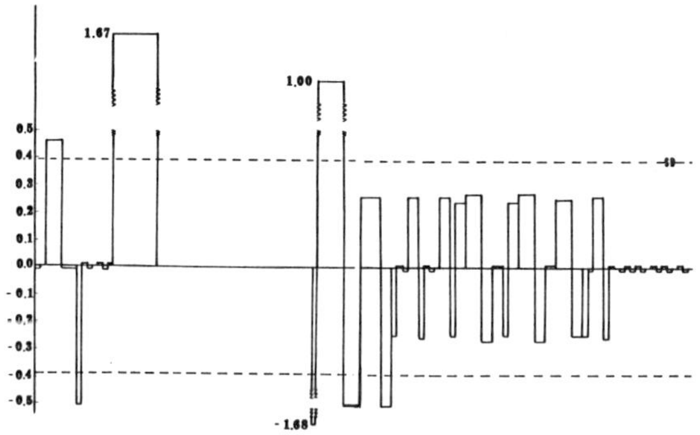

FIGURE 5b. As for Fig. 5a but for Embrapa.

convective and multilayered rainfall, and 0.50 mm/min as the boundary between multilayered and single layer rainfall. Table 3a shows the rainfall at the 0.25 mm/min resolution of the measurement. Table 3b shows the three classes each as a percentage of the total.

TABLE 3a. RAIN RATES AND THE ASSOCIATED PERCENTAGE OF THE TOTAL RAINFALL IN EACH RATE INTERVAL

Station	Total Rain	.25	.50	.75	≥ 1.0
Embrapa	32.01	31.0%	17.4%	14.3%	37.3%
Carapana	33.02	21.6%	10.8%	6.9%	60.7%
ZF-1	15.49	47.5%	9.9%	29.5%	13.1%

The percentage amounts ascribed to the convective class ranges between 40 and 70%, single layered cloud 20 and 50%, the remaining 10 to 20% is placed in the multilayered category.

The forward propagation of the squall line is measured from satellite enhanced infrared images to be 51 kph. The width of the

TABLE 3b. RAIN RATES FOR THE THREE CLASSES OF CONVECTIVE (C $\geq$ .75 mm/min), MULTILAYERED (0.5 $\leq$ M < .75) AND SINGLE LAYERED (S < 0.5)

Station	Total Rain	S	M	C
Embrapa	32.01	31.0%	17.4%	51.6%
Carapana	33.02	21.6%	10.8%	67.6%
ZF-1	15.49	47.5%	9.9%	42.6%

system normal to its direction of propagation as measured from the leading edge to the corresponding grey scale on the opposite side of the squall line remained nearly constant at 290 km.

The width of the precipitating areas can be estimated from the time occupied by the rain rate class multiplied by the speed of propagation of the system. The results are shown in Fig. 6 and summarized in Table 4.

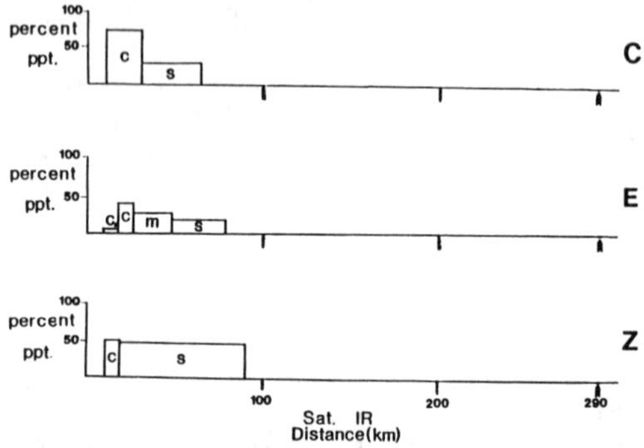

FIGURE 6. Percent of total rain in each rain class (C_p = pre-convective, C = convective, M = multilayered, and S = single layered) for each station (C = Carapana, E = Embrapa, and Z = ZF-1). The horizontal axis represents the width of the system (290 km) as obtained from the IR image. The width of each precipitating class is obtained from (time of class rate) x 51 kph.

4. CONCLUSIONS

The rain rates of 0.75 mm/min or greater associated with the convective core of this squall line are significantly higher than rates ascribed to tropical oceanic squall lines (0.25 mm/min). Rates as low as 0.25 mm/min would not appear to correspond to the highly variable region of the squall line where high variability might be equated to convective rainfall.

Spatial variability is also, however, high and at two stations a convective stratiform classification could apply yielding 50 to 70% of the rain in the convective region and 30 to 50% in the stratiform

TABLE 4. RAIN DISTRIBUTIONS FROM TIME SPACE CONVERSION

	Pre-Conv	Conv	Multilayered	Single Layered
Carapana	0 km	20.4 km (71%)	0	33.9 km (29%)
Embrapa	8.5 km	8.5 km (41.3%)	22.1 km (33.3%)	29.7 km (29.0%)
ZF-1	0 km	10.2 km (51%)	0	89.1 km (49%)

region. We believe that the convective core produces between 40 and 70% of the total rainfall.

The remaining 30 to 60% of the rainfall may occur at relatively high rates in the multi-cloud layered region of the storm and at low rates in the single cloud layered region. This variability (30 to 60%) is both spatial and temporal. Any indirect sensing technique employed with the hope of estimating rainfall with an accuracy of greater than 20% will have to deal with these distributions.

5. REFERENCES

Gamache, J.F., and R.A. Houze, Jr., 1982: Mesoscale air motions associated with a tropical squall line. Mon. Wea. Rev., 110, 118-135.

________, and ________, 1983: Water budget of a mesoscale convective system in the tropics. J. Atmos. Sci., 40, 1835-1850.

________, and ________, 1985: Further analysis of the composite wind and thermodynamic structure of the 12 September GATE squall line. Mon. Wea. Rev., 113, 1241-1259.

Harriss, R.C., S.C. Wofsy, M. Garstang, L.C.B. Molion, R.S. McNeal, J. M. Hoell, R.J. Bendura, S.M. Beck, R.L. Navarro, J.T. Riley, and R.C. Shell, 1988: The Amazon Boundary Layer Experiment (ABLE 2A): Dry Season 1985. J. Geophys. Res., 93, 1351-1360.

Houze, R.A., Jr., 1977: Structure and dynamics of a tropical squall line system. Mon. Wea. Rev., 105, 1540-1567.

Houze, R.A., Jr., 1981: Structure of atmospheric circulation: A global survey. Radio Science, 16, 671-689.

Leary, C.A., and R.A. Houze, Jr., 1979: Melting and evaporation of hydrometeors in precipitation from the anvil clouds of deep tropical convection. J. Atmos. Sci., 36, 669-679.

Riehl, H., 1954: Tropical Meteorology. McGraw-Hill Book Company, Inc., New York, 392 pp.

Smull, B.F., and R.A. Houze, Jr., 1987: Dual-Doppler radar analysis of a midlatitude squall line with a trailing region of stratiform rain. J. Atmos. Sci., 44, 2128-2148.

Wei, T., and R.A. Houze, Jr., 1987: The GATE squall line of 9-10 August 1974. Adv. Atmos. Sci., 4, 85-92.

TRMM GROUND TRUTH IN A MONSOON ENVIRONMENT: DARWIN, AUSTRALIA

T.D. Keenan, G.J. Holland, and M.J. Manton
Bureau of Meteorology Research Centre
Melbourne 3001, Australia

ABSTRACT

The establishment of a TRMM ground truth station consisting of a Doppler radar and associated mesoscale rainfall network in the monsoon environment of the "maritime continent" of northern Australia is described. The aims are to provide a high quality monsoon regime rainfall data set suitable for pre-TRMM studies and to undertake detailed case studies of the weather systems affecting northern Australia.

1. INTRODUCTION

The TRMM ground truth mission (Thiele, 1987) includes the objective of determining the three-dimensional structure of tropical precipitation and rainfall amounts from spatial scales less than 1 km up to a 10^5 km^2 area for a one month time period. To date, ground truth data on tropical rainfall has been limited to short-period field experiments, eg. GATE, AMEX/EMEX, and at locations not representative of the entire global regime. To this end, a TRMM ground truth station has been established in Darwin, Australia in a collaborative venture between NASA, NOAA and BMRC. The objective is to obtain a climatology of convection and total rainfall indicative of the monsoon climate and the "maritime continent". This will provide a unique opportunity to collect data on the wide range of tropical oceanic and continental weather phenomena affecting the climate of northern Australia.

2. CLIMATOLOGY OF THE NORTHERN AUSTRALIAN MONSOON

Northern Australia lies at the southern edge of the "maritime continent", where a large proportion of the total atmospheric latent heat release occurs. It experiences a classical monsoonal climate, described in detail by Holland (1986), with a sudden wind flow reversal from low-level easterlies during the winter to summertime westerly flow. The onset of the "active period", when rainfall and low-level equatorial westerly flow occur, is normally during late December, with the retreat to easterly flow and dry conditions usually during late February or early March. At upper levels, the flow is reversed, with a transition from westerly to easterly flow during the summer "active period". "Break periods", when low-level subtropical easterly flow is again established also occur during the summer period. On average

TROPICAL RAINFALL MEASUREMENTS
John S. Theon and Nobuyoshi Fugono (Eds.)

ISBN 0-937194-14-4

20% of the season experiences this "break" condition and there are two to three "active" westerly flow phases per season, with a mean period between active phases of 40 days. There is a close relationship between low-level westerly flow at Darwin and rainfall over northern Australia (Holland, 1986). In the Darwin region, however, thunderstorm and squall line activity during the "transition" season contributes up to 30% of the total rainfall (Nicholls et al., 1982). Considerable interannual variability in the strength and duration of the summer monsoon also occurs at Darwin.

The dominant rain-producing systems, during both the "transition" and "wet" season, include oceanic monsoon cloud clusters and equatorial cloud bands, "maritime continent" island thunderstorms, "pop-corn" cumuli and continental tropical mesoscale convective complexes, north Australian cloud lines and tropical squall lines, monsoon depressions and tropical cyclones. These continental and maritime systems contain a range of convective and mesoscale structure and undergo distinct lifecycles ranging from the diurnal time scale to that associated with synoptic-scale organization. Subtropical interactions appear to be particularly important (Keenan and Brody, 1988). At Darwin, there is considerable variability in the occurrence of many of these systems from season to season.

3. THE DARWIN OBSERVING SYSTEM AND OBSERVATIONAL PROGRAM

The Bureau of Meteorology Research Centre (BMRC) Darwin ground truth site shown in Fig. 1 consists of the National Oceanic and Atmospheric Administration (NOAA)/Tropical Oceans and Global Atmosphere (TOGA) Doppler radar and a mesoscale network of raingauges. Within

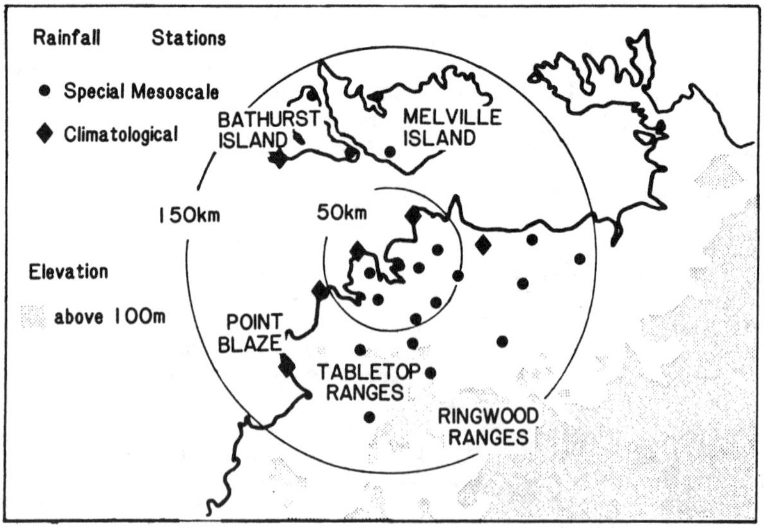

Fig. 1 The Darwin radar site and associated raingauge network.

150km of the site there is approximately 48% coverage of the ocean, 43% coverage of the mainland and 9% coverage of tropical islands. The radar and associated climatological rainfall network will be in operation for the 1987/88 and 1988/89 "wet seasons" commencing 1 November each year and ceasing on 30 April. In addition, full upper tropospheric thermodynamic soundings will be obtained every 12 hours and wind soundings every 6 hours. Supplementary data will also be available from the routine surface network of the Bureau of Meteorology.

The observational program will consist of continuous climatological data collection supplemented with Special Observational Periods (SOP) in support of the TRMM objectives. During SOPs the radar will be available to support specific observational programs designed by project principal investigators, subject to the constraint that the climatological program is not compromised. BMRC plans to have two such programs during 1987/88. One for the study of "maritime continent" thunderstorms during early December and one for monsoon system studies during January and February, 1988. Additional SOPs are planned by NASA Goddard for November and February. During the SOPs the radar observational practice will be optimized for the particular phenomenon but will aim to collect as much high resolution data as possible. It is planned to coordinate the SOPs and climatological program to provide as much attended operation of the radar as possible.

Radar-derived rainfall rates will be calibrated using the raingauge network shown in Fig. 1. The climatological networks of six gauges will operate throughout the entire observing period. The upgraded network of 26 stations will be utilised during SOPs.

Long-term calibration of the radar will be essential and will be maintained by real-time monitoring and quick post-analysis procedures. In real-time, electronic bench calibrations will be used to maintain quality in both Doppler and reflectivity measurements. Comparison of radar echo strength with raingauge measurements and analysis of Doppler VAD-derived divergence fields will be part of the on-going post-analysis quality control procedures.

All data collected during the observational programs will be forwarded to NASA Goddard Space Flight Center for archive.

4. SOME RADAR CHARACTERISTICS OF TROPICAL CONVECTION AT DARWIN

The mean radar returns at the Darwin site obtained during AMEX Phase II, as shown in Fig. 2, exhibit considerable spatial and temporal variability. Echo frequency maxima exist over Bathurst and Melville Islands and over the oceanic and coastal region to the south-west of Darwin. The island maximum was an indication of the "maritime continent" thunderstorm activity. The maximum to the southwest of Darwin was due to the presence of major convective bands associated with monsoon depressions in the Joseph Bonaparte Gulf. In addition, interaction between the local land and sea breeze circulations and the monsoon westerly flow, analogous to that observed by Houze et al. (1981), are thought to be responsible for the maximum to the south-west. The offshore maximum occurs during the early morning dissipating by midday. The onshore component of the echo frequency maximum on the Cox Peninsula occurs during the early afternoon. An evening maximum also occurs on the ranges to the south-east of Darwin. Similar local circulations are thought to be responsible for diurnal variations in land/sea echo frequency around Bathurst and Melville Islands.

% OCCURRENCE OF REFL LEVELS >15dBZ

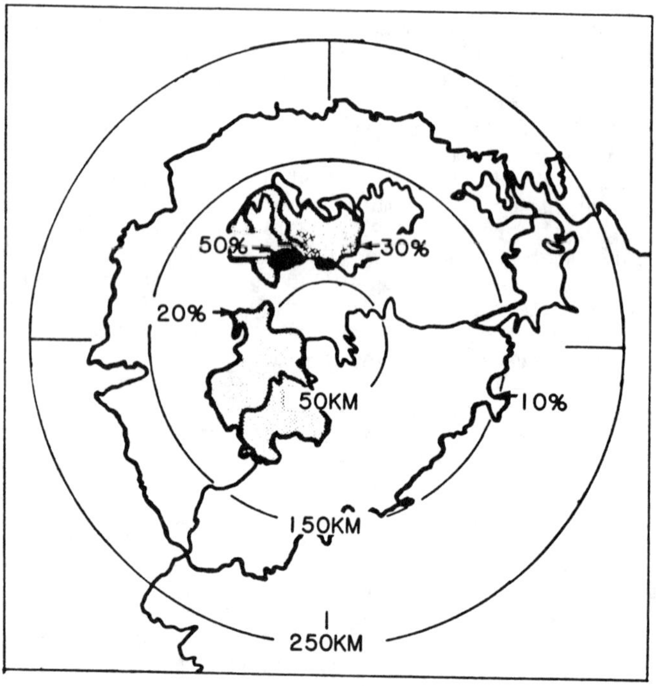

Fig. 2 Percentage occurrence of radar echoes greater than 15dBZ during Phase II AMEX at Darwin

The rainfall at Darwin undergoes a distinct diurnal variation, as summarized in Fig. 3 where the data were obtained from radar ranges between 50 and 250 km. The oceanic domain extended from 225 to 300 degrees azimuth and the land domain from 90 to 210 degrees azimuth. Oceanic convective echoes underwent a 2:1 variation and had a broad peak between 1800 GMT (0330 LST) and 0000 GMT (0930 LST), ie. the morning. The associated stratiform component underwent a 2.5:1 variation and was at a maximum at 0000 GMT (0930 LST). Over land, the maximum in convective activity occurred near noon at 0300 GMT (1230 LST) and the minimum at 1800 GMT (0330 LST). Stratiform precipitation over the land had a smaller diurnal cycle with a much broader peak. It was at a minimum during the early morning period between 1800 GMT and 0000 GMT.

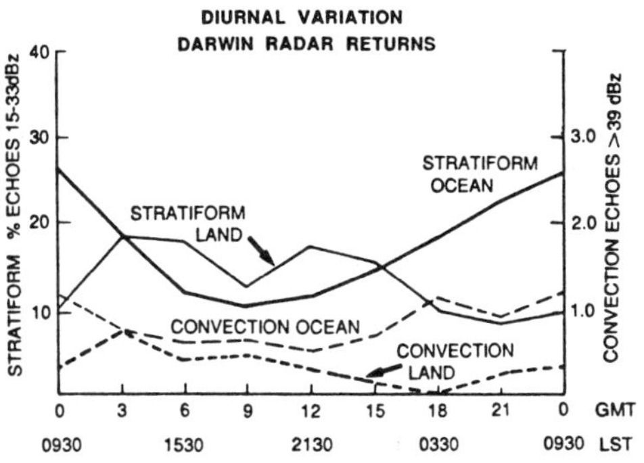

Fig. 3 Diurnal variation of oceanic and land echoes observed with the Darwin radar during AMEX Phase II

5. RESEARCH PROGRAM

The BMRC research program will focus on the following :

5.1 TROPICAL PRECIPITATION STUDIES

This will include SOP radar/mesoscale raingauge calibration studies to optimise the accuracy of the "ground truth" data, assessment of the time and space scales of convective and stratiform precipitation, the relationship between radar signatures and the observed precipitation. These studies will be useful for specifying TRMM sampling requirements, comparing the Darwin rainfall regime with data sets from other regions and providing a basis for validation of satellite-derived precipitation estimates.

5.2 INVESTIGATION OF ISLAND THUNDERSTORMS

BMRC, in collaboration with Monash University Geophysical Fluid Dynamics Laboratory, the NASA Goddard Laboratory for Atmospheres and the National Center for Atmospheric Research, plans to obtain a comprehensive data set on the generation and evolution of tropical island thunderstorms. These data will be used in analytical and numerical modelling studies of tropical convection and in a comparative test of a number of existing numerical models for convective clouds.

As shown by Fig. 2, island thunderstorms occur regularly over Bathurst and Melville Islands (north of Darwin). These systems are representative of the large number of similar systems that occur throughout the maritime continent. They are topographically fixed, occur during known periods, and are thus suitable for detailed observational studies.

5.3 CASE STUDIES OF AUSTRALIAN TROPICAL WEATHER SYSTEMS

Data will be collected for a range of "transition" and "monsoon" season systems affecting northern Australia. Particular attention will be directed at the "maritime" continent thunderstorm and squall lines during the "transition" season. During the "monsoon" season the structure of oceanic monsoon cloud clusters, differences between continental and oceanic mesoscale convective complexes, land-based monsoon depressions and tropical cyclones will be studied. Local mesoscale/diurnal/topographically driven circulations and hazardous low-level wind shear will also be investigated throughout the observational period.

6. SUMMARY

The two-year observational program of the NOAA/TOGA Doppler radar and the associated mesoscale observational network at Darwin will provide a unique opportunity to study a wide range of tropical weather systems representative of the monsoon climate of the "maritime continent". This is a NASA/NOAA/BMRC collaborative program that should result in a better appreciation of the structure, dynamics and relationship between latent heat release and observed precipitation within these systems. All these factors are important design considerations for TRMM.

REFERENCES

Holland, G.J., 1986: Interannual variability of the Australian summer monsoon at Darwin: 1952-82, Mon. Wea. Rev., 114, 594-604.

Houze, R.A., S.G. Geotis, F.D. Marks and A.K. West, 1981: Winter monsoon convection in the vicinity of North Borneo. Part I: structure and time variation of the clouds and precipitation, J. Atmos. Sci., 109, 1595-1614.

Keenan, T.D. and L.R. Brody, 1988: Synoptic scale modulation of convection during the Australian summer monsoon, Mon. Wea. Rev., 116, 71-85.

Nicholls, N., J.L. McBride and R.J. Ormerod, 1982: On predicting the onset of the Australian wet season at Darwin, Mon. Wea. Rev., 110, 14-17.

Thiele, O.W., 1987: On requirements for a satellite mission to measure tropical rainfall, NASA Reference Publication, 1183, 49pp.

TWO IMPROVED Z-R RELATIONSHIPS: THE PROCESS RECURRENT METHOD AND THE VARIABLE COEFFICIENT METHOD

Peicai Yang, JinLi Liu, Hai Lin, and Daren Lü*
Institute of Atmospheric Physics
Academia Sinica
Beijing, PRC

ABSTRACT

In this paper two radar precipitation measurement methods are established. They are: (1) The process recurrent Z-R relationship implying climatic information and (2) The variable coefficient Z-R relationship simulating the combined radar-radiometer system. Their measurement errors in three light precipitation processes are discussed and compared with the usual Z-R relationship with constant coefficients. The results show that the two improved methods have higher measurement accuracy.

1. INTRODUCTION

Since radar can provide continuous precipitation information in time and space, it is a valuable rainfall measurement tool. In many measurement methods the most prevalent one is the empirical relationship based on raindrop-size spectrum. Its expression is

$$Z=AR^{B} \tag{1}$$

where Z and R are radar reflectivity factor(mm^6/m^3) and rainfall rate (mm/hr) respectively, both A and B are constants. However, many investigations have shown that A and B have a very wide variable range.They not only change with area, season and types of precipitation but with time and space, even though in identical precipitation processes. This can result in considerable measurement error. Therefore, how to select appropriate A and B values becomes the key improving the measurement accuracy.

In this paper, we have presented two modified Z-R relationships to try to reduce the method error. Our experiment results show that the methods are quite useful. The details are given below.

2. DATA

The data used in this paper are from three light precipitations which happened in the Beijing area on July 24, Aug. 12 and Aug. 15, 1980.

The rainfall data were obtained from 12 weather stations, which are located 20-80km northwest of the radar station. The time resolution of the rainfall data is ten minutes. The corresponding

*Present affiliation: National Natural Science Foundation of China.

ISBN 0-937194-14-4

errors in rainfall rates are not greater than ±0.1mm/hr.

The echo data are collected by a C-band radar which is in Xianghe county lying southeast of Beijing. In order to insure consistency with the resolution of the precipitation data, the sampling period of the radar is taken as ten minutes. Each sample consists of echo power P_r on the ten different radial paths, each path passing through one or two raingauges.

All the above rainfall processes are light convective precipitations. The rainfall rate in all these cases is smaller than 10 mm/hr. The numbers of the samples are 22 on July 24, 55 on Aug. 12, and 110 on Aug. 15, 1980.

3.METHOD

In order to establish method 1, considering the difference of the sample numbers among the above three precipitation processes, we have first deduced the Z-R relationship by means of the actual data obtained with radar and raingauges on Aug. 15, 1980(written as M.1a) and applied it to estimate the rainfall rate on Aug. 12, 1980. Then we use the actual data on Aug.12, 1980 to modify the M.1a and generate a new empirical relationship M.1b. Finally we apply M.1b to measure the rainfall rate on July 24. It is apparent that the above order should be inverted in practice. The M.1a is obtained through a linear regression. The expression is as follows,

$$Z=294R^{2.2} \tag{2}$$

and its relative root mean square deviation is 50.2%. The M.1b is established based on Eq.(2) and through reselecting the coefficient A. If $\{Z_i, R_i\}$ expresses the data set on Aug. 12, according to the following error equation,

$$\varepsilon^2=\Sigma[\alpha z_i^{\beta_0}-R_i]^2$$

we have that the coefficient α, which minimizes ε^2, should be

$$\alpha=\Sigma R_i z_i^{\beta_0}/\Sigma z_i^{2\beta_0}$$

where $\beta_0=1/2.2$ and $\alpha=A^{-\beta_0}$. Then we obtain α=0.05901(or A=506). Thus the M.1b may be expressed as

$$Z=506R^{2.2} \tag{3}$$

In fact, the climate information is implied in Eq.(3). If the empirical relationship is successively modified, the information on season is included naturally.

In method 2, the information amount will have a large expansion because of inputting a lot of echo data on the radial paths of radar beam. It is different from the previous Z-R relationship (i.e. M.1a and M.1b). The coefficient A, here, is not a constant but a variable which changes with the distribution of the reflectivity factor Z on the radial paths. The deduction of this method is given in the following paragraph.

From the relationship between Z and R (Eq.(1)) and the empirical relationship between the attenuation coefficient σ and R

$$\sigma=CR^{D} \tag{4}$$

eliminating R, we have

$$\sigma=EZ^{F} \tag{5}$$

where σ is in km^{-1} and C and D are empirical constants. In this paper, the values of A and B are taken to be 4.41×10^{-4} and 1.11(Lü Daren and Lin Hai, 1980). We isometrically take the values of the reflectivity factor along the radial paths and write them as $\{Z_{i,k}\}$, $i=1,2,\cdots,N$; $k=1,2,\cdots,M$; Subscripts i and k represent different places on one radial path and on different paths, respectively.

Substituting Z_{ik} into (5), we can obtain σ_{ik}. Let the mean values of σ_{ik} on the radial path be

$$\bar{\sigma}_k=\frac{1}{N}\sum_{1}^{N}\sigma_{ik}$$

Substituting it into (4), we have the mean rainfall rate. It is

$$\bar{R}_k=(\bar{\sigma}_k/C)^{1/D} \tag{6}$$

However, $\bar{R}_k$ are also expressed as

$$\bar{R}_k=\frac{1}{N}\sum_{1}^{N}R_{ik}=\frac{1}{N}\sum_{1}^{N}\left(\frac{Z_{ik}}{A}\right)^{1/B} \tag{7}$$

From (6) and (7), as B,C and D are given, simply obtain

$$A_k=\left[\sum_{1}^{N}Z_{ik}^{1/D} \,/\, N(\bar{\sigma}_k/C)^{1/D}\right]^{B} \tag{8}$$

After A_k are calculated, from the values of Z over the rainfall area, the corresponding rainfall rates may be calculated by the formula

$$R=(Z/A_k)^{1/B} \tag{9}$$

In fact, the mean attenuations of rainfall area on the radial paths are easy to measure by the microwave radiometer. Therefore, if the microwave radiometer can be used in the precipitation measurement system, it will be possible to increase the accuracy of the precipitation measurement.

4.RESULTS

In order to examine the measurement accuracy of the above two methods, we have treated the data on July 24 and Aug. 12, 1980 as the test samples and compared them with the measurement results of usual Z-R relationship. For the latter, we have adopted the expression given by the Chinese Meteorological Institute (1977). It is as follows:

$$Z=264R^{1.39} \tag{10}$$

For convenience, we refer to it as Method M.1c. We have also given the following two definitions of error: (1) The relative error of the total rainfall amount,

$$\delta H=\left|\frac{\Sigma\hat{H}_i-\Sigma H_i}{\Sigma H_i}\right|=\left|\frac{\Sigma\hat{R}_i-\Sigma R_i}{\Sigma R_i}\right| \tag{11}$$

and (2) Relative root mean square error of the rainfall rate,

$$\delta R=\sqrt{\Sigma(\hat{R}_i-R_i)^2/\Sigma R_i^2} \tag{12}$$

where H stands for the total rainfall amount and sign "∧" represents the estimated values of precipitation given by radar.

The calculation results are as follows.

First, we discuss the results of the usual Z-R relationship. Using Eq.(10) to calculate R, the measurement error of the above two groups of the test samples are, respectively, δH=280%, δR=336% (for July 24) and δH=19%, δR=140% (for Aug. 12). These results show that, in these two cases, the M.1c is a bad empirical relationship, the accuracy of which is hardly acceptable in practice.

Utilizing the M.1a, measurement accuracy improves greatly. Measurement errors are reduced to δH=89%, δR=96% (for July 24) and δH=8%, δR=65% (for Aug. 12), much lower than those of the M.1c.

When the M.1b is applied to the test samples on July 24, the measurement error becomes δH=47% and δR=58%. Compared with the M.1a, the measurement accuracy increases by approximately a factor of two. The comparison of the regression line, based on the tested samples of July 24 with empirical relationships (2), (3) and (10), is in logarithmic coordinate and given in Fig.1.

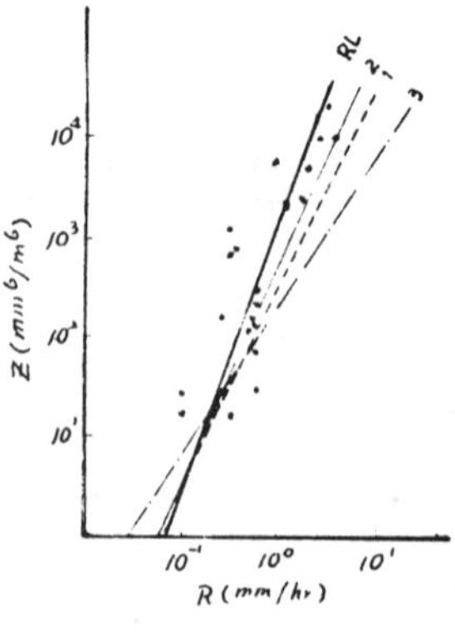

Fig.1 The comparison of the regression line (RL) based on the tested samples of July 24 with lines shown by formulae (2), (3) and (10), which are represented as 1, 2 and 3, respectively.

M.2 has been discussed in three cases according to the exponent B. When B is taken as the same as in M.1c (written as M.2c), for the test samples on July 24, all the 16 values Of A_k on the different radial paths at different times fall in the range from 229 to 275. The rainfall measurement errors are δH=272% and δR=365% Although this result is not good, it is better than the M.1c. For the test samples on Aug. 12, we have similar results. When B is taken as the same value as the M.1a (written as M.2a), still using the samples of July 24 as an example, all A_k are between 289 and 349. The measurement errors are δH=77% and δR=84%. This measurement accuracy is better than the M.1a When the coefficients A and B are both taken as the corresponding values in M.1b (written as M.2b), the A_k for the samples of July 24 ranges from 501 to 605, and δH and δR decreases down to 34% and 48%, respectively. Compared with the M.1c, the worst case, this result is undoubtedly a great improvement. The above results are listed in Table 1.

TABLE 1. THE PRECIPITATION MEASUREMENT ERRORS OF SCHEMES 1,2 AND 3

DATA	M.1a		M.1b		M.1c		M.2a		M.2b		M.2c	
	δH (%)	δR (%)	δH (%)	δR (%)	δH (%)	δR (%)	δH (%)	δR (%)	δH (%)	δR (%)	δH (%)	δR (%)
Aug. 15	/	/	/	/	131	385	/	/	/	/	115	357
Aug. 12	8	65	/	/	18	152	15	62	/	/	7	138
July 24	89	96	47	58	307	413	77	84	34	48	272	365

5. CONCLUSION

The above analysis leads to the following conclusions.

(1) Compared with the usual Z-R relationship, in the light precipitation situation discussed in this paper, the method based on the actual data observed with radar and raingauges has two advantages as follows: (a) It has higher measurement accuracy, and (b) It is easier to expand the sample number by taking samples from other precipitation processes. Especially, the latter advantage may gain from the feedback of climatic information.

(2) Utilizing radar information on radial paths can make the coefficient A or B in Z-R relationship become a variable of time and space; offers improvements in measurement accuracy and is simple to derive.

The Methodology here described may also be regarded as a simulation of a system consisting of radar and radiometers. It shows that the above precipitation measurement methods are feasible in theory and practice.

REFERENCES

Lü Daren and Lin Hai, 1980, Comparisons and combined uses of radar and radiometer in the remote sensing of rainfall distributions, Scientia Atmospherica Sinica, 4, pp.30-39

Meteorological Institute, National Meteorology Administration, 1977, The errors of several selected Z-R relationships, In Collected Papers of Radar Meteorology, Meoteorol. Ins. NMA(Ed), Beijing, pp.95-103

BOUNDARY LAYER MEASUREMENTS OF A CONVECTIVE RAIN BAND IN TROPICAL CYCLONE IRMA

B.F. Ryan and R.C. O'Brien
CSIRO Division of Atmospheric Research
Aspendale, Victoria, Australia

ABSTRACT

As part of the Equatorial Mesoscale Experiment (EMEX), the COSSA Fokker F-27 aircraft, together with the NOAA WD-P3 and NCAR Electra aircraft, investigated a rainband associated with tropical cyclone Irma. The low level structure of the rainband is deduced by compositing the F-27 aircraft data collected during transects of the convective band. Of particular interest is the calculation of rainfall rate from the vapour mixing ratio and the total water mixing ratio. The deduced rainfall rates are compared with those derived from the Bureau of Meteorology radar sited at Gove in northern Australia and good agreement is found. The measurements show that the total water instrument, in conjunction with standard environmental instruments, has the potential for providing ground truth measurements over the ocean during Tropical Rainfall Measuring Mission.

1. INTRODUCTION

The rainfall associated with storms and cyclones is an important component of the hydrological cycle, but the measurement of that rainfall over the ocean is very difficult. Within a tropical cyclone there are mesoscale regions of intense convective activity giving rise to very heavy rainfall rates interspersed with larger regions of steady stratiform rain. The convective rain and the stratiform rain are both important components of the water cycle and need to be measured in any tropical rainfall measuring mission. In weaker tropical cloud clusters Leary and Houze (1979) have shown that after the systems reach maturity the stratiform rain becomes an increasing fraction of the total rainfall. In the formative and intensifying stages of the systems life cycle convective rain dominates.

During January-February 1987 the Equatorial Mesoscale Experiment (EMEX) and the Australian Monsoon Experiment (AMEX) were conducted over northern Australia. The aircraft component of EMEX consisted of the NOAA P3-WD aircraft, the NCAR Electra, and the CSIRO Office of Space Science Applications (COSSA) Fokker F-27. The role of the COSSA F-27 was to make measurements in the boundary layer beneath the two US aircraft.

On 18 January a mission took place to study an outer rainband that was part of tropical cyclone Irma. The instrumentation on the F-27 aircraft included a device developed by Coulman and Parker (1982) for measuring the total water in cloud. The aim of this paper is to show that this total water instrument is effective for measuring moderate to heavy rainfall. Furthermore it is shown

ISBN 0-937194-14-4

that the dynamic and thermodynamic structure of the boundary layer containing the rainband can be deduced from the overall suite of instruments on the F-27.

2. F-27 INSTRUMENTATION

The thermodynamic instrumentation carried by the COSSA F-27 is summarized in Table 1. An important feature of the instrumentation is its redundancy. In particular, the vapour mixing ratio can be calculated from the Rosemount reverse flow temperature sensor and the CSIRO wet bulb temperature sensor or from the EG&G hygrometer.

The total water instrument was designed by Coulman and Parker (1982). It evaporates impacting cloud drops in a labyrinth of heated plates. The mixing ratio of the air containing the evaporated water is then measured psychrometrically. In clear air the total water instrument behaves as a third instrument to measure vapour mixing ratio. The vapour mixing ratio determined from the reference instruments (ie. the Rosemount dry bulb sensor and the wet bulb sensor) and the total water instrument follow a regression line of the form

$$q_t = 0.064q_v + 0.392 \quad . \tag{1}$$

where q_t is the vapour mixing ratio from the total water instrument and q_v is the vapour mixing ratio from the reference temperatures. The scatter about the regression line is of the order of o.2 g/kg^{-1}. This relation is used to remove the systematic differences between the mixing ratios determined by the two techniques. That is, there is a systematic difference between the calibration of the two instruments. On the assumption that the reference instruments are likely to have the better calibration, the total water instrument mixing ratio, q_t was scaled in accordance with Equation 1. A similar relation is used to remove systematic differences between the vapour mixing ratios found using the total water instrument and the dew point instrument.

TABLE 1. AIRCRAFT THERMODYNAMIC SENSORS

Sensor	Parameter	Response Time	Accuracy
Rosemount Reverse Flow Themometer	dry bulb temp.	1s	0.05 C
CSIRO Wet Bulb Sensor	wet bulb temp.	4s	0.05 C
E G&G Aircraft Hygrometer System	dew point	2s	0.5 C
Coulman-Parker Total Water Sensor	total water	4s	0.2 g/kg
King Hot Wire	liquid water	0.1	0.2 g/kg

Figure 1 shows the total water mixing ratio, the vapour mixing ratio from the reference thermometers and the saturation mixing ratio for a pass through the rainband associated with tropical cyclone Irma. Outside the rainband the air is subsaturated and at that time the total water mixing ratio and vapour mixing ratio measurements agree to within the noise level, that is to within 0.2 g kg^{-1}. However, within the rainband itself, the total water mixing ratio exceeds the vapour mixing ratio by more than 2 g kg^{-1}. At 2125 GMT there is evidence for wetting of the reference thermometer since the mixing ratio exceeds the saturation mixing ratio. A comparison between the vapour mixing ratios, derived from the dew point instrument and the reference instruments, confirm that there was wetting of the reverse flow temperature sensor between 2124.30 and 2125.30 GMT.

After passing through the rainband (see Fig.1) the aircraft entered a region of stratiform rain at which point the rain water

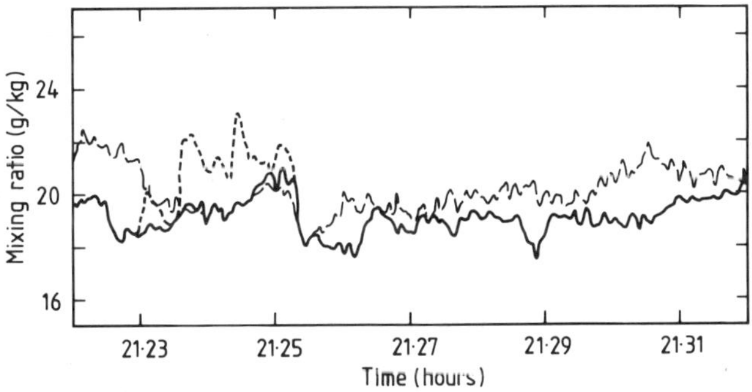

Figure 1. Pass through the rainband in tropical cyclone Irma between 2124 GMT and 2132 GMT on the 18th November 1987. The solid line is the vapour mixing ratio from the reference instruments, the pecked line is the total water mixing ratio and the dashed line is the saturation mixing ratio.

component of the total water was in the noise level of the total water instrument. During heavy rain, in this and other passes, the liquid water as measured by the King hot wire instrument was consistently less than that found by the total water instrument, thus showing that the hot wire was failing to evaporate the drops completely.

In the moderate to heavy rainfall associated with the convective rainbands the signal from the total water instrument unambiguiously shows presence of rainwater. However, in the stratiform rain the contribution of the rain water to the total water is less than 0.2 g kg^{-1}. The light stratiform rain is not able to be resolved by the total water instrument.

An independent assessment of the effectiveness of the total water instrument can be obtained by comparing the aircraft observations with the radar observations from Gove. For comparison with radar observations, it is necessary to convert the rain water mixing ratio to a reflectivity (dBz).

If the Marshall-Palmer model relates to the rain drop size distribution, N(D) with drops of diameter D viz,

$$N(D) = N_o \exp(-\Lambda D) \quad . \qquad (2)$$

We assume this to be valid with $N_0 = 8*10^6 \; N_o \; M^{-3} \; M^{-1}$ (Srivastava 1971) and if Λ is related to the rainfall by the following equation (see Doviak and Zrnic 1984)

$$\Lambda = 4.1 \; R^{0.21} \; mm^{-1} \quad . \qquad (3)$$

where R is the rainfall rate in mm/hr^{-1} all three parameters, namely the rainwater mixing ratio, reflectivity and rainfall then can be readily calculated since the radar reflectivity and liquid water content are known functions of N(D) and D.

It can be shown that the total water instrument should be capable of detecting rainfall rates greater than 3 mm/hr^{-1} and a comparison between the total water instrument and a radar should be valid for reflectivities exceeding 30dBz.

3. THE STRUCTURE OF THE BOUNDARY LAYER OF A TROPICAL CYCLONE RAINBAND

The rainband echoes are clearly identifiable on the PPI sweeps at 2255 GMT from the Australian Bureau of Meteorology radar sited at Gove in the Northern Territory. The Fokker F-27 made a number of passes at different flight levels on the 115° radial from Gove at a distance of 100 to 200 km from Gove. The rainband studied was determined to be moving at -5 m/s^{-1} (i.e. to the east) and all aircraft analyses have been made in this frame of reference. At 2155 the origin of the flight track (o in Fig.2) is approximately 145 km from Gove.

Figure 2a shows the flight track through the rainband between 2122 and and 2236. The 30 second averages of rain water (i.e. water content of the evaporated rain drops) are shown and reflectivity contours in dBz are drawn. Using the relative speed of -5 m/s^{-1} a consistent picture of the rainband has emerged and a new band ahead of the main rainband appears to be in the process of formation.

Figure 2b shows that the rainband has cooled the lower boundary layer air and that the newly developing rainband is at the leading edge of the cold outflow. There are two convergence zones with updraft regions associated with the two rainbands. However the region of highest radar reflectivity corresponds to a strongly divergent region where downdrafts are found.

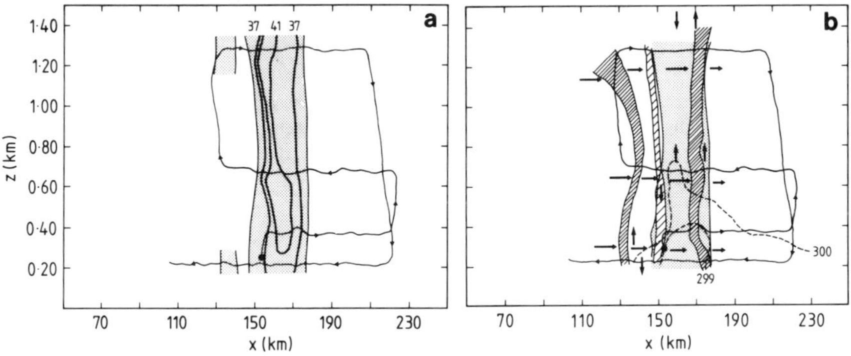

Figure 2(a). Shows the relative aircraft rack assuming a rainband speed of -5 m/s between 2122 and 2236 GMT on the 18th November 1987. The 30 second average values of rain water exceeding 0.25 g/kg are marked on the flight track. Contours of reflectivity are drawn through the rainwater areas.

Figure 2(b). Shows the position of the rainband (stippled), there are two regions of convergence (hatched) and a region of divergence on the western boundary of the rainband (hatched). The 300 and 299 K potential temperature contours are shown. The horizontal arrows correspond to the normal wind velocity component and the vertical arrows to updraft and down draft region

Figure 3 shows two RHI sections taken by the Gove radar. Figure 3a is at 2125 GMT and shows the rainband some 160 to 180 km from Gove in the early stages of development. By 2155 GMT (Fig.3b) the rainband has substantially developed with the main convective development some 160 km from Gove. The radar return at levels above the Fokker F-27 are less than 39 dBz (rainfall rates less than 13 mm/hr^{-1}) whereas the aircraft observations suggest rainfall rates in excess of 18mm/hr^{-1} with peaks exceeding 40 mm/hr^{-1}.

A detailed interpretation of the results and a comparison with results from the few other mesoscale observations of tropical cyclone rainbands (eg Barnes et al., 1983; Barnes and Stossmeister, 1986) will be published elsewhere.

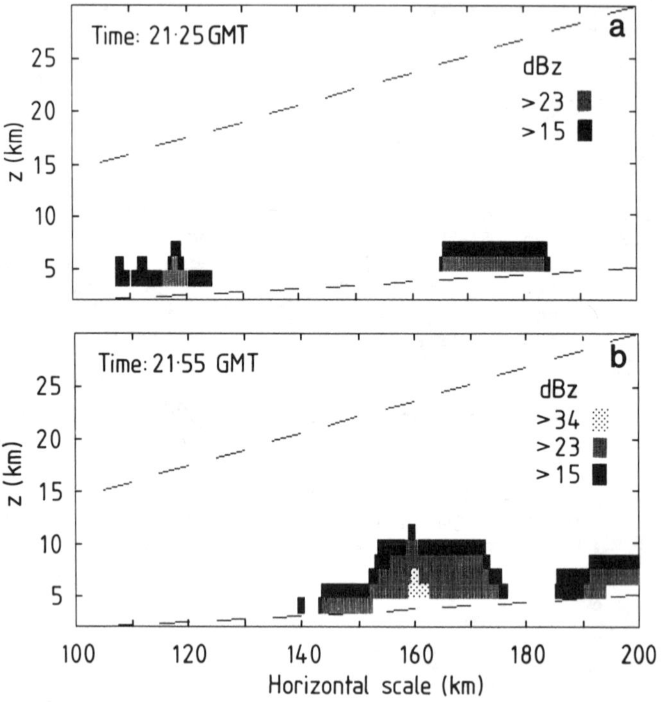

Figure 3(a). Gove radar RHI section along radial 115° from Gove at 2125 GMT on the 18th November 1987.

(b). Gove radar RHI section along radial 115° from Gove at 2155 GMT on the 18th November 1987.

4. CONCLUSIONS

It has been shown that by using the total water instrument, together with the standard thermodynamic instruments on the COSSA Fokker F-27 aircraft, it is possible to deduce the rainfall in the convective regions of a tropical cyclone. The deduced reflectivities are consistent although systematically stronger than observed at higher levels by the Gove radar. With the addition of a Knollenberg 2-D probe to characterize the raindrop spectra an even more accurate evaluation of the total water instrument could have been made.

It would appear that the total water instrument is one of the few instruments that is capable of measuring heavy rain while it is over the ocean and out of radar range. As such it may be able to be exploited as a ground truth tool in the development of any proposed satellites for the Tropical Rain Measuring Mission.

REFERENCES

Barnes, G.M., E.J.Zipser, D Jorgensen and F. Marks, 1983 : Mesoscale and convective structure of a hurricane rainband, J.Atmos.Sci., 40, 2125-2137.

Barnes, G.M. and G.J. Stossmeister, 1986 : The structure and decay of a rainband in hurricane Irene. Mon.Wea.Rev., 114, 2590-2601.

Coulman, C.E. and M.A.Parker, 1982 : On the calibration and performance of an instrument for measuring total water mixing ratio in cloud. J.Appl.Met., 21, 695-702.

Doviak, R.J. and D.S. Zrnic, 1984 : Doppler radar and weather observations. Academic Press, Orlando, Fla. 458 pp.

Leary, C.A. and R.A. Houze Jr., 1979 : The structure and evolution of convection in a tropical cloud cluster. J.Atmos.Sci., 36, 473-457.

Srivastava, R.C., 1971 : Size distribution of raindrops generated by their breakup and coalescence. J.Atmos.Sci., 28, 410-415.

RADIO LINK MEASUREMENTS FOR TRMM GROUND TRUTH

T. Ihara and K. Okamoto
Radio Research Laboratory
Ministry of Posts and Telecommunications
Koganei-shi, Tokyo 184, Japan

ABSTRACT

Capability of radio link technique for obtaining TRMM ground truth data is numerically examined in connection with the mission objectives of the TRMM rain radar. It is shown that radio links at frequencies near 35 GHz can provide most satisfactory capability for obtaining path-averaged rainfall rate. This capability can be further improved by a combined use of much higher frequency systems, especially for retrieving the weak rainfall intensity.

1. INTRODUCTION

Acquiring the ground truth data for rainfall is an indispensable step toward a suitable interpretation of data obtainable by the planned TRMM rain radar. Rainfall rate near the ground, obtainable by TRMM rain radar having a beam footprint of 4 km in diameter at nadir and a range resolution of 0.25 km, is essentially an area-averaged rainfall rate. Since spatial non-uniformity of rainfall usually exists in such an extended area, point measurement by a single raingage is obviously insufficient as TRMM ground truth. Ground-based rain radar, a densely distributed raingage network, radio link techniques, etc. should be cooperatively used to get comprehensive ground truth data.

Within such a variety of techniques, the present paper discusses the capability of a radio link technique for TRMM ground truth.

2. THEORETICAL CONSIDERATION

2.1 BASIC THEORY

The basic idea of the radio link technique is to derive path-averaged rainfall rate from wave attenuation measurements. Attenuation A (dB) which radiowaves undergo during rain through traveling along a path of length L (km), is represented as follows:

$$A = 4343 \int_0^{D_{max}} Q_t(D)\ \overline{N}(D)\ dD \quad L \qquad (1)$$

where D is the diameter of raindrop (mm), $Q_t(D)$ is the total cross section of raindrop (m^2), $\overline{N}(D)$ is the path-averaged drop size distribution ($m^{-3}\ mm^{-1}$), and D_{max} is the maximum diameter of the raindrops (mm).

Similarly, the path-averaged rainfall rate $\overline{R}$ (mm/h) is expressed by using the path-averaged drop size distribution $\overline{N}(D)$ as follows:

$$\overline{R} = 6\ 10^{-4}\ \pi \int_0^{D_{max}} D^3\ v(D)\ \overline{N}(D)\ dD \quad L \qquad (2)$$

ISBN 0-937194-14-4

where v(D) is the fall velocity of the raindrops (m/sec) as given by Gunn and Kinzer (1949).

In principle, there are two ways to derive $\overline{R}$ from attenuation measurements: one is to derive $\overline{N}(D)$ from multifrequency attenuation measurements and then, using the derived $\overline{N}(D)$, calculate $\overline{R}$ through Eq. (2), and the other is to derive $\overline{R}$ from attenuation measurement directly through a semi-theoretical and semi-empirical relationship between A and $\overline{R}$. In the latter case, the choice of wave frequency is essentially important, and the frequency should be chosen so that functional dependence of $Q_t(D)$ on D becomes similar to that of $D^3 \cdot v(D)$ as closely as possible. If $Q_t(D)$ is directly proportional to $D^3 \cdot v(D)$, then A is also directly proportional to $\overline{R}$, hence measurement of A enables the derivation of $\overline{R}$ in a straightforward way by merely taking account of a multiplicative factor. It is widely accepted that frequencies in the vicinity of 35 GHz roughly satisfy the condition mentioned above (Atlas and Ulbrich, 1977; Manabe et al., 1984). In the following, a potential of the technique based on the A-$\overline{R}$ relation is reviewed, especially considering the use of frequencies near 35 GHz. The technique using inferred drop size distribution is discussed elsewhere in detail (Furuhama and Ihara, 1981; Ihara et al., 1984).

2.2 NUMERICAL CONSIDERATION

In Fig. 1, the value of $Q_t(D)$ in m^2 is plotted against the drop diameter D for 11.5, 34.5, 50.4, 81.8, and 245.5 GHz. The values of $Q_t(D)$ are calculated by using Mie scattering theory on the assumption of spherical drop shape and dielectric constants of water at 20 C. In the same figure, the value of $D^3 \cdot v(D)$ is also shown as a function of D. In the calculation, data of v(D) given by Gunn and Kinzer (1949) are used. The ordinate scale for $D^3 \cdot v(D)$ is expressed in arbitrary units.

The functional similarity between $Q_t(D)$ at 34.5 GHz and $D^3 \cdot v(D)$ clearly exists for the range of D roughly less than 3.5 mm. This similarity places a theoretical basis for the radio link at frequency around 35 GHz to work as a path-averaged raingage. However, the two quantities compared show a growing discrepancy as the drop diameter increases beyond 3.5 mm. If such large drops predominantly contribute to rainfall rate, capability of 35 GHz link as the path-averaged raingage may significantly deteriorate.

First order approximation of the functional form of the raindrop size distribution is a negative exponential one, i.e., $N_0 \exp(-\Lambda D)$. In the following, the function of radio link at about 35 GHz is further examined by assuming this type of drop size distribution. In Fig. 2, the normalized attenuation coefficient r (dB/km/mm h^{-1}), i.e., attenuation coefficient per unit rainfall rate, is shown against Λ for 11.5, 34.5, and 81.8 GHz (Manabe et al., 1984). The value of N_0, not shown in this figure, is adjusted so as to normalize the attenuation coefficients with respect to rainfall rate. The curve of r for 34.5 GHz reveals a broad peak centered around Λ=2, taking nearly a constant value 0.24 for the range of Λ 1.8 to 2.5 (mm^{-1}). This range of Λ corresponds to rain rates between 10 and 50 mm/h for the case of Marshall and Palmer distribution (1948), implying that this range of

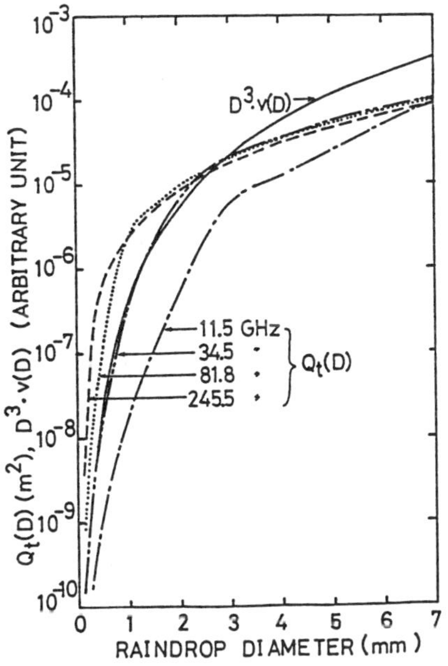

FIGURE 1. Functional dependences of $Q_t(D)$ and $D^3v(D)$ on drop diameter D.

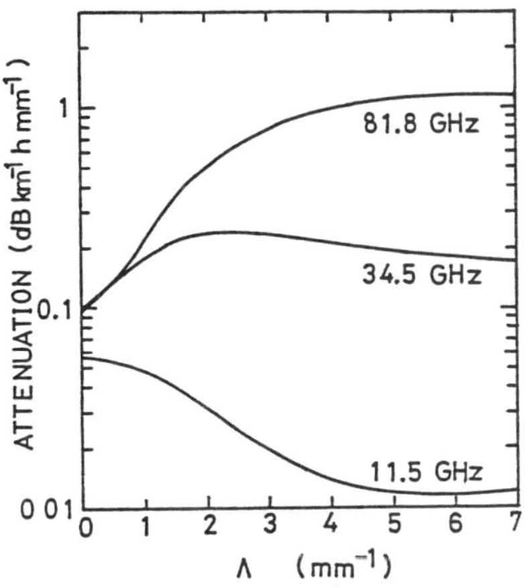

FIGURE 2. Normalized attenuation coefficient v.s. Λ.

rain rates would be estimated accurately from attenuation measurements at 34.5 GHz by using r=0.24. At both sides of the peak, the value of r decreases with the change of Λ: the slope for the lefthand side of the peak is much steeper than that for the opposite side because large drops play much more important role in the former case. Unified use of r=0.24 tends to underestimate the rain rate for the ranges of Λ outside the interval of 1.8 to 2.5. For example, the underestimates of rain rates are -11, -4, and -14 % respectively for Λ= 1.56 (100mm/h), 2.9 (5mm/h), and 4.1 (1mm/h), where the rain rates corresponding to the Marshall and Palmer distribution are parenthesized. Since, as described in the next section, fractional contributions to annual accumulation of rainfall from rainfall rates less than 1 mm/hr and greater than 100 mm/hr are less than about 10%. Some 10% of the underestimated rainfall rate in the ranges less than 1 mm/hr and greater than 100 mm/hr produces an overall error of less than a few percent in the estimate of the annual rainfall accumulation. From the above discussion, radio link technique using frequencies near 35 GHz is expected to work quite well as the path-averaged raingage.

3. SOME PRACTICAL CONSIDERATIONS FOR TRMM GROUND TRUTH

3.1 REQUIREMENT FOR THE RANGE OF RAINFALL RATE

One of the most important mission objectives of the TRMM rain radar is to provide a knowledge of monthly accumulation of rainfall averaged over the area of 360,000 km^2 with observation errors less than 20 % (Keating, 1987). From this mission objective, we can understand the range of rainfall rates for which TRMM radar measurement must accurately be made, and hence the ground truth data must be obtained. In the following, requirement for the range of rainfall rates is examined. At present, the adequate monthly data base is not easy to access, so that the examination is made based on the annual rainfall data.

Fig. 3 shows a curve of fractional rainfall accumulation, typical of tropical regions, for the annual time base: the ordinate represents the contribution from rainfall rates less than the value shown in the abscissa. This curve is obtained from the annual cumulative distribution data of rainfall rates tabulated for rain climatic zone N, typical of tropical rain climate, in the classification by International Radio Consultative Committee (CCIR, 1986). It should be noted that relatively weak rainfalls of less than 10 mm/h produce a significant contribution to annual accumulation of rainfall, nearing 50 % of the total accumulation. Although the importance of weak rain is emphasized, the contribution of strong rain should also be adequately taken into account: information on rainfalls with intensities at least up to 50 mm/h is needed to account for the 80 % of annual rainfall accumulation.

The above discussion provides an estimate of the range of rainfall rates for which ground truth measurements must be accurately made.

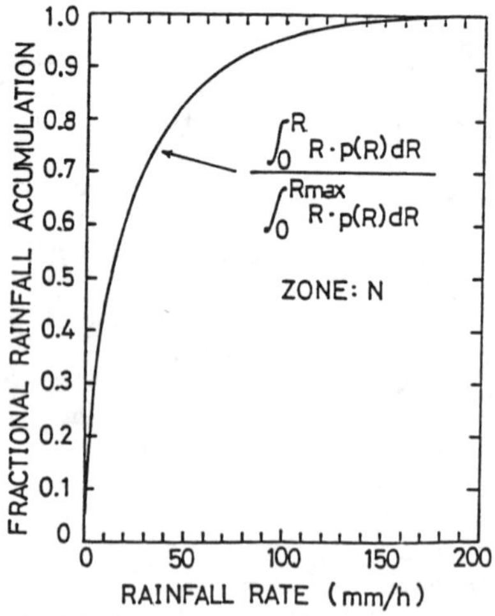

FIGURE 3. Fractional rainfall accumulation v.s. rainfall rate.

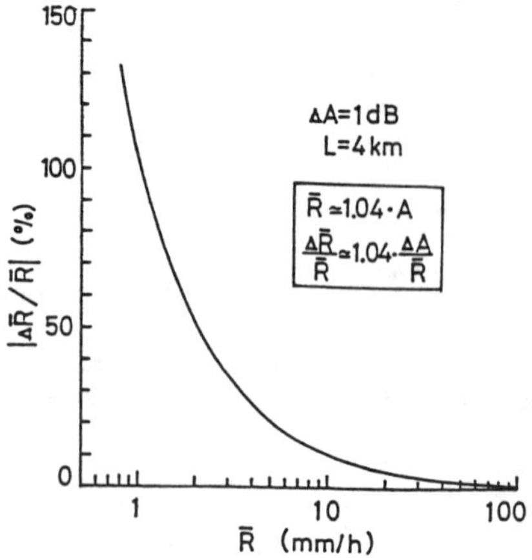

FIGURE 4. Effect of attenuation measurement error on estimate of rain rate.

3.2 EFFECT OF ATTENUATION MEASUREMENT ERROR

In this section, the effect of error involved in the attenuation measurement is discussed. Assuming that radio link experiment is performed at 34.5 GHz on a 4 km path (corresponding to the size of radar beam footprint), path-averaged rainfall rate is approximately related to attenuation through the following equation:

$$\overline{R} = 1.04 A \quad (3)$$

where r=0.24 is employed. Then, the error $\Delta\overline{R}$ in the estimate of $\overline{R}$ is related to the error ΔA in the measurement of A as follows:

$$\Delta\overline{R}/\overline{R} = 1.04 \Delta A/\overline{R} \quad (4)$$

In Fig. 4, the value of $|\Delta\overline{R}/\overline{R}|$ is plotted against the value of $\overline{R}$ for the case of ΔA = 1 dB. Values of $|\Delta\overline{R}/\overline{R}|$ reach 100 %, 40 %, 20 %, and 10 % respectively at 1mm/h, 2.5 mm/h, 5 mm/h, and 10 mm/h. Such magnitudes of errors should not be ignored because rainfalls of intensities less than about 10 mm/h play a very important role in the total accumulation of rainfall as mentioned in Sec. 3.2.

One of the possible means to mitigate this problem is a combined use of radio system operated at higher frequency, which can enhance the sensitivity of total radio link to weak rain. For example, attenuation coefficients at 81.8 GHz become roughly four or five times as large as those at 34.5 GHz for rainfall rates less than 5 mm/h, as seen in Fig. 2. This means that $|\Delta\overline{R}/\overline{R}|$ can be reduced to roughly fourth or fifth of that for the case of 34.5 GHz, when both frequencies are employed in a common radio path and identical measurement errors are expected for both frequencies.

Use of multifrequencies also has an advantage of providing the capability of inferring the path-averaged drop size distribution (Furuhama and Ihara, 1981; Ihara et al., 1984), which may be helpful for calibration of Z-R relationship used in the radar equation.

4. CONCLUSION

Capability of radio link technique for obtaining TRMM ground truth data is numerically examined in connection with the mission objectives of the TRMM rain radar. It is shown that radio links at frequencies near 35 GHz can provide most satisfactory capability for obtaining the path-averaged rainfall rate. This capability can be further improved by a combined use of higher frequency systems, especially for retrieving the weak rainfall intensity.

Despite the advantages of radio link technique mentioned above, there still remains a question to be answered about whether the path-averaged rainfall rate can be fully representative of the area-averaged rainfall rate, which is ultimately required for TRMM ground truth.

REFERENCES

Atlas, D., and C.W. Ulbrich, 1977: Path- and area-integrated rainfall

measurement by microwave attenuation in the 1-3 cm band, J. Appl. Meteor., 16, 1322-1331.
CCIR, 1986: Radiometeorological data, Recommendations and Reports of the CCIR, Rep. 563-3.
Furuhama, Y., and T. Ihara, 1981: Remote sensing of path-averaged raindrop size distributions from microwave scattering measurements, IEEE Trans. Ant. Prop., AP-29, 275-281.
Gunn, K.L.S., and Kinzer, G.D., 1949: The terminal velocity of fall for water droplets in stagnant air, J. Meteor., 16.
Ihara, T., Y. Furuhama, and T. Manabe, 1984: Inference of raindrop size distribution from rain attenuation statistics at 12, 35, and 82 GHz, Trans. IECE Japan, E67, 211-217.
Keating, T., 1987: Tropical rainfall measuring mission (TRMM), presented at EASCON Symp.
Manabe, T., T. Ihara, and Y. Furuhama, 1984: Inference of raindrop size distribution from attenuation and rain rate measurements, IEEE Trans. Ant. Prop., AP-32, 474-478.
Marshall, J.S., and W.M. Palmer, 1948: The distribution of raindrop size with size, J. Meteor., 5, 165-166.

AN ESTIMATION OF PRECIPITATION INTENSITY FROM GMS DATA BY USING DIGITAL RADAR ECHO INTENSITY DATA

Shingo Osano, Toshihiro Motoki, and Kazufumi Suzuki
Meteorological Satellite Center
Japan Meteorological Agency
Kiyose-shi, Tokyo 204, Japan

ABSTRACT

By combining GMS (Geostationary Meteorological Satellite) data and digital radar data, a new scheme of satellite-derived rainfall intensity estimation was developed. Correlation of the reflectivity in the visible channel with the radar echo intensity was stronger than that of the brightness temperature of the infrared channel. Applying a regression analysis, an estimation of radar echoes was made.

The correlation coefficient between radar echo intensity data and estimates (tentatively named SP) was 0.62 in the case of combination of Tbb and Vis, while 0.50 in the case of only Tbb.

1. INTRODUCTION

Detailed rainfall estimation data are desired by the forecasters, especially in the nowcast field. To support them, digital estimates of rainfall data should be made with fine mesh and communicated in a timely manner.

The digital radar network developed by the Japan Meteorological Agency (JMA), that covers major portions of Japanese Islands, enables us to make a quantitative comparison with the GMS data over a considerable area, and to develop a new scheme of rainfall estimation (Osano et al., 1986).

2. METHOD

Input data sets are the digitized GMS data (wave lengths of infrared channel and visible channel are 10.5-12.5 um and 0.55-0.75 um respectively) and the digitized radar echo intensities at or below 2 km height, with ground echoes eliminated, processed so as to be representative of each of 0.1 lat. X 0.1 lon. squares within 30 N to 41 N and 130 E to 145 E.

Radar observations with 5 km resolution from the six radar sites were changed into the 0.1 lat. and 0.1 lon. resolution, and merged to fill a part of 111 x 151 grid-point data. Radar echo intensities are denoted by R hereafter. An example of the composite radar echo intensity chart, 00 GMT 06 AUG 1985, is shown in Fig. 1, in which shading indicates the areas where beam heights are greater than 2 km. The radar observations with beam height 2 km or below were utilized here.

The GMS infrared (Tbb) and visible (Vis) data sets were also changed into (111 x 151) mesh data of 0.1 x 0.1 resolution. Vis were standardized after Davis et al. (1974).

TROPICAL RAINFALL MEASUREMENTS
John S. Theon and Nobuyoshi Fugono (Eds.)

ISBN 0-937194-14-4

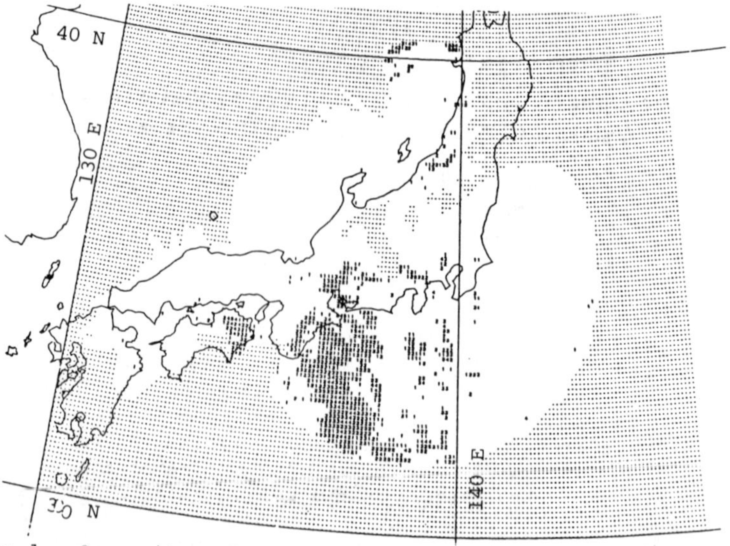

Figure 1. Composite radar echo intensity chart, 00GMT 06 AUG 1985. Digits indicate radar echo intensity levels,i.e. 1:0< 1 mm/hour, 2:1< <2, 3:2< <4, 4:4< <8, 5:8< <16, 6:16< <32, 7:32< <64, 8:64< <128, 9:128< . The dots in the study area show where beam height is above 2.0 km.

called Tbb and Vis respectively hereafter. Vis were standardized after the method of Davis et /1).

Least-squares multivariate linear regression was applied to obtain estimated R in terms of the ten levels from GMS data.

A regression equation is assumed:

$$GR = a_1 Tbb + a_2 Vis + a_3 Tbb^2 + a_4 Tbb \times Vis + a_5 Vis^2 + a_6 Tbb^3 + a_7 Tbb^2 \times Vis + a_8 Tbb \times Vis^2 + a_9 Vis^3 + C \quad (1)$$

where Tbb is equivalent black body temperature,
Vis is reflectivity in the visible channel,
GR is the GMS based, estimated R,
a_i and C are constants.

The least-squares method was applied to GR and R.

$$dGR = d\left(\sum_{i,j}\left(R_{i,j} - GR_{i,j}\right)^2\right) = 0 \quad (2)$$

where i and j are grid point numbers.

Once an empirical formula of Eq. (1) is made, a look-up table is set up for obtaining GR values directly from digitized GMS observations. Accuracy of GR values was evaluated by correlation coefficient.

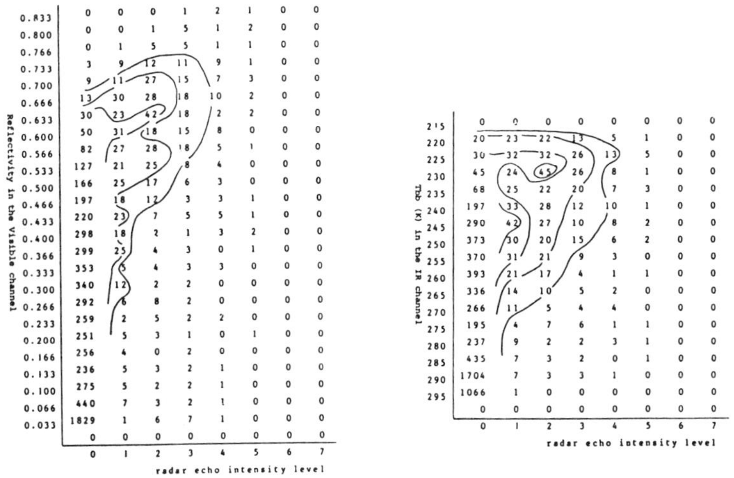

Figure 2. Two-dimensional histogram between reflectivity in visible channel (Vis) or equivalent black body temperature (Tbb) and radar echo intensity (R) for 00 GMT 06 AUG 1985.

3. SENSITIVITY OF VIS AND TBB TO R

Two-dimensional histogram analysis on Vis, Tbb and R using the data of 00 GMT 06 AUG 1985 is shown in Fig. 2, in which numerals are the number of data points in association with Vis (or Tbb) and R values. Let us call the R of zero intensity level as R0 and R of levels "1" to "9" as Rr for convenience. It is seen in Fig. 2 that data points of Rr begin to increase from Vis 0.4 and, as Vis increases beyond 0.6, Rr jumps while R0 drops. Rr increases as Tbb becomes colder than 265 K.

It appears that, in comparing Fig. 2, Vis responds more sharply to the change in the R level than Tbb does.

4. ASSESSMENT OF ESTIMATES

Accuracy of GR values are assessed for each of different data sets: Tbb alone and combination of Tbb and Vis.

The changes in value of correlation coefficients with different input data sets were assessed. Two kinds of the GMS data sets of 00 GMT 06 AUG 1985 were tested, i.e., Tbb and combination of Tbb and Vis. The radar chart, GR values and GMS pictures (visible and infrared) are shown in Figs. 1,3-5, respectively. In the case of Tbb alone (Fig. 6), GR points are seen near A where corresponding R points do not exist in Fig. 3. As cirri-form clouds near A had very low temperatures, as do CB-clusters causing heavy rainfall, GR values were not correct. Comparing Fig. 3 (Vis and Tbb), general pattern of GR points and locations of weak/strong GR values are very similar. Correlation coefficient between R and GR reached 0.62 in the case of Tbb-Vis, while only 0.50 for Tbb alone.

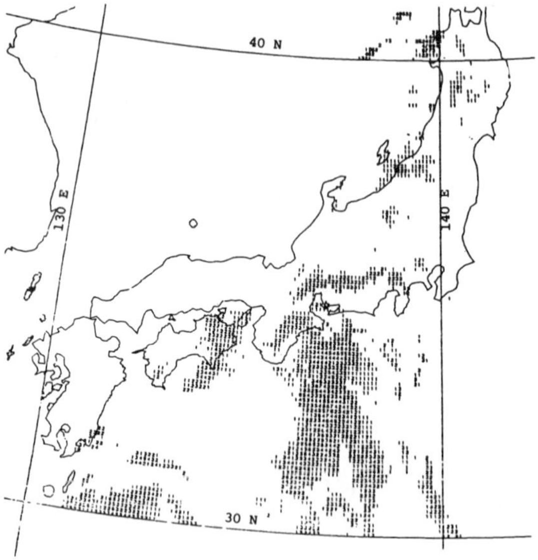

Figure 3. GR chart from combination of Vis and Tbb, 00 GMT 06 AUG 1985. Each grid-point value was multiplied by 2.

Figure 4. Visible picture, 00 GMT 06 AUG 1985.

Figure 5. Infrared picture, 00 GMT 06 AUG 1985.

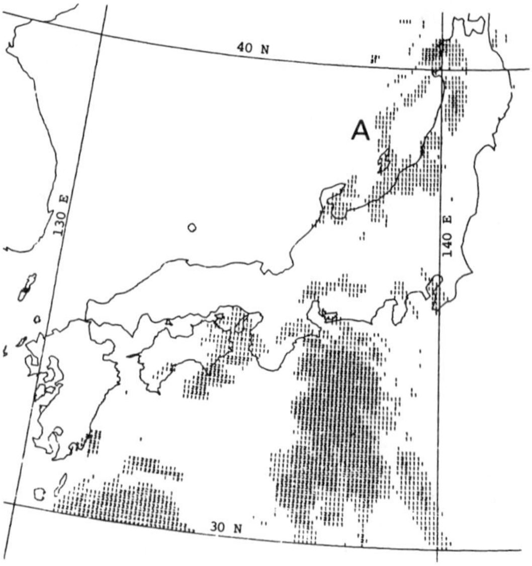

Figure 6. Same as Fig. 3, except for Tbb alone.

In this case, a tropical air mass covered the research area. Over the area northeast of typhoon center(28 N, 132 E), GR values and distribution agreed with both the echo pattern and the location of strong echoes associated with CB-clusters.

5. CONCLUSIONS

Radar echo intensities greater than level "1" (0.0< <1.0 mm/hour) were associated with reflectivity in the visible channel greater than 0.4. The number of radar echo data points rapidly increased when the reflectivity in the visible channel exceeded 0.6.

As to the Tbb data, radar echo intensities greater than level "1" appeared when Tbb became colder than 265 K. As Tbb became colder than 245 K, the number of strong echo data points increased, while that of level zero remained unchanged.

It is seen that Vis responds more sharply to the change in the R level than Tbb does.

With regards to the estimates, the correlation coefficient between radar echo intensity data and GR was 0.62 in the case of combination of Tbb and Vis, while 0.50 in the case of Tbb alone.

There is a difficulty in the use of this technique at night time and further investigation based on the one channel data set of Tbb will be required.

ACKNOWLEDGMENTS

The authors would like to express their sincere thanks to Mr. Michio Hituma and Mr. Yukiou Takemura, JMA, for their helpful suggestions. Sincere appreciation is expressed to Mr. Toshiaki Yasuda, JMA, for reviewing the manuscript.

REFERENCES

Davis, P. A. et al.,1974: Infrared emissivities and upper tropospheric cloud motion, SRI Project ERU-2014, Stanford Research Insitute, 7-10.

Osano, S., T. Motoki and K. Suzuki, 1986: An estimation of precipitation intensity from GMS data by using digital radar echo intensity, Preprint for the 23rd Conference on Radar Meteorology and Cloud Physics, JP46-49.

ESTIMATION OF LAGGED SPACE-TIME CORRELATIONS IN RAIN DATA[1]

S. Nakamoto, J.-T. Wang, D.A. Short[2], *and G.R. North*
Texas A&M University
College Station, Texas 77843, USA

ABSTRACT

Lagged space-time correlations of rain field data are needed in many applications pertinent to mission planning of rain rate measuring satellites. The estimator for such a correlation calculation can take one of several forms each of which is correct in the limit of large data sets. However, for a finite amount of data, most estimators are biased. We examine two such estimators of special interest to TRMM planning and show which has the lesser bias by direct simulation and by an asymptotic theory.

1. INTRODUCTION

The lagged space-time correlations of rain rates are needed in the construction of random rain field models (Bell, 1987), in the estimation of sampling errors in finite rain gauge networks (Gurney, private communication), and in the estimation of bias due to beam filling (Chiu et al., 1987). An intuitive sense of the spatial autocorrelation function is that its e-folding length is a measure of the *range of influence* of a point gauge. For a given time-averaging interval, if the circles of influence of several neighboring point gauges overlap, then we can assume that sampling errors in estimating area average rain rate will be small.

The lagged autocovariance function can be estimated by the formula

$$\hat{C}(m) = \frac{1}{N} \sum_{n=1}^{N-m} (x_{n+m} - \bar{x})(x_n - \bar{x})$$

where the overbar represents an average through the data included in the N sum. The lagged autocorrelation is

---

[1]Supported in part by a grant from NASA Headquarters

[2]Permanent Address: Code 613, NASA Goddard Space Flight Center.

TROPICAL RAINFALL MEASUREMENTS
John S. Theon and Nobuyoshi Fugono (Eds.)

ISBN 0-937194-14-4

$$\rho(m) = \hat{C}(m) / \hat{C}(0)$$

The formula above is a *consistent estimator* of the autocovariance in the sense that the limit of large N does not lead to a value different from the true ensemble average. On the other hand in rain rate data one often has a finite strip of data N, but many different independent realizations of the strip. In this case, which is the better estimator to use in applications? Method A): Use estimates of C(0) and $\bar{X}$ from the individual strips to form estimates of the autocorrelation, $\hat{\rho}(m)$; then average across all the $\hat{\rho}(m)$ estimates to form an estimate of $\rho(m)$. Method B): Form the estimates of $\bar{X}$ and C(0) from all the strips and proceed to calculate C(1) as above. Trenberth (1984) has considered these two estimators with the application of large-scale meteorological fields in mind. He found that method B was the better for that application.

We decided to reconsider the problem because rain fields are so distinctly nongaussian that this might bear strongly on the choice of estimator. Our approach was to build a "time" series using an AR1 process so that we could have an exact theoretical answer to compare with our estimation results.

2. AR1 PROCESS

The AR1 process

$$x_{n+1} - \mu = a(x_n - \mu) + z_n$$

where x_n is the value of the variate (rain rate) at time (or space) step n and z_n is a white noise forcing. The theoretical value of the lagged autocorrelation is a^m where m is the lag. This last result is independent of the pdf of z_n. We can, for example, imagine a process where z_n is chosen from a binomial distribution (p,1-p) where p is very small. This kind of choice for the pdf of z_n leads to a white noise series that is mostly zeroes. Similarly, the x_n so generated will have long lists of zeroes (or at least very small values) provided a is small compared to one. Since the spatial autocorrelation functions for rain fields look qualitatively like this functional form, and because the series has the character of being dominated by zeroes, we have an excellent candidate for testing the competing estimators.

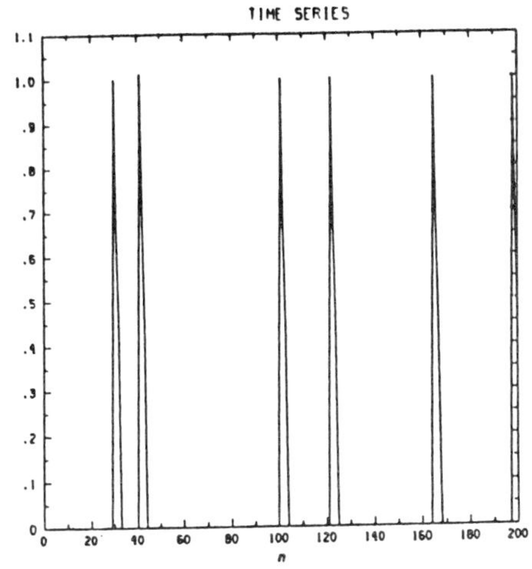

Figure 1. *Time series of the binomially driven AR1 process with p=0.03 and a = 0.667.*

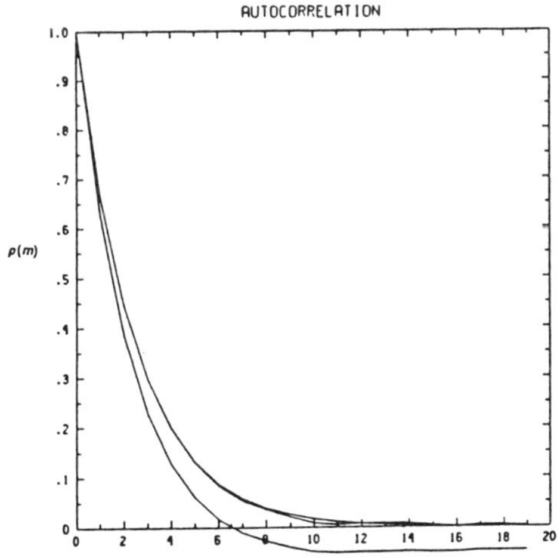

Figure 2. *Spatial auto-correlation functions for the AR1 process shown in Figure 1. The upper curves (nearly coincident) show the theoretical value and Method B. The lower curve shows Method A with strips of length 70.*

3. NUMERICAL SIMULATIONS

Figure 1 shows a typical time series of simulated rain rates for a value of p = 0.03 (in the white noise forcing) and a = 0.667. This case corresponds roughly to the case encountered in the GATE spatial autocorrelation case and is the one needed by Bell (1987). It corresponds to an autocorrelation length of 12km and a probability of rain of about 0.08. For a strip in GATE data, N=70. Figure 2 shows the autocorrelation functions for this case. The upper curve is method B and the lower is method A. Method B is so close to the *exact answer* , a^m , that the two are nearly coincident. While the time series would look very different for the gaussian versus the binomially driven process, it is remarkable that the autocorrelation function depends only upon a; this latter also holds to a good approximation for the estimators making use of either method.

4. ASYMPTOTIC THEORY

By looking at the estimators in the limit of large N (details to be published elsewhere) we find a single form that encompasses both cases except for a different interpretation of the symbol N:

$$b = \frac{(1+a)(a^m - 1)}{(1-a)N}$$

where b is the bias for lag m, a is the AR1 parameter, and for Method A: N is the number of members of the *strip*. Whereas, for Method B, N is the number of members of the *entire* data set including all strips. Since the autocorrelation "time" for the series is $1/(1-a)$, the product $N(1-a)$ is just twice the number of independent sampling intervals in the aggregate N.

Bell's method of estimating the autocorrelation function in space (it originated with A. McConnell, private communication) seems to be the correct one in this case.

REFERENCES

Bell, T. L., 1987: A space time stochastic model of rainfall for satellite remote-sensing studies, J.Geophys.Res., D8, 9631-9645.

Chiu, L., G. R. North, A. McConnell and D. Short, 1987: Estimation of bias due to nonuniformly filled FOV in satellite rain rate sensing. (to be submitted; also see Chiu and North, this conference).

Trenberth, K. E., 1984: Some effects of finite sample size and persistence on meteorological statistics. Part I: Autocorrelations. Mon.Wea.Rev., 112, 2359-2368.

RAINFALL ESTIMATE FROM DIGITAL RADARS IN TAIWAN AREA

Tai-Chi Chen Wang, Long-Nan Chang, and Pay-Liam Lin
National Central University
Chung-Li, Taiwan, R.O.C.

ABSTRACT

Two digital S-band radars together with a proper raingauge network provided an excellent opportunity for a rainfall estimate study in the Taiwan area.

Rainfall distributions in space and time from composite radar data were compared to those obtained from surface gauge stations.

An optimizing technique suggested by Brandes (1982) was utilized to calibrate the radar rainfall estimate with raingauge measurements. Barne's objective analysis (1973) was applied in the technique to interpolate both radar and gage data onto common cartesian grids, a more accurate rainfall estimate can be obtained with a variable g/r ratio on every grid.

By 1989, a real-time monitoring network combining three digital radars and automatic raingauges around the island will be implemented. The present study will serve as a guide for the design of this network and may also provide a ground truth information for the TRMM mission.

1. INTRODUCTION

The heavy rainfall during typhoon and "Mei-Yu" seasons sometimes brings extensive damage to the Taiwan area. On the other hand, rainfall is a very important water resource to the island. A raingauge network consisting of 74 surface weather stations and 126 self-recording raingauges was set up by the Central Weather Bureau. Five conventional radars constitute the radar network. Two S-band radars have provided digital data since summer 1985. Currently, only qualitative estimates of rainfall are made for operations. How to combine the gauge and radar observations to provide a more realistic quantitative rainfall estimate is a major interest of meteorologists in Taiwan.

Radar data are collected at the first ten minutes of every hour unless special experiments are being conducted. A statistical comparison between radar estimate and gauge measurement is not reasonable with one hour time resolution of radar data. Rainfall characteristics, such as duration, coverage and intensity of different types of rain, will be considered in order to decide a proper radar data acquisition scheme for the future.

TROPICAL RAINFALL MEASUREMENTS
John S. Theon and Nobuyoshi Fugono (Eds.)

ISBN 0-937194-14-4

A case was selected to show the variation of G/R ratio. An optimizing technique suggested by Brandes (1974) was adopted to estimate the radar rainfall with raingauge measurements. The results indicate that a more homogeneous raingauge distribution in some areas is needed to provide a better estimate. The proper Z-R relations of various raintypes also need more comparisons in the future.

2. GAUGE NETWORK AND RADAR NETWORK

Fig. 1 and Fig. 2 show the locations of surface weather stations and self-recording raingauges as of this date. About one-third of stations are located at the northern part of Taiwan near the capital of Taipei (one gauge per 100-km area approximately), the rest are located on the west side of the Central Mountain Ridge in a very inhomogeneous distribution. The topography of Taiwan island is shown in Fig. 3. The steep mountains and short rivers increase the possibility of flash flood, hence more gauges were placed at the upstream of a watershed and dam area. In the remote mountain area only a few gauges were installed. On the south west plain there are some places with very sparse gauge distributions (one gauge in 500-km area).

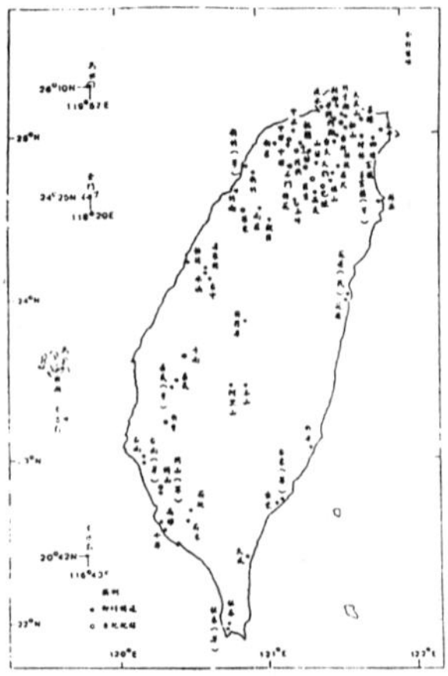

FIGURE 1. Locations of surface station

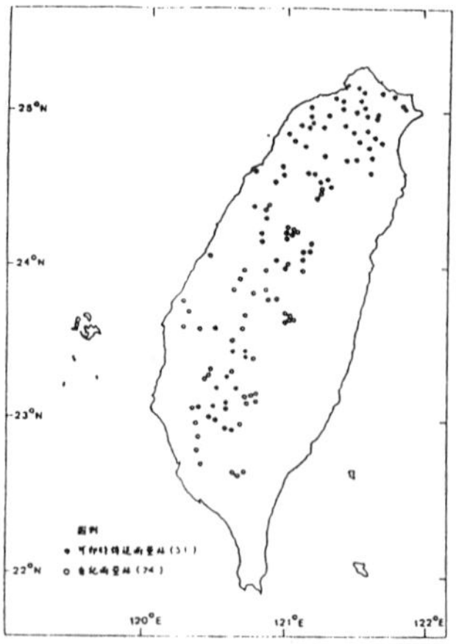

FIGURE 2. Locations of self-recording rainguages

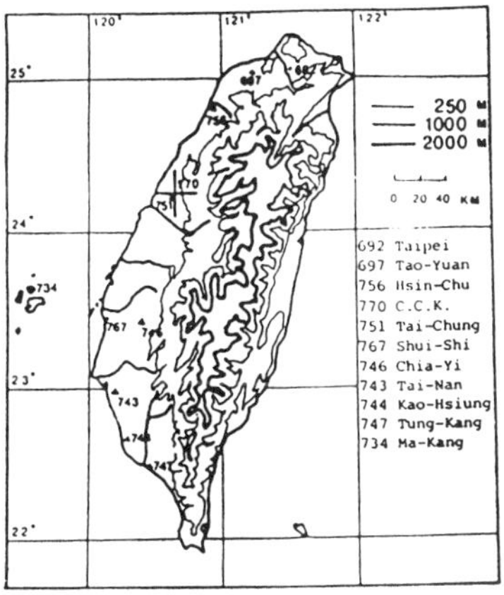

FIGURE 3. Topographic map of Taiwan

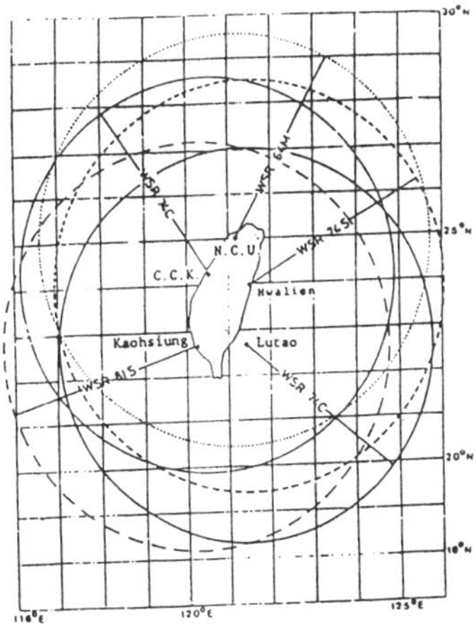

FIGURE 4. Location and coverage of radars

Specification	Hualien/Kaoshiung WSR-74S/81S	C.C.K./Lutau WSR-74C	NCU WSR-64M
Wavelength	10.35 -- 11.1 cm	5.3 cm	10.35 -- 11.1 cm
Frequency	2700-2900 MHz	5600-5650 MHz	2700-2900 MHz
Peak transmitting Power	500 KW	250 KW	500 KW
Pulse Repitition and frequency	539 & 164 PPS	266 PPS	545 & 164 PPS
Pulse width	0.5 & 4 sec 10%	3.0 sec	0.5 & 4.5 sec.
Antenna diameter	12 feet	8 feet	12 feet
Beam width	2.25 degrees	1.65 degrees	2 degrees
PPI display range	50 km, 125 km 250 km, 500 km	100 km, 230 km 450 km	25 nm, 50 km 125 nm, 250 km
RHI display range	250 km, 500 km 50 km, 125 km	450 km 100 km, 230 km	25 nm, 50km 125 nm, 250 km

TABLE 1. Specification of radars

The location and coverage of each radar is shown in Fig. 4 and the radar specifications are listed in Table 1. Although Taiwan island and its vicinity are within the range of the network, due to the topography of Central Mountain Ridge, a great deal of land area is blocked.

Two radar stations (Hwalien WSR 74/S and Kaohsiung WSR 81/S) can currently provide digital data storage on tapes. From June 1985 to April 1987, only the Cartesian interpolated grid data called CV (Column Vector), at one hour intervals are available. Each bin consists of the maximum reflectivity values within each vertical column. Horizontal resolution is 4 km x 4 km. The dbz value above noise level, ranging from 0 db to 63 db in every 0.5 db, were recorded. Since May 1987, only the CA (constant altitude) data sets for selected levels at 30-minute intervals are available upon request.

3. RAINFALL CHARACTERISTICS

A few rain types can be distinguished from a statistical study (Chen and Wu, 1978). In the northern part of Taiwan, the widespread persistent frontal rain dominates the winter and the spring, while in the central and southern part, from October to April the weather is rather dry. May to June is the "Mei-Yu" season, and the rainfall over the whole island reaches a peak. The complex rain types in this season are still not well-recognized. Some deep convective systems may develop and dump heavy rainfall. From July to September, typhoons also bring a lot of rainfall to the whole island. The daily average rainfall from Taipei and Kaohsiung are shown in Fig. 5. It illustrates the rainfall characteristics of northern and southern Taiwan.

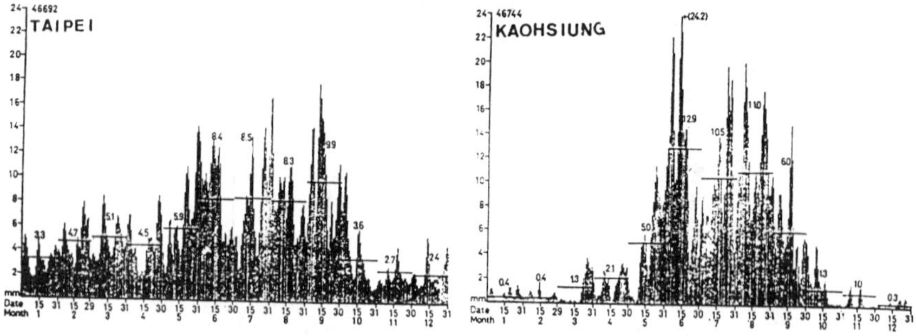

FIGURE 5. Daily rainfall characteristics (1951-1970) of Taipei and Kaohsiung (adopted from Chen and Wu, 1978)

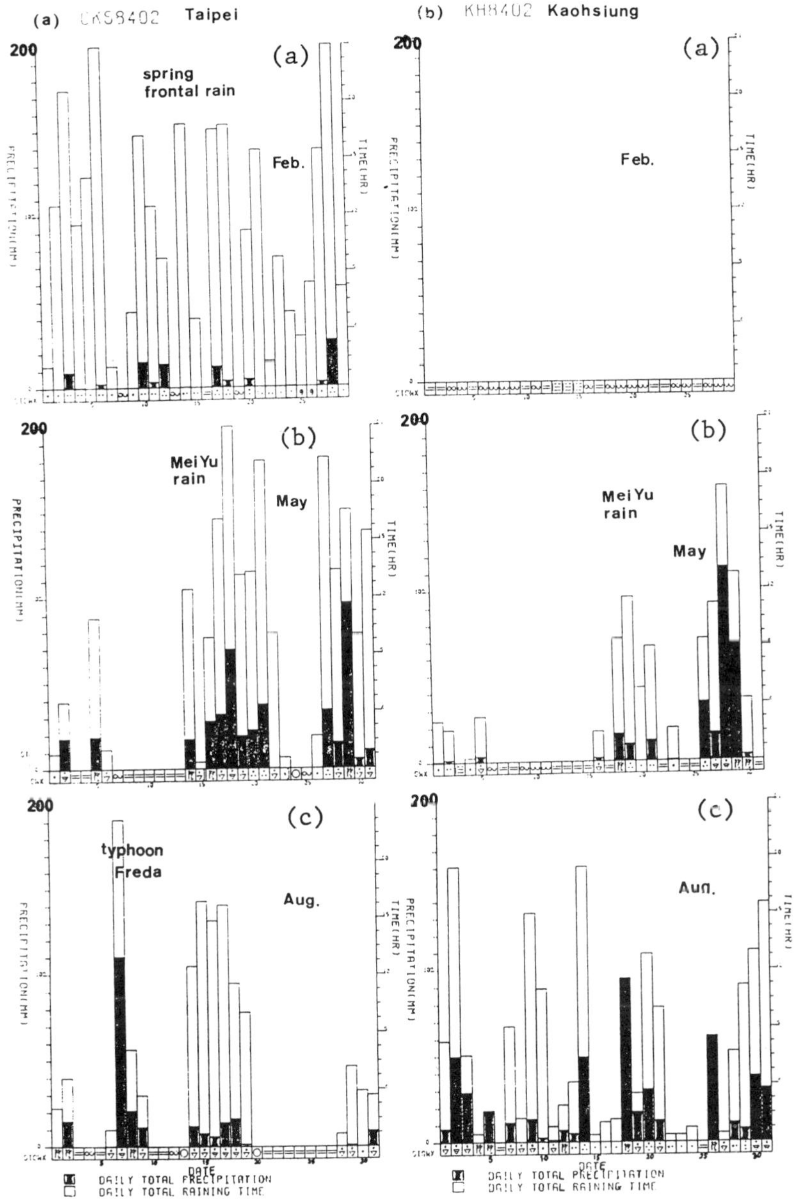

FIGURE 6. Composite histogram of Taipei daily rainfall amount and precipitation hours with additional symbol of the most significant weather marked - (a) February, 1984 (b) May, 1984 (c) August, 1984. (Adopted from CAA annual report)

FIGURE 7. Same as Figure 6 except for Kaohsiung

Recently, some radar analysis (Liao et al., 1984; Chen 1984) showed summer convection enhanced by topography is also a very important rain type. Actually the unique topography of Taiwan influences the spatial distribution of rainfall of every type (Liu and Chu, 1986).

As an example of the duration of various raintypes, Fig. 6 and Fig. 7 showed the daily rainfall and total rain time of February, May, and August 1984 of the Taipei and Kaohsiung (locations are marked on Fig. 3).

4. COMPARISON BETWEEN DIGITAL RADAR ESTIMATE AND GAUGE MEASUREMENT

For the past three years pictures of the Kaohsiung digital radar were collected and compared to the hourly gauge data. A few problems were revealed. Because of the location of Kaohsiung radar, it is very difficult to observe the winter and spring rain in northern Taiwan. Due to the earth curvature effect, only the upper part of a very strong echo (above 8 km) could be seen by this radar. Secondarily, the ground and sea clutter are very serious due to its 338.1 m station height and 2.25 beam width. It is only near the central part of the west plain that radar pictures can be compared with surface gauge data in Mei-Yu and typhoon seasons. Since radar data were collected during the first ten minutes in one hour intervals, the gauge rainfall rates (mm/hr) were collected at the same time (first ten minutes accumulation multiplied by 6) for comparison. Radar rainfall estimates were made using the Marshall and Palmer (1948) Z=200R relation. The results indicate a positive correlation in time series of most gauge stations. However, our sample cases were too few to make a statistical comparison. The results also show the radar estimates tend to underestimate the rainfall rate in heavy rain cases. Many systematic error sources could contribute to this discrepancy (Brandes, 1982; Austin, 1987).

In May and June 1986, a field experiment called "TAMEX phase I" was held in the Taiwan area as a testing field program for 1987 TAMEX (Taiwan Area Mesoscale Experiment). For this period, 0.5-hour interval digital radar and 10 minute gauge data at the major events are available. From May 20 to May 21 1986, a precipitation system was observed across the Taiwan Strait which later brought a rainfall accumulation of approximately 200 mm in two days in some major cities in central and northern Taiwan. From 2300 May 20 to 0700 May 21, sixteen reflectivity distributions for every 30 minutes were carefully analyzed. In Fig. 8, the echo pattern of six different times are selected to show the evolution of the precipitation system.

Gauge measurements in 10-minute intervals for Wuchi and Taichung are illustrated in Figs. 9a and 9b. The locations of these two cities are marked in Fig. 8. The gauge accumulations for every 10 minutes were multiplied by a factor of 6 to provide the unit of rainfall rate, mm/hour. The radar estimate of the same location is also plotted in the same figure in a 30-minute-interval time series. Comparisons between reflectivity and gauge data show a positive correlation. From Fig. 9 we notice that the rainfall rate changes rapidly. Intense heavy rain could fall for less than 30 minutes then get weak suddenly. According to Adler (1981) a time resolution of 15 minutes or less is needed for the radar scan to observe the temporal change of a strong convective thunderstorm case. Fig. 10 shows the G/R ratio in 30-minute time series for WuChi and Taichung. The gauge rainfall rate and the radar estimate at H+10 and H+40 were used to calculate the ratio. Serious underestimates of the radar measurement were observed during the heavier rain. The G/R ratios were greater than 3.0 when gauge measurements were greater than 10 mm/hour. On the other hand overestimates happened when gauge measurements were less than 2 mm/hour. The discrepancy is probably due to the strong spatial gradient of the real rainfall distribution. The large resolution volume of radar data tends to smooth out the sharp gradient. Another possibility could be that the Z=200 R relation was not proper for this strong convective precipitating system.

The spatial distributions of the G/R ratios within a 160 km x 200 km domain (Box I in Fig. 8) are illustrated in Fig. 11 at the selected time 0500. A well-organized NE-SW oriented rainband had already passed Wuchi and Taichung at this time (Fig. 8). The heavy rain core was located east of Taichung. The reflectivity was 30 dbz at Taichung and 20 dbz at Wuchi. Gauge measurements were 10.2 mm/hour at Taichung and 0.6 mm/hour at Wuchi. An underestimate (G/R 4.2) was found near Taichung, and an underestimate (G/R 0.5) was found near Wuchi. This sharp gradient of calibration factor could be the consequence of the large variation of real rainfall in both space and time.

The rainfall optimizing scheme suggested by Brandes was applied in this case. Within the domain of Box I, the G/R ratios were interpolated onto each grid by the objective analysis (Barnes' scheme 1973). In the gauge data void area, an average G/R ratio was assigned to the grid. Fig. 12a shows the smoothed radar estimate (using Z=200 R) over a 40-km x 40-km window (Box II in Fig. 11), Taichung was located at the center of the domain (43,10) and Wuchi was located at (41,12). After the G/R ratios were interpolated, the radar estimates were multiplied by the gridded ratio to provide the "corrected" rainfall estimates. From the results of Fig. 12b, the objective analysis was able to bring up the radar estimate near heavy rain core and retain the spatial variation of the sharp gradient of rainfall. The rainfall estimate of the heavy rain core was about 45 mm/hour. This cell actually passed through Taichung from 0400 to 0500 (Fig. 8). The accumulation over Taichung in one hour was 37 mm. This result indicates the "corrected" estimate seems much more correct than the first estimate. However, much denser gauge network data is necessary to verify this result. Currently such data are not available.

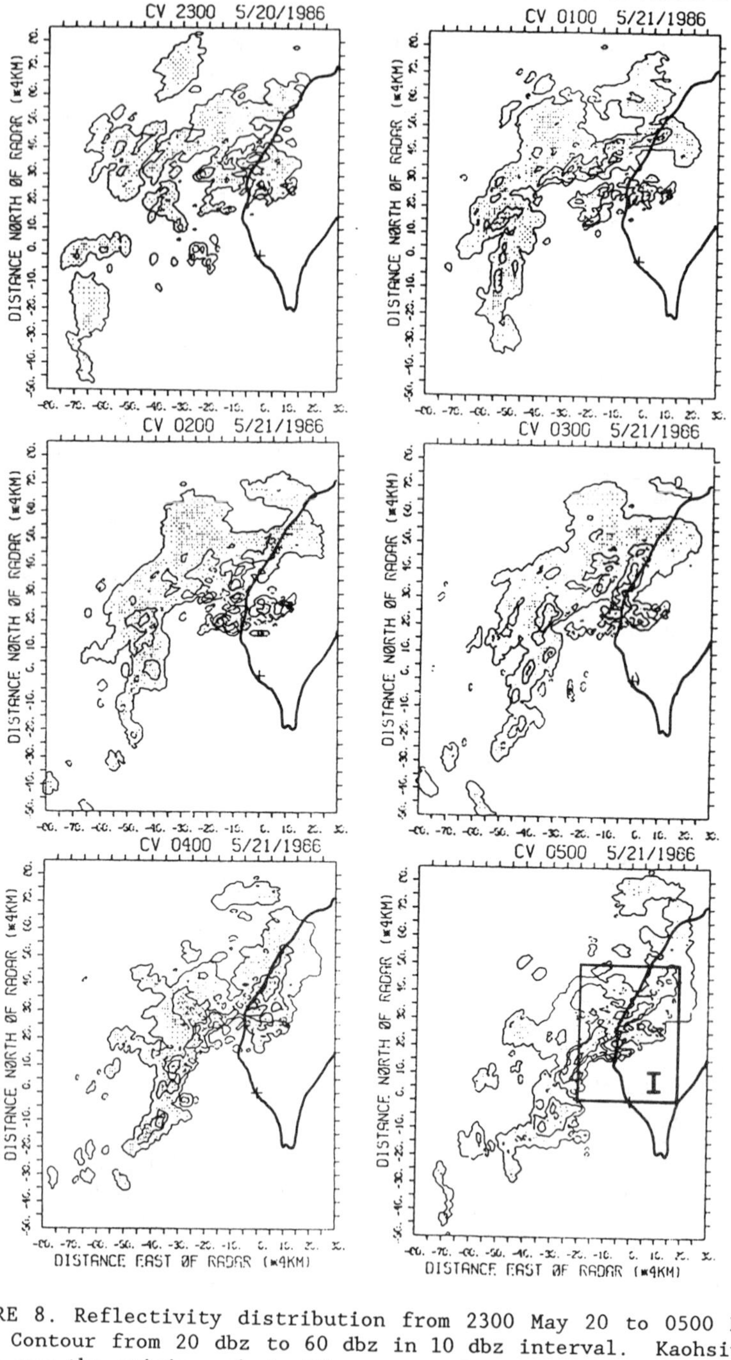

FIGURE 8. Reflectivity distribution from 2300 May 20 to 0500 May 21. Contour from 20 dbz to 60 dbz in 10 dbz interval. Kaohsiung radar was the origin marked with a cross sign.

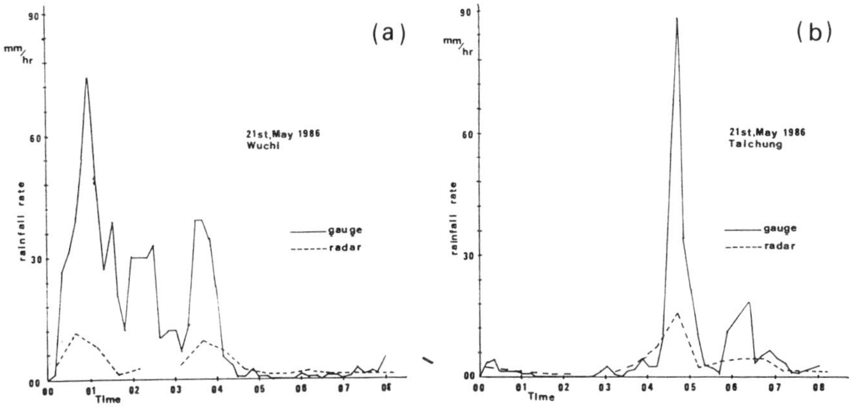

FIGURE 9. Gauge measurement in 10 minute intervals. Radar in 30 minute intervals of Wuchi (9a) and Taichung (9b).

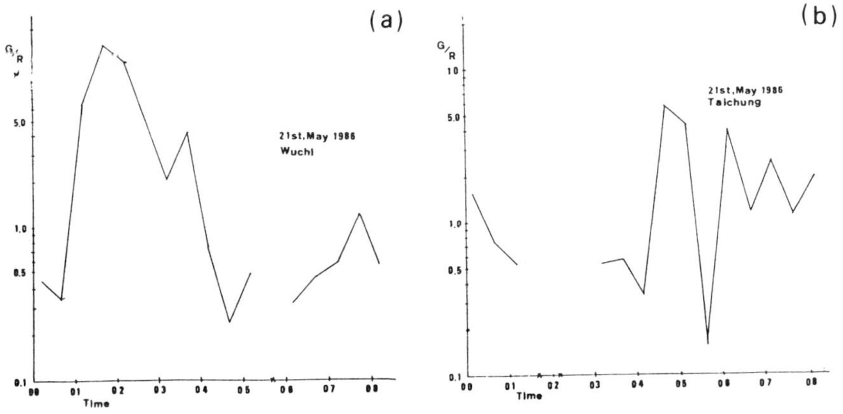

FIGURE 10. G/R ratio [30 minute interval for Wuchi (10a) and Taichung (10b)]

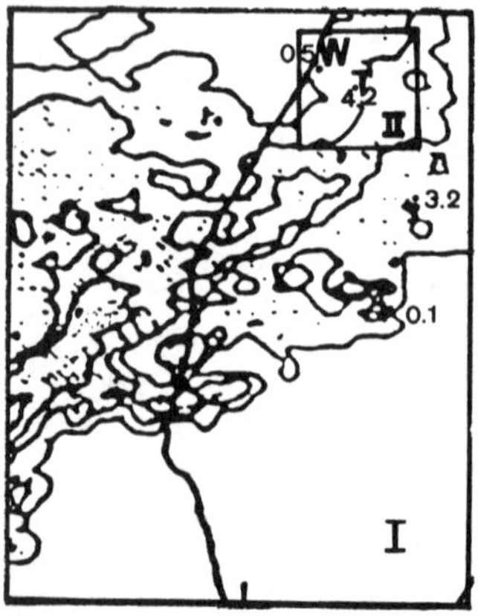

FIGURE 11. Spatial distribution of G/R ratio. (The relative position of optimizing technique calculation domain is marked in Fig. 8).

CV 0500 5/21/1986
smoothed uncalibrated radar estimate

	5	6	7	8	9	10	11	12	13	14	15	
48	.9	.9	.9	1.0	1.2	1.5	1.9	2.3	2.8	3.7	5.2	48
47	.9	.9	.9	1.0	1.2	1.6	2.4	3.0	3.4	4.3	5.1	47
46	.9	.9	1.0	1.2	1.4	2.2	3.6	4.5	4.6	4.8	5.1	46
45	.9	1.0	1.2	1.4	1.8	2.5	3.7	4.6	4.8	4.4	4.2	45
44	.9	1.0	1.2	1.6	2.0	2.4	2.9	3.7	4.6	4.7	4.4	44
43	.9	.9	1.1	1.4	1.8	2.4	2.9	3.8	5.2	6.0	5.9	43
42	.9	1.0	1.2	1.3	1.6	2.4	3.5	5.4	7.4	7.7	7.1	42
41	1.3	1.4	1.4	1.5	1.8	2.5	4.1	7.1	9.5	9.1	7.4	41
40	2.3	2.3	2.5	3.1	3.5	3.6	4.6	6.6	7.9	7.1	5.8	40
39	5.2	5.0	4.6	5.1	6.0	6.2	5.7	5.1	4.6	4.0	3.6	39
38	10.0	9.3	7.5	7.2	7.7	7.9	6.6	4.3	3.0	2.7	2.7	38

CV 0500 5/21/1986
optimizing calibration factor rainfall

	5	6	7	8	9	10	11	12	13	14	15
48	.0	.0	.2	.5	1.1	2.2	4.2	7.0	10.1	5.5	7.6
47	.0	.1	.4	.8	1.6	3.2	6.5	10.3	13.7	18.8	7.5
46	.0	.3	.6	1.3	2.5	5.5	11.6	17.4	19.7	21.9	24.4
45	.2	.5	1.1	2.2	4.0	7.6	13.4	19.0	21.5	20.7	20.3
44	.3	.7	1.6	3.1	5.5	8.5	11.7	16.2	21.1	22.7	21.8
43	.5	1.0	2.0	3.6	6.0	9.2	12.3	17.1	24.5	29.2	29.7
42	.9	1.6	2.6	4.0	6.0	9.7	15.7	25.4	35.8	38.3	35.5
41	1.6	2.7	3.9	5.4	7.3	10.7	19.1	34.0	46.4	44.7	36.4
40	3.9	5.9	8.0	11.7	14.9	16.5	21.6	31.6	38.0	33.3	24.9
39	7.5	15.1	16.7	21.0	26.4	28.5	26.8	23.7	20.2	15.5	12.7
38	14.7	13.5	29.8	31.2	35.1	36.2	28.5	16.8	10.8	9.1	9.1

FIGURE 12. (a) Radar estimate using $Z=200R^{1.6}$ relations in mm/hour. (b) Corrected radar estimate using Brandes' (1974) optimizing technique in mm/hour

Comparisons between gauge and radar data of different raintypes and synoptic conditions need to be accomplished in the future. The validity of optimizing schemes also needs to be carefully verified for different kinds of situations.

6. FUTURE OUTLOOK

For such a complex situation of rainfall estimates, it was agreed by many meteorologists that both radar data and gauge data are needed to better estimate rainfall. In Taiwan we are in the beginning stage of learning how to utilize digital radar data.

From September 1987, a Doppler radar belonging to the CAA (Civil Aeronautics Administration, Republic of China) started to operate at Chiang Kai-Shek airport. Before 1991 another digital radar will be installed by the Central Weather Bureau near Taipei. The other three conventional radars will also be modified to provide digital data storage in the near future. Since May 1987, a careful calibration has been carried out for the preparation of the TAMEX experiment. Since then, the radars have been calibrated with the same procedure every two weeks. A common network connecting these seven radars and real-time gauges was proposed and approved in CWB. This plan will also be carried out soon. In the meantime, more research should be done with available data sets. Currently, a suggestion has been made to CWB and CAA for a radar scanning scheme based on this preliminary study.

A 2 km x 2 km bin is suggested for data storage once the precipitating system is within 200 km of the radar. For the wide spread stratiform rain in winter and spring near northern Taiwan, data at a very low elevation angle should be collected every 30 minutes. For strong convective systems in Mei-Yu and typhoon seasons, the 30-minute volume scan data should be collected every 15 minutes. Also, more rain gauges should be installed on the islands and in the strait to provide comparisons offland.

In conclusion, the equipment in Taiwan should be able to provide a ground truth observation for TRMM. However more work needs to be done by meteorological organizations and meteorologists in the next few years to achieve the goal.

7. ACKNOWLEDGMENTS

Special thanks go to CWB of R.O.C. for providing the grant for this study. The authors would like to express their appreciation to many individuals in CWB for their assistance.

REFERENCE

Austin, P.M., 1987: Relation between measured radar reflectivity and surface rainfall. Mon. Wea. Rev., 115, 1053-1070.

Adler, R.F., 1981: Severe storm requirements for precipitation information, Precipitation Measurements from Space, P 31-34.

Barnes, S.L., 1973: Mesoscale objective map analysis using weighted time-series observations, NOSS Tech. Memo. ERL NSSL-62, Norman, Oklahoma, 60pp.

Brandes, E., 1974: Radar rainfall pattern optimizing technique. NOAA Tech. Memo. ERL NSSL-67, Norman, Oklahoma, 16pp.

Brandes, E., 1982: Rainfall measurement with Radar, Proceedings of the NEXRAD Doppler Radar Symposium/Workshop sponsored by the CIMMS, Univ. of Oklahoma-NOAA, Sept. 22-24, 1982, p. 144-163.

Chen, T.J., G. and C.-C. Wu, 1978: On the climatological characteristics at five cities in Taiwan. Atmos. Sci. vol. 5, No. 2, the Meteorological Society of the Republic of China, 1-16.

Chen, C.-S., 1984: The preliminary investigation of terrain effect on the movement of summertime convective systems. Bullentin of Geophysics, 26, 133-156, National Central Univ., Chung-Li, Taiwan, R.O.C.

Liao, S.-Y., C.-K. Lo, S.-S. Hong, C.-S. Chen, H.-M. Jiang, L.-C. Kung, C.-H. Shiao, Y.-M. Jong, C.-C. Liu and C.-H. Liu, 1984: Radar analysis of mesoscale weather system (II) Further investigation, Tech. Report, pp. 56, NSC73-0202-M008-10.

Liu, K.Y. and K.-K. Chu, 1986: The effects of the Taiwan terrain on the rainfall distribution. Tech. Report, pp. 72, NSC74-0202-M-072-09.

Marshall, J.S. and W.M. Palmer, 1948: The distribution of raindrops with size. J. Meteor., 5, 165-166.

POLARIMETRIC RADAR CHARACTERIZATION OF TRMM MICROPHYSICS*

A.R. Jameson
Applied Research Corporation
Landover, Maryland 20785, USA

ABSTRACT

The capability to distinguish among different types of precipitation (eg., snow, graupel, hail, rain) using radar polarization measurements is undergoing rapid development. It is argued that such measurements can greatly aid the development of the next generation of radiometric rainfall retrieval algorithms required for the next decade.

1. INTRODUCTION

Observations of the earth from space are intrinsically oriented towards phenomena of global scale. It was proposed ten years ago (Wilheit et al., 1977), that radiometric measurements of upwelling radiation could be used to estimate rainfall rates over the ocean. Because of the low emissivity of the oceans and of icy hydrometeors, the measured radiation (normally expressed as brightness temperatures) should be due almost entirely to the rainmass along the path between the earth and the satellite. From the brightness temperature the rainfall rate could be estimated provided the height over which the rain extended could also be estimated or measured.

While sound in principle, complications always arise to challenge scientists. Radiometric measurements from aircraft (Wilheit et al., 1982) revealed that icy hydrometeors were not as transparent as originally hoped. While it is true that their emissivity is low, they can still be effective scatterers of both the upwelling radiation and the isotropic radiation from space. Consequently, overlying canopies of icy hydrometeors can substantially reduce the brightness temperatures.

This conclusion was refined by the computations of Wu and Weinman (1984) who found that the obscuring effects of ice increased substantially at frequencies above 21 GHz and that aspherical ice induced a dependence of brightness temperature on polarization. Since satellite based radiometric measurements are proposed using frequencies above about 37 GHz, icy hydrometeors are likely to influence the brightness temperatures measured from satellite.

---

*This work was supported under grants ATM85-96007 from the National Science Foundation and grant NAS5-27796 from the Goddard Space Flight Center of the National Aeronautics and Space Administration.

ISBN 0-937194-14-4

The scattering by icy hydrometeors can vary substantially, however, depending upon the size, density (index of refraction), shape, the particle orientation with respect to the plane of propagation, and the wavelength at which observations are made. Pristine ice crystals, aggregate snow, graupel, and hail do not scatter alike. In addition, radiometric measurements are sensitive to the depths and areal extent of these different types of icy hydrometeors. These factors, in turn, are functions of the climate and season around the globe.

While the ultimate implementation of global satellite precipitation measurements may require undertaking a global-scale microphysics climatology program (GSMCP), the TRMM objectives are focused on the tropics. Even so, our knowledge of the temporal and spatial distribution of the different types of hydrometeors, even in this more restricted domain, is practically non-existent.

In order to develop the next generation of radiometric rainfall retrieval algorithms required by the 1990's, this void in our knowledge should be eliminated. One possible approach is to use numerical simulations. The microphysical output of model simulations of tropical system dynamics can be used as input into radiative transfer models. While this approach is potentially useful and certainly better than what is now available, the microphysical processes in numerical models are usually highly parameterized. While parameterization is necessary, some important precipitation processes are not sufficiently understood to formulate realistic parameterizations. Factors such as shape, which affects the scattering properties of hydrometeors, are often not even included in the parameterizations. In addition, computations are usually restricted to only a few categories of hydrometeor sizes and types. This is especially true as the computations are expanded to larger and larger domains. Consequently, the correspondence between the results from numerical simulations and reality is not entirely clear.

Observations of the areal extent, temporal duration, and depth of different types of hydrometeors in different tropical systems are, therefore, required. Not only can such observations check the numerical simulations and improve their performance, but they can also provide information which may never be simulated adequately. The classical method of gathering such measurements is by aircraft. Although aircraft observations are attractive because one sees what is actually up there, aircraft can not provide the necessary temporal and spatial resolution over extended time frames. In addition, aircraft are unable to penetrate into the regions of most intense precipitation. Presently the only viable hope for obtaining measurements of hydrometeor type over extended periods and regions in the next ten years is by radar.

2. RADAR TECHNIQUES FOR HYDROMETEOR IDENTIFICATION

2.1 REFLECTIVITY FACTOR

The simplest radar measures the power backscattered at one frequency and one polarization. Usually this backscattered power is reexpressed in terms of the reflectivity factor (Z) which is only a function of meteorological quantities. While the measurement of Z is convenient for locating precipitation, it is not very useful by itself for discriminating among hydrometeor types or for time-resolved quantitative rainfall measurements. Because Z is a function of particle sizes, concentrations, and types, it is impossible to separate the influence of each factor using only the single measurable Z.

2.2 DOPPLER VELOCITIES

With the development of more sophisticated radars, more variables have become available. The trick, however, is to identify measurables which can be associated with particular physical characteristics unique to each type of hydrometeor. One of the first improvements to the simplest radars was to make them coherent. By pointing such radars vertically, it is possible to measure the Doppler velocity spectra of the falling hydrometeors. Using a VHF radar (Wakasugi et al., 1986) or a combination of VHF-UHF radar (Larsen and Röttger, 1987), it is apparently possible to separate vertical air speed from the hydrometeor terminal fallspeed. This information can probably be used to discriminate between some types of hydrometeors, given a basic knowledge of the stratification of the atmosphere. There is, however, significant overlap between the fallspeeds of different hydrometeors such as raindrops and graupel. The main limitation of this approach, though, is that only a limited two dimensional view is obtained as the precipitation moves past the radar.

To some extent this limitation can be overcome using the VAD (Lhermitte and Atlas, 1961) technique as extended by Srivastava et al.(1986). While applicable to homogeneous stratiform precipitation, it cannot, however, be used in perturbed convective conditions.

While it is conceivable, in principle, to use dual-Doppler measurements from an airborne radar, in practice the derived estimates of particle fallspeeds have such large errors that they are useless (Bohne and Srivastava, 1975).

2.3 DUAL-WAVELENGTH TECHNIQUES

Although a dual-wavelength system can be used to detect hail (Atlas and Ludlam, 1961; Westcott et al., 1984), its potential use in the tropics would be to distinguish liquid from icy hydrometeors and to provide estimates of attenuation rate which can be used to estimate rainfall (Atlas and Ulbrich, 1974, 1977; Jameson, 1983b). Hydrometeors attenuate microwaves by scattering and absorption. Since microwave absorption is much greater for liquid than for icy hydrometeors, significant attenuation might often be used to successfully identify liquid precipitation. There are shortcomings, however.

Icy hydrometeors can cause considerable loss by scattering when the circumferences of the hydrometeors are comparable to the wavelength. Attenuation is not always synonymous with absorption. In addition, the selection of wavelengths exhibiting attenuation sufficient to provide resolution along a radar radial will usually only yield measurements restricted to short distances. The role of dual-wavelength measurements for discrimination of hydrometeors, therefore, is probably very limited.

2.4 POLARIZATION

The only other property of electromagnetic waves which is available for hydrometeor discrimination is polarization. The plane wave backscattered to a radar can always be decomposed into a component with the same sense of polarization as that transmitted (the co-polarized component) and a component with the sense of polarization orthogonal to that transmitted (the cross-polarized component). If a radar is equipped with two receivers, it can receive both components simultaneously. If it has only one receiver, then polarization measurements can only be obtained by transmitting and receiving alternatively between two polarizations, usually between the two orthogonal polarizations. (Because the transmitted and received polarizations are identical, this pair of polarizations is referred to as a pair of orthogonal co-polarizations. A good example is alternating horizontal and vertical linear polarizations.)

Parameters derived from measurements at different polarizations are functions of the shapes, indices of refraction, and the orientations and asymmetries of the hydrometeors. The challenge is to identify unique combinations of these parameters which correspond only to a particular type of hydrometeor. This is probably best done by using the parameters to describe the distribution of hydrometeor shapes and the degree of orientation, conveniently expressed as the spread of the distribution of canting angles with respect to the local vertical direction.

In this paper, only the horizontal-vertical orthogonal pair of linear polarizations will be considered. At the wavelengths used for most meteorological radars, however, the parameters defined by this polarization pair can also be measured using circular polarizations (Jameson and Dave', 1988) and probably other elliptical polarizations as well.

2.4.1 Distribution of Hydrometeor Shapes

Because of the enhanced index of refraction of water over that for ice and because of the greater gravitational orientation of liquid hydrometeors, rain constitutes a highly oriented medium (McCormick et al., 1972) compared to other types of precipitation (Hendry et al., 1976, 1987). Consequently since a raindrop is approximately oblate as it falls (Pruppacher and Beard, 1970), the power backscattered at horizontal polarization is greater than that at vertical polarization. The ratio of the power measured at horizontal polarization to that at vertical is the differential reflectivity (Seliga and Bringi, 1976). In oriented icy (Jameson, 1985) or liquid (Jameson, 1983a) mediums, it provides a measure of mean particle shape (axis ratio) provided the hydrometeors are small compared to the radar wavelength. For liquid precipitation the differential reflectivity is usually substantially between 1.25 and 4.

In rain a measure of the mean axis ratio can be converted into a mean drop size which can then be used to estimate rainfall (Seliga and Bringi, 1976; Jameson, 1983b; Jameson and Dave', 1988). In icy hydrometeors, however, differential reflectivity is small because of the reduced index of refraction of ice compared to liquid water and because icy hydrometeors exhibit greater canting as they fall. This contrast in the differential reflectivity between liquid and icy hydrometeors can be used to discriminate between these two classes of particles (Hall et al., 1980, 1984).

While differential reflectivity can often be used to discriminate rain from ice, it is always near unity regardless of the type of icy hydrometeor. Differential reflectivity is, therefore, of little use in distinguishing among various icy hydrometeors through the measurement of mean shape unless the effects of particle canting can be removed. This appears to be possible (Jameson and Mueller, 1985) provided some measure of canting can be found. Since all icy hydrometeors do not behave identically as they fall, it is also likely that a parameter for measuring canting could also be used for discriminating among some different solid hydrometeors.

2.4.2 Distribution of Hydrometeor Canting Angles

It is well known that when the hydrometeors are small compared to the wavelength (so-called Rayleigh-Gans scatterers) , the electric field induced within a particle by an incident radar wave can be expressed in terms of the dipole moments along each of three principal axes (Gans, 1912; Atlas et al., 1953; van de Hulst, 1957). However, the magnitude of the induced dipole is a function of the electric field "seen" by the hydrometeor. This field, in turn, depends upon the orientation of the particle with respect the plane of propagation defined by the oscillation of the components of the incident electromagnetic wave. For horizontal and vertical linear polarizations, it is the tilt or canting of the axis of symmetry of a hydrometeor from vertical that produces backscattered power associated with a wave which is cross-polarized to that transmitted. The ratio of this power to that in the co-polarized channel is called the linear depolarization ratio (LDR).

For this reason LDR is a 'natural' parameter for measuring hydrometeor canting. Unfortunately, the magnitude of LDR also depends upon the shapes of the hydrometeors. This dependence can be eliminated, however, by combining measurements of LDR, differential reflectivity, and the magnitude of the cross-correlation function between horizontal and vertical co-polarizations (Jameson, 1985).

With this canting parameter, it is possible to estimate the variance of the distribution of hydrometeor canting angles.

As the dimension and mass of a hydrometeor increase, it wobbles and spirals more and more as it falls. It is reasonable, therefore, to expect the canting parameter to increase as measurements are made first in snow, then graupel and finally hail (Jameson, 1985).

3. THE PRESENT AND FUTURE

3.1 EXAMPLES

At this time the only radar in the world capable of measuring the canting parameter is at DFVLR (Deutsche Forshungs- und Versuchanstalt für Luft- und Raumfahrt e.V.) in Oberpfaffenhofen, F.R.G. While it is anticipated that initial measurements of the canting parameter will be available soon, only measurements of LDR are presently available. Nevertheless, data from COHMEX (Cooperative Huntsville Meteorological Experiment) and MAYPOLE (May Polarization Experiment) are shown in Herzegh and Jameson (1988) which illustrate systematic structure in LDR data especially in the icy regions of clouds, structure which is independent of that for the reflectivity factor and for differential reflectivity. LDR (and, hence, probably the canting parameter) provides unique information not available from any other measurable.

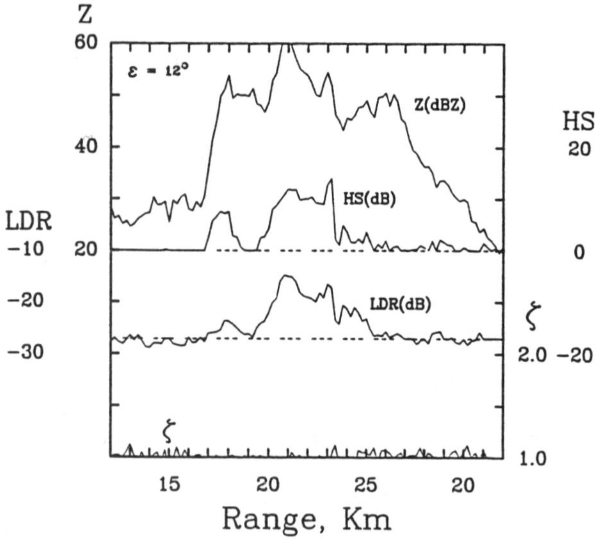

FIGURE 1. Plot of the radar reflectivity factor at a wavelength of 10.71 cm and horizontal polarization (Z), the dual-wavelength (10.71, 3.21 cm) hail signal (HS), the differential reflectivity (ζ), and the attenuation corrected linear depolarization ratio (LDR) at 3.21 cm wavelength as a function of radar range through an intense hailstorm. In this plot the maxima of LDR and HS are 4.2 km above ground (about 1 km above the melting level).

While there is insufficient room for all the data to be presented in this brief paper, one example is shown in Fig. 1. It is a plot of radar data along one radial as a function of distance from the radar. These data are from the severe Denver hailstorm on June 13, 1984 collected during MAYPOLE. These data are unique because simultaneous polarization and dual-wavelength measurements were also collected. In Fig. 1, the coincidence of substantial LDR exactly where the dual-wavelength hail signal indicates the presence of hail (larger than 1.25 cm diameter) is striking especially since these measurements are entirely independent of one another. While LDR exceeded -15 dB, as expected in ice the differential reflectivity (ζ) remains small ($<$ 0.8 dB) and exhibits no apparent structure. The background level of LDR (about -27 dB) is characteristic of aggregate snow (Hendry et al., 1987) although no confirmatory in situ measurements are available.

The existence of coherent structure in LDR as exemplified in Fig. 1 is apparent throughout the depth of the storm. The collocation of significant dual-wavelength hail signals with substantial LDR is consistent with theoretical expectations (Jameson, 1985). While at present work of this type is still in the research domain, measurements from the DFVLR radar in the next few years are expected to lead to the development of operational tools for the radar characterization of hydrometeors before TRMM. A polarization radar with the capabilities required for hydrometeor identification will also be able to estimate rainfall simultaneously in more than one way thereby providing, at a minimum, a means for better tuning relationships between radar reflectivity and rainfall in different locations.

REFERENCES

Atlas, D., and F.H. Ludlam, 1961: Multi-wavelength radar reflectivity of hailstorms. Quart. J. Roy. Meteor. Soc., 87, 523-534.

Atlas, D., and C.W. Ulbrich, 1974: The physical basis for attenuation-rainfall relationships and the measurement of rainfall parameters by combined attenuation and radar methods. J. Rech. Atmos., 8, 275-298.

Atlas, D., and C.W. Ulbrich, 1977: Path- and area-integrated rainfall measurement by microwave attenuation in the 1-3 cm band. J. Appl. Meteor., 16, 1322-1331.

Atlas, D., M. Kerker, and W. Hitschfeld, 1953: Scattering and attenuation by non-spherical atmospheric particles. J. Atmospheric Terrest. Phys., 3, 108-119.

Bohne, A., and R.C. Srivastava, 1975: Random errors in wind and precipitation fall speed measurement by a triple Doppler radar system. Rpt. No. 37, Lab. for Atmos. Probing, Dept. Geophys. Sci., Univ. of Chicago, 20pp.

Gans, R., 1912; Über die Form ultramikroshopischer Goldteilchen., Ann. Phys., 37, 881-900.

Hall, M.P.M., S.M. Cherry, J.W.F. Goddard, and G.R. Kennedy, 1980: Raindrop sizes and rainfall rate measured by dual-polarization radar. Nature, 285, 195-198.

Hall, M.P.M., J.W.F. Goddard, and S.M. Cherry, 1984: Identification of hydrometeors and other targets by dual-polarization radar. Radio Sci., 19, 132-140.

Hendry, A., G. C. McCormick and B. L. Barge, 1976: The degree of common orientation of hydrometeors observed by polarization diversity radars. J. Appl. Meteor., 15, 633-640.

Hendry, A., Y.M.M. Antar, and G.C. McCormick, 1987: On the relationship between the degree of preferred orientation in precipitation and dual-polarization radar echo characteristics. Radio Sci., 22, 37-50.

Herzegh, P. H., and A. R. Jameson, 1988: Observing precipitation structure through dual-polarization radar measurements., Bull. Amer. Meteor. Soc., 69, submitted.

Jameson, A. R., 1983a: Microphysical interpretation of multi-parameter radar measurements in rain. Part I: Interpretation of polarization measurements and estimation of raindrop shapes. J. Atmos. Sci., 40, 1792-1802.

Jameson, A. R., 1983b: Microphysical interpretation of multi-parameter radar measurements in rain. Part II: Estimation of raindrop distribution parameters by combined dual-wavelength and polarization measurements. J. Atmos. Sci., 40, 1803-1813.

Jameson, A. R., 1985: On deducing the microphysical character of precipitation from multiple-parameter radar polarization measurements. J. Climate Appl. Meteor., 24, 1037-1047.

Jameson, A. R., and E. A. Mueller, 1985: Estimation of differential phase shift from sequential orthogonal linear polarization radar measurements. J. Atmos. Oceanic Technol., 2, 133-137.

Jameson, A.R., and J.H. Dave', 1988: An interpretation of circular polarization measurements affected by propagation differential phase shift., J. Atmos. Oceanic Technol., 5, in press.

Larsen, M.F., and J. Röttger, 1987: Observations of thunderstorm reflectivities and Doppler velocities measured at VHF and UHF. J. Atmos. Oceanic Technol., 4, 151-159.

Lhermitte, R.M., and D. Atlas, 1961: Precipitation motion by pulse-Doppler radar. Proc. 9th Conf. Radar Meteorology, Boston, Amer. Meteor. Soc., 218-223.

McCormick, G. C., A. Hendry and B. L. Barge, 1972: The anisotropy of precipitation media. Nature, 238, 214-216.

Pruppacher, H.R., and K.V. Beard, 1970: A wind tunnel investigation of the internal circulation and shape of water drops falling at terminal velocity in air. Quart. J. Roy. Meteor. Soc., 96, 247-256.

Seliga, T. A., and V.N. Bringi, 1976: Potential use of radar differential reflectivity measurements at orthogonal polarizations for measuring precipitation. J. Appl. Meteor., 15, 69-76.

Srivastava, R.C., T.J. Matejka, and T.J. Lorello, 1986: Doppler radar study of the trailing anvil region associated with a squall line. J. Atmos. Sci., 43, 356-377.

Van de Hulst, H. C., 1957: Light Scattering by Small Particles. Wiley and Sons, 470pp.

Wakasugi, K., A. Mizutani, Matsuo, M., S. Fukao, and S. Kato, 1986: A direct method for deriving drop-size distribution and vertical air velocities from VHF Doppler radar spectra. J. Atmos. Oceanic Technol., 3, 623-629.

Westcott, N.E., D.B. Johnson, and A.R. Jameson, 1984: Multi-parameter radar measurements of a severe convective storm in CCOPE. Proc. 22nd Conf. Radar Meteorology, Boston, Amer. Meteor. Soc., 103-107.

Wilheit, T.T., A.T.C. Chang, M.S. Rao, E.B. Rodgers, and J.A. Theon, 1977: A satellite technique for quantitatively mapping rainfall rates over the ocean. J. Appl. Meteor., 16, 551-560.

Wilheit, T.T., A.T.C. Chang, J.L. King, E.B. Rodgers, R.A. Nieman, B.M. Krupp, A.S. Milman, J.S. Stratigos, and H. Siddalingaiah, 1982: Microwave radiometric observations near 19.35, 92 and 183 GHz of precipitation in tropical storm Cora. J. Appl. Meteor., 21, 1137-1145.

Wu, R., and J.A. Weinman, 1984: Microwave radiances from precipitating clouds containing aspherical ice, combined phase and liquid hydrometeors. J. Geophys. Res., 89, 7170-7178.

ON THE COMBINING OF RAINGAUGE, RADAR AND SATELLITE ESTIMATES OF RAINFALL

G.L. Austin
McGill Radar Weather Observatory
Macdonald College
Ste. Anne de Bellevue, Que
Canada H9X 1CO

ABSTRACT

Extreme spatial and temporal variability in the observed rainfall fields suggests that great care needs to be exercised in the calibration of rainfall estimating schemes from gauge networks and other surface data. The development of analytical tools to combine several not very precise rainfall estimating techniques to yield optimum rainfall estimates is reviewed. It is concluded that a large data base of simulated 3-D rainfall data, which can be regarded as accurate, is necessary. These data can then be used to obtain the required optimum integration scheme and error estimates based on a knowledge of the physics of the instrumental measuring biases, noise and sampling characteristics.

1. INTRODUCTION

One of the most significant problems associated with the remote sensing of rainfall is the development of strategies to combine data derived from different rainfall measuring techniques to yield the optimum estimate of areal precipitation. An equally urgent and related problem is to estimate the likely impact of sampling, instrumental and interpretational errors in the remote-sensed precipitation data.

Some attention has been given to the former as it relates to the combining of raingauge and radar data (Austin, 1987; Harrold et al., 1973; Wilson and Brandies, 1979). The literature on techniques for the combining of gauge and satellite data is also quite extensive (see Barrett and Martin, 1981 for a review).

2. RAINFALL VARIABILITY

Some of these authors have recognized a major problem in dealing with precipitation data sets - the presence of extreme variability and intermittency (Schertzer and Lovejoy, 1987). The effect of this variability is that schemes designed to produce areal rainfall fields have low accuracy, even for quite dense gauge networks, and particularly in convective situations. Damant et al. (1983), for example, obtain errors of approximately 70% for total storm accumulations over a 4,800 km^2 catchment using 10 raingauges. Bellon and Austin (1986) estimate mean point total storm accumulation differences from gauges at different spacing and obtained 60% differences at a distance of 10 km and 100% difference at a separation of 100 km.

ISBN 0-937194-14-4

These results have considerable impact on the accuracy with which a remote-sensing rainfall measuring system can be calibrated with sparse gauge networks. The calibration of large footprint estimating schemes, particularly those having non-linear averaging characteristics with a sparse gauge data set is clearly hazardous. The use of well-calibrated weather radar data with an estimated ± 30% uncertainty over small catchments for storm totals (Austin (1987) and Bellon and Austin (1986)) then becomes particularly attractive. In order to achieve these accuracies in the radar estimates, specific account has to be taken of the temporal sampling errors (frequencies greater than five minutes are sometimes required), and the variability in the parameters of the non-linear relationship between rainfall and radar reflectivity. The author believes that some of the published accuracy claims for satellite rainfall estimating schemes are unduly optimistic, since they are certainly higher than the accuracy with which the rainfall could be estimated by the sparse calibrating gauge network.

The important point is that there is no existing accurate way to measure the space/time distribution of precipitation. This means that the problem of combining rainfall estimates becomes one of combining several wrong estimates of rainfall to get a best estimate of the true rainfall pattern.

3. POSSIBLE STRATEGY

Since no direct estimate of the 'correct' value of the rainfall pattern exists, it is not clear that it is possible to devise the optimum strategy or estimate its accuracy directly from the wrong fields themselves, unless we make considerable assumptions about the statistical properties of the 'correct' field.

If a rainfall field in space and time which could have existed was simulated stochastically, then simulated rainfall estimates based on the known physics of the measuring system can be generated. This accounts, in part at least, for the great deal of activity in recent years directed toward the stochastic modelling of rainfall patterns. This type of activity has resulted recently in a special issue of the Journal of Geophysical Research (August 1987) and even the establishment of a new journal (Journal of Stochastic Hydrology) to discuss the problems.

An alternative strategy, which to some extent avoids the difficulties associated with producing stochastic rainfall models with correct statistical properties, is to take a series of 3-D radar data sets and argue that although they do not represent the actual rainfall pattern that existed in space at the indicated time, they do represent a pattern that could have existed. This pattern can then be modified by the theoretically-derived sampling, non-linear integration and noise contamination characteristics inherent in each of the measurement techniques. This procedure has the enormous advantage that since we start from a 'correct' rainfall pattern we can calculate the statistical properties inherent in each of the measurement schemes, and presumably the optimum combining strategies.

The great advantage of the procedure is that no assumptions have to be made about the statistical properties of the rainfall fields except at space and time scales smaller than the resolution of the radar (perhaps 1 km and 5 minutes). This is clearly likely to be a much less drastic assumption than those which assume properties of homogeneity and independence on the large scale. Some would argue, however, that much of the observed discrepancy between radar and gauge data is due to real sub-kilometer variability and that in fact the effective area of representativeness for a raingauge is little more than a circle of radius 4". What has to be assumed is that the physics of the rainfall measuring procedure used by the instrument is well understood. This author, at least, believes that to assume we understand the measuring process in our instruments is generally much safer than to make assumptions about the statistical properties of rainfall patterns which have only been poorly measured. This basic concept has already been used in Damant et al. (1983) in the investigation of the sampling errors of raingauges.

4. METHOD

The problem can be set up following the notation used in Krajewski (1987) in his discussion of cokriging. If $X(k)$ is the true rainfall pattern, where k is a position vector, then one of the remote-sensed estimates of the rainfall at the point k will be given by:

$$R_k = \frac{1}{A} \int_A F_{Rk}\, X(k)\, dk \; + E_{Rk}$$

where A is the measurement cell area, F_{Rk} is the bias in the estimate, and E_{Rk} is the random error. On the other hand, the gauge estimate at the points where gauges exist, k^1 is

$$6_k = F_{6k}\, X(k^1) + E_{6k}$$

where F_{6k} is the gauge bias and E_{6k} the sampling noise.

The problem is to find the best estimate $Y^x(k)$, of

$$Y(k) = \frac{1}{A} \int_A X(k)\, dk$$

In the cokriging approach it is normal to now assume a possible solution of the form

$$Y^x(k) = \sum_{i=1}^{N_6} \lambda_{6i}\, G_i(k^1) + \sum_{i=1}^{N_R} \lambda_{Ri}\, R_i(k) + \ldots.$$

where the λ's are distance weighting functions which may be determined from the data fields.

5. RESULTS

Preliminary results for this type of work have been obtained for summer rainfall patterns in Montreal, where it has been found that more than 5 km from the raingauge a radar estimate of rainfall is more accurate and that more than 40 km from a gauge then a simple VIS/IR threshold technique yields better rain estimates. A substantial quantity of Kennedy Space Center weather radar data is now being analyzed with a view to finding the appropriate parameters for a more representative tropical rainfall field.

6. CONCLUSIONS

Extreme spatial and temporal variability in rainfall patterns produce considerable problems for the calibration of remote-sensed rain estimation schemes, the development of optimum combining systems for several techniques and error analysis. Work on mid-latitude rain patterns in Montreal, for example, give the results that an estimate of rainfall at a point is better made by radar than a gauge about 5 km away and by satellite using a simple VIS/IR thresholding technique if the gauge is more than about 40 km away.

When we consider systems of remote-sensing devices such as are proposed in the TRMM satellite, then in order to establish procedures for producing combined estimates of precipitation together with their likely accuracy, a considerable simulation based on large 3-D data sets is required.

REFERENCES

Austin, P.M., 1987: Relationship between measured radar reflectivity and surface rainfall, Mon. Wea. Rev., 115, 1053-1070.

Barrett, E.C. and D.W. Martin, 1981: The use of satellite data in rainfall monitoring, Academic Press, 340 pp.

Bellon, A. and G.L. Austin, 1986: On the relative accuracy of satellite and raingauge rainfall measurements over middle latitudes during daylight hours, J. Climate and Appl. Meteor., 25, 1712-1724.

Damant, C., G.L. Austin, A. Bellon and R.S. Broughton, 1983: Errors in the Thiessen technique for estimating areal rain amounts using weather radar data. J. Hydrol., 62, 81-94.

Harrold, T.W., E.J. English and C.A. Nichols, 1973: The measurement of area precipitation using radar, Weather, 28, 332-338.

Krajewski, W.F., 1987: Cokriging radar-rainfall and rain gage data. J. Geophy. Res., 92, 9571-9580.

Schertzer D. and S. Lovejoy, 1987: Physical modeling and analysis of rain and clouds by anisotropic scaling multiplicative processes, J. Geophys. Res., 92, 9693-9714.

Wilson, J.W. and E.A. Brandies, 1979: Radar measurement of rainfall - A summary, Bull. Amer. Met. Soc., 60, 1048-1058.

VHF RADAR REMOTE SENSING OF RAINFALL RATE AND VERTICAL PROFILE OF LIQUID WATER CONTENT OVER TAIWAN DURING THE TIME PERIOD OF TYPHOON WAYNE*

R.J. Hung and Y.D. Tsao
The University of Alabama in Huntsville
Huntsville, Alabama 35899, USA

and

A.J. Chen and L.-N. Chang
National Central University,
Chung Li, Taiwan

ABSTRACT

The VHF radar located at the subtropical and low latitude observing site of Taiwan has been used to observe atmospheric parameters during the time passage of Typhoon Wayne in August 1986. Time-dependent wind velocities with three dimensional profiles are plotted by processing the recorded data of the backscattered echo power and the Doppler spectral width of the signal returns from the VHF radar. The VHF radar can be used to detect time-dependent vertical profiles of the liquid water content. The present study shows that the equivalent rainfall rate calculated from the VHF radar is in good agreement with the ground-based measurement of rainfall rate in the same area.

1. INTRODUCTION

Recently, the National Central University (NCU) of Taiwan has installed a VHF radar at their subtropical and low latitude observing site. This VHF radar is a phase coherent radar that measures the amplitude and Doppler shift of radio waves scattered back to the receiving antennas. The application of the VHF radar in the observations of the troposphere, stratosphere and mesosphere has been reviewed by Balsley and Gage (1980), Rottger (1981), and others.

Gravity waves play an important role in the coupling between the lower and upper atmospheres. Gravity waves observed in the ionosphere have been traced to sources mostly in the troposphere (Baker and Davies, 1969; Hung et al., 1978a; Hung and Smith, 1979).

Sources of gravity waves can be determined by using ray tracing computations (Hung et al., 1978b; Hung and Smith, 1978; Hung et al., 1979a, and 1979b).

---

* The research is supported by NASA through Grant No. NAGW-1165, and the National Science Council of Taiwan, ROC, through Grant Nos. NSC75-0202-M008-21, and NSC76-0202-M008-23.

ISBN 0-937194-14-4

Gravity waves are responsible for the perturbation of atmospheric density. Short-term density variability is generated by the propagation of gravity waves, which are detected by the backscattered echo power and the Doppler spectral width of the signal, returns of the VHF radar (Gage et al., 1981). Gravity waves observed at different heights can be used to trace the vertical profile of the liquid water content in the troposphere. Equilibrium and perturbation quantities of density are the summation of the density of dry air, the density of water vapor, and the density of liquid water. Since the densities of dry air and water vapor can be determined from rawinsonde data and by other means, the vertical profile of liquid water content can, thus, be determined from the VHF radar observation. Results have been compared with ground-based rain gauge measurements with good agreement.

Atmospheric disturbances over Taiwan during the time period of Typhoon Wayne were investigated in this study. Figure 1 shows the locations and the times of the Typhoon Wayne storm track on 21-22 August, 1986. Figure 2 shows the satellite imagery of Wayne with the storm track right on Taiwan at 0000 GMT, 22 August, 1986.

2. REMOTE SENSING OF WIND VELOCITY

In this study, the Doppler method, which measures the Doppler shift of the scattered radar echoes at oblique incidence, is used in the analysis. The complex autocorrelation or the spectral analysis yields radial velocities v' at different directions. Assuming that the antenna beam points at a zenith angle δ and at an azimuth angle α , directed clockwise from the north, then

$$v' = v_x \sin\alpha \cdot \sin\delta + v_y \cos\alpha \cdot \sin\delta + v_z \cos\delta \qquad (1)$$

for a constant zenith angle, and assuming the wind field is uniform, will vary sinusoidally with the azimuth angle α. The Fourier analysis of the data series directly yields the eastward wind v_x and the northward wind v_y from the Fourier coefficients, as well as the vertical velocity v_z . This is the so-called VAD method (velocity azimuth display) (Battan, 1973).

Three arrays with 64 yagi antennas each are installed in Chung Li, Taiwan. These antennas are arranged in such a way that three arrays are all 17° from zenith vertically, one faced toward the north, the second faced toward the east, and the other faced toward the west. The radial velocities of the winds observed from the three arrays are, then, fitted into Equation (1), and the three dimensional wind speeds are computed. Table 1 shows the three-dimensional wind speeds observed from the VHF radar at the selected heights of 3, 5.1 and 9.6 km, from 0455 to 0603 GMT, 22 August 1986. The observations are made for a time resolution with a data point approximately every 2 minutes for this case.

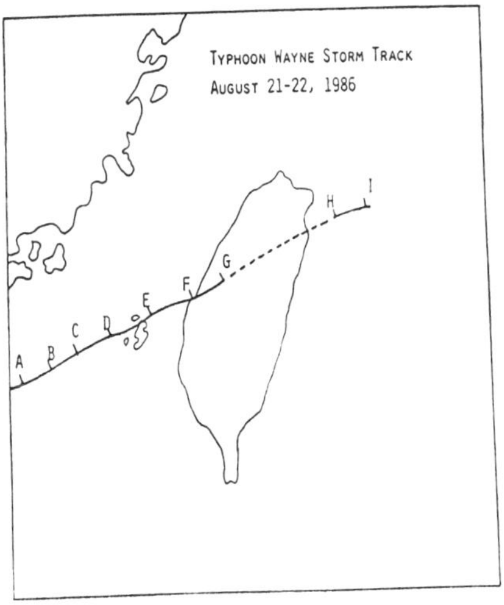

A: 1100 GMT Aug. 21, 1986
B: 1300 GMT Aug. 21, 1986
C: 1500 GMT Aug. 21, 1986
D: 1800 GMT Aug. 21, 1986
E: 2030 GMT Aug. 21, 1986
F: 2300 GMT Aug. 21, 1986
G: 0100 GMT Aug. 22, 1986
H: 0900 GMT Aug. 22, 1986
I: 1000 GMT Aug. 22, 1986

Figure 1. Locations and times of Typhoon Wayne storm track on 21-22 August, 1986

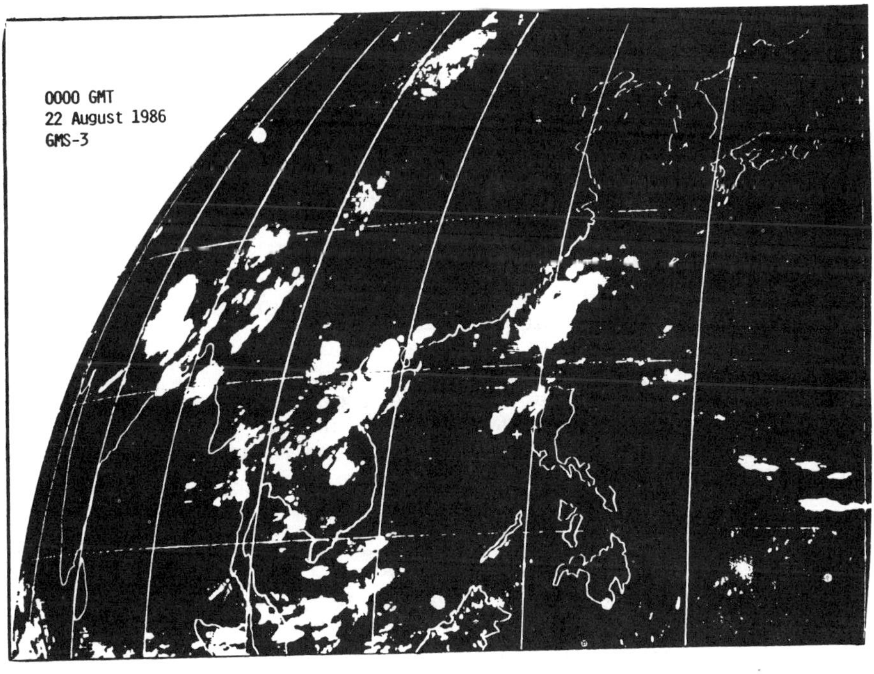

Figure 2. Satellite imagery of Typhoon Wayne with the storm track right on Taiwan at 0000 GMT, 22 August 1986.

To show a time-dependent three-dimensional wind profile, which covers the entire heights from 1.5 to 13.2 km, Figures 3 and 4 illustrate the horizontal and vertical wind profiles, respectively. Figure 3 shows the time-dependent horizontal wind from 0455-0603 GMT, 22 August 1986. The maximum wind is indicated with the longest arrow and in the figure corresponds to 13.2 m/s. This figure shows that the horizontal wind demonstrated a clockwise circulation between the heights of 4 to 8 km during the time period of 0500 to 0540 GMT, and between the heights of 3 to 5 km during the time period of 0550 to 0600 GMT. This introduced the down-draft motion of the vertical wind. This result is in agreement with the vertical wind observation.

Figure 4 shows the time dependent profile of the vertical wind with the same heights and time period of observation coverage as Figure 3. A description of the symbols is included in the figure legend.

During this time period, most heights were in the region of the up-draft motion, except the 2 to 3 km wide zone which was in the region of the down-draft motion. This interesting zone of the down-draft motion moved downward as time passed. This indicates the increasing convective motion and concentration of water vapor. The result is in agreement with the calculated result of the liquid water content to be shown in Section 4 of this paper.

3. DETECTION OF GRAVITY WAVES

Frequency, horizontal wavelength, vertical wavelength, and phase velocity of gravity waves can be determined from the Fourier power spectrum analysis of the horizontal and vertical wind velocities from the multiple beams of the VHF radar and their cross-correlation analysis (Chanin and Hauchecorne, 1981; Dunkerton, 1982b).

Gravity waves associated with tornado activity (Hung et al., 1978a; Hung and Smith, 1975) and a hurricane (Hung and Kuo, 1978) have been observed. By using a ray tracing technique, Hung et al. (1979 a; 1980) have shown that the enhanced convection-initiated gravity waves associated with severe storms were generated by thunderheads embedded in a squall line and/or isolated cloud with intense convection.

High wind shear, flows across topographical barriers, and convectively unstable flows are believed to be the major causes of the excitation of gravity waves. Of the three dominant cases, convection as a source of gravity waves is the least understood. In the case studied here, wave-like disturbances were observed in three-dimensional wind data for this stormy day. Detailed descriptions of the data analysis techniques for the deduction of gravity waves are given by Hung and Kuo (1978), and Hung and Smith (1981).

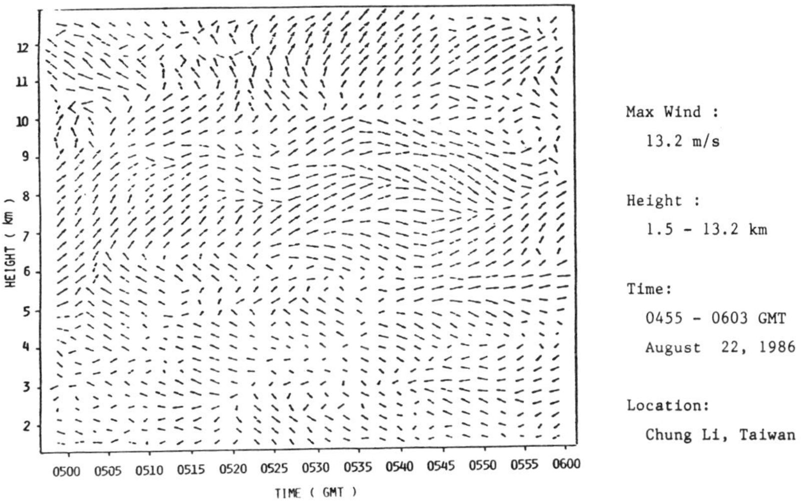

Figure 3. Time dependent variation of horizontal wind profile for heights from 1.5 to 13.2 km, during the time period of 0455 to 0603 GMT, 22 August 1986 observed from the VHF radar at Chung Li, Taiwan. The maximum wind indicated with the longest arrow corresponds to 13.2 m/s.

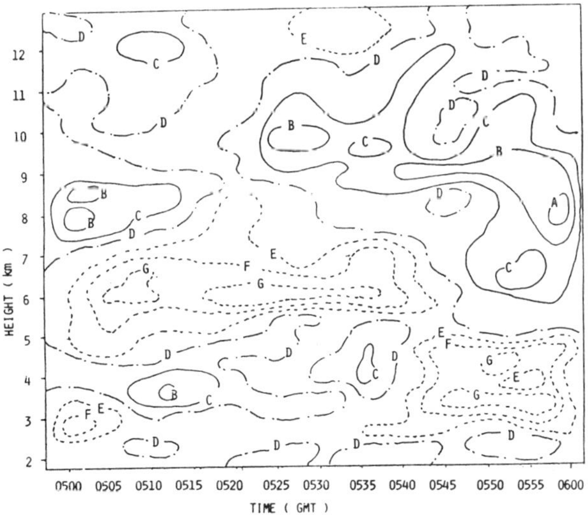

Figure 4. Time dependent variation of the vertical wind profile for the heights of 1.5 to 13.2 km, during the time period of 0455 to 0603 GMT, 22 August 1986 observed from the VHF radar at Chung Li, Taiwan. Symbol A = 1.5 m/s, B = 1.0 m/s, C = 0.5 m/s, D = 0 m/s, E = -0.5 m/s, F = -1.0 m/s, and G = -1.5 m/s.

Table 1. Three Dimensional Wind Speeds Observed.
Time : 0455 - 0603 GMT, August 22, 1986.
Location : Chung Li, Taiwan

Time (GMT)	Height 3 km Wind Speed Vx (m/s)	Vy (m/s)	Vz (m/s)	5.10 km Wind Speed Vx (m/s)	Vy (m/s)	Vz (m/s)	9.6 km Wind Speed Vx (m/s)	Vy (m/s)	Vz (m/s)
0459	-1.43	-1.40	-0.47	-0.28	-0.97	-0.27	-2.50	-0.20	0.04
0501	-3.16	-1.98	-0.51	0.32	-0.66	-0.58	-2.10	0.06	0.07
0503	-4.76	-3.04	-0.26	0.41	0.70	-1.00	-2.77	0.29	0.03
0506	-5.94	-4.80	0.11	-0.50	0.82	-1.11	-4.04	1.15	-0.08
0508	-5.79	-5.55	0.57	-1.74	0.95	-0.96	-5.20	1.29	-0.25
0510	-5.31	-6.12	0.97	-2.79	1.35	-0.89	-5.58	0.53	-0.12
0512	-3.53	-6.41	1.07	-3.70	1.38	-0.74	-5.22	0.14	-0.10
0514	-1.88	-7.22	0.82	-4.01	1.26	-0.66	-4.31	-0.32	-0.12
0516	-2.40	-7.57	0.55	-4.35	2.36	-0.89	-2.69	-0.56	0.08
0518	-2.23	-7.74	0.11	-4.89	1.97	-0.74	-1.68	-0.54	0.05
0521	-2.99	-9.90	0.12	-5.10	0.93	-0.57	-0.64	-0.72	0.27
0523	-3.38	-10.20	0.14	-5.37	0.04	-0.29	0.78	-1.11	0.66
0525	-2.73	-10.99	0.21	-5.31	-2.47	0.11	1.37	-1.69	0.94
0527	-2.06	-10.87	0.00	-4.68	-3.10	0.07	1.64	-1.97	1.15
0529	-1.65	-10.61	0.12	-3.43	-3.64	0.16	0.72	-2.03	1.04
0531	-1.60	-10.23	0.31	-2.16	-2.21	-0.03	-1.80	-1.95	0.64
0533	-0.92	-8.39	-0.05	-1.38	-1.77	-0.15	-2.43	-1.92	0.47
0536	-1.74	-6.66	0.07	-0.77	-0.68	-0.05	-2.26	-1.75	0.37
0538	-2.59	-1.13	-0.06	-1.26	-0.40	-0.16	-2.22	-1.71	0.32
0540	-1.65	0.79	-0.52	-1.61	-1.03	-0.17	-0.14	-1.94	0.54
0542	-1.62	2.48	-0.90	-2.16	-0.98	-0.11	0.06	-1.24	0.37
0544	-0.84	3.29	-1.36	-2.72	-0.94	-0.27	-0.72	-0.48	0.08
0546	0.29	1.52	-1.94	-2.94	-1.29	0.02	1.04	-0.91	0.39
0548	0.23	0.70	-1.82	-3.04	-3.68	0.61	2.17	-1.64	0.70
0551	0.50	0.00	-1.86	-1.70	-3.30	0.67	3.01	-2.00	0.84
0553	0.95	-1.01	-1.77	-0.54	-4.16	0.75	2.99	-2.34	0.77
0555	1.78	-2.52	-1.68	0.65	-5.14	0.51	1.01	-2.30	0.36
0557	3.62	-3.05	-1.76	0.47	-6.70	0.24	-1.58	-2.29	0.20
0559	4.45	-6.74	-0.59	-0.44	-9.65	0.20	-4.20	-1.85	0.27

Table 2. Characteristics of Gravity Waves Observed in Troposphere
Sampling Time : 0455 - 0603 GMT August 22, 1986.
Location : Chung Li, Taiwan

Height (km)	Wave Period (Min)	Wavelength Horizontal (km)	Wavelength Vertical (km)	Horizontal Propagation Direction (degree)	Phase Velocity Horizontal (m/s)	Phase Velocity Vertical (m/s)
3	7.2	9.45	1.59	-55	21.8	3.7
5.1	9.5	7.66	1.84	66	13.5	3.2
9.6	11.7	26.4	1.91	62	36.8	2.7

Table 2 shows the characteristics of the gravity waves detected at the selected heights of 3, 5.1, and 9.6 km. The background information is the same as the three-dimensional wind velocity shown in Table 1. The horizontal propagation direction of the gravity waves is measured clockwise from the north. The ray tracing of the wave source indicates that the gravity waves were excited by the cloud bands of Typhoon Wayne which broke into three parts affected by the high topographical barriers after it hit Taiwan. Table 2 shows the gravity waves with wave periods in the range of 7.2 to 11.7 min; horizontal wavelength, 7.6 to 26.4 km; vertical wavelength, 1.59 to 1.91 km; horizontal phase velocity, 13.5 to 36.8 m/s; and vertical phase velocity, 2.7 to 3.7 m/s were detected in the troposphere.

4. REMOTE SENSING OF LIQUID WATER AND CLOUD ICE CONTENTS

Gravity waves play a major role in causing the atmospheric density perturbations. The total density of the atmosphere, ρ , is the summation of the equilibrium value of the density, ρ_0 , and the density perturbation, ρ' , i.e.,

$$\rho = \rho_0 + \rho' \tag{2}$$

Let us consider acoustic-gravity waves progagating in the terrestrial atmosphere which is assumed to be compressible, nonrotating and adiabatic. Detailed derivation of atmospheric density perturbations caused by the propagation of acoustic-gravity waves is illustrated in Hung et al. (1987). The expression for an atmospheric density perturbation can be shown as follows:

$$\frac{\rho'}{\rho_0} = -\left\{ 1 + \frac{\left(\frac{N}{\omega}\right)^2 - \left(\frac{g\Gamma_d}{T_0 \omega^2}\right)}{\left[\frac{(v_p)_H \, v_H \cos\theta_p}{2\,(v_p)_v \, v_v} - 1\right]} \right\}^{-1}$$

$$\left\{ (\gamma - 1)\left[\frac{v_H}{(v_p)_H}\cos\theta_p + \frac{v_v}{(v_p)_v}\right] + \frac{\gamma\,(v_p)_H \, v_H \cos\theta_p}{2C^2} \right\} \tag{3}$$

Here ρ' denotes the perturbed quantity of the atmospheric density; ρ_0, the equilibrium quantity of the atmospheric density; g, the gravitational acceleration; γ ,the ratio of the constant pressure specific heat to the constant volume specific heat; T_O, the equilibrium quantity of the atmospheric temperature; C, the speed of sound; v_H, the horizontal wind velocity; $(v_p)_H$, the horizontal phase velocity of the gravity waves; v_V, the vertical wind velocity; $(v_p)_v$,the vertical phase velocity of the gravity waves; and Θ_p, the angle between the horizontal phase velocity of the gravity waves and the horizontal wind velocity.

The Brunt-Vaisalla frequency, N, is defined by

$$N^2 = \frac{g}{T_0}\left[\Gamma_d + \frac{dT_0}{dz}\right] \tag{4}$$

where

$$\Gamma_d = \frac{g}{C_p} \tag{5}$$

is the adiabatic lapse rate, and C_p denotes the constant specific heat.

On the other hand, the total density of the atmosphere is the summation of the density of dry air ρ_a, the density of water vapor ρ_v, and the density of liquid water and/or cloud ice ρ_e, i.e.,

$$\rho = \rho_a + \rho_v + \rho_e \tag{6}$$

Since ρ_a and ρ_v can be observed from rawinsonde data and by other means, the liquid water and/or cloud ice contents can, thus, be determined.

Figure 5 shows the time-dependent liquid water content at the selective heights of 3.0 and 5.1 km, and that of the cloud ice content at the selective height of 9.6 km during the time period of 0500 to 0600 GMT, 22 August 1986. It is shown that the maximum liquid water content was 62 g/m^3, and that of the cloud ice content was 8 g/m^3. Both the liquid water and the cloud ice contents started to increase rapidly after 0555 GMT.

Figure 6 shows the ground-based measurements of the hourly rainfall rate at the same observing site during the time period of 1700 GMT, 21 August 1986 to 1600 GMT, 22 August 1986. It is shown that the moderate rainfall rates were observed during the time periods of 0400 to 0600 GMT, and from 0800 to 1100 GMT, while heavy rainfall rates occurred during the time of 0600 to 0800 GMT. VHF radar observations were taken during the time period of 0500 to 0600 GMT which was in the middle of the transition from a moderate rainfall rate to a heavy rainfall rate. The results of the VHF radar observations, shown in Figure 5, are in good agreement with the ground-based observation.

Raindrop spectra may extend to drop diameters as large as 6 mm (Blanchard and Spencer, 1970). Such large drops are rather rare since they are found only in very heavy rain with rainfall rate, $R >$ 100 mm/hr. Most of the time, at smaller rainfall intensities, rainfall spectra usually extend only to drop diameters of 2 to 3 mm at the ground (Pruppacher and Klett, 1978). Larger drops tend to break up as a result of collision with other drops. Cloud drop size

Liquid Water and Cloud Ice Contents

Location : Chung Li, Taiwan

Time : 0500 - 0600 GMT, August 22, 1986

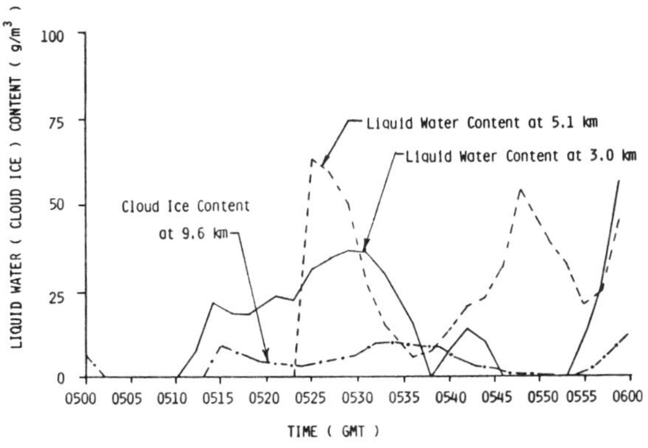

Figure 5. Time dependent variation of liquid water content at the selective heights of 3.0 and 5.1 km, and cloud ice content at the selective height of 9.6 km for the observation time of 0500 to 0600 GMT, 22 August 1986, based on the VHF radar observation at Chung Li, Taiwan.

Observed Hourly Rainfall Rate

Location : Chung Li, Taiwan

Time : 1700 GMT, 21 August 1986
to 1600 GMT, 22 August 1986

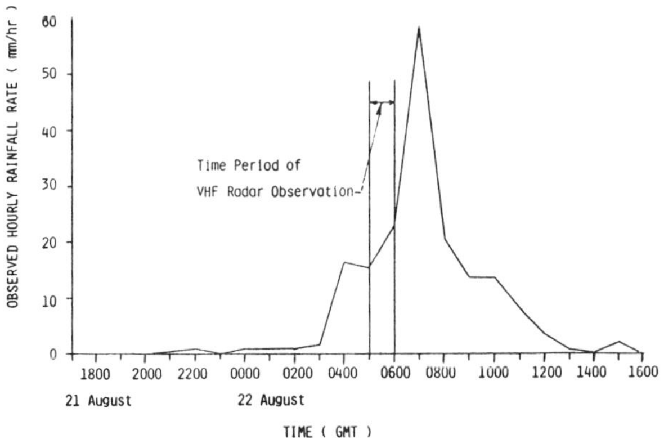

Figure 6. Ground-based observation of the hourly rainfall rate for the observation time of 1700 GMT, 21 August 1986 to 1600 GMT, 22 August 1986, at Chung Li, Taiwan.

distribution is a function of the development stage of a cloud. Battan and Reitan (1957) show that fair weather cumuli have relatively narrow drop size spectra, while the spectra of tropical cumuli which have reached the more mature development stage of a cumulus congestus are considerably broader. The computed liquid water content, based on the VHF radar observation, can be converted to an equivalent hourly rainfall rate with the assumption that the average radius of water drops at 3.0 km height are to be 0.2 mm, and that of the water drops at 5.1 km height are 0.1 mm. By using the values of the average size of water drops, the average terminal speed of the water drops at 3.0 km height is 0.25 m/s, and that of the water drops at 5.1 km height is 0.1 m/s (Mason, 1971). Figure 7 shows the computed time-dependent equivalent rainfall rate during the time period of 0500 to 0600 GMT, 22 August 1986. In reality, this result corresponds to the instantaneous rainfall rate. On the other hand, the ground-based measurement is an hourly rainfall measurement. In order to make the comparison more accurate, the following computation was used.

$$R_a = \frac{1}{t_1 - t_2} \int_{t_1}^{t_2} R_i \ (t) \ dt \qquad (7)$$

Here, R_i denotes the instantaneous rainfall rate, shown in Figure 7; R_a, the time-average rainfall rate. The integration is carried over from time t_1 to t_2. This calculation shows that the time-average rainfall rate for the VHF radar observation at the height of 3.0 km is 16 mm/hr, and that of the rainfall rate at the height of 5.1 km is 10 mm/hr. The ground-based measurement of hourly average rainfall rate at the same period of time is 17 mm/hr. This shows that the VHF radar observation gives an excellent agreement with the ground-based observations.

5. DISCUSSION AND CONCLUSION

The VHF radar located at the subtropical area of Taiwan has been used to observe the atmospheric disturbances caused by the propagation of gravity waves excited by the enhanced convective cloud bands of Typhoon Wayne, on 22 August 1986.

From the backscattered echo power and the Doppler spectral width of the signal returns from the VHF radar, we have determined the three-dimensional wind velocities at different heights. With fast algorithms and computers this could be done in real time from the troposphere to middle atmospheric heights.

The VHF radar observation in the troposphere can contribute to the measurement of the vertical profile of the liquid water and cloud ice contents. This vertical profile of the liquid water

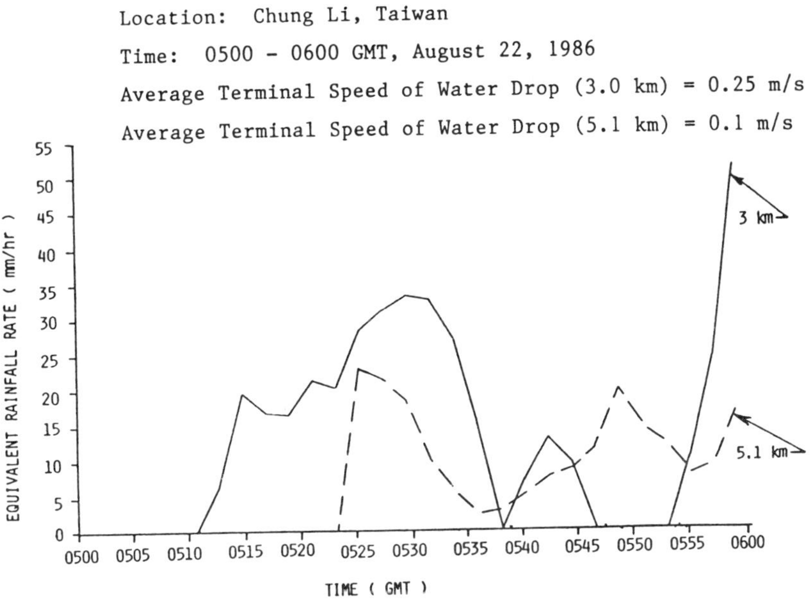

Figure 7. Time dependent variation of the equivalent rainfall rate at the selective heights of 3.0 and 5.1 km, converted from Figure 10, based on the VHF radar observation at Chung Li, Taiwan.

content can be converted to an equivalent rainfall rate, and compared with the ground-based measurement of the hourly recording raingauge. In this study, both the vertical profiles of the liquid water and the cloud ice contents are investigated from the VHF data. A comparison between the equivalent rainfall rate calculated from the VHF data and the ground-based hourly measurement of the rainfall rate at the same location shows an excellent agreement.

REFERENCES

Baker, D. M., and Davies, K., 1969, F2-region acoustic waves from severe weather. J. Atmos. Terr. Phys., 31, 1345.

Balsley, B. B., and Gage, K. S., 1980. The MST radar technique: Potential for middle atmospheric studies. Pure Appl.Geophys., 118, 452.

Battan, L. J., 1973, Radar Observation of the Atmosphere, The University of Chicago Press, Chicago, Ill., USA.

Battan. L. J., and Reitan, C. H., 1957: Artificial Stimulation of Rain, pp. 184, Pergamon Press, New York

Blanchard, D. C., and Spencer, A. D., 1970: Experiments on the generation of raindrop-size distributions by drop breakup. J. Atmos. Sci., 27, 101.

Chanin, M. L., and Houchecorne, A., 1981, Lidar observation of gravity and tidal waves in the middle atmosphere. J.Geophys. Res., 86, 9715.

Gage, K. S., and Green, J. L., 1982, An objective method for the determination of tropopause height from VHF radar observations. J. Appl. Meteor., 21,1150.

Hung, R. J., and Kuo, J. P., 1978, Ionospheric observations of gravity waves associated with Hurricane Eloise. J.Geophys., 45, 67.

Hung, R. J., Phan, T., Lin, D. C., Smith, R. E., Jayroe, R. R., and West, G. S., 1980, Gravity waves and GOES IR data study of an isolated tornadic storm on 29 May 1977. Mon.Weath. Rev., 108, 456.

Hung, R. J., Phan, T., and Smith, R. E., 1978a, Observation of gravity waves during the extreme tornado outbreak of April 3, 1974. J. Atmos. Terr. Phys., 40, 831.

Hung, R. J., Phan, T., and Smith, R. E., 1978b, Ionosphere Doppler sounder for detection and prediction of severe storms. AIAA Jl, 16, 763.

Hung, R. J., Phan, T., and Smith, R. E., 1979a, Case studies of gravity waves associated with isolated tornadic storms on January 13, 1976. J. Appl. Met., 18, 460.

Hung, R. J., Phan, T., and Smith, R. E., 1979b, Coupling of ionosphere and troposphere during the occurrence of isolated tornadoes of November 20, 1973. J. Geophys.Res., 84, 1261.

Hung. R. J., and Smith, R. E., 1975, Observation of severe weather activities by Doppler sounder array. J. Appl. Meteor., 14, 1611.

Hung, R. J., and Smith, R. E., 1978, Ray tracing of gravity waves as a possible warning system for tornadic storms and hurricanes. J. Appl. Met., 17, 3.

Hung, R. J., and Smith, R. E., 1979, Dynamics of severe storms through the study of thermospheric-tropospheric coupling. J. Geomag. Geoelect., 31, 183.

Hung, R. J., and Smith, R.. E., 1981, Ionospheric remote sensing of medium scale gravity waves and tornadic storms. Nuovo Cim. C, 4, 339.

Hung, R. J., and Tsao, Y. D., 1987, VHF radar remote sensing of atmospheric parameters over Taiwan during the time period of typhoon Wayne. Int. J. Remote sensing, 8, in press.

Larsen, M. F., Kelley, M. C., and Gage, K. S., 1982, Turbulence spectra in the upper troposphere and lower stratosphere between 2 hours and 40 days. J. Atmos. Sci., 39, 1035.

Larsen, M. F., and Rottger, J., 1983, Comparison of tropopause height and frontal boundary locations based on radar and radiosonde data. Geophys. Res. Lett., 10, 325.

Lilly, D. K., 1983, Stratified turbulence and the mesoscale variability of the atmosphere. J. Atmos. Sci., 40, 749

Mason, B. J., 1971, The Physics of Clouds, Clarendon Press, Oxford, England.

Pruppacher, H. R., and Klett, J. D., 1978: Microphysics of Clouds and Precipitation, pp. 714, D. Reidel Publish. Co., Dordrecht, Holland.

Rottger, J. 1981, Investigations of lower and middle atmosphere dynamics antenna drifts radars. J. Atmos. Terr. Phys., 43, 277.

Schoeberl, M. R., Strobel, D. F., and Apruzese, J. P., 1983, A numerical model of gravity wave breaking and stress in the mesosphere. J. Geophys. Res., 88, 5249.

Uccellini, L. W., 1975, A case study of apparent gravity waves initiation of severe convective storms. Mon. Weath. Rev., 103, 497.

Vincent, R. A., and Rottger, J., 1980, Spaced antenna VHF radar observations of tropospheric velocities and irregularities. Radio Sci., 15, 319.

INDIRECT TECHNIQUE FOR MEASURING RAINFALL OVER THE OCEAN

Paul C. Etter and Joseph J. Tamul, Jr.
SYNTEK Engineering & Computer Systems, Inc.
Rockville, Maryland 20852, USA

and

Peter J. Lamb
Illinois State Water Survey
Champaign, Illinois 61820, USA

ABSTRACT

This paper describes an indirect technique for measuring rainfall over the surface of the ocean for the purpose of verifying satellite data. The approach is based on a freshwater budget method organized to solve for monthly, spatially averaged rainfall as a residual. The Gulf of Mexico is selected as the initial experiment location due to its semi-enclosed configuration. Preliminary results are presented to demonstrate the merits of this technique.

1. INTRODUCTION

Precipitation is a key climatic index as well as a major component of both the global energy and water budgets. It is both a cause and an effect of air-sea interaction, since it involves thermodynamic changes of state that influence the energy available to the coupled ocean-atmosphere system. The difference between oceanic evaporation (E) and precipitation (E-P) is proportional to the salt (particularly) and heat fluxes at the sea surface.

A recent survey (Atlas and Thiele, 1981) indicated that precipitation measurements used in conjunction with investigations of the global climate need an accuracy of 10-25%, a horizontal resolution of 200-500 km, and a temporal resolution of 1 week -1 month. For energy flux calculations over the oceans, the accuracy needed is about the same as that for oceanic evaporation ($\sim$1 cm per month) over a (500 km)^2 grid.

The need for describing precipitation on a multitude of time and space scales is fundamental to advancing the state-of-the-art in weather and climate predictability. Remote sensing of oceanic precipitation shows great promise in providing the needed information, and satellites, in particular, can provide global coverage. It has been noted that the lack of accurate area-wide ground-truth precipitation data, particularly over the oceans, has impeded the full development and operation of a spaceborne precipitation measurement capability (Atlas and Thiele, 1981). To date, proposed

ISBN 0-937194-14-4

ground-truth measurement systems have included various types of radars as well as measurements of the underwater acoustic noise associated with oceanic rainfall (Nystuen, 1986). At present, rainfall data from ships and island stations can exhibit considerable differences and are believed to contain large biases. It has been suggested that coastal and island stations could be used to compute time series of precipitation for comparison to a time series of satellite estimates for the particular ocean area in question. Such an investigation would be feasible, it is argued, since most ocean climatic areas have reporting meteorological stations from which data are readily available.

Large-scale, long-term portraitures of oceanic precipitation patterns have historically been accomplished by a combination of two methods: (1) extrapolation from coastal and island precipitation records, or (2) extrapolation from shipping reports of precipitation occurrence. Figure 1 presents a comparison of the annual mean precipitation patterns over the Gulf of Mexico derived from two recent studies. The differences in both the magnitudes and morphologies are striking. Thus, the mere availability of coastal and island precipitation data, at least in this instance, does not appear to promote reconciliation among the differing approaches.

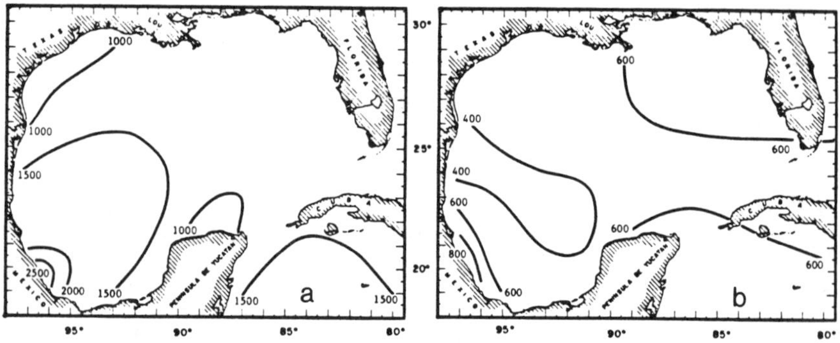

FIGURE 1. Comparison of mean annual precipitation (mm) over the Gulf of Mexico according to: (a) Jaeger, 1976, and (b) Dorman and Bourke, 1981.

A new application of the water budget method may provide data of sufficient quality to serve as ground-truth for satellite measurements of rainfall over the ocean. This method is described in the following sections.

2. FRESHWATER BUDGET

The governing freshwater budget equation is derived and arranged to solve for precipitation as a residual.

The water balance of the Gulf of Mexico can be expressed as (Etter, 1983; Etter, et al., 1987)

$$\frac{dF}{dt} + (E - P) - R = (I - O) \quad (1)$$

where: F = mass of water in the Gulf of Mexico
E = rate of evaporation
P = rate of precipitation (rainfall)
R = net rate of mass inflow due to river discharge
I = net rate of inflow into basin
O = net rate of outflow out of basin
t = time

Neglecting the small contributions due to flow of water through the basin floor and surface runoff in coastal areas, and assuming stationary conditions, Eq. (1) reduces to

$$(E - P) - R = (I - O) \quad (2)$$

The quantity (I - O) is equated with the freshwater flux (F_W) which is defined as the rate of water exchange between the ocean and the rest of the earth through evaporation, precipitation, and river discharge. A direct calculation of F_W can be accomplished by using volume transport and salinity data at the Yucatan Channel and the Straits of Florida. Specifically,

$$F_W = \frac{- M_F (S_F - S_Y)}{S_Y} \quad (3)$$

where: F_W = freshwater flux
M_F = volume transport through the Florida Straits
S_F = vertically averaged salinity at the Florida Straits
S_Y = vertically averaged salinity at the Yucatan Channel

M_F, S_F and S_Y are evaluated to a set depth; ideally, this would be the bottom depth, but in practice could be much shallower.

Combining Eqs. (2) and (3), rearranging, and solving for P, we obtain

$$P = E - R - F_W = E - R + M_F \frac{(S_F - S_Y)}{S_Y} \quad (4)$$

Thus, given monthly estimates of E, R, M_F, S_Y and S_F, spatially averaged estimates of monthly precipitation (P) can be calculated. In this context, a negative value for M_F indicates motion out of the system (the Gulf of Mexico).

3. PRELIMINARY RESULTS

In order to demonstrate the merits of this technique, preliminary results are presented. Related research (Etter, 1983; Etter, et al., 1987) has already dealt with heat and freshwater budgets of both the Gulf of Mexico and the Caribbean Sea. These particular investigations were concerned with monthly, multi-annual time scales spatially averaged over regions on the order of $(300\ km)^2$. The application proposed here entails similar spatial scales, but with individual monthly time periods; otherwise, the methods are quite similar. The burden on data acquisition is greatly increased since reliance can no longer be made on multi-annual archival data sources.

In the Gulf of Mexico, it has been shown that the terms P, E, R and F_W are of similar orders of magnitude (Etter, et al., 1987). Thus, reasonable estimates of E, R and F_W should provide reasonable, but indirect, monthly estimates of P. Precipitation data from coastal and island stations in the Gulf of Mexico can also be used to "calibrate" the indirect results, although, as shown earlier, this option does not guarantee conclusive results.

TABLE 1. SUMMARY OF PRELIMINARY PRECIPITATION RATES OVER THE GULF OF MEXICO IN THE SUMMER SEASON[a]

Parameter	$cm\ season^{-1}$
Evaporation	33[b] - 36[c]
River Discharge	13
Freshwater Flux	16
Precipitation (residual)	4 - 7
(calculated)	16[d] - 37[e]

Notes:

(a) summer season: June - August
(b) Hastenrath and Lamb (1987)
(c) Bunker (unpublished)
(d) Dorman and Bourke (1981)
(e) Jaeger (1976)

Using historical data from the summer season, the application of the proposed technique can be better illustrated. This particular example blends data of differing base years; thus, these results must be considered only preliminary. Furthermore, where more than one source of supporting calculations is available, ranges of values are indicated to show potential uncertainties. These results are summarized in Table 1.

These preliminary results, which are based on data of differing base periods, are not in good agreement. However, we do believe that they serve to demonstrate that the freshwater budget technique may hold some promise in providing collaborating data useful in ground-truthing satellite measurements.

ACKNOWLEDGMENT

We are grateful to Karen Fox for her efforts in preparing this manuscript. This study was partially supported by NSF Grant ATM 85-20877.

REFERENCES

Atlas, D., and O.W. Thiele, Eds., 1981: Precipitation Measurements from Space: Workshop Report. National Aeronautics and Space Administration, Goddard Space Flight Center.

Dorman, C.E., and R.H. Bourke, 1981: Precipitation over the Atlantic Ocean, 30° S to 70° N. Mon. Weath. Rev., 109, 554-563.

Etter, P.C., 1983: Heat and freshwater budgets of the Gulf of Mexico. J. Phys. Oceanogr., 13, 2058-2069.

Etter, P.C., P.J. Lamb, and D.H. Portis, 1987: Heat and freshwater budgets of the Caribbean Sea with revised estimates for the Central American Seas. J. Phys. Oceanogr., 17, 1232-1248.

Hastenrath, S.L., and P.J. Lamb, 1978: Heat Budget Atlas of the Tropical Atlantic and Eastern Pacific Oceans. University of Wisconsin Press, 104 pp.

Jaeger, L., 1976: Monatskarten des Niederschags für die ganze Erde. Ber. Dtsch. Wetterdienstes, 139, 38 pp., 13 charts.

Nystuen, J.A., 1986: Rainfall measurements using underwater ambient noise. J. Acoust. Soc. Am., 79, 972-982.

WEATHER MONITORING SYSTEM FROM EARTH OBSERVATION SATELLITES

T. Sakata, H. Shimoda, and Y. Matsumae
Tokai University Research & Information Center
Tokyo 151, Japan

ABSTRACT

An earth observation satellite data receiving station named Tokai Space Information Center (TSIC) has been established within Tokai University. In this center data from MOS-1, NOAA and GMS are received, archived and processed for monitoring global change. A semi-real time global cloud monitoring system has been developed in Tokai University Research and Information Center (TRIC) using data from TSIC in addition to data from GOES and METEOSAT via tele-communication network.

1. INTRODUCTION

Monitoring of weather conditions was limited to local areas until the launch of artificial satellites. With the launch of the geostationary meteorological satellite system, humankind has obtained this global weather monitoring capability. However, the utilization of those meteorological satellites had been focused mainly on short-term weather forecasting until recent years. During recent years, it has become clear that long term meteorological changes may greatly affect the lives of humankind. One major example of this effect is the case of the African drought. Precipitation changes on the African continent which may have been caused by El Nino have resulted in a tragedy for people living on that continent.

It is now an urgent matter to monitor global meteorological changes in order to prevent reoccurrences of such tragedies. The only way to achieve such monitoring in a timely manner is the utilization of earth observation satellites. In this paper, efforts in Tokai University Research & Information Center to realize monitoring of global meteorological changes are described.

2. SATELLITE DATA RECEIVING SYSTEM

Tokai University Space Information Center (TSIC) was established in November, 1986, at Kumamoto, Japan. The purposes of TSIC are: data collection from earth observation satellites such as MOS-1,

TROPICAL RAINFALL MEASUREMENTS
John S. Theon and Nobuyoshi Fugono (Eds.)

ISBN 0-937194-14-4

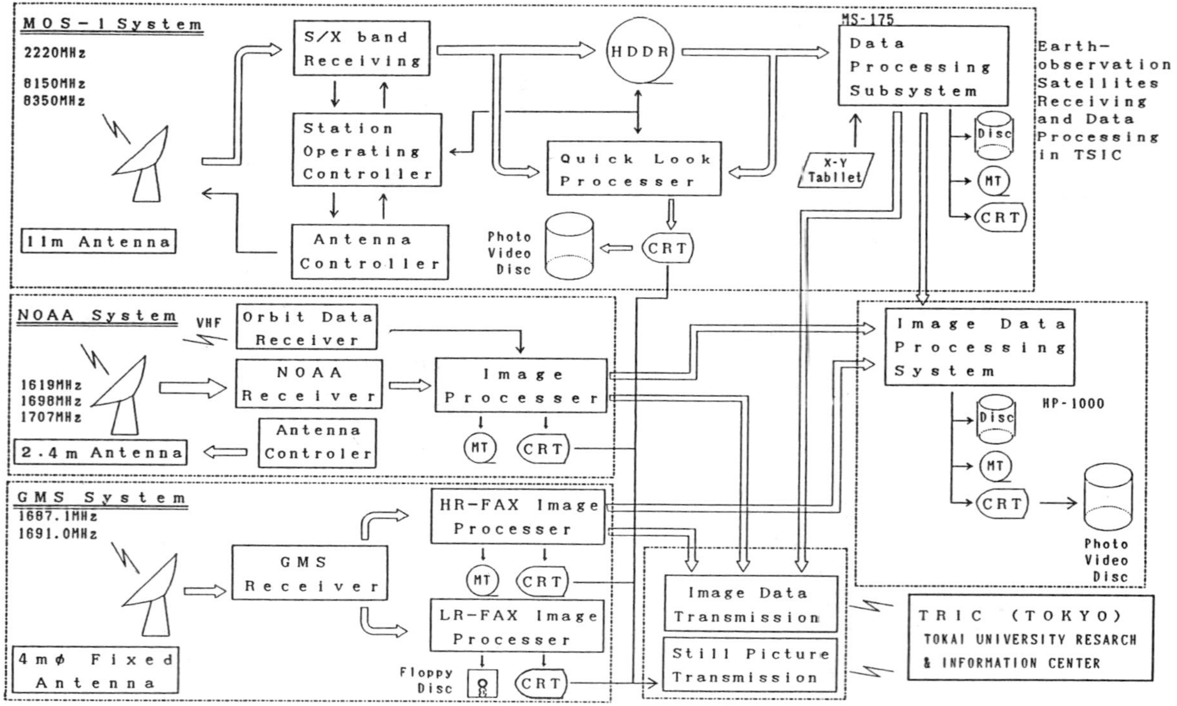

FIGURE 1. Block diagram of satelliteimage data receiving and processing system in TSIC.

NOAA, GMS, etc. as well as from several communication or direct broadcasting satellites; and processing and analysis of this data and dissemination of collected information. TSIC will be the center of an advanced integrated information network system which is under development by Tokai University.

2.1 OUTLINE OF THE SYSTEM

The block diagram of the satellite image data receiving system in TSIC is shown in Figure 1. This system is composed of the following subsystems:

a) Antenna and antenna control units

Three antennas are used to receive MOS-1, NOAA, and GMS satellites, respectively. MOS-1, MESSR and VTIR data are received by a 11m X/S-band antenna. It is an AZ-EL, Cross-EL mount antenna and has a full coverage as an X-Y mount antenna. NOAA HRPT data are received by a 2.4m S-band antenna which has an AZ-EL mount. GMS HR-FAX data are now received by a 4m antenna, but will receive stretched VISSR data after next March.

b) Data receiving units with quick look

The data receiving unit has several functions, i.e. satellite orbit calculations, automatic tracking of satellites based on an operation schedule, remote control, and monitoring of the reception related units, data receiving and demodulation, and data storage.

The system can easily be operated by one person. Personal computers are used as the host of this unit. First, the satellite orbit is calculated based on six orbital elements, and then, the antenna is controlled according to the calculated results. In the automatic operation mode, the following controls are performed based on the predetermined operation plan. They are schedule management, readiness checks, initialization of each unit, satellite tracking, control of data recording, logging operational history and preparation of the next path.

Received data is then stored in storage units. Simultaneously, it is displayed on quick look monitors. MOS-1 data is stored on high density digital tapes (HDDT) while NOAA and GMS data are stored on magnetic disks.

c) Data processing units

The MOS-1 data processing unit generates MESSR and VTIR CCTs from the data stored in High Density Digital Tape (HDDT). The radiometric and geometric corrections of MESSR/VTIR images are performed in this unit. The processing levels vary from level 0 to level 4. Level 0 corresponds to uncorrected data, while level 1 corresponds to radiometrically corrected, level 2 to system corrected, level 3 to precisely corrected, and level 4 to precisely registered. The host computer of this unit is a MS-175 minicomputer.

NOAA AVHRR data, TOVS data and GMS HR-FAX data are formatted and recorded on CCTs, but in this case, no corrections are performed in the data processing units.

d) Data archive unit

In order to retrieve MOS-1 data effectively, an archive unit is installed in this system. In the archive unit, image data are stored in optical disks which can store 24,000 images, while attribute data, which is necessary in retrieval operations, are stored in the hard disk of a personal computer as a data base. These attributes are sensor names, received days and times, path and row numbers, cloud coverage, processing levels, HDDT and/or CCT numbers, etc. Retrievals can be done with any logical combination of attributes and retrieved images are shown on the display monitor.

As for NOAA and GMS data, similar archive systems will be installed in the near future.

e) Data analysis units

All the received data can be transferred directly to an image analysis system. Fig. 2 shows the block diagram of the image analysis system. As for software, a general purpose image processing software package called TIPE (TIAS Image Processing Executive) is installed on the system. Therefore, most of the imaging processing and analysis tasks can be performed within TSIC.

f) Data dissemination units

All the received data can be transmitted in three ways through communication lines as shown in Fig. 3. They are through a telephone line with modems, x.25 packet networks and facsimile lines. The first two units are already installed while the last one will be installed within this year and will be used as routine data broadcasting for users. Usage of communication satellites for data transmission are now under consideration.

3. GLOBAL CLOUD MONITORING SYSTEM

Geostational meteorological satellite data is one of the most important information resources for weather forecasting. For regional utilization, it is sufficient to use one of these satellites. However, recently it has become clear that the integration of global information is important for the study of global area phenomena such as meteorology or oceanology. We tried to make the world cloud cover image by using meteorological satellite data in semi-real time. In order to clearly display cloud locations, a base image was first generated.

The base image of earth was made by using NOAA data. At first, the vegetation index data, which is a polar-stereographic projection, was transformed to longitude and latitude coordinates. Thus, a cloud-free base image in longitude and latitude coordinates was obtained.

The second step was to make the world cloud cover image. In this study, we used the data from four satellites: GMS, GOES, METEOSAT and INSAT. The data of GOES, METEOSTAT and INSAT were transferred as facsimile signals through overseas telephone lines. These data were geometrically transformed to the longitude and latitude coordinate system. By combining four images using digital

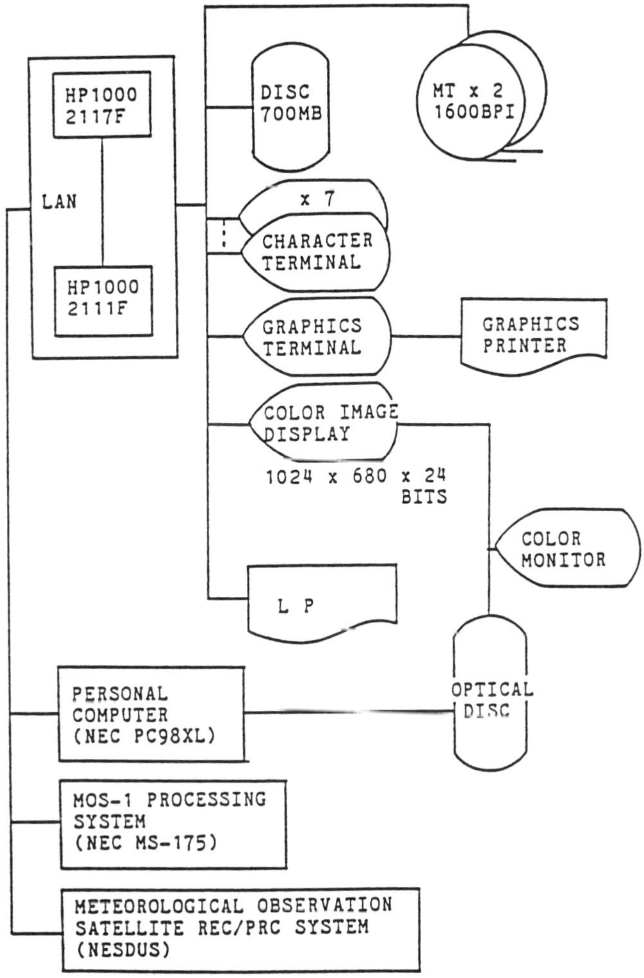

FIGURE 2. Block diagram of the image analysis system

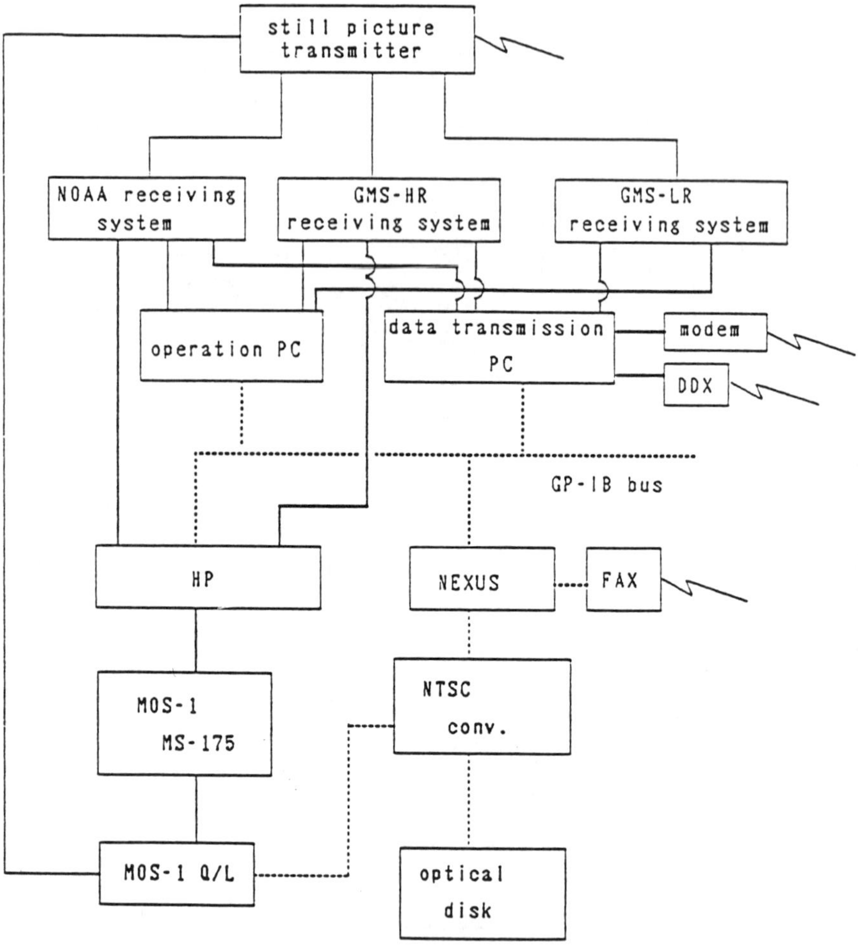

FIGURE 3. Data transmission system in TSIC

mosaicing technique, a world cloud cover image was obtained. This image was overlayed on the base image produced by NOAA data.

The third step was to transform the coordinate system from longitude and latitude to orthographic equator projection. Thus, we could make the picture of earth viewed (from infinite distance) on the equator. The longitude of the viewing direction was shifted one degree by one degree. Thus, 360 images were generated. Finally, these images were transferred to a video recorder through an NTSC encoder, frame by frame, and the rotating earth with real-time cloud cover was shown on a monitor.

Several other kinds of image processing for meteorological applications, such as sea surface temperature extractions, animation of cloud movement, 3-D display of clouds, etc. were tried in our center.

LIST OF AUTHORS

SUBJECT INDEX